Methods in Molecular Biology™

Series Editor
John M. Walker
School of Life Sciences
University of Hertfordshire
Hatfield, Hertfordshire, AL10 9AB, UK

For further volumes:
http://www.springer.com/series/7651

Ion Channels

Methods and Protocols

Second Edition

Edited by

Nikita Gamper

School of Biomedical Sciences, University of Leeds, Leeds, UK

Humana Press

Editor
Nikita Gamper
School of Biomedical Sciences
University of Leeds
Leeds, UK

ISSN 1064-3745 ISSN 1940-6029 (electronic)
ISBN 978-1-4939-6009-5 ISBN 978-1-62703-351-0 (eBook)
DOI 10.1007/978-1-62703-351-0
Springer New York Heidelberg Dordrecht London

Softcover re-print of the Hardcover 2nd edition 2013

Cover illustration: Taken from Figure 1 of Chapter 16

Printed on acid-free paper

Humana Press is a brand of Springer
Springer is part of Springer Science+Business Media (www.springer.com)

Preface

Every cell in our body contains a great variety and number of permeability pathways for various organic and inorganic ions, water, metabolites, nutrients, and signaling molecules. Maintenance and precise control of gating within these pathways are fundamental principles of life as these underlie basic cellular functions such as communication, contractility, and metabolism. This book focuses on the strategies, approaches, methods, and protocols for studying a large family of proteins that form ionic channels in the plasma membrane and intracellular membranes of cells. Like other permeability pathways of biological membranes, ion channels are essential for life as they generate action potentials and regulate synaptic transmission in neurons and muscle cells, underlie intracellular Ca^{2+} signalling, and contribute to the charge separation across plasma membranes. Not surprisingly, genetic deficiencies or acute deregulations of ion channel activity, trafficking, or degradation often cause or contribute to severe human disorders (often called "channelopathies") and pathologies, e.g., arrhythmias, epilepsies, chronic pains, deafness, diabetes, and many others. Conversely, ion channels are increasingly recognized as therapeutic targets.

Slightly over half a century ago, ion-selective channels in the plasma membrane were postulated by Alan Hodgkin and Andrew Huxley as a purely theoretical concept. Now, at the beginning of the twenty-first century, hundreds of ion channel genes are cloned, and the currents conducted by many of them are exhaustively characterized. Some ion channels are assigned with clear physiological functions while some are linked to human diseases, and, for a handful of them, functional structures are proposed. This is tremendous progress, yet there is even more that we do not know. The aim of the present book is twofold: firstly, using practical examples from the cutting-edge current research, we will take a look back at the major methods and approaches that allowed us to progress to our current understanding of ion channel function, structural design, and biological roles; and secondly, we will try to look forward and identify approaches that will lead us to future discoveries.

This book will be of interest to specialists in academia and industry looking for specific methodology in studying ion channels. It will be helpful for lecturers and advanced students in the university classroom as well as for anyone interested in the state-of-the art biomedical toolkit.

Leeds, UK *Nikita Gamper*

Contents

Contributors

KARIN ABARCA-HEIDEMANN • *Department of Biochemistry, Temple University School of Medicine, Philadelphia, PA, USA*

ARMEN N. AKOPIAN • *Department of Endodontics, Dental School, University of Texas Health Science Center at San Antonio, San Antonio, TX, USA*

ALESSANDRO ALAIMO • *Unidad de Biofísica (CSIC-UPV/EHU), Universidad del País Vasco, Leioa, Spain*

KERMAN ALORIA • *Proteomics Core Facility-SGIKER, Universidad del País Vasco, Leioa, Spain*

PILAR ARESO • *Faculty of Medicine and Odontology, Department of Pharmacology, Universidad del País Vasco, Leioa, Spain*

STEPHEN A. BALDWIN • *Astbury Centre for Structural Molecular Biology, School of Biomedical Sciences, University of Leeds, Leeds, UK*

DAVID J. BEECH • *Faculty of Biological Sciences, Multidisciplinary Cardiovascular Research Centre, University of Leeds, Leeds, UK; Faculty of Medicine and Health, Multidisciplinary Cardiovascular Research Centre, University of Leeds, Leeds, UK*

ANAMIKA BHARGAVA • *Department of Cardiovascular Sciences, National Heart and Lung Institute, Imperial College London, London, UK*

SONYA M. BIERBOWER • *Department of Physiology, University of Texas Health Science Center at San Antonio, San Antonio, TX, USA*

NINA BOIKO • *Department of Physiology, University of Texas Health Science Center at San Antonio, San Antonio, TX, USA*

VLADISLAV BUGAY • *Department of Physiology, University of Texas Health Science Center at San Antonio, San Antonio, TX, USA*

XINGJUAN CHEN • *The Key Laboratory of Neural and Vascular Biology, Ministry of Education, Shijiazhuang, Hebei Province, China; The Key Laboratory of Pharmacology and Toxicology for New Drugs, Ministry of Education, Shijiazhuang, Hebei Province, China; Department of Pharmacology, Hebei Medical University, Shijiazhuang, China*

CHRISTOPHER J. COCKCROFT • *Faculty of Biological Sciences, School of Biomedical Sciences, University of Leeds, Leeds, UK*

BERTRAND COSTE • *Centre de Recherche en Neurobiologie et Neurophysiologie de Marseille, UMR 7286, CNRS, Aix-Marseille Université, Marseille, France*

MARCEL CREST • *Centre de Recherche en Neurobiologie et Neurophysiologie de Marseille, UMR 7286, CNRS, Aix-Marseille Université, Marseille, France*

PATRICK DELMAS • *Centre de Recherche en Neurobiologie et Neurophysiologie de Marseille, UMR 7286, CNRS, Aix-Marseille Université, Marseille, France*

ELKE DUCHARDT-FERNER • *Institute for Molecular Biosciences, Center for Biomolecular Magnetic Resonance (BMRZ), Johann Wolfgang Goethe University, Frankfurt am Main, Germany*

BENJAMIN A. EATON • *Department of Physiology, University of Texas Health Science Center at San Antonio, San Antonio, TX, USA*

NIKITA GAMPER • *Faculty of Biological Sciences, School of Biomedical Sciences, University of Leeds, Leeds, UK*
SYLVAIN GIGOUT • *Faculty of Biological Sciences, School of Biomedical Sciences, University of Leeds, Leeds, UK*
JULIA GORELIK • *Department of Cardiovascular Sciences, National Heart and Lung Institute, Imperial College London, London, UK*
ALEKSANDRA GRUSLOVA • *Department of Physiology, University of Texas Health Science Center at San Antonio, San Antonio, TX, USA*
BINGCAI GUAN • *The Key Laboratory of Neural and Vascular Biology, Ministry of Education, Shijiazhuang, Hebei Province, China; The Key Laboratory of Pharmacology and Toxicology for New Drugs, Ministry of Education, Shijiazhuang, Hebei Province, China; Department of Pharmacology, Hebei Medical University, Shijiazhuang, China*
JIZHE HAO • *Centre de Recherche en Neurobiologie et Neurophysiologie de Marseille, UMR 7286, CNRS, Aix-Marseille Université, Marseille, France*
JAMES HEWINSON • *School of Physiology and Pharmacology, University of Bristol, Bristol, UK*
STEPHAN M. HUBER • *Department of Radiation Oncology, University of Tübingen, Tübingen, Germany*
DARIA V. ILATOVSKAYA • *Department of Physiology, Medical College of Wisconsin, Milwaukee, WI, USA; Institute of Cytology, Russian Academy of Sciences, St. Petersburg, Russia*
HARALD JANOVJAK • *Institute of Science and Technology, Klosterneuburg, Austria*
LIN-HUA JIANG • *Faculty of Biological Sciences, School of Biomedical Sciences, University of Leeds, Leeds, UK*
SERGEY KASPAROV • *School of Physiology and Pharmacology, University of Bristol, Bristol, UK*
HANNAH M. KIRTON • *Faculty of Biological Sciences, School of Biomedical Sciences, University of Leeds, Leeds, UK*
WOJCIECH A. KOSIBA • *Department of Physiology, University of Texas Health Science Center at San Antonio, San Antonio, TX, USA*
VOLODYMYR KUCHER • *Department of Physiology, University of Texas Health Science Center at San Antonio, San Antonio, TX, USA*
DAUNIA LAURENTI • *Illawarra Health and Medical Research Institute, School of Biological Sciences, University of Wollongong, Wollongong, NSW, Australia*
AMELIA LESIUK • *Astbury Centre for Structural Molecular Biology, School of Biomedical Sciences, University of Leeds, Leeds, UK*
JOHN Y. LIN • *Department of Pharmacology, University of California San Diego, San Diego, CA, USA*
JOHN E. LINLEY • *Faculty of Biological Sciences, School of Biomedical Sciences, University of Leeds, Leeds, UK*
JONATHAN D. LIPPIAT • *Faculty of Biological Sciences, School of Biomedical Sciences, University of Leeds, Leeds, UK*
COVADONGA MALO • *Unidad de Biofísica (CSIC-UPV/EHU), Universidad del País Vasco, Leioa, Spain*

MYKOLA MAMENKO • *Department of Integrative Biology and Pharmacology, University of Texas Health Science Center at Houston, Houston, TX, USA*
GREGORY I. MASHANOV • *MRC National Institute for Medical Research, London, UK*
CATHERINE MCKENZIE • *Institute of Science and Technology, Klosterneuburg, Austria*
OSCAR MILLET • *Proteomics Unit, CIC bioGUNE CIBERehd, Derio, Spain*
CAROL J. MILLIGAN • *Florey Institute of Neuroscience and Mental Health, Melbourne Brain Centre, Kenneth Myer Building, Royal Parade, Melbourne, VIC, Australia*
ELENA MIRONOVA • *Department of Physiology, University of Texas Health Science Center at San Antonio, San Antonio, TX, USA*
CLEMENS MÖLLER • *Life Sciences Faculty, Albstadt-Sigmaringen University, Sigmaringen, Germany*
JUSTIN E. MOLLOY • *MRC National Institute for Medical Research, London, UK*
JACQUELINE NAYLOR • *Xention Limited, Cambridge, UK*
NGUYEN V. NGUYEN • *Department of Radiology, University of Miami School of Medicine, Miami, FL, USA*
ROGER G. O'NEIL • *Department of Integrative Biology and Pharmacology, University of Texas Health Science Center at Houston, Houston, TX, USA*
LEZANNE OOI • *Illawarra Health and Medical Research Institute, School of Biological Sciences, University of Wollongong, Wollongong, NSW, Australia*
JULIAN F.R. PATON • *School of Physiology and Pharmacology, University of Bristol, Bristol, UK*
LOUISA PETTINGER • *Faculty of Biological Sciences, School of Biomedical Sciences, University of Leeds, Leeds, UK*
OLEH POCHYNYUK • *Department of Integrative Biology and Pharmacology, University of Texas Health Science Center at Houston, Houston, TX, USA*
VINCENT G.L. POSTIS • *Astbury Centre for Structural Molecular Biology, School of Biomedical Sciences, University of Leeds, Leeds, UK*
ANDREA E. RAWLINGS • *Astbury Centre for Structural Molecular Biology, School of Biomedical Sciences, University of Leeds, Leeds, UK; School of Physics and Astronomy, University of Leeds, Leeds, UK*
TIBOR ROHACS • *UMDNJ, New Jersey Medical School, Newark, NJ, USA*
KIRSTIN E. ROSE • *Faculty of Biological Sciences, School of Biomedical Sciences, University of Leeds, Leeds, UK*
BRAD S. ROTHBERG • *Department of Biochemistry, Temple University School of Medicine, Philadelphia, PA, USA*
YANN ROUDAUT • *Centre de Recherche en Neurobiologie et Neurophysiologie de Marseille, UMR 7286, CNRS, Aix-Marseille Université, Marseille, France*
JÉRÔME RUEL • *Centre de Recherche en Neurobiologie et Neurophysiologie de Marseille, UMR 7286, CNRS, Aix-Marseille Université, Marseille, France*
SARAH SCHWARZER • *Barts and The London School of Medicine & Dentistry, Queen Mary, University of London, London, UK; William Harvey Heart Centre, Queen Mary, University of London, London, UK*
MALA M. SHAH • *Department of Pharmacology, UCL School of Pharmacy, University College London, London, UK*

MARK S. SHAPIRO • *Department of Physiology, University of Texas Health Science Center at San Antonio, San Antonio, TX, USA*
FRANK J. SMITH • *Department of Biochemistry, Temple University School of Medicine, Philadelphia, PA, USA*
ALEXANDER STARUSCHENKO • *Department of Physiology, Medical College of Wisconsin, Milwaukee, WI, USA*
JAMES D. STOCKAND • *Department of Physiology, University of Texas Health Science Center at San Antonio, San Antonio, TX, USA*
STEPHANIE SZOBOTA • *Department of Neurology, Yale University School of Medicine, New Haven, CT, USA*
ANDREW TINKER • *Barts and The London School of Medicine & Dentistry, Queen Mary, University of London, London, UK; William Harvey Heart Centre, Queen Mary, University of London, London, UK*
ALVARO VILLARROEL • *Unidad de Biofísica (CSIC-UPV/EHU), Universidad del País Vasco, Leioa, Spain*
BIN WANG • *Department of Physiology, University of Texas Health Science Center at San Antonio, San Antonio, TX, USA*
JENS WOEHNERT • *Institute for Molecular Biosciences, Center for Biomolecular Magnetic Resonance (BMRZ), Johann Wolfgang Goethe University, Frankfurt am Main, Germany*
DAVID C. WRIGHTON • *Faculty of Biological Sciences, School of Biomedical Sciences, University of Leeds, Leeds, UK*
WEI YANG • *Department of Neurobiology, Zhejiang University School of Medicine, Hangzhou, China; Key Laboratory of Medical Neurobiology of the Ministry of Health, Hangzhou, China; Faculty of Biological Sciences, School of Biomedical Sciences, University of Leeds, Leeds, UK*
OLEG ZAIKA • *Department of Integrative Biology and Pharmacology, University of Texas Health Science Center at Houston, Houston, TX, USA*
ELEONORA ZAKHARIAN • *Department of Cancer Biology and Pharmacology, University of Illinois College of Medicine, Peoria, IL, USA*
HAILIN ZHANG • *Department of Pharmacology, Hebei Medical University, Shijiazhuang, China; The Key Laboratory of Neural and Vascular Biology, Ministry of Education, Shijiazhuang, Hebei Province, China; The Key Laboratory of Pharmacology and Toxicology for New Drugs, Ministry of Education, Shijiazhuang, Hebei Province, China*

Part I

Cloning and Heterologous Expression of Ion Channel Genes

Chapter 1

Approaches to Cloning of Pain-Related Ion Channel Genes

Armen N. Akopian

Abstract

Molecular pain research is a relatively new and rapidly expanding field that represents advancement in conventional pain research. One of the fundamentals of molecular pain involves the cloning of genes and especially the ion channels specifically involved in nociceptive processing at the periphery and centrally. A variety of approaches were used to isolate these critically important genes. Cloning of these genes involved innovative strategies based on existing molecular approaches. This review will discuss well-utilized cloning approaches and their exploitation in molecular pain research.

Key words Molecular pain research, Subtractive cloning, Differential cloning, Functional cloning, Expression cloning, Homology cloning

1 Introduction

Substantial advances in our understanding of pain mechanisms during the past two decades can be traced to the application of molecular biology in pain research (1, 2). This new period, which can be labeled as molecular pain research, gave rise to a fully recognized category in the pain field, leading to the creation of dedicated and specialized journals including *Molecular Pain*. In regard to the field of pain, molecular biology can offer several powerful approaches, including methods and tools to identify genes and the underlying molecular mechanisms specifically involved in variety of physiological and pathophysiological acute and chronic pain conditions. Thus, modern molecular biology has following main approaches:

1. *Cloning*: Cloning methods were developed through a string of seminal discoveries in the 1970s and early 1980s (3, 4). The application of modern cloning approaches and methods to pain research led to the identification of new transcription factors, cytoplasm proteins, adapter proteins, receptors, and ion channels involved in generating, modulating, and propagating

Nikita Gamper (ed.), *Ion Channels: Methods and Protocols*, Methods in Molecular Biology, vol. 998,
DOI 10.1007/978-1-62703-351-0_1, © Springer Science+Business Media, LLC 2013

action potentials along the nociceptive pathways (5–13). Cloning of the genes involved in certain physiological conditions is considered a primer and ultimately a starting point for all further functional studies on the underlying molecular mechanisms. Eventually these functional studies, which also utilize molecular biology approaches, result in a true appreciation of the importance of cloned genes (1).

2. *Identification of regulatory pathways*: Molecular biology approaches are utilized to identify transcriptional, posttranscriptional (i.e., splice variants), translational, and posttranslational changes in gene expressions after pathological painful conditions such as tissue damage, inflammation, and nerve damage (14, 15). These alterations underlie neural reorganization (i.e., "plasticity") in the nociceptive signal transduction pathways, which include sensory neurons, spinal cord and brain stem neurons, and certain brain regions including the hypothalamus. It is presumed that chronic pain conditions are manifestations of neuronal plasticity and therefore, uncovering the genes and molecular mechanisms contributing to the generation of this plasticity could support discoveries of novel analgesic targets and promote designs of novel anti-pain therapeutic strategies (16).
3. *Identification of mutated genes using genetics*: Molecular genetics tools are widely used to identify genes responsible for particular phenotypes in defined animal lines or genetic mutations in humans leading to diseases, such as congenital insensitivities to pain (17, 18). As the human genome project approaches completion, "new" genes involved in specific pain conditions or chronic pain diseases may be uncovered (19, 20). Moreover, molecular genetics approaches can help to identify determinants of individual differences in pain and analgesia in animals and humans (21, 22).
4. *Generation of genetically modified animals*: Molecular biology provided advanced tools for pharmacological investigations. Molecular genetics manipulations allow for the generating of animal lines with deleted genes or gene segments, called conventionally "knockout" mutations. Knockout mice are useful in investigating the contributions of particular proteins to nociception. Thus, functions of many important proteins for the pain pathway were assessed using these knockout mice (23–26). One of the main drawbacks of knockout technology is that the gene of interest is ablated in every cell type where it is expressed. To bypass this drawback, the conditional knockout technology (i.e., Cre–lox null-mutants) was developed. This method allows for the ablation of genes in specific cell types. The specificity of ablation is defined by expression pattern of Cre-recombinase. Recently, mouse lines expressing

Cre-recombinase only in subsets of sensory neurons were generated (27, 28).

5. *Gene therapy and gene delivery technologies*: Molecular biology provides anticipation for pain control through the application of gene therapy. Gene therapy can correct genetic defects by replacing or substituting the defective gene with a new, functional copy. Studies exploring gene therapy in pain control are just beginning (29, 30). Hopefully, in the near future, gene therapy using the antisense strategy or delivery of genetic material using viral vectors into the pain pathway may be applied clinically.

The listed molecular biology approaches contribute to our understanding of the molecular mechanisms underlying nociceptive transmission. However, cloning of novel genes involved in nociceptive transmission is a critically important step in studies of molecular mechanisms of nociception. This step provides the required tools (i.e., the genes themselves) for a plethora of studies, including molecular, biochemical, pharmacological, and physiological research. In this chapter, I will review a variety of strategies and methods used in pain research to clone genes, especially those encoding ion channels which play vital roles in the nociceptive processing. Among multiple cloning strategies, three stand out: subtractive, differential, and functional cloning. The use of these strategies allowed researchers to not only identify new genes but also characterize previously recognized genes who then became founding members of novel gene families. Accordingly, we will review these strategies in detail.

2 Subtractive and Differential Cloning

There are many variations of the subtractive and differential cloning strategies. However, every modification follows the same principle: obtaining gene(s) which is expressed in a chosen cell type but not in other selected cell type(s). The original method was developed and utilized by Davis and colleague to make a seminal finding—cloning of the elusive antigen receptor expressed specifically on the surface of T lymphocytes (31, 32). The basic sequence of steps for the subtractive cloning strategy is as follows: single-stranded cDNA generated from mRNA isolated from a cell type of interest is hybridized with mRNA from cell type(s) in which the targeted gene should not be expressed; un-hybridized cDNA is then separated and is used for cDNA library construction. In an ideal case, every clone in the library could represent the specific gene for a chosen cell type. In reality, an additional step is required; the cDNA library needs to be differentially screened for final identification of specific clones. These approaches have been used in pain research to identify several critically important ion channels and other types of proteins: TTX-resistant voltage-gated sodium channel $Na_V1.8$ (originally known as SNS; (5)), sensory neuron-specific ATP-gated channel (P2X3; (6)); sensory neuron-specific proton-gated channel (ASIC; (33)); TRPV1 adapter protein *Pirt* clone B4 (12, 34); and mechano-gated Piezo1 and Piezo2 channels (35).

Figure 1 shows schematically one of the approaches for isolation of these genes (34). The main aim of this study was to isolate genes that are expressed in dorsal root ganglia (DRG) sensory neurons, but not in kidney, liver, cerebellum, and cortex (34). Accordingly, single-stranded cDNA from DRG mRNA (polyA$^+$ RNA fraction) was synthesized using oligo (dT)/NotI primer-adapter. This cDNA was purified from unincorporated nucleotides and primers (Fig. 1). This purification step is essential if the cDNA is used for PCR

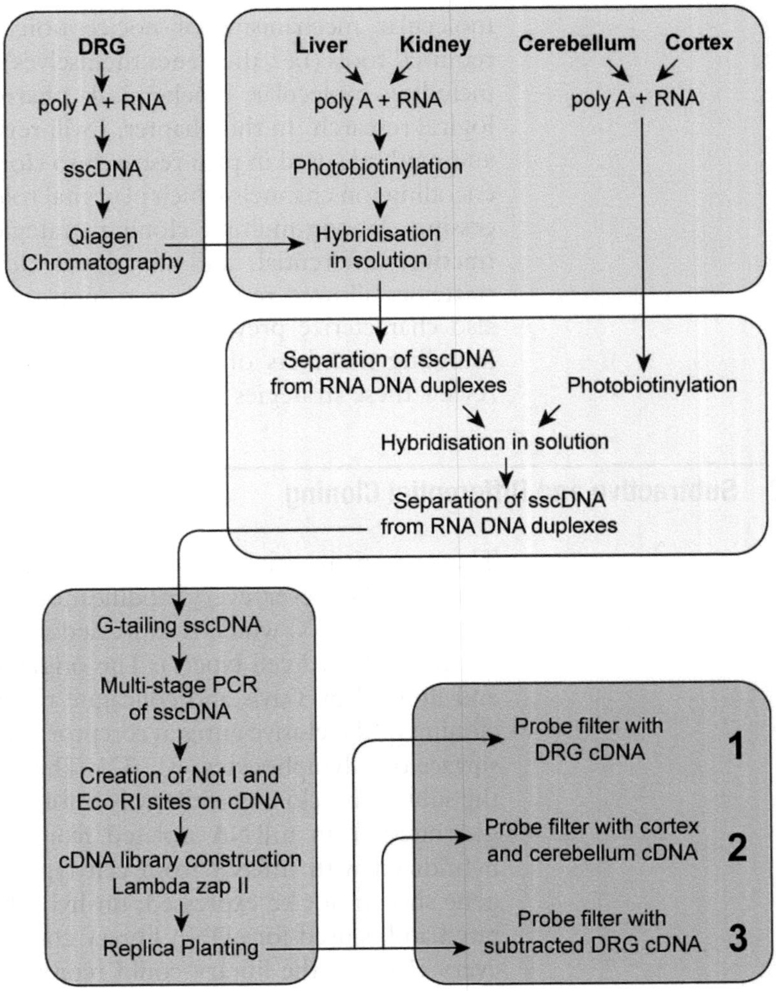

Fig. 1 Schematic representation of subtractive cloning and differential screening that has yielded many sensory neuron-specific genes, including P2X3 and $Na_V1.8$ (aka SNS) channels. Different critical steps of procedure are illustrated as separate blocks

amplification in the following steps. The DRG cDNA was hybridized in two rounds with 100-fold excess amounts of photobiotinylated mRNA from liver, kidney, cerebellum, and cortex (Fig. 1). Photobiotinylation was an innovative step in the subtractive cloning procedure (34). This allowed for the effective separation of unhybridized DRG cDNA on a streptavidin–avidin column. Amounts of remaining enriched DRG-specific cDNA were too small for direct construction of a cDNA library. Therefore, several PCR amplification cycles were used to accumulate a sufficient amount of double-stranded cDNA for construction of subtractive DRG cDNA library. Unfortunately, small-sized cDNA are naturally PCR amplified more effectively. Therefore, a majority of long cDNA, which encode many channels, will be lost on this stage. To counteract this phenomenon, during every cycle of cDNA amplification, the cDNA pool was enriched with long-sized amplified cDNA (34). This step is the cornerstone of the procedure represented in Fig. 1. Sufficient insert lengths of DRG-specific cDNA library cannot be achieved without the enrichment of cDNA pool with long-sized cDNA molecules during multistep PCR procedure. Finally, the constructed DRG-specific cDNA library was deferentially hybridized with extremely high P^{32}-radio-labeled DRG, cortex + cerebellum, or liver + kidney cDNAs (Fig. 1). Replica clones that showed hybridization with labeled DRG cDNA, but not other labeled cDNAs were selected for further analysis (34). Sequence analysis revealed that isolated clones contained many known sensory neuron-specific genes such as peripherin, gamma-synuclein and villin, and one unknown gene with homology to voltage-gated sodium channel, which later was revealed to be $Na_V1.8$ (i.e., SNS; (5)), and six totally unknown genes, four of which have become known as P2X3, DRASIC and ASIC-β channels, and Pirt (6, 12, 33, 36). This strategy (Fig. 1) contained many previously described and utilized steps; however, photobiotinylation and differential amplification of enriched cDNA were innovative approaches, promoting the generation of subtractive cDNA library from miniscule amounts of tissue (34).

Differential cloning is a simplified version of the subtractive cloning strategy. In this approach, a subtractive cDNA library is not constructed. In the original differential cloning experiments, direct differential screening of conventional cDNA library has been utilized (37). However, this approach in its original version is not very sensitive, and only abundantly expressing proteins can be cloned. In addition, differential screenings of conventional cDNA libraries generate hundreds of positive signals, which make analysis cumbersome and time-consuming. To tackle these two obvious disadvantages, the differential cloning approach has substantially been redeveloped. First, a conventional cDNA library is not constructed anymore; libraries are now represented by large pools of single-stranded oligonucleotides attached to solid matrixes (named "chips" or "microarray"; (38)). Positions of each defined oligonucleotide on chips are known. Second, differential screenings of chips are performed with fluorophore-labeled single-stranded cDNAs, and hybridization results are revealed by use of scanning. Third, analyses of differential screenings are performed by specialized software, designed for these particular types of chips. These developments simplified differential cloning tremendously. Thus, the full experiment can be performed within 1–2 weeks, unlike the original differential cloning approach that consumed at least 4–6 months of extensive bench-work.

Modern differential cloning approaches exploiting oligonucleotide-containing chips has been employed to isolate two of the most coveted channels in pain research—mechano-gated Piezo1 and Piezo2 channels (35). First, cell lines expressing and non-expressing functional mechano-gated channel(s) were identified. Then, differential cloning using chips was performed between these lines. A set of genes which are specific for cell lines containing functional mechano-gated channel(s) was isolated. Finally, full-length cDNA clones for unknown genes were isolated and expressed in a heterologous system and whole-cell and single-cell mechano-gated currents were recorded from one of the clones named Piezo-1. Altogether, this differential cloning yielded two proteins, Piezo1 and Piezo2, having very large sizes and belonging to an unknown family of genes (35). Beside this classical work, modern differential cloning on chips is widely used to identify genes up- or down-regulated during some physiological or pathophysiological acute or chronic pain conditions (39).

3 Functional Cloning

Functional cloning could be divided into three separate approaches: expression cloning, cloning of interacting protein(s), and isolation of mutated genes using genetic mapping. Each of these approaches will be reviewed separately.

3.1 Expression Cloning

The expression cloning strategy to isolate genes of interest was widely utilized in pain research. Notable discoveries of ATP-gated P2X2 (40), capsaicin-/heat-gated TRPV1 (7), and menthol/cold-gated TRPM8 channels (9) were made using expression cloning strategies. To successfully accomplish expression cloning, three parameters should be defined: (a) decision on the use of a particular expression system needs to be made; (b) agonists activating proteins of interests have to be known; and (c) methods for detection of protein activation should be selected. A majority of expression cloning experiments were performed either by injection of mRNA in *Xenopus* oocytes (40) or transfection of cell lines (such as HEK, COS-7, or CHO) with expression cDNA libraries (7). However, occasionally these direct approaches could not be fruitful. Thus, mechano-gated channels Piezo1 and Piezo2 are very large proteins; therefore, it is almost impossible to generate full-length cDNA library that will express functional Piezo1 or Piezo2 proteins. To overcome this difficulty, differential cloning on cell lines naturally expressing and not expressing Piezo1 and Piezo2 were utilized (see above and (35)). Selection of ligands appears straightforward for a bulk of expression cloning experiments. However, in some cases, difficulties have been met, and imaginative adjustments and presumptions have been made. Thus, to clone a cold-gated channel, cool-like sensation produced by menthol has been selected as a "ligand" to activate a putative cold-gated channel in an expression system (9). Decision on selection of a mechanical stimulus was required to clone a mechano-gated channel, which is distinct from stretch- and osmotic pressure-activated channels. Coste and colleagues (35) selected a pinching device driven by precisely graded and calibrated piezo elements, which has previously been utilized to characterize mechano-gated currents in sensory neurons (41). Selection of methods for detection of protein activation is a most critical step for successful accomplishment of expression cloning. Expression cloning is naturally not a sensitive approach, as the activities of the proteins of interests need to be detected in a large pool containing thousands of different cDNAs or mRNAs. For example, an ATP-gated current was recorded from *Xenopus* oocytes injected with whole polyA$^+$ RNA fraction from smooth muscle cells (40). In contrast, attempts to record capsaicin-gated current after injection of *Xenopus* oocytes with polyA$^+$ RNA fraction from DRG did not yield the desired result. Hence, Caterina and colleagues employed Ca^{2+}-imaging to detect capsaicin-evoked Ca^{2+} accumulations in individual HEK cells transfected with DRG cDNA pools (7). This innovative and imaginative approach increased sensitivity of the expression cloning experiment by 50–100-fold, as an entire DRG cDNA library was divided into >20 pools. In addition, Ca^{2+} imaging is a sensitive detection method for activation of Ca^{2+} channels.

Figure 2 illustrates schematically the approach used for isolation of the capsaicin-/heat-gated channel, TRPV1 (7). The same approach was used to clone menthol/cold-gated channel, TRPM8 (9). At the start, conventional DRG cDNA library was constructed using an expression vector (Fig. 2; Step 1). Whole cDNA library was divided into >20 pools each of which contains 5,000–10,000 independent clones. Each pool was transfected into a mammalian expression system, such as HEK or CHO cells (Fig. 2; Step 2), and then transfected cells were probed with capsaicin and Ca^{2+}-accumulation was registered (Fig. 2; Step 3). HEK and CHO cells are capable of uptaking thousands of independent cDNA clones, but each transfected cell contains a unique pool of transfected cDNA. This, in turn, increased detection sensitivity, as each cell expresses lesser clones than in an initial cDNA pool (7). A positive

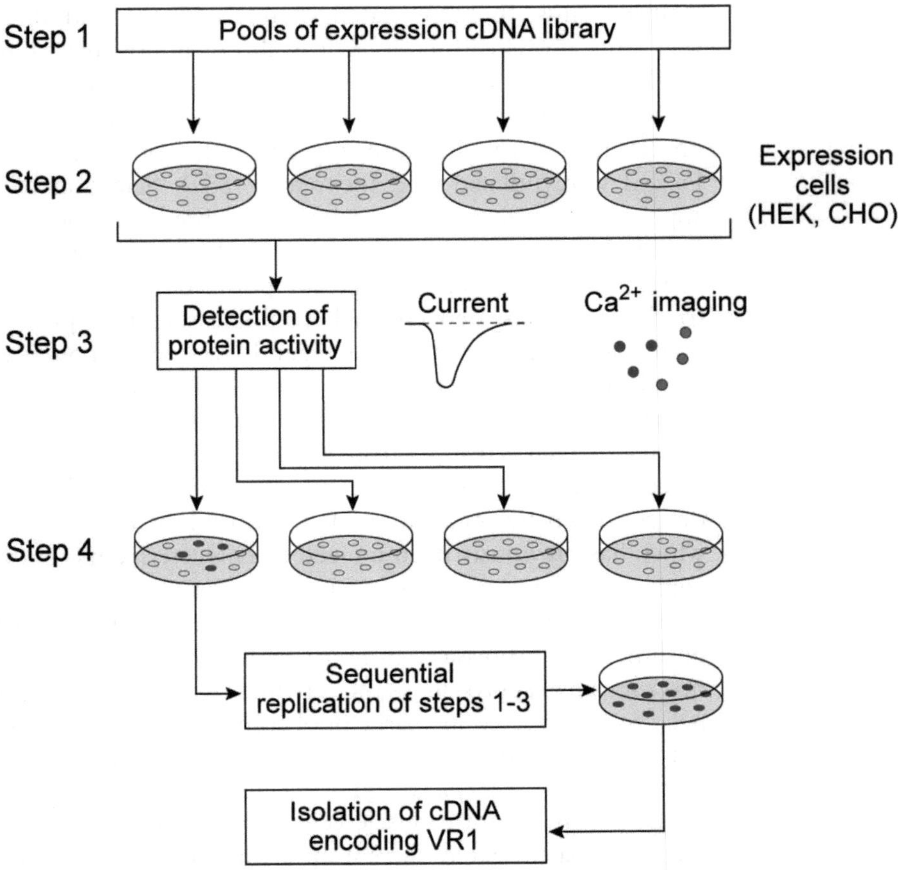

Fig. 2 Schematic representation of expression cloning employed to isolate TRPV1 (aka VR1). Positive clones are marked as *dark circles* on dishes

pool was again divided into ten cDNA pools, and expression screening procedure was repeated (Fig. 2; Step 4). Step-by-step, positive cDNA pool diminished (in increment of ten) and only one capsaicin or menthol-sensitive clone remained. The capsaicin-sensitive clone became the founding member of a novel gene family, the TRPV-family (7), while the menthol-sensitive clone expressed a novel member (TRPM8) of the TRPM-family (9).

3.2 Isolation of Mutated Genes Using Genetic Mapping

There are two separate groups of genetic diseases, which are characterized by alterations in pain perceptions in patients. One group is congenital insensitivity to pain (CIP), which is a rare syndrome with various clinical expressions, characterized by a dramatic *reduction* (or ablation) in pain perception since birth. According to some classifications, there are five major types of CIP (42). CIP types I–IV are manifested by hereditary sensory and autonomic degeneration (i.e., neuropathy; HSAN) involving the Aδ and C-fiber nociceptors (43). However, some CIP patients show normal morphology for nerve biopsies (44). Another group including inherited erythromelalgia (IEM) and paroxysmal extreme pain disorder (PEPD) are characterized with *severe episodes of pain* (45)).

Identification of the molecular basis of CIP, IEM, and PEDP could lead to a better understanding of the mechanisms involved in the functioning, development, and survival of nociceptors. Determining the genetic bases of certain disorders is performed in two steps. First, the loci responsible for a disorder are established by the genetic linkage analysis relatively to known genetic markers (46). Second, molecular analysis of the identified genetic loci is performed to discover mutation(s) (i.e., deletion, missense, or additional incorporation of nucleotides) in particular gene(s). This classical and now standard approach has been used to define molecular bases of several CIP, IEM, and PEDP. Hereditary sensory radicular neuropathy (aka HSAN type I) has been associated with a mutation in the *SPTLC1* gene which encodes for serine palmitoyltransferase—the enzyme for the synthesis of the sphingolipids, ceramide, and sphingomyelin (47, 48). Congenital sensory neuropathy (aka HSAN type II) gene is located on chromosome 12, but its molecular nature is not very clear as yet. Familial dysautonomia (FD)/Riley–Day syndrome (aka HSAN type III) has been linked to a single point mutation on the *IKBKAP* gene encoding IKAP, a subunit of the highly conserved complex involved in transcriptional elongation (49). CIP with anhidrosis or partial anhidrosis (aka HSAN type IV) is caused by mutations in the *NTRK1* (trkA) gene which is the receptor for NGF (17). Recently, mutations of the gene coding for the sodium channel $Na_v1.7$—a voltage-gated sodium channel expressed preferentially on nociceptors and sympathetic neurons—have been found to be the cause of CIP in patients with a normal nerve biopsy, IEM, and PEDP. CIP with intact sensory nerves is associated to loss-of-function mutations of $Na_v1.7$ (44). Other sensory modalities and the remainder of the central and peripheral nervous systems in these patients were preserved (44). IEM is associated with several "gain-of-function" mutations of the same ion channel (45, 46). Finally, a different set of "gain-of-function" mutations that impair inactivation of $Na_v1.7$ lead to PEDP (19).

3.3 Cloning of Genes Encoding Interacting Proteins

Channels are often presented in cells as multi-subunit proteins. Some channel-associated proteins promote functional channel expressions (50). Thus, $Na_V1.8$ channel transfection into superior cervical ganglion (SCG) neurons shows exactly the same biophysical properties as those observed in DRG neurons (23). In contrast, the $Na_V1.8$ channel cannot be effectively and properly expressed in popular mammalian cell lines such as HEK and CHO, even in the presence of auxiliary β-subunits (51). Therefore, it could be suggested that $Na_V1.8$ channel requires an adapter protein or auxiliary subunit for correct functioning.

There are several cloning strategies to isolate adapter proteins, additional or auxiliary subunits for defined channels. In 1980s and early 1990s, a biochemical approach was the most popular. This approach relies on direct isolation of interacting proteins using affinity chromatography (52, 53). A major disadvantage of the method is low sensitivity, as substantial amounts of tissue are required for purification of interacting proteins. To increase sensitivity and efficiency of this cloning procedure, it was suggested that approaches involving expression cloning and detection of interacting proteins could be combined. This basic idea eventually brought to discovery and development the two-hybrid cloning system for identification of genes via protein–protein interactions (54). This approach was improved within the last two decades; however, general principles remained the same (Fig. 3). In normal yeast cells, *Gal4* transcription factor consists of two subunits (BD and AD). Interaction of these subunits leads to binding of an activation sequence (AS) and transcription of genes. This way *Gal4* can activate a reporter gene such as *LacZ* (Fig. 3a). A yeast line without *Gal4* and with LacZ reporter was generated. A critical step is the creation of an interaction domain on the protein of interest called the "bait." Bait fused to BD subunit of *Gal4* does not produce activation of LacZ (Fig. 3b). A cDNA library is constructed in a specialized vector containing the AD subunit of *Gal4*. Transfection of the yeast line with the cDNA–AD library should not produce LacZ activation (Fig. 3c). However, the cDNA–AD library contains a candidate clone, named "prey" (the one which is the object of identification), which can interact with the "bait." The yeast line is co-transfected with cDNA–AD library and "bait"–BD construct. Hopefully, in some yeast cells, "pray"–AD and "bait"–BD will copresent, and this will result in activation LacZ reporter (Fig. 3d). The LacZ activation can be detected by a staining procedure, which will turn yeast cells blue. The yeast clone will be selected and "prey"–AD cDNA can be isolated, sequenced, and analyzed. This approach was used to clone an adapter protein for $Na_V1.8$ channel (51). The two-hybrid system is now widely used. However, it has two deficiencies: (1) cloning transmembrane proteins is problematic with this approach and (2) false-positive clones can be identified and, thus, these need to be filtered out with further detailed biochemical studies. Nevertheless, it offers the possibility to clone proteins that control channels' activities in the nociceptive pathways. These proteins could provide potentially attractive and novel analgesic targets, because channel activities could be regulated by disrupting interactions of channels with newly discovered adapter proteins or auxiliary subunits.

3.4 Biochemical Approach to Functional Cloning

Although fairly outdated now, in the early days of molecular cloning, many classical channels were cloned using biochemical approaches. The biochemical approach is based on cloning genes encoding

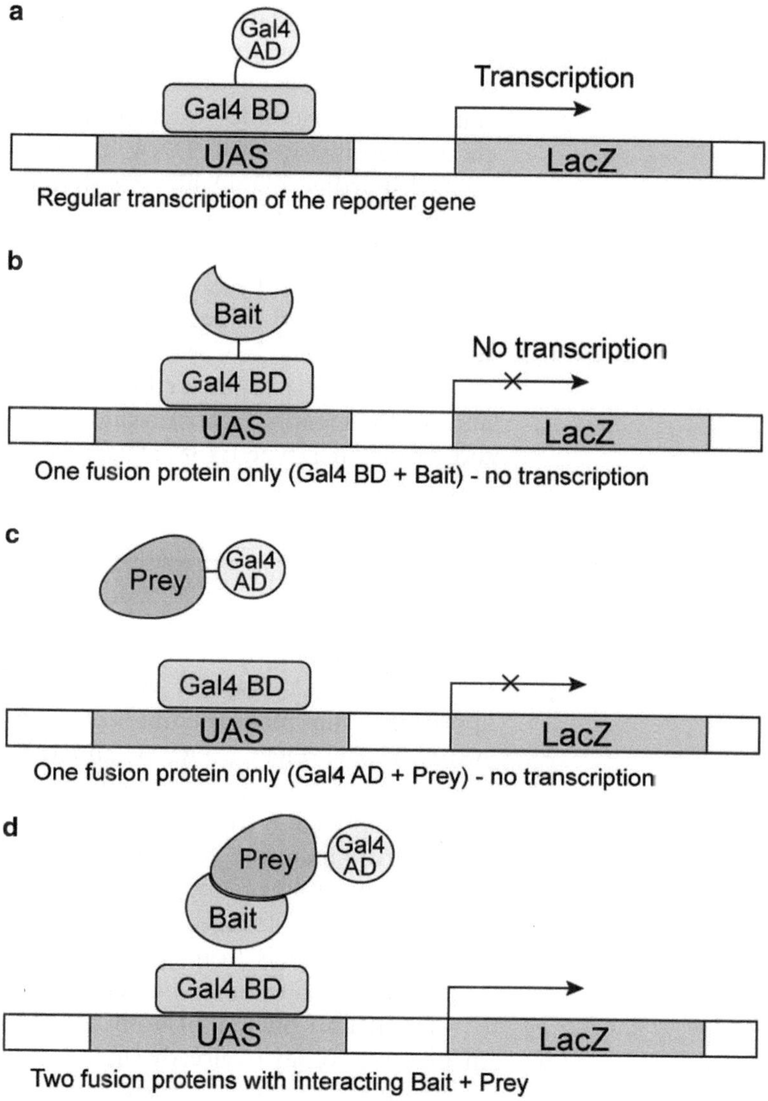

Fig. 3 Schematic representation of two-hybrid cloning of interacting proteins. This approach was utilized to clone adapter protein for $Na_v1.8$ sodium channel

particular purified proteins. Thus, the first step of this approach is to isolate and purify the required amount of protein of interest. The next step involves short (10–20 aa) sequences of a purified protein from the start codon (Met) as well as any 2–3 regions within the protein. To achieve this, a peptide map of the protein is resolved using a variety of proteases, as well as several peptide fragments of the purified protein are generated, isolated, and sequenced (10–20 aa) from N-terminal parts. It is expected that enough of the purified protein is generated for these procedures. On the basis of the protein sequences, degenerate oligonucleotides (20–40 p) are synthesized. In this case, degenerate oligonucleotides are a mixture of all possible oligonucleotides that will encode a determined protein sequence. Finally, the degenerated oligonucleotides are labeled and used for conventional screening of cDNA library. Such screening with degenerated oligonucleotides offers low sensitivity. Hence, to achieve enough sensitivity, small cDNA libraries (up to 100,000 clones) are utilized. Further, cDNA libraries should be full length as much as possible, because the protein sequence encoded by degenerated oligonucleotides belongs to the deep N-terminal part of open reading frames. To filter off false clones, isolated clones are probed with other degenerated primers encoding other determined protein sequences (usually 2–3 belonging to a variety of parts of the protein). This approach was used to clone such classic channels as subunits of the nicotinic acetylcholine receptor (55–57), voltage-gated sodium channel (type I; (58, 59), and voltage-gated calcium channel (L-type; (60, 61)).

4 Homology Cloning

Homology cloning is an extremely productive approach for cloning of genes. A majority of channels were cloned using this approach. Thus, many members of channel families including P2X, ASIC, TRPV, TRPC, TRPM, and voltage-gated sodium, potassium, and calcium channel families were cloned using homology cloning approaches (62–73). The original cloned genes could have origins from mammal, fly, or worm. For example, the original nicotinic acetylcholine receptor and voltage-gated sodium channels have been cloned from squid neurons (55, 74), and voltage-gated potassium channels from flies (75). Their mammalian orthologs were isolated using homology cloning approaches (58, 63, 76).

The main principle of homology cloning is to isolate gene(s) that has homology to already cloned genes. Homology cloning can be conducted two ways: screening conventional cDNA library with homologous probes or finding homologous genes using bioinformatics software and genome sequences databases generated by the Human Genome Project.

To screen cDNA library with homologous probes, several parameters should be taken into account. First, it should be defined in what tissue to look for homologous cDNA. Base on this presumption (or information), a cDNA library is constructed from mRNA isolated from this particular, selected tissue (or cell lines). Second, one of the most critical steps for successful homology cloning is to select (or decide on) probe(s) for homology screening and decide on hybridization temperature during screening. Such probes can cover a conservative domain (62) or a part of gene(s) (76, 77). All in all, homology screening can involve trial and error to succeed in cloning.

Homology cloning based on the bioinformatics approach is straightforward. Genome sequences available from different data bases are screened for homology to particular sequences using specialized software. Here, a critical step is the algorithm for software used to locate homologous sequences in whole genomes. Often, several software packages with different algorithms are used to find homologous sequences in genomes. Homologous sequences usually do not cover an entire cDNA. Therefore, this partial cDNA sequence is used for direct screening of cDNA libraries. Alternatively, there are software packages which are able to identify intron–exon structure of an entire gene. In this case, full-length cDNA can be PCR amplified from total mRNA. This procedure is much faster and more cost-effective. Homology cloning based on bioinformatics approaches has been employed to clone such notable channels as TRPV3 (78) and TRPA1 (8).

5 Conclusion

It is well accepted that molecular biology revolutionized many research fields. Molecular pain research is not the exception to this rule. Cloning of genes involved in physiological and pathophysiological processes requires molecular biology approaches. Initially, many genes important for nociceptive processing were cloned with other goals in mind. Take for example the TTX-sensitive sodium channels type III and type VI and subunits of nicotinic acetylcholine receptors. They play critical roles in many neuronal processes including nociceptive signal processing. ATP-gated channel, P2X3, could be considered the first gene cloned specifically for nociceptive processing (6). The following decades produced many notable discoveries related to cloning of genes involved in nociceptive processing in periphery and centrally: PN1, SNS, and NaN voltage-gated sodium channels; TRPV1, TRPM8, and TRPA1 temperature-gated channels; Piezo1 and 2 mechano-gated channels; and ASIC 1 and 3 proton-gated channels (5, 7, 8, 19, 33–36, 44, 67, 79). Almost all existing cloning approaches have been used to clone pain genes. Although the application of molecular cloning techniques to discovery of pain-specific genes has probably passed its peak, the discussed approaches can be used to pinpoint genes involved in a variety of acute and chronic pain conditions in animals and humans (17, 44). Furthermore, advances in gene therapy could eventually make examination of molecular mechanisms utilizing the above techniques invaluable.

Acknowledgements

I would like to thank to members of my lab Dr. Belugin, Mr. Patil, Ms. Salas, and Ms. Phoebe as well as collaborators Drs. Wood, Hargreaves, Henry, Jeske, Diogenes, Gamper Staruschenko,

Brooks, Dube, Nikita Ruparel, Shivani Ruparel, and Patwardhan for helping me in contribution to molecular pain research. Supporting grants are DE014928 and DE019311.

References

1. Akopian AN, Abson NC, Wood JN (1996) Molecular genetic approaches to nociceptor development and function. Trends Neurosci 19:240–246
2. Muir WW 3rd, Woolf CJ (2001) Mechanisms of pain and their therapeutic implications. J Am Vet Med Assoc 219:1346–1356
3. Berg P (2008) Meetings that changed the world: Asilomar 1975: DNA modification secured. Nature 455:290–291
4. Berg P, Baltimore D, Brenner S, Roblin RO 3rd, Singer MF (1975) Asilomar conference on recombinant DNA molecules. Science 188:991–994
5. Akopian AN, Sivilotti L, Wood JN (1996) A tetrodotoxin-resistant voltage-gated sodium channel expressed by sensory neurons. Nature 379:257–262
6. Chen CC, Akopian AN, Sivilotti L, Colquhoun D, Burnstock G, Wood JN (2005) A P2X purinoceptor expressed by a subset of sensory neurons. Nature 377:428–431
7. Caterina MJ, Schumacher MA, Tominaga M, Rosen TA, Levine JD, Julius D (1997) The capsaicin receptor: a heat-activated ion channel in the pain pathway. Nature 389:816–824
8. Story GM, Peier AM, Reeve AJ, Eid SR, Mosbacher J, Hricik TR, Earley TJ, Hergarden AC, Andersson DA, Hwang SW et al (2003) ANKTM1, a TRP-like channel expressed in nociceptive neurons, is activated by cold temperatures. Cell 112:819–829
9. McKemy DD, Neuhausser WM, Julius D (2002) Identification of a cold receptor reveals a general role for TRP channels in thermosensation. Nature 416:52–58
10. Waldmann R, Champigny G, Bassilana F, Heurteaux C, Lazdunski M (1997) A proton-gated cation channel involved in acid-sensing. Nature 386:173–177
11. Chen CL, Broom DC, Liu Y, de Nooij JC, Li Z, Cen C, Samad OA, Jessell TM, Woolf CJ, Ma Q (2006) Runx1 determines nociceptive sensory neuron phenotype and is required for thermal and neuropathic pain. Neuron 49: 365–377
12. Kim AY, Tang Z, Liu Q, Patel KN, Maag D, Geng Y, Dong X (2008) Pirt, a phosphoinositide-binding protein, functions as a regulatory subunit of TRPV1. Cell 133:475–485
13. Ravenall SJ, Gavazzi I, Wood JN, Akopian AN (2002) A peripheral nervous system actin-binding protein regulates neurite outgrowth. Eur J Neurosci 15:281–290
14. Woolf CJ, Costigan M (1999) Transcriptional and posttranslational plasticity and the generation of inflammatory pain. Proc Natl Acad Sci U S A 96:7723–7730
15. Woolf CJ, Salter MW (2000) Neuronal plasticity: increasing the gain in pain. Science 288: 1765–1769
16. Befort K, Costigan M, Woolf CJ (2001) Differential gene expression—how to find new analgesic targets. Curr Opin Investig Drugs 2:396–398
17. Indo Y, Tsuruta M, Hayashida Y, Karim MA, Ohta K, Kawano T, Mitsubuchi H, Tonoki H, Awaya Y, Matsuda I (1996) Mutations in the TRKA/NGF receptor gene in patients with congenital insensitivity to pain with anhidrosis. Nat Genet 13:485–488
18. Drenth JP, Waxman SG (2007) Mutations in sodium-channel gene SCN9A cause a spectrum of human genetic pain disorders. J Clin Invest 117:3603–3609
19. Fertleman CR, Baker MD, Parker KA, Moffatt S, Elmslie FV, Abrahamsen B, Ostman J, Klugbauer N, Wood JN, Gardiner RM et al (2006) SCN9A Mutations in paroxysmal extreme pain disorder: allelic variants underlie distinct channel defects and phenotypes. Neuron 52:767–774
20. Weiss J, Pyrski M, Jacobi E, Bufe B, Willnecker V, Schick B, Zizzari P, Gossage SJ, Greer CA, Leinders-Zufall T et al (2011) Loss-of-function mutations in sodium channel Nav1.7 cause anosmia. Nature 472:186–190
21. Mogil JS, Yu L, Basbaum AI (2000) Pain genes?: natural variation and transgenic mutants. Annu Rev Neurosci 23:777–811
22. Diatchenko L, Slade GD, Nackley AG, Bhalang K, Sigurdsson A, Belfer I, Goldman D, Xu K, Shabalina SA, Shagin D et al (2005) Genetic basis for individual variations in pain perception and the development of a chronic pain condition. Hum Mol Genet 14:135–143
23. Akopian AN, Souslova V, England S, Okuse K, Ogata N, Ure J, Smith A, Kerr BJ, McMahon SB, Boyce S et al (1999) The tetrodotoxin-resistant sodium channel SNS has a specialized function in pain pathways. Nat Neurosci 2:541–548
24. Caterina MJ, Leffler A, Malmberg AB, Martin WJ, Trafton J, Petersen-Zeitz KR, Koltzenburg

M, Basbaum AI, Julius D (2000) Impaired nociception and pain sensation in mice lacking the capsaicin receptor. Science 288:306–313
25. Kwan KY, Allchorne AJ, Vollrath MA, Christensen AP, Zhang DS, Woolf CJ, Corey DP (2006) TRPA1 contributes to cold, mechanical, and chemical nociception but is not essential for hair-cell transduction. Neuron 50:277–289
26. Souslova V, Cesare P, Ding Y, Akopian AN, Stanfa L, Suzuki R, Carpenter K, Dickenson A, Boyce S, Hill R et al (2000) Warm-coding deficits and aberrant inflammatory pain in mice lacking P2X3 receptors. Nature 407:1015–1017
27. Zurborg S, Piszczek A, Martinez C, Hublitz P, Al Banchaabouchi M, Moreira P, Perlas E, Heppenstall PA (2011) Generation and characterization of an Advillin-Cre driver mouse line. Mol Pain 7:66–69
28. Stirling LC, Forlani G, Baker MD, Wood JN, Matthews EA, Dickenson AH, Nassar MA (2005) Nociceptor-specific gene deletion using heterozygous Nav1.8-Cre recombinase mice. Pain 113:27–36
29. Meunier A, Pohl M (2009) Lentiviral vectors for gene transfer into the spinal cord glial cells. Gene Ther 16:476–482
30. Ro LS, Chang KH (2005) Neuropathic pain: mechanisms and treatments. Chang Gung Med J 28:597–605
31. Hedrick SM, Cohen DI, Nielsen EA, Davis MM (1984) Isolation of cDNA clones encoding T cell-specific membrane-associated proteins. Nature 308:149–153
32. Hedrick SM, Nielsen EA, Kavaler J, Cohen DI, Davis MM (1984) Sequence relationships between putative T-cell receptor polypeptides and immunoglobulins. Nature 308:153–158
33. Chen CC, England S, Akopian AN, Wood JN (1998) A sensory neuron-specific, proton-gated ion channel. Proc Natl Acad Sci U S A 95:10240–10245
34. Akopian AN, Wood JN (1995) Peripheral nervous system-specific genes identified by subtractive cDNA cloning. J Biol Chem 270: 21264–21270
35. Coste B, Mathur J, Schmidt M, Earley TJ, Ranade S, Petrus MJ, Dubin AE, Patapoutian A (2010) Piezo1 and Piezo2 are essential components of distinct mechanically activated cation channels. Science 330:55–60
36. Waldmann R, Bassilana F, de Weille J, Champigny G, Heurteaux C, Lazdunski M (1997) Molecular cloning of a non-inactivating proton-gated Na+channel specific for sensory neurons. J Biol Chem 272:20975–20978
37. Rougeon F, Chambraud B, Foote S, Panthier JJ, Nageotte R, Corvol P (1981) Molecular cloning of a mouse submaxillary gland renin cDNA fragment. Proc Natl Acad Sci USA 78:6367–6371
38. Fodor SP, Rava RP, Huang XC, Pease AC, Holmes CP, Adams CL (1993) Multiplexed biochemical assays with biological chips. Nature 364:555–556
39. Diogenes A, Patwardhan AM, Jeske NA, Ruparel NB, Goffin V, Akopian AN, Hargreaves KM (2006) Prolactin modulates TRPV1 in female rat trigeminal sensory neurons. J Neurosci 26:8126–8136
40. Brake AJ, Wagenbach MJ, Julius D (1994) New structural motif for ligand-gated ion channels defined by an ionotropic ATP receptor. Nature 371:519–523
41. Drew LJ, Wood JN, Cesare P (2002) Distinct mechanosensitive properties of capsaicin-sensitive and -insensitive sensory neurons. J Neurosci 22:RC228
42. Axelrod FB, Hilz MJ (2003) Inherited autonomic neuropathies. Semin Neurol 23: 381–390
43. Axelrod FB, Gold-von Simson G (2007) Hereditary sensory and autonomic neuropathies: types II, III, and IV. Orphanet J Rare Dis 2:39
44. Cox JJ, Reimann F, Nicholas AK, Thornton G, Roberts E, Springell K, Karbani G, Jafri H, Mannan J, Raashid Y et al (2006) An SCN9A channelopathy causes congenital inability to experience pain. Nature 444:894–898
45. Fischer TZ, Waxman SG (2010) Familial pain syndromes from mutations of the Nav1.7 sodium channel. Ann N Y Acad Sci 1184: 196–207
46. Yang Y, Wang Y, Li S, Xu Z, Li H, Ma L, Fan J, Bu D, Liu B, Fan Z et al (2004) Mutations in SCN9A, encoding a sodium channel alpha subunit, in patients with primary erythermalgia. J Med Genet 41:171–174
47. Bejaoui K, Wu C, Scheffler MD, Haan G, Ashby P, Wu L, de Jong P, Brown RH Jr (2001) SPTLC1 is mutated in hereditary sensory neuropathy, type 1. Nat Genet 27: 261–262
48. Bejaoui K, Uchida Y, Yasuda S, Ho M, Nishijima M, Brown RH Jr, Holleran WM, Hanada K (2002) Hereditary sensory neuropathy type 1 mutations confer dominant negative effects on serine palmitoyltransferase, critical for sphingolipid synthesis. J Clin Invest 110:1301–1308
49. Slaugenhaupt SA, Blumenfeld A, Gill SP, Leyne M, Mull J, Cuajungco MP, Liebert CB, Chadwick B, Idelson M, Reznik L et al (2001) Tissue-specific expression of a splicing mutation in the IKBKAP gene causes familial dysautonomia. Am J Hum Genet 68:598–605

50. Isom LL (2001) Sodium channel beta subunits: anything but auxiliary. Neuroscientist 7: 42–54
51. Okuse K, Malik-Hall M, Baker MD, Poon WY, Kong H, Chao MV, Wood JN (2002) Annexin II light chain regulates sensory neuron-specific sodium channel expression. Nature 417: 653–656
52. Vaghy PL, Striessnig J, Miwa K, Knaus HG, Itagaki K, McKenna E, Glossmann H, Schwartz A (1987) Identification of a novel 1,4-dihydropyridine- and phenylalkylamine-binding polypeptide in calcium channel preparations. J Biol Chem 262:14337–14342
53. Hartshorne RP, Catterall WA (1984) The sodium channel from rat brain. Purification and subunit composition. J Biol Chem 259:1667–1675
54. Fields S, Song O (1989) A novel genetic system to detect protein–protein interactions. Nature 340:245–246
55. Noda M, Takahashi H, Tanabe T, Toyosato M, Furutani Y, Hirose T, Asai M, Inayama S, Miyata T, Numa S (1982) Primary structure of alpha-subunit precursor of Torpedo californica acetylcholine receptor deduced from cDNA sequence. Nature 299:793–797
56. Noda M, Takahashi H, Tanabe T, Toyosato M, Kikyotani S, Furutani Y, Hirose T, Takashima H, Inayama S, Miyata T et al (1983) Structural homology of Torpedo californica acetylcholine receptor subunits. Nature 302: 528–532
57. Noda M, Takahashi H, Tanabe T, Toyosato M, Kikyotani S, Hirose T, Asai M, Takashima H, Inayama S, Miyata T et al (1983) Primary structures of beta- and delta-subunit precursors of Torpedo californica acetylcholine receptor deduced from cDNA sequences. Nature 301:251–255
58. Noda M, Ikeda T, Kayano T, Suzuki H, Takeshima H, Kurasaki M, Takahashi H, Numa S (1986) Existence of distinct sodium channel messenger RNAs in rat brain. Nature 320:188–192
59. Noda M, Ikeda T, Suzuki H, Takeshima H, Takahashi T, Kuno M, Numa S (1986) Expression of functional sodium channels from cloned cDNA. Nature 322:826–828
60. Tanabe T, Takeshima H, Mikami A, Flockerzi V, Takahashi H, Kangawa K, Kojima M, Matsuo H, Hirose T, Numa S (1987) Primary structure of the receptor for calcium channel blockers from skeletal muscle. Nature 328: 313–318
61. Tanabe T, Beam KG, Powell JA, Numa S (2002) Restoration of excitation-contraction coupling and slow calcium current in dysgenic muscle by dihydropyridine receptor complementary DNA. Nature 336:134–139
62. Xu H, Ramsey IS, Kotecha SA, Moran MM, Chong JA, Lawson D, Ge P, Lilly J, Silos-Santiago I, Xie Y et al (2002) TRPV3 is a calcium-permeable temperature-sensitive cation channel. Nature 418:181–186
63. Kayano T, Noda M, Flockerzi V, Takahashi H, Numa S (1988) Primary structure of rat brain sodium channel III deduced from the cDNA sequence. FEBS Lett 228:187–194
64. Clapham DE (2003) TRP channels as cellular sensors. Nature 426:517–524
65. MacKenzie AB, Surprenant A, North RA (1999) Functional and molecular diversity of purinergic ion channel receptors. Ann N Y Acad Sci 868:716–729
66. North RA, Barnard EA (1997) Nucleotide receptors. Curr Opin Neurobiol 7:346–357
67. Toledo-Aral JJ, Moss BL, He ZJ, Koszowski AG, Whisenand T, Levinson SR, Wolf JJ, Silos-Santiago I, Halegoua S, Mandel G (1997) Identification of PN1, a predominant voltage-dependent sodium channel expressed principally in peripheral neurons. Proc Natl Acad Sci U S A 94:1527–1532
68. Guler AD, Lee H, Iida T, Shimizu I, Tominaga M, Caterina M (2002) Heat-evoked activation of the ion channel, TRPV4. J Neurosci 22:6408–6414
69. Xu XZ, Moebius F, Gill DL, Montell C (2001) Regulation of melastatin, a TRP-related protein, through interaction with a cytoplasmic isoform. Proc Natl Acad Sci USA 98:10692–10697
70. Waldmann R, Champigny G, Lingueglia E, De Weille JR, Heurteaux C, Lazdunski M (1999) H(+)-gated cation channels. Ann N Y Acad Sci 868:67–76
71. Miljanich GP, Ramachandran J (1995) Antagonists of neuronal calcium channels: structure, function, and therapeutic implications. Annu Rev Pharmacol Toxicol 35:707–734
72. Waxman SG, Cummins TR, Dib-Hajj SD, Black JA (2000) Voltage-gated sodium channels and the molecular pathogenesis of pain: a review. J Rehabil Res Dev 37:517–528
73. Kubo Y (1994) Towards the elucidation of the structural-functional relationship of inward rectifying K^+ channel family. Neurosci Res 21(2):109–117
74. Noda M, Shimizu S, Tanabe T, Takai T, Kayano T, Ikeda T, Takahashi H, Nakayama H, Kanaoka Y, Minamino N et al (1984) Primary structure of Electrophorus electricus sodium channel deduced from cDNA sequence. Nature 312:121–127
75. Papazian DM, Schwarz TL, Tempel BL, Jan YN, Jan LY (1987) Cloning of genomic and complementary DNA from shaker, a putative potassium channel gene from drosophila. Science 237:749–753

76. Boulter J, Evans K, Goldman D, Martin G, Treco D, Heinemann S, Patrick J (1986) Isolation of a cDNA clone coding for a possible neural nicotinic acetylcholine receptor alpha-subunit. Nature 319: 368–374
77. Caterina MJ, Rosen TA, Tominaga M, Brake AJ, Julius D (1999) A capsaicin-receptor homologue with a high threshold for noxious heat. Nature 398:436–441
78. Smith GD, Gunthorpe MJ, Kelsell RE, Hayes PD, Reilly P, Facer P, Wright JE, Jerman JC, Walhin JP, Ooi L et al (2002) TRPV3 is a temperature-sensitive vanilloid receptor-like protein. Nature 418:186–190
79. Dib-Hajj SD, Tyrrell L, Black JA, Waxman SG (1998) NaN, a novel voltage-gated Na channel, is expressed preferentially in peripheral sensory neurons and down-regulated after axotomy. Proc Natl Acad Sci USA 95:8963–8968

Chapter 2

Mammalian Expression Systems and Transfection Techniques

Daunia Laurenti and Lezanne Ooi

Abstract

To delineate the function of a single ion channel subtype amongst the multitude that normally constitutes a signalling pathway, it is often insightful to study the function or signalling of that one ion channel in the absence of the others. Mammalian cell lines that do not normally express the gene of interest can be manipulated to do so via plasmid DNA expression vectors. However, large and highly charged molecules like DNA cannot passively diffuse through cell membranes. Therefore introducing nucleic acid into mammalian cells may involve introducing pores transiently into the cell membrane to allow the passage of circular plasmid DNA. This is relatively easily carried out using cationic lipids that form liposomes around the DNA and fuse with the cell membrane to introduce the DNA inside the cell. Alternatively, a highly successful mechanism for introduction of DNA involves utilizing viral vectors. These take advantage of the molecular mechanisms that viruses have evolved to efficiently transport their genome inside cells. Lipid-based transfection techniques and adenoviral delivery of plasmids encoding large genes (such as ion channel genes) for expression in mammalian cells are the focus of this chapter.

Key words Transient transfection, Stable transfection, Adenovirus, Transfection efficiency, Reporter gene, Bicistronic vector

1 Introduction

Since many ion channels are regulated by diverse posttranslational modifications (such as phosphorylation, glycosylation, disulfide bond formation) that can differ between species, in order to study mammalian ion channel function in a heterologous system, it is beneficial to use a mammalian expression system. Mammalian cells can be transfected with plasmid DNA encoding ion channel genes. Plasmid DNA may either be introduced into the cells in a transient fashion or may integrate into the cell genome. Cells that have incorporated the plasmid DNA into their genome can be selected via the use of antibiotic resistance genes to generate stable clones that express a protein of interest. This chapter focuses on the methods of transient and stable transfection by lipid-based

Nikita Gamper (ed.), *Ion Channels: Methods and Protocols*, Methods in Molecular Biology, vol. 998,
DOI 10.1007/978-1-62703-351-0_2, © Springer Science+Business Media, LLC 2013

transfection and transient expression by adenoviral delivery. Three methods of assessing transfection efficiency and identifying transfected cells are also discussed.

1.1 Using Reporters to Analyze the Efficiency of Transfection

The efficiency of transfection is a major concern for many applications and should be monitored using a reporter, such as a fluorescent protein, e.g., enhanced green fluorescent protein (eGFP), to confirm the transfer of nucleic acid to the cell. This can be achieved by (1) transfecting two separate plasmids, one of which encodes eGFP. This is the weakest of the three options presented here; although it is likely that if one plasmid was able to transit through a particular cell membrane then both plasmids were able to do so, it cannot be guaranteed (see Note 1). (2) Generation of a fusion protein, using a plasmid such as pEGFP-N1 or pEGFP-C1 (Clontech), in which your protein of interest is tagged with eGFP (as one continuous protein). Careful consideration should be given as to the position of the eGFP relative to the protein of interest (i.e., N-terminal or C-terminal) since the 27 kDa eGFP may affect folding, localization, or function of your target protein. (3) Another option would be to use a bicistronic vector (generates a single mRNA that encodes two proteins), such as pIRES2-eGFP (Clontech). This vector contains an internal ribosomal entry site (IRES) sequence, a nucleotide sequence that promotes translation initiation in the middle of an mRNA and thus allows the simultaneous expression of the protein of interest and eGFP (as a separate protein) from the same mRNA transcript. With this method eGFP expression can be used to determine efficiency of transfection and identify those cells that are positively transfected (since all cells expressing the eGFP were successfully transfected) without the eGFP affecting the folding of the protein of interest. One other important consideration is that the expression of any reporter may affect the expression of other proteins or the function of the cell, thus control transfections should be carried out in which the eGFP (in the absence of your protein of interest) is expressed (see Note 2).

1.2 Transient Versus Stable Transfection

The type of transfection method used, the size of the DNA plasmid to be transfected, the amount of DNA transfected relative to the amount of reagent, and the number of cells plated will all affect the efficiency of transfection. Many ion channel genes are relatively large, often leading to low-transfection efficiencies in mammalian cells. Particularly in cases where the efficiency of transfection is an issue, or to standardize experiments, you may want to consider generating a stable cell line, in which the plasmid has integrated into the genome of the cell. This means that the cell line stably expresses your protein(s) of interest through successive rounds of division. With transient transfections, expression of the proteins will reduce as the plasmid is lost through cell division.

For this reason, cells are often used for experiments 24–48 h after transfection (division approximately once every 24 h in many mammalian cell lines). Although more time consuming initially, generating stable cell lines can save a large amount of time and money throughout the project and may also lead to more consistent results. Plasmids such as pEGFP-N1 and pIRES2-GFP contain a neomycin resistance gene for selection of stably transfected cells. Selection of stable clones is achieved by incubating the cells with G418 (also known as geneticin) over a period of weeks. Cells that have not integrated the plasmid into the genome die following extended exposure to G418. Individual cells expressing the reporter gene can then be picked and diluted to single cells that then undergo clonal expansion.

1.3 Adenoviral Delivery

Viral vectors permit high levels of transgene expression in many cell types, often without affecting cell viability. Since the adenoviral DNA does not integrate into the genome and is not replicated during cell division, adenoviral delivery mediates transient expression of transgenes. The adenoviral DNA vector contains a portion of the viral genome plus a gene of interest cloned into a multiple cloning site inserted into the early region 1A (E1A) of the genome. Recombinant adenoviral vectors are generated using this viral DNA vector and a packaging cell line (e.g., HEK293, which have been stably transfected with the E1A region of the adenoviral genome). Via a method adapted from He and colleagues (1), we have previously successfully prepared adenoviral vectors to transiently express genes and reporters in cell lines (2, 3) and primary cells, including neurons (4) and cardiac myocytes (5). The gene of interest is cloned into the pAdTrack-CMV vector, linearized, and transformed into an *Escherichia coli* strain BJ5183 that expresses pAdEasy adenoviral vector (pAdEasy-1). The recombinant adenoviral construct is then transfected into a packaging cell line (e.g., HEK293) for virus production. During the generation of the HEK293 cell line, normal human embryonic kidney cells were transformed with adenovirus five DNA and this cell line can be utilized for the propagation of adenoviral vectors. Since adenoviruses are pathogenic it is important to use adenoviral vectors in which certain genes are deleted so that the viruses are unable to replicate after entering a cell. HEK293 can be safely used in combination with the adenoviral vector pAdEasy, in which the E1 and E3 genes have been deleted (1). Using this system viral production and transgene expression can be monitored with the reporter eGFP.

In this chapter, we outline the lipid-based methods used to transfect mammalian cells transiently and following on from this how to generate stable cell lines. We then outline how to prepare adenovirus particles, which can be used to infect mammalian cells to express genes of interest.

2 Materials

1. Cell culture medium: Dulbecco's Modified Eagle Medium (DMEM), 10% fetal bovine serum, 2 mM glutamine, streptomycin (10 g/L), and penicillin (10 g/L); pre-warmed to 37°C.
2. Serum-free media (e.g., Opti-MEM).
3. Lipid transfection reagent [e.g., Lipofectamine 2000, Lipofectamine LTX (Invitrogen)].
4. Cell culture plates.
5. Phosphate-buffered saline (PBS).
6. Sterile, round bottomed 14 mL centrifuge tubes.
7. Purified plasmid DNA (e.g., pIRES2-GFP, pAdTrack-CMV).
8. Cell line for transfection, e.g., HEK293, CHO, Neuro2A

   For stable transfection:
9. G418 (Sigma).
10. 96-well cell culture plates.

    For adenovirus preparation:
11. Restriction enzymes *Pme*I, *Pac*I.
12. Bacterial strain BJ5183 containing pAdEasy-1.
13. Bacterial strain XL1 Blue.
14. Electroporation cuvettes, electroporator (option to use chemically competent cells and heat shock for electroporation).
15. Luria-Bertani Media (LB Broth: Tryptone 1.0%, Yeast Extract 0.5%, NaCl 1.0%)+kanamycin (10 μg/mL) or ampicillin (25 μg/mL). For plates, 10% agar.
16. Tris–HCl pH 8.0 0.1 M.
17. Sodium deoxycholate 5%.
18. Magnesium chloride 2 M.
19. DNaseI (100 mg DNaseI in 10 mL of 20 mM Tris–HCl pH 7.4, 50 mM NaCl, 0.1 mM DTT, 0.1 mg/mL BSA, and 50% glycerol).
20. Cesium chloride 1.5 g/mL (90.8 g CsCl and 109.2 g 10 mM Tris pH 8).
21. Cesium chloride 1.35 g/mL (70.4 g CsCl and 129.6 g 10 mM Tris pH 8).
22. Cesium chloride 1.25 g/mL (54 g CsCl and 146 g 10 mM Tris pH 8).
23. Polyallomer tubes.
24. Dialysis buffer (10 mM Tris–HCl pH 8, 10% glycerol).
25. Slide-a-lyzer dialysis cassette.

3 Methods

3.1 Transient Transfection

1. Seed cells at a density of 1.5×10^6 cells per well of a 6-well plate 24 h prior to transfection.
2. Dilute 10 μg of plasmid in 250 μL of Opti-MEM (mix A).
3. Dilute 30 μL of Lipofectamine 2000 in 250 μL Opti-MEM (mix B).
4. Incubate at room temperature for 20 min.
5. Add mix A to mix B.
6. Incubate for 25 min at room temperature.
7. Wash the cells with PBS.
8. Add the mix A + B to cells.
9. Incubate for 1 min with gentle agitation.
10. Add 500 μL of Opti-MEM medium.
11. Incubate for 1 h at 37°C.
12. Add 1.5 mL of DMEM: 10% fetal bovine serum and glutamine added—NO ANTIBIOTICS.
13. Incubate the cells for 24 h at 37°C 5% CO_2.
14. Change medium or add 1 mL of DMEM: 10% fetal bovine serum and glutamine WITH antibiotics (if required).
15. After 24 h change medium and check fluorescence under microscope.

3.1.1 Optimization of Transfection Efficiency

Cell lines that transfect easily and show high levels of transfection efficiency (>80%) using this method include human embryonic kidney 293 (HEK293) and Chinese hamster ovary (CHO). Other cell lines that show reduced transfection efficiencies may be required for use, e.g., the mouse neuroblastoma cell line Neuro2A. In this case, it is extremely important to optimize transfection; some hints for this are included below. We tested Lipofectamine LTX (which is suggested by the manufacturer for high-efficiency transfection of Neuro2A cells) and Lipofectamine 2000 (which is not specifically recommended by the manufacturer for transfection of Neuro2A cells). Based on the original protocol from both products two different transfections were tested. For each, a control transfection was carried out with the "empty plasmid" pEGFP-N1 (4.7 kb) (control plasmid) and the test plasmid containing the cloned gene of interest in pEGFP-GOI (11 kb).

A. Lipofectamine 2000, following the manufacturer's protocol for general transfection of 1×10^6 murine cells in a 6-well plate. The protocol was as above except for:

 1. Transfection with 4 μg DNA and 10 μL Lipofectamine 2000.

2. Incubate mix A and B at room temperature for 5 min.
3. Incubate mix A+B with cells at room temperature for 1 h.
4. Add 1 mL of DMEM 10% FBS+glutamine without antibiotics.
5. Incubate cells for 24 h at 37°C 5% CO_2.
6. After 24 h add 1 mL of DMEM with antibiotics.

B. Lipofectamine LTX, following the manufacturer's protocol for transfection of 2.5×10^5 Neuro2A cells in a 6-well plate.

1. Aspirate media and wash cells with PBS.
2. Add 1.5 mL DMEM: 10% FBS and glutamine without antibiotics.
3. Incubate 1.25 μg DNA in 500 μL of Opti-MEM.
4. Incubate at room temperature for 5 min.
5. Add 20 μL Lipofectamine LTX to the mix.
6. Incubate at room temperature for 25 min.
7. Add mix to cells.
8. Incubate cells for 24 h at 37°C 5% CO_2.
9. Change medium after 24 h with DMEM with antibiotics.

The transfection efficiencies achieved with these two protocols of Neuro2A cell transfection with pEGFP-N1 and pEGFP-GOI are given in Table 1. Increased efficiencies for transfection were identified by optimization of the following steps:

1. Increased number of cells seeded (2×10^6 cells in a 6-well plate) (step 1 of Subheading 3.1).
2. Increased Lipofectamine 2000 (30 μL was the optimal volume; the results were similar with 35 μL but lower with 25 μL) (step 3 of Subheading 3.1).
3. Increased time of separate incubation for mix A and mix B in Opti-MEM (up to 25 min) (step 4 of Subheading 3.1).
4. Increased time of incubation after transfection (2–3 h) (step 11 of Subheading 3.1).

Table 1
Transfection efficiencies vary with plasmid size and reagent used

Lipofectamine + plasmid	Efficiency of transfection (%)
Lipofectamine 2000 + plasmid pEGFP-N1 4.7 kb (empty, control)	20
Lipofectamine 2000 + plasmid pEGFP-GOI 11 kb	15
Lipofectamine LTX + plasmid pEGFP-N1 4.7 kb (empty, control)	15
Lipofectamine LTX + plasmid pEGFP-GOI 11 kb	5

Table 2
Transfection efficiencies vary with ratio of DNA:lipophilic reagent and incubation time

Plasmid size (kb)	Incubation time	DNA (μg)	Lipofectamine (μL)	Efficiency of transfection (%)
4.7	10	10	30	10–15
4.7	20	10	30	15–20
4.7	25	10	30	20–30
4.7	30	10	30	20–30
4.7	35	10	30	10–15
11	10	10	30	5–10
11	20	10	30	15–20
11	25	10	30	20–30
11	30	10	30	30–35
11	35	10	30	30–35

The efficiencies of transfection for the 11 and 4.7 kb plasmid were both increased to 40% with these changes (see Table 2 and Note 3).

3.2 Stable Transfection and Clone Selection

1. After transfection of Neuro2A cells with Lipofectamine 2000 as above, add G418 to a final concentration of 400 μg/mL.
2. Incubate for 2 days with G418.
3. Change medium and incubate without G418 for 4–5 days.
4. After 5 days change medium to include G418 to a final concentration of 600 μg/mL.
5. Increase the concentration of G418 up to a final concentration of 800 μg/mL for a further 2 weeks (see Note 4).
6. Trypsinize cells and dilute to the appropriate volume to yield single cells in a 96-well plate and allow clones to expand (Fig. 1). To do this perform dilutions in a 96-well plate, starting with 2×10^5 cells in A1 then serially diluting in the first column of the plate by a dilution of 1:2. Perform a second dilution series, beginning from column 1 and serially diluting by 1:2 in columns 2–12. Incubate the 96-well plates in a 5% CO_2 incubator at 37°C.
7. Identify wells containing single clones (generally rows F–H columns 10–12), transfer each clone to a single well in a 24-well plate, and incubate with G418 at a final concentration of 600 μg/mL for 2 weeks.

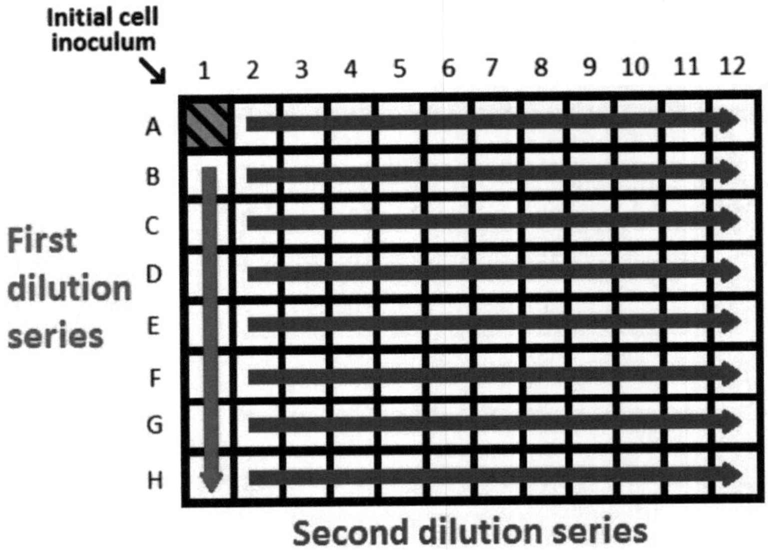

Fig. 1 Dilution of transfected cells to isolate single clones. Perform dilutions in a 96-well plate, starting with 2×10^5 cells in A1 then serially diluting in the first column of the plate by a dilution of 1:2. Perform a second dilution series, beginning from column 1 and serially diluting by 1:2 in columns 2–12. Wells containing single cells are generally within the bottom right-hand corner of the plate, e.g., rows F–H, columns 10–12

8. Transfer single clones to a single well in a 24-well plate and allow to proliferate for 5 days (see Note 5). Treat with G418 at 400 μg/mL to maintain stably transfected cells.

3.3 Adenovirus Particle Preparation

3.3.1 pAdEasy Recombination

1. Digest 0.5–1.0 μg of pAdTrack-CMV plasmid with *Pme*I and purify by agarose gel purification. Resuspend DNA in 50 μL purified water.
2. Set up 5 mL overnight culture of BJ5183 (pAdEasy-1) in LB + ampicillin at 37°C shaking at 225 rpm on an orbital shaker.
3. Grow for several hours until reaching an OD of 0.6–0.8.
4. Pellet cells at 2,000 × *g* 10 min at 4°C.
5. Resuspend in 10 mL ice cold water.
6. Pellet cells at 2,000 × *g* 10 min at 4°C.
7. Resuspend in 1 mL of ice cold water and transfer to Eppendorf tube.
8. Pellet cells in microfuge for 10 s.
9. Wash pellet two times with ice cold water.
10. Pellet cells in microfuge for 10 s.
11. Resuspend cells in 25 μL of digested DNA.

12. Warm 1 mL SOC media per transformation.
13. Place cell suspension in 2 mm cuvettes.
14. Electroporate at 2.5 kV, 200, and 25 μF in a Bio-Rad Gene Pulser T.
15. Immediately transfer cells to 1 mL warm SOC.
16. Incubate at 37°C with shaking for 1 h.
17. Plate onto LB + kanamycin plates and incubate at 37°C overnight.
18. Pick a colony and grow overnight in 5 mL LB + kanamycin at 37°C and prepare DNA by standard miniprep kit (e.g., Qiagen). Identify pAdEasy clones and transform into XL1 Blue.

3.3.2 Transfection of pAdEasy

1. Prepare pAdEasy recombinant plasmid DNA from a 50 mL culture from a colony from step 18 (XL1 Blue transformed).
2. Digest 5 μg of DNA with *Pac*I in 100 μL volume at 37°C for 1 h.
3. Purify DNA and resuspend in 10 μL water.
4. Transfect HEK293 cells with purified DNA, as per Subheading 3.1 above.
5. Culture cells for 7–10 days, until the majority of cells are floating and expressing GFP (the adenovirus infected cells will also change morphology and become rounded). Harvest free-floating cells by centrifugation.
6. Resuspend in 1 mL PBS and transfer to Eppendorf tube.
7. Freeze/thaw to lyse cells by transferring from −70 to 37°C for three cycles and vortex.
8. Pellet cell debris at 18,000 × *g* for10 min at 4°C.
9. Transfer supernatant to fresh tube and store lysate at −70°C.

3.3.3 Amplification and Purification of Virus (See Note 6)

1. Take the lysate from step 9 of the previous section and use to infect 2 × 10 cm tissue culture dishes of HEK293.
2. After 7–10 days the majority of cells will be floating and expressing GFP. Harvest free-floating cells by centrifugation. Resuspend cell pellet in 1 mL PBS per 10 mL media.
3. Freeze/thaw three times in −70°C/37°C, vortexing each time.
4. Pellet cell debris at 18,000 × *g* 10 min at 4°C.
5. Transfer supernatant to fresh tube and store at −70°C.
6. Use the second lysate to infect 20 × 10 cm dishes of HEK293. Lyse cells as above.
7. Use the third lysate to infect 40 × 15 cm dishes of HEK293 and harvest and purify adenoviral particles as below.

8. Collect cells in 400 mL centrifuge tubes and pellet by centrifugation at 10,000 × *g* for 20 min at 4°C.
9. Resuspend cells in 10 mL 0.1 M Tris pH 8 per 15 cm dish.
10. Pellet cells by centrifugation at 10,000 × *g* for 20 min at 4°C.
11. Resuspend cells in 0.5 mL 0.1 M Tris–HCl pH 8 per 15 cm dish.
12. Add 100 μL of 5% Na deoxycholate per mL of cell lysate.
13. Incubate 30 min at room temperature to produce a clear, highly viscous suspension.
14. Add 10 μL 2 M $MgCl_2$ and 5 μL DNase (10 mg/mL) per mL of lysate.
15. Mix and incubate 37°C for 60 min, agitating every 10 min. Viscosity should reduce.
16. Centrifuge at 18,000 × *g* for 20 min at 4°C.
17. Prepare a discontinuous cesium chloride (CsCl) gradient in a polyallomer centrifuge tube (do not disturb gradients once formed). To do this, add the following volumes of CsCl of different densities slowly down the side of the tube in this order:
    (a) Add 0.5 mL of 1.5 mg/mL CsCl.
    (b) Overlay with 3.0 mL of 1.35 mg/mL CsCl.
    (c) Overlay with 3.0 mL of 1.25 mg/mL CsCl.
18. Apply 5 mL of supernatant to the CsCl gradient. Prepare further gradients for the remainder of the supernatant.
19. Centrifuge at 28,000 × *g* at 10°C for 1 h. Do not use the brake.
20. Collect viral bands which should be visible as white/purplish bands:
    (a) Put the tube in a stand in the tissue culture hood above a beaker.
    (b) Remove several mL of CsCl from the top of the tube to reduce the pressure when you puncture the tube.
    (c) Place a piece of tape vertically on the side of the tube.
    (d) Using an 18 gauge needle on a 2 mL syringe, carefully puncture the tube just below the viral band with the bevel of the needle facing upwards (make sure to only puncture one side of the tube, do not push through the other side, insert needle tip to the middle of the tube, directly underneath the viral band).
    (e) Extract the viral band (approx 1–1.5 mL volume).
21. Transfer the pooled virus to a fresh tube and top up with 1.35 g/mL CsCl.
22. Centrifuge at 28,000 × *g* for 16–24 h at 10°C. Do not use brake.

23. Collect viral band as above in about 0.5–1.0 mL volume and transfer to a slide-a-lyzer dialysis cassette.
24. Dialyse for 3 × 1 h in 1 L 10 mM Tris–HCl pH 8 10% glycerol in a beaker with a magnetic stirrer.
25. Aliquot and store at −70°C.
26. Titrate virus in cell culture media to infect cells. Test dilutions at 1:50–1:50,000. Incubate cells for 24 h at 37°C 5% CO_2 then monitor GFP expression.

4 Notes

1. Consider increasing the amount of the plasmid encoding your protein of interest compared to the plasmid encoding eGFP (e.g., 2:1 or 3:1) to increase the likelihood of expression of your protein of interest in eGFP-positive cells. This may be especially important if the plasmid encoding your protein of interest is much larger than the plasmid encoding eGFP. For efficient DNA transfection into cells, preparation of high-quality plasmid DNA is essential. It is best to prepare DNA using a commercial kit based on ion exchange chromatography. Removal of endotoxins during the plasmid isolation procedure also enhances transfection efficiency and many companies supply plasmid isolation kits that will remove endotoxin contamination.
2. Transfection can result in changes in gene expression and consequently in altered cellular responses and/or morphology. It is always advisable to compare responses to control transfected cells. In the case of transfection with liposomes, the transfection reagent on its own can affect the cell membrane (and any membrane proteins, such as ion channels); it may therefore also be necessary to include a "transfection reagent" control in experiments, in which transfection reagent is added as per the above protocol but in the absence of plasmid DNA.
3. The incubation time of the single mixes affects transfection efficiency (6). The "empty plasmid" pEGFP-N1 (4.7 kb) (control plasmid) was negatively affected by long incubation times but the larger test plasmid containing the cloned gene of interest pEGFP-GOI (11 kb) showed improved efficiency by increasing the incubation time. Lipofectamine volume was chosen based on the amount of cells transfected and the toxicity to cells (confluence of cells compared with the initial confluence). Finally, decreasing the amount of plasmid reduced efficiency.
4. With the use of G418 for selection of resistant clones it may be important to titrate the antibody concentration depending on the

cell line and plasmid used; concentrations between 200 μg/mL and 1,400 μg/mL are recommended for testing. We also found that G418 reduces proliferation and at low densities the cells were particularly susceptible to death. Therefore we found it unnecessary to split the cells more than once per 2–4 weeks during the time of treatment.

5. We have found that treating the cells with G418 in this single cell state (even very low concentrations) results in high levels of cell death. It is advisable to allow the cells to proliferate before repeating treatments with G418. G418 treatment should be used periodically to maintain selection, since even plasmids that were integrated into the genome can be lost.

6. Exposure to adenovirus (even replication-deficient adenovirus) may cause acute respiratory illness. Adenoviruses are biohazards and appropriate biosafety procedures should be adhered to. Work with virus in a certified Class II biosafety cabinet. Discard of all virus in bleach (1% sodium hypochlorite, made fresh) or similar for at least 15 min and disinfect all plasticware contaminated with virus in bleach for at least 15 min, followed by autoclaving.

References

1. He T-C, Zhou S, da Costa LT, Yu J, Kinzler KW, Vogelstein B (1998) A simplified system for generating recombinant adenoviruses. Proc Natl Acad Sci U S A 95:2509–2514
2. Ooi L, Belyaev ND, Miyake K, Wood IC, Buckley NJ (2006) BRG1 chromatin remodeling activity is required for efficient chromatin binding by repressor element 1-silencing transcription factor (REST) and facilitates REST-mediated repression. J Biol Chem 281:38974–38980
3. Johnson R, Gamblin RJ, Ooi L, Bruce AW, Donaldson IJ, Westhead DR, Wood IC, Jackson RM, Buckley NJ (2006) Identification of the REST regulon reveals extensive transposable element-mediated binding site duplication. Nucleic Acids Res 34:3862–3877
4. Mucha M, Ooi L, Linley JE, Mordaka P, Dalle C, Robertson B, Gamper N, Wood IC (2010) Transcriptional control of KCNQ channel genes and the regulation of neuronal excitability. J Neurosci 30:13235–13245
5. Bingham AJ, Ooi L, Kozera L, White E, Wood IC (2007) The repressor element 1-silencing transcription factor regulates heart-specific gene expression using multiple chromatin-modifying complexes. Mol Cell Biol 27:4082–4092
6. Dalby B, Cates S, Harris A, Ohki EC, Tilkins ML, Price PJ, Ciccarone VC (2004) Advanced transfection with Lipofectamine 2000 reagent: primary neurons, siRNA, and high-throughput applications. Methods 33:95–103

Chapter 3

Use of *Escherichia coli* for the Production and Purification of Membrane Proteins

Vincent G.L. Postis, Andrea E. Rawlings, Amelia Lesiuk, and Stephen A. Baldwin

Abstract

Individual types of ion channels and other membrane proteins are typically expressed only at low levels in their native membranes, rendering their isolation by conventional purification techniques difficult. The heterologous over-expression of such proteins is therefore usually a prerequisite for their purification in amounts suitable for structural and for many functional investigations. The most straightforward expression host, suitable for prokaryote membrane proteins and some proteins from eukaryotes, is the bacterium *Escherichia coli*. Here we describe the use of this expression system for production of functionally active polytopic membrane proteins and methods for their purification by affinity chromatography in amounts up to tens of milligrams.

Key words Membrane protein, Protein expression, Nonionic detergent, Protein purification, Affinity chromatography, Size exclusion chromatography

1 Introduction

Twenty to thirty percent of genes in most living organisms encode membrane proteins (1), which play critical roles in many aspects of biology, ranging from cell signalling to nutrition. Moreover, such proteins represent the targets of more than 50% of currently used therapeutic drugs (2). Isolation of membrane proteins for structural and functional analysis is therefore of great biological and medical importance. However, individual types of ion channels, transporters, and other membrane proteins typically represent less than 1% of the total protein content of natural biological membranes and this low level of expression hinders their purification. Moreover, the nonionic detergents required to solubilize the proteins can interfere with purification by conventional approaches, such as ion-exchange chromatography, by shielding some of the charged groups on the protein surface within the hydrophilic head

Nikita Gamper (ed.), *Ion Channels: Methods and Protocols*, Methods in Molecular Biology, vol. 998,
DOI 10.1007/978-1-62703-351-0_3, © Springer Science+Business Media, LLC 2013

groups of the detergents in the protein-detergent micelle. Typically, isolation of membrane proteins in amounts sufficient for structural and other investigations therefore requires their over-expression, bearing an affinity tag, in a suitable heterologous expression system. Choice of the type of affinity tag and the location of its attachment must be carefully considered, because these can influence the synthesis and insertion of the protein in the membrane (3). Choice of expression system is also important. For example, while some human membrane proteins can be functionally expressed in bacteria (4), high-level expression usually requires a eukaryotic host (5). Additionally, the choice of detergent used for protein solubilization is often critical in order that the protein not only remains in a soluble, non-aggregated state for prolonged periods of time but also that it retains its native structure and function.

Here we describe methods that can be employed for the expression of membrane proteins in *Escherichia coli* and for their subsequent purification. Whilst most applicable to prokaryote membrane proteins, including homologues of human ion channels and transporters, the procedures involved can be adapted for the purification of proteins expressed in eukaryotic systems such as mammalian and insect cells or yeasts.

2 Materials

2.1 Reagents and Buffers for Bacterial Cell Culture and Membrane Preparation

1. *E. coli* host strains: BL21-gold(DE3) (Stratagene), BL21 Star™ (DE3) (Invitrogen), and C43(DE3) (Lucigen Corporation), all harboring the plasmid pRARE2 (Novagen) if required (see Note 1).
2. Expression constructs: These should contain the open reading frame (ORF) for the protein of interest, bearing appropriate N- and/or C-terminal affinity tags, under the control of a lactose-inducible *tac* promoter (i.e., in a derivative of the vector pTTQ18 (6)) or another suitable promoter (see Note 2).
3. 1 M isopropyl-β-D-thiogalactoside (IPTG): Dissolve 2.38 g IPTG in a final volume of 10 mL H_2O, sterilize by passage through a 0.22 μm filter, and store at −20°C.
4. 100 mg/mL carbenicillin: Dissolve 500 mg carbenicillin in a final volume of 5 mL H_2O, sterilize by filtration through a 0.22 μm filter, and store in the dark at −20°C.
5. 30 mg/mL chloramphenicol: Dissolve 150 mg chloramphenicol in 5 mL 100% ethanol and store at −20°C.
6. 50× 5052: 25% (w/v) glycerol, 2.5% (w/v) glucose, 10% (w/v) α-lactose monohydrate. Weigh 25 g glycerol into a beaker and then add 73 mL H_2O, 2.5 g glucose, and 10 g α-lactose. Stir until dissolved (see Note 3) then sterilize by filtration through a 0.22 μm filter.

7. 20× NSPC: 0.5 M Na_2HPO_4, 0.5 M KH_2PO_4, 0.1 M Na_2SO_4, 1 M NH_4Cl. Dissolve 7.1 g Na_2HPO_4, 6.8 g KH_2PO_4, 1.42 g Na_2SO_4, and 5.35 g NH_4Cl in 75 mL H_2O. Adjust to pH 7.0 using NaOH then make volume up to 100 mL. Sterilize by autoclaving.
8. 1 M $MgSO_4$: Dissolve 24.65 g $MgSO_4 \cdot 7H_2O$ in a final volume of 100 mL H_2O then sterilize by autoclaving or filtration through a 0.22 μm filter.
9. 1 M $CaCl_2$: Dissolve 14.7 g $CaCl_2 \cdot 2H_2O$ in a final volume of 100 mL H_2O then sterilize by autoclaving or filtration through a 0.22 μm filter.
10. 50% (w/v) glycerol: Dissolve 100 g glycerol in H_2O to give a final volume of 200 mL. Sterilize by autoclaving.
11. 40% (w/v) glucose: Dissolve 40 g glucose in H_2O to give a final volume of 100 mL. Sterilize by filtration through a 0.22 μm filter.
12. 20× M9 salts: Dissolve 120 g Na_2HPO_4, 60 g KH_2PO_4, 10 g NaCl, and 20 g NH_4Cl in 800 mL H_2O and adjust pH to 7.4 with 10 M NaOH. Make up volume to 1 L then sterilize by autoclaving.
13. 20% (w/v) casamino acids: Dissolve 20 g casamino acids in H_2O to give a final volume of 100 mL then sterilize by autoclaving.
14. Standard M9 medium (for IPTG induction): To 924 mL sterile H_2O add 50 mL 20× M9 salts, 0.2 mL 1 M $CaCl_2$, 2 mL 1 M $MgSO_4$, 20 mL 20% casamino acids, and 4 mL 50% glycerol. Add carbenicillin (100 μg/mL) and chloramphenicol (30 μg/mL) as appropriate before use (see Note 4).
15. M9 medium for autoinduction: Dissolve 6 g Na_2HPO_4, 3 g KH_2PO_4, 1 g NH_4Cl, 1 g casamino acids, and 0.003 g $CaCl_2 \cdot 2H_2O$ in a final volume of 1 L H_2O then sterilize by autoclaving.
16. Standard Lysogeny Broth (LB) medium: Dissolve 10 g tryptone, 5 g yeast extract, and 10 g NaCl (omit NaCl if the medium is to be used for autoinduction experiments) in 800 mL H_2O and adjust to pH 7.4 by addition of 1 M NaOH. Make up volume to 1 L then sterilize by autoclaving. Add carbenicillin (100 μg/mL) and chloramphenicol (30 μg/mL) as appropriate before use (see Note 4).
17. Non-inducing LB medium ($LB_{glucose}$): Mix 92.4 mL LB medium (lacking NaCl) with 100 μL 1 M $MgSO_4$, 2.5 mL 40% glucose, and 5 mL 20× NPSC. Add carbenicillin (100 μg/mL) and chloramphenicol (30 μg/mL) as appropriate before use (see Note 4).
18. Standard Superbroth (SB) medium: Dissolve 32 g tryptone, 20 g yeast extract, and 5 g NaCl (omit NaCl if the medium is to be used for autoinduction experiments) in 800 mL water

and adjust pH to 7.4 by addition of 1 M NaOH. Make up volume to 1 L with H_2O then sterilize by autoclaving. Add carbenicillin (100 μg/mL) and chloramphenicol (30 μg/mL) as appropriate before use (see Note 4).

19. Complete media for autoinduction ($M9_{auto}$, LB_{auto}, and SB_{auto}) (per L): To 929 mL M9 medium for autoinduction, or LB or SB medium lacking NaCl, add, in the following order, 1 mL 1 M $MgSO_4$, 20 mL 50× 5052, and 50 mL 20× NSPC (see Note 5). Add carbenicillin (100 μg/mL) and chloramphenicol (30 μg/mL) as appropriate before use (see Note 4).
20. 10× Phosphate-buffered saline, pH 7.4 (10× PBS): 100 mM Na_2HPO_4, 18 mM KH_2PO_4, 1,370 mM NaCl, 40 mM KCl, pH 7.4. Dissolve 80 g NaCl, 3 g KCl, 14.4 g Na_2HPO_4, and 2.4 g KH_2PO_4 in 800 mL water, adjust pH to 7.4, and make up the volume to 1 L. Sterilize by autoclaving for long-term storage.
21. Cell lysis solution: 50 mM HEPES, 5 mM $MgCl_2$, 1% (v/v) Triton X-100, 25% sucrose, 10 U/mL OmniCleave endonuclease, 0.1 mg/mL lysozyme, pH 8.0. Dissolve 1.192 g HEPES, 25 g sucrose, 0.048 g $MgCl_2$, and 1 mL Triton X-100 in 80 mL H_2O, adjust pH to 8 with 5 M NaOH, then make up volume to 100 mL. Store in aliquots at −20°C. Just before use, add 10 U/mL OmniCleave endonuclease (Epicentre Biotechnologies) and 0.1 mg/mL lysozyme.
22. 200 mM Tris–HCl, pH 8.0: Dissolve 2.42 g Tris in 800 mL H_2O, adjust pH to 8.0 with HCl, and then make up volume to 1 L.
23. Sucrose buffer: 200 mM Tris–HCl, 1 mM EDTA, 1 M sucrose, pH 8.0. Dissolve 0.037 g $Na_2EDTA{\cdot}2H_2O$ and 34.23 g sucrose in 200 mM Tris–HCl, pH 8.0, to give a final volume of 100 mL.
24. Denaturing solution for dot blotting: 100 mM Tris–HCl, 8 M guanidinium chloride, pH 8.0: Dissolve 76.4 g guanidine hydrochloride in 50 mL 200 mM Tris–HCl, pH 8.0, plus sufficient H_2O to give a final volume of 100 mL (see Note 6).
25. 10× Tris-buffered saline (10× TBS): 500 mM Tris–HCl, 1.5 M NaCl, pH 7.5. Dissolve 60.57 g Tris and 87.66 g NaCl in 900 mL H_2O, adjust pH to 7.5 with HCl, and then make up volume to 1 L.
26. Tris-buffered saline Tween (TBST): 50 mM Tris–HCl, 150 mM NaCl, 0.1% Tween 20, pH 7.5. Dilute 100 mL of 10× TBS and 1 mL Tween 20 to 1 L with H_2O.
27. Blocking buffer: 3% bovine serum albumin (BSA) in TBST. Dissolve 3 g BSA in 100 mL TBST.
28. Bicinchoninic acid (BCA) reagent (Thermo Scientific).
29. cOmplete, EDTA-free Protease Inhibitor Cocktail Tablets (Roche Diagnostics Ltd.).

2.2 Reagents and Buffers for Purification of His-Tagged Proteins by Immobilized Metal Affinity Chromatography

1. HisPur™ cobalt resin (Thermo Scientific).
2. *n*-dodecyl-β-D-maltoside (DDM) (GLYCON Biochemicals GmbH): 25% (w/v) solution in H_2O. Store at −80°C.
3. 3 M imidazole, pH 7.4: Dissolve 20.42 g imidazole in ~80 mL H_2O, adjust pH to 7.4 with 1 M NaOH then make up to a final volume of 100 mL. For making buffers with low concentrations of imidazole, make a 1 M sub-stock by dilution in water.
4. 3 M NaCl: Dissolve 87.66 g NaCl in H_2O to give a final volume of 500 mL.
5. 2× Immobilized metal affinity chromatography (IMAC) solubilization buffer: 20 mM Na_2HPO_4, 3.6 mM KH_2PO_4, 8 mM KCl, 574 mM NaCl, 15 mM imidazole, 20% glycerol, protease inhibitor cocktail without EDTA, pH 7.4. Dissolve 1 tablet protease inhibitor cocktail without EDTA in a mixture of 10 mL 50% glycerol, 2.5 mL 3 M NaCl, 375 μL 1 M imidazole, 5 mL 10× PBS plus sufficient H_2O to give a final volume of 25 mL.
6. IMAC wash buffer 1: 10 mM Na_2HPO_4, 1.8 mM KH_2PO_4, 4 mM KCl, 287 mM NaCl, 7.5 mM imidazole, 10% glycerol, 0.05% DDM, pH 7.4. Dilute 10 mL 50% glycerol, 2.5 mL 3 M NaCl, 375 μL 1 M imidazole, 5 mL 10× PBS, and 100 μL 25% DDM to a final volume of 50 mL with H_2O.
7. IMAC wash buffer 2: 10 mM Na_2HPO_4, 1.8 mM KH_2PO_4, 4 mM KCl, 287 mM NaCl, 15 mM imidazole, 10% glycerol, 0.05% DDM, pH 7.4. Dilute 10 mL 50% glycerol, 2.5 mL 3 M NaCl, 250 μL 3 M imidazole, 5 mL 10× PBS, and 100 μL 25% DDM to a final volume of 50 mL with H_2O.
8. IMAC elution buffer: 10 mM Na_2HPO_4, 1.8 mM KH_2PO_4, 4 mM KCl, 287 mM NaCl, 250 mM imidazole, 10% glycerol, 0.05% DDM, pH 7.4. Dilute 10 mL 50% glycerol, 2.5 mL 3 M NaCl, 4.17 mL 3 M imidazole, 5 mL 10× PBS, pH 7.4 and 100 μL 25% DDM to a final volume of 50 mL in H_2O.
9. IMAC dialysis buffer: 10 mM Na_2HPO_4, 1.8 mM KH_2PO_4, 4 mM KCl, 137 mM NaCl, 10% glycerol, 0.03% DDM, pH 7.4. Dilute 200 mL 50% glycerol, 100 mL 10× PBS, pH 7.4 and 1.2 mL 25% DDM to a final volume of 1 L in H_2O.

2.3 Reagents and Buffers for Purification of Strep-Tagged Proteins by Strep-Tactin Affinity Chromatography

1. Strep-Tactin® Superflow® resin (IBA GmbH).
2. D-Desthiobiotin.
3. 1 M 2-(*p*-Hydroxyphenylazo)benzoic acid (HABA): Dissolve 2.422 g HABA in a final volume of 10 mL H_2O. Store at 4°C.
4. cOmplete, Mini Protease Inhibitor Cocktail Tablets (Roche Diagnostics Ltd.).

5. 1 M Tris–HCl, pH 7.4: Dissolve 60.57 g Tris in 400 mL H_2O. Adjust the pH to 7.4 with HCl then make up to a final volume of 500 mL.
6. 0.5 M EDTA, pH 7.4: Dissolve 9.307 g $Na_2EDTA.2H_2O$ in about 40 mL H_2O. Warming will probably be necessary. Cool to room temperature, adjust pH to 7.4 by addition of 1 M NaOH, then make up volume to 50 mL. Store at 4°C, or frozen for extended periods.
7. 2× Strep-Tactin affinity chromatography (SAC) solubilization buffer: 100 mM Tris–HCl, 200 mM NaCl, 2 mM EDTA, 20% glycerol, protease inhibitor cocktail, pH 7.4. Dissolve 1 tablet protease inhibitor cocktail in a mixture of 2.66 mL 3 M NaCl, 4 mL 1 M Tris–HCl, pH 7.4, 160 μL 0.5 M EDTA, pH 7.4 and 16 mL 50% glycerol plus sufficient H_2O to give a final volume of 40 mL in H_2O.
8. SAC wash buffer: 50 mM Tris–HCl, 100 mM NaCl, 1 mM EDTA, 10% glycerol, 0.05% DDM, pH 7.4. Dilute 5 mL 1 M Tris–HCl, pH 7.4, 3.33 mL 3 M NaCl, 200 μL 0.5 M EDTA, pH 7.4, 20 mL 50% glycerol, and 200 μL 25% DDM to a final volume of 100 mL in H_2O.
9. SAC elution buffer: 50 mM Tris–HCl, 100 mM NaCl, 1 mM EDTA, 2.5 mM D-Desthiobiotin, 10% glycerol, 0.05% DDM, pH 7.4. Dissolve 5.4 mg D-Desthiobiotin in 10 mL SAC wash buffer.
10. SAC dialysis buffer: 50 mM HEPES, 100 mM NaCl, 1 mM EDTA, 5% glycerol, 0.05% DDM, pH 7.5. Dissolve 11.915 g HEPES in ~800 mL H_2O. Add 33.3 mL 3 M NaCl and 2 mL 0.5 M EDTA, pH 7.4 then adjust pH to 7.5 using NaOH. Add 100 mL 50% glycerol and 2 mL 25% DDM then make up volume to 1 L.
11. SAC regeneration buffer: 50 mM Tris–HCl, 100 mM NaCl, 1 mM EDTA, 1 mM HABA, pH 7.4. Dilute 5 mL 1 M Tris–HCl, pH 7.4, 3.333 mL 3 M NaCl, 200 μL 0.5 M EDTA, pH 7.4 and 0.1 mL 1 M HABA to a final volume of 100 mL in H_2O.

2.4 Equipment and Materials

1. Temperature-controlled orbital shaker for multiwell plates.
2. 24 Deep-well and 96-well plates.
3. Breathable seals for deep-well plates.
4. Temperature-controlled orbital incubator for flasks.
5. Mechanical homogenizer, e.g., T18 basic ULTRA-TURRAX® homogenizer (IKA).
6. Cell disruptor, e.g., TS series continuous cell disruptor (Constant Systems, UK).

7. Refrigerated bench centrifuge capable of centrifuging 96-well plates.
8. Refrigerated high-speed centrifuge.
9. Refrigerated ultracentrifuge.
10. Handheld glass/Teflon homogenizer.
11. UV spectrophotometer.
12. Chromatography columns.
13. Dialysis tubing.
14. Centrifugal concentrators.
15. Apparatus for SDS-PAGE and for electrophoretic transfer of proteins to nitrocellulose membranes for Western blotting.
16. Äkta Fast Protein Liquid Chromatography (FPLC) system (GE Healthcare) or equivalent.

3 Methods

Design of constructs for expression of membrane proteins in *E. coli* needs to take into account both the origin of the gene to be expressed and the topology of protein. For example, for proteins from organisms in which codon usage differs from that of *E. coli* use of host strains harboring a plasmid encoding tRNAs for rarely used codons can sometimes improve expression levels (7) and is the method described here. Alternatively, the corresponding gene can be produced synthetically, a process that allows not only codon optimization but also control of other features, such as mRNA secondary structure and the beneficial presence of rare codon clusters, which may influence membrane protein biogenesis (8). Knowledge of the topology of the membrane protein of interest is also useful in deciding how the sequence should be tagged; if the topology has not yet been established experimentally it can be predicted with reasonable success using a range of algorithms such as TOPCONS (9). For the majority of membrane proteins, which have cytoplasmic C-termini (10), addition of a C-terminal octahistidine tag typically allows good expression and affinity purification. An added advantage of C-terminal tagging is that, in combination with N-terminal sequencing, the detection of the tag by Western blotting indicates that the protein produced is full length. Co-purification of endogenous histidine-rich *E. coli* proteins, such as AcrB (11), can be avoided by incorporating a protease cleavage site between the tag and the membrane protein, and rechromatographing the protein after protease treatment (12). However, addition of an oligohistidine tag to a protein terminus normally located on the periplasmic/extracellular side of the membrane typically results in misfolding and/or reduced levels of expression of membrane proteins, probably because the positively charged nature of the tag interferes with topogenesis (3). To circumvent this problem a C-terminal *Strep*-tag II affinity tag (Trp-Ser-His-Pro-Gln-Phe-Glu-Lys), which bears a much smaller net charge, can instead be used (3). Alternatively, for proteins with a periplasmic N-terminus, the latter can be fused to a maltose binding protein tag bearing a signal sequence (13).

In our laboratory expression vectors derived from the plasmid pTTQ18, bearing a *tac* promoter, have been successfully used to express many membrane proteins (3, 14). Expression using these constructs can be achieved either using induction with isopropyl-β-D-thiogalactoside (IPTG) or by autoinduction with lactose (15). Alternatively, pET vectors (Novagen), bearing a T7 promoter, and *E. coli* (DE3) host strains can be employed. Use of this strong promoter can sometimes result in misfolding of a proportion of the expressed protein, but expression can be "tuned" using the Lemo21(DE3) strain of *E. coli* in order to maximize the amount of correctly folded protein (16).

3.1 Initial Optimization of Protein Expression

The following procedure is designed for rapid screening of a range of host strains and culture media, using autoinduction (15), to find those in which expression is optimal.

1. Freshly transform *E. coli* strains BL21-gold(DE3), BL21 Star™ (DE3), and C43(DE3) with a pTTQ18-derived expression construct encoding an affinity-tagged ORF of the protein of interest.
2. Inoculate 2 mL samples of $LB_{glucose}$ in a 24 deep-well plate with single colonies from plates streaked with the above transformants. Seal the plate with breathable seal then incubate overnight at 37°C in a humidified plate shaker at 1,300 rpm.
3. Inoculate 3 mL samples of $M9_{auto}$, LB_{auto}, and SB_{auto} in a 24 deep-well plate with 15 μL samples of the overnight cultures. Seal the plate with breathable seal then incubate for 24 h at 37°C in a humidified plate shaker at 1,300 rpm.
4. Transfer 400 μL samples of the autoinduced expression cultures to a 96-well plate and pellet the cells by centrifugation at $2{,}000 \times g_{av}$ for 15 min at 4°C.
5. Pour off the medium and tap on clean dry tissue to remove any remaining liquid.
6. Freeze the cell pellets at −80°C for 1 h then add 100 μL cell lysis reagent per well and incubate on a plate shaker at 1,000 rpm for 30 min at room temperature.
7. Take duplicate 5 μL samples for assay of protein concentration by the bicinchoninic acid (BCA) assay (17).

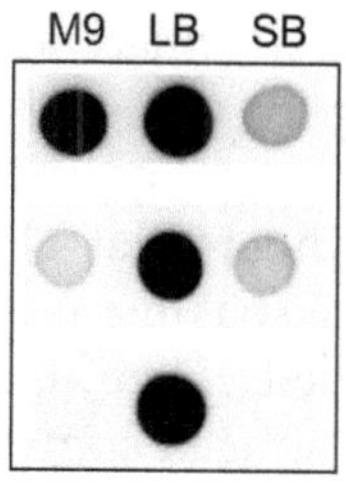

Fig. 1 Effect of host strain and culture medium on expression of a C-terminally His_8-tagged construct of the *Escherichia coli* nucleoside transporter NupG. The *E. coli* strains indicated, harboring a pTTQ18-based expression construct of NupG, were cultured at 37°C for 18 h in the autoinduction media shown. Following cell lysis samples (4 μg protein) were spotted onto nitrocellulose membrane and then the His_8-tagged protein visualized by incubating with a horseradish peroxidase conjugate of a monoclonal antibody against oligohistidine, followed by chemiluminescent detection

8. Mix 12.5 μL of cell lysate with 37.5 μL denaturing solution for dot blotting, incubate for 1 h at room temperature, and then spot 3 μL samples onto nitrocellulose membrane.
9. Block free protein binding sites on the membrane by incubation with blocking buffer for 1 h at room temperature or in a cold room overnight and then detect the affinity-tagged protein by standard Western blotting procedures, using an antibody appropriate for the tag (see Note 7).

An example of typical results is shown in Fig. 1.

3.2 Further Optimization of Protein Expression

The small-scale culture and dot blotting procedures described above should provide an initial idea of the best expression host and growth medium required to achieve the desired expression level of the protein of interest. However, before embarking on large-scale expression it is wise to determine the optimum time of induction and to compare yields obtained using autoinduction with those obtained using IPTG induction, by performing cultures in flasks as follows (see Note 8):

1. Inoculate 50 mL $LB_{glucose}$ in a 250 mL baffled flask with single bacterial colony from a freshly streaked plate and incubate overnight at 37°C with orbital shaking at 200 rpm.
2. Measure the D_{600nm} of the overnight culture and inoculate a series of 250 mL baffled flasks, each containing 50 mL of the appropriate autoinduction medium identified as detailed in Subheading 3.1 and supplemented with antibiotics as appropriate, to give a theoretical D_{600nm} value of 0.05. Incubate at 37°C with orbital shaking at 200 rpm for 24, 36, 48, and 60 h then prepare cytoplasmic membranes as detailed in Subheading 3.3 for analysis of expression levels by Western blotting.

3. In parallel, inoculate a series of 250 mL baffled flasks, each containing 50 mL of the appropriate standard medium (M9, LB, or SB) and supplemented with antibiotics as appropriate, to give a theoretical D_{600nm} value of 0.05. Incubate at 37°C with orbital shaking at 200 rpm until a D_{600nm} value of 0.7 has been reached. Then, either harvest the cells immediately for use as un-induced controls or induce expression by addition of 25 μL 1 M IPTG to give a final concentration of 0.5 mM. Continue incubation of the latter samples at 37°C with orbital shaking at 200 rpm for 3, 6, and 16 h then prepare membranes from these cultures and from the un-induced controls as detailed in Subheading 3.3 for analysis of expression levels by Western blotting.

3.3 Small-Scale Preparation of Membranes and Analysis of Expression Levels by Western Blotting

1. Take 50 mL of an IPTG-induced culture and collect the cells by centrifugation at 12,000 × g_{av} for 10 min at 4°C (see Note 9).
2. Resuspend the pellet in 10 mL 200 mM Tris–HCl, pH 8.0, using a handheld homogenizer, then shake in an orbital incubator at 250 rpm for 20 min at 25°C.
3. At time zero add 4.85 mL sucrose buffer and briefly mix by vortexing.
4. At 1.5 min add 65 μL 10 mg/mL lysozyme (freshly prepared in the sucrose buffer) and briefly mix by vortexing.
5. At 2 min add 9.6 mL H_2O, vortex briefly, and continue shaking for 20 min to allow spheroplast formation (see Note 10).
6. Collect the spheroplasts by centrifuging at 25,000 × g_{av} for 20 min at 4°C.
7. Discard the supernatant then resuspend the spheroplasts in 15 mL H_2O using a handheld homogenizer and allow to stand at room temperature for 30 min to ensure complete lysis.
8. Collect the resultant membranes by centrifuging as in step 6.
9. Wash the membranes three times by resuspending each time in 15 mL 1× PBS, pH 7.4, using a handheld homogenizer, then centrifuging as in step 6.
10. Resuspend the final membrane pellet in 0.2–1.0 mL 1× PBS, pH 7.4.
11. Measure the protein concentration using the BCA assay (17) then snap-freeze and store in aliquots at −80°C.
12. Assess the expression level of the protein of interest by Western blotting, using an appropriate antibody (see Note 11).

3.4 Large-Scale Expression and Membrane Preparation

Once the optimal combination of host strain, medium, induction method, temperature, and time of induction has been determined, expression can be scaled up either in flask culture or, if available, in a fermenter (14) to enable sufficient material to be produced for protein purification (see Note 12). As an example, the following method is employed in our laboratory for production of *E. coli* membranes containing the peptide transporter $PepT_{So}$ from *Shewanella oneidensis* (18):

1. Transform *E. coli* strain BL21-gold(DE3) with a pTTQ18-derived plasmid encoding $PepT_{So}$ bearing a C-terminal hexa-histidine tag. Streak onto an LB-agar plate supplemented with carbenicillin (100 μg/mL) and incubate overnight at 37°C.
2. Inoculate 50 mL LB medium, supplemented with carbenicillin (100 μg/mL) and 1% glucose, with a single colony from the plate and incubate overnight at 37°C with orbital shaking at 200 rpm in a 250 mL baffled flask.
3. Measure the D_{600nm} of the overnight culture and inoculate the required number of 2.5 L baffled flasks, each containing 500 mL LB medium supplemented with carbenicillin (100 μg/mL), with a volume sufficient to give a theoretical D_{600nm} value of 0.05, and then incubate at 37°C with orbital shaking at 200 rpm. Monitor the D_{600nm} and when it reaches 0.7 induce protein expression by adding IPTG to give a final concentration of 0.5 mM (see Note 13). Incubate for a further 3 h under the same conditions.
4. Harvest the cells by centrifugation at 9,000 × g_{av} for 40 min at 4°C then resuspend the cell pellets (6 mL buffer/g cells) in ice-cold 1× PBS, pH 7.4, containing 1 tablet protease inhibitor cocktail without EDTA (see Note 14) per 50 mL.
5. Homogenize the cell suspension using a mechanical homogenizer to ensure that no lumps are present and then lyse the cells by passage through a cell disruptor operating at 30 kPSI and 4°C (see Note 15).
6. Pellet the cell debris by centrifugation at 14,000 × g_{av} for 45 min at 4°C.
7. Collect the membranes by centrifugation of the supernatant at 100,000 × g_{av} for 2 h at 4°C.
8. Resuspend the membranes in a minimal volume (a few mL) of 1× PBS, pH 7.4, or other buffer appropriate for the required downstream processing (e.g., 1× IMAC solubilization buffer, without detergent), using a handheld homogenizer. Take a small sample for measurement of the protein concentration using the BCA assay (17), then snap-freeze and store at −80°C (see Note 16). The typical yield is 150 mg membrane protein per liter of culture.

3.5 Purification of His-Tagged Proteins by IMAC

A necessary first step in membrane protein purification is solubilization of the membrane in a nonionic or zwitterionic detergent. Commonly used detergents include alkyl glucosides (e.g., octyl

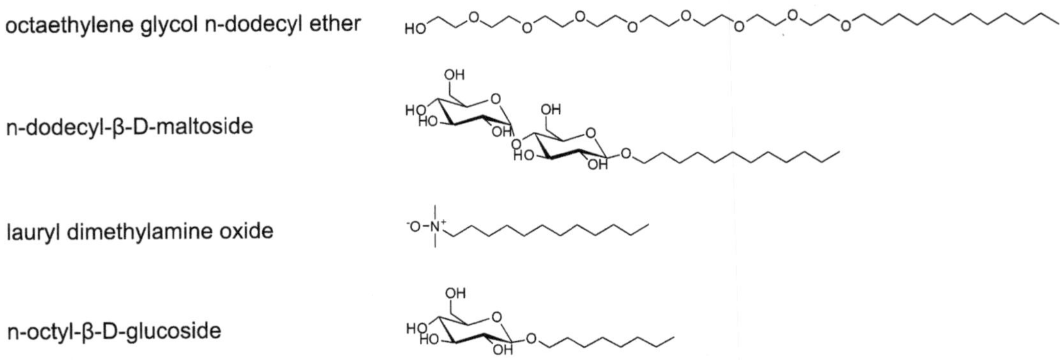

Fig. 2 Structures of detergents commonly used for membrane protein purification

glucoside) and maltosides (e.g., *n*-dodecyl-β-D-maltoside; DDM), polyoxyethylene glycols (e.g., octaethylene glycol *n*-dodecyl ether; $C_{12}E_8$), and lauryl dimethylamine oxide (LDAO) (Fig. 2). For efficient solubilization of membranes, not only must the detergent concentration be greater than the critical micellar concentration (CMC) but also a sufficiently high ratio of detergent to membrane concentration must be employed. The optimal detergent and its concentration for solubilization of a particular membrane protein must be determined empirically by incubating membranes, typically at a protein concentration of 2 mg/mL, on ice with a range of detergent concentrations (e.g., 0.5, 1, 1.5, and 2% (w/v)) for 1 h. After centrifuging at 100,000 × g_{av} to remove insoluble material the amount of the desired protein in the supernatant can then be assessed by SDS-polyacrylamide gel electrophoresis (SDS-PAGE) and Western blotting. In the case of His_6-tagged $PepT_{So}$, which is used as an example below, the optimal detergent for solubilization is DDM. Indeed, in our hands DDM solubilizes many prokaryote membrane proteins efficiently.

All purification steps should be done on ice or at 4°C as appropriate.

1. Dilute 100 mg of membrane protein with 10 mL ice-cold 2× IMAC solubilization buffer (see Note 17) then add ice-cold H_2O to give a final volume of 19.2 mL. Homogenize using a handheld homogenizer, then add 0.8 mL 25% DDM, to give a final detergent concentration of 1% and a protein concentration of 5 mg/mL; incubate with gentle mixing for 1 h (see Note 18). Take a small sample (e.g., 100 μL), snap-freeze, and store at −80°C for subsequent SDS-PAGE/Western blotting to assess the total amount of $PepT_{So}$ in the starting material.
2. Pellet the insoluble fraction in an ultracentrifuge at 100,000 × g_{av} for 1 h. Carefully remove the supernatant from the pellet and keep this on ice. If desired, solubilize the pellet in SDS-PAGE sample buffer (it is hard to get an even resuspension if ordinary

buffer is used) and snap-freeze for subsequent SDS-PAGE/ Western blotting to assess the amount of $PepT_{So}$ that has not been solubilized. Similarly, take a small sample (e.g., 100 μL) of the supernatant to assess the amount of $PepT_{So}$ that has been solubilized.

3. Pre-equilibrate 1.2 mL HisPur™ cobalt resin (50% slurry, i.e., 0.6 mL packed resin) (see Note 19) in a 50 mL plastic tube by washing three times with 10 mL H_2O and then three times with 3 mL IMAC wash buffer 1. Do this by gently inverting the tube then centrifuging at $700 \times g_{av}$ for 5 min each time.
4. Add the supernatant from the ultracentrifugation step to the resin and allow protein binding to occur for 2 h with gentle mixing on a roller mixer (see Note 20).
5. Centrifuge the resin slurry again at $700 \times g_{av}$ for 5 min, and remove and discard the supernatant (but first keep a sample for SDS-PAGE/Western blotting to assess the amount of $PepT_{So}$ that has remained unbound). Wash the resin in the same fashion twice with 6 mL IMAC wash buffer 1, then resuspend in 3 mL IMAC wash buffer 1.
6. Next, pack the resin into a column of appropriate dimensions with a tap on the outlet, then wash the resin under gravity with 18 mL (i.e., 30 column volumes (CV)) of IMAC wash buffer 1, followed by 18 mL IMAC wash buffer 2 (see Note 21).
7. Close the column outlet and gently add 0.6 mL IMAC elution buffer, without disturbing the top of the resin. Open the column again and collect ~0.5 mL of the eluate (see Note 22) before closing the column outlet again.
8. With the column outlet still closed, add a further 1 mL elution buffer, resuspend the resin with a pipet tip, cap the column, and incubate for 10 min on a rotary mixer.
9. Return the column to the vertical position, open the outlet and collect the eluate, plus that generated by subsequent gentle addition of 0.6 mL elution buffer, in a single tube. Close the outlet again. Measure the A_{280nm} of the eluted fraction to provide an indication of protein concentration, using elution buffer as the blank.
10. Repeat steps 8 and 9 until all the protein has eluted. Pool the peak fractions, as indicated by their absorbance (see Note 23).
11. To remove the imidazole, dialyze overnight against at least 2×500 mL IMAC dialysis buffer (see Note 24).
12. Assess the progress of the purification and the purity of the final material by SDS-PAGE and Western blotting (see Note 11).
13. If necessary, concentrate the dialyzed sample using a centrifugal concentrator with a 100 kDa molecular weight cutoff by centrifugation at $3{,}800 \times g_{av}$ (see Note 25). The homogeneity of

the purified protein can then be investigated by size exclusion chromatography, as discussed for the protein MPSIL0347 in Subheading 3.6.

14. Assess the functional state of the purified protein using a suitable assay (see Note 26).

3.6 Purification of Strep-Tagged Proteins by SAC

In cases where protein topology precludes the use of an oligohistidine affinity tag, we have successfully used an attached *Strep*-tag II sequence for membrane protein purification (3). This approach circumvents the problem of co-purification of the protein with endogenous His-rich proteins, sometimes seen in IMAC purifications (11), although the relatively low affinity of the tag for the Strep-Tactin affinity resin can result in low yields for some proteins, especially if they are monomeric. The following method is used in our laboratory for purification of MPSIL0347, a homologue of the mammalian bestrophin family of chloride channels (19), encoded by the gene alr2987 from the cyanobacterium *Nostoc* sp. PCC 7120.

All purification steps should be done on ice or at 4°C as appropriate.

1. Solubilize membranes as detailed in step 1 of Subheading 3.5, except using 2× SAC solubilization buffer and incubating with gentle mixing overnight.
2. Separate the detergent-soluble from the insoluble protein by ultracentrifugation as detailed in step 2 of Subheading 3.5.
3. Pre-equilibrate 1 mL of a 50% slurry of Strep-Tactin® Superflow® resin slurry (i.e., 0.5 mL packed resin [see Note 27]) in a 50 mL plastic tube by washing twice with 10 mL SAC wash buffer. Do this by gently inverting the tube then centrifuging at $700\times g_{av}$ for 5 min each time.
4. Add the supernatant from the ultracentrifugation step to the resin and allow protein binding to occur for 1 h with gentle mixing.
5. Centrifuge the resin slurry again at $700\times g_{av}$ for 5 min, remove and discard the supernatant (but first keep a small sample, e.g., 1 mL, for SDS-PAGE/Western blotting to assess the amount of MPSIL0347 that has remained unbound), then gently resuspend the resin in 5 mL of SAC wash buffer.
6. Next pack the resin into a column of appropriate dimensions, then wash the resin under gravity with 20 mL (i.e., 40 CV) of SAC wash buffer, collecting the eluate for further analysis if required.
7. Elute the bound protein by successive addition of ten 0.5 mL volumes of SAC elution buffer, collecting 0.5 mL fractions.

8. Measure the A_{280nm} of the eluted fractions to provide an indication of protein concentration, using elution buffer as the blank. Pool the peak fractions, as indicated by their absorbance.
9. Assess the progress of the purification and the purity of the final material by SDS-PAGE and Western blotting. Results for a typical purification are shown in Fig. 3.
10. To remove the desthiobiotin, dialyze overnight against 2 × 500 mL SAC dialysis buffer.
11. If necessary, concentrate the dialyzed sample to a volume of ~250 μL using a centrifugal concentrator with a 100 kDa molecular weight cutoff by centrifugation at 3,800 × g_{av} for 30 min.
12. As an additional purification step, and to assess the oligomeric state of the purified protein, the dialyzed sample can be subjected

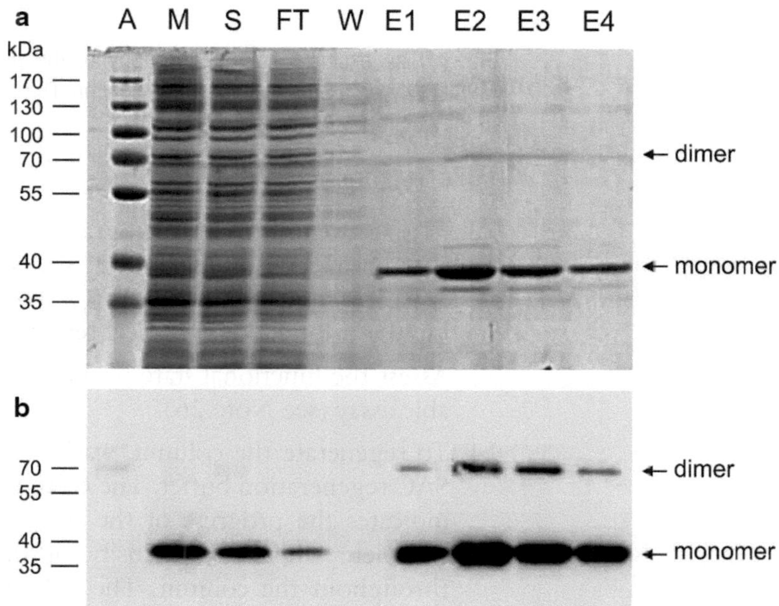

Fig. 3 (**a**) Coomassie blue-stained SDS-polyacrylamide gel and (**b**) corresponding Western blot stained with an antibody against the *Strep*-tag II, illustrating the purification of a *Strep*-tag II modified form of the bestrophin homologue MPSIL0347 using SAC, as detailed in the text. Lanes were loaded with 10 μL samples from the following stages of the purification process: *M Escherichia coli* membranes used as starting material, *S* supernatant following incubation of the membranes with the detergent DDM and subsequent ultracentrifugation, *FT* flow through fraction that remained unbound following passage of the supernatant through a column of Strep-Tactin® Superflow® resin, *W* wash fraction resulting from passage of SAC wash buffer through the column, *E1–E4* fractions eluted using a buffer containing 2.5 mM D-Desthiobiotin. *Lane A* contained protein markers with the indicated molecular masses

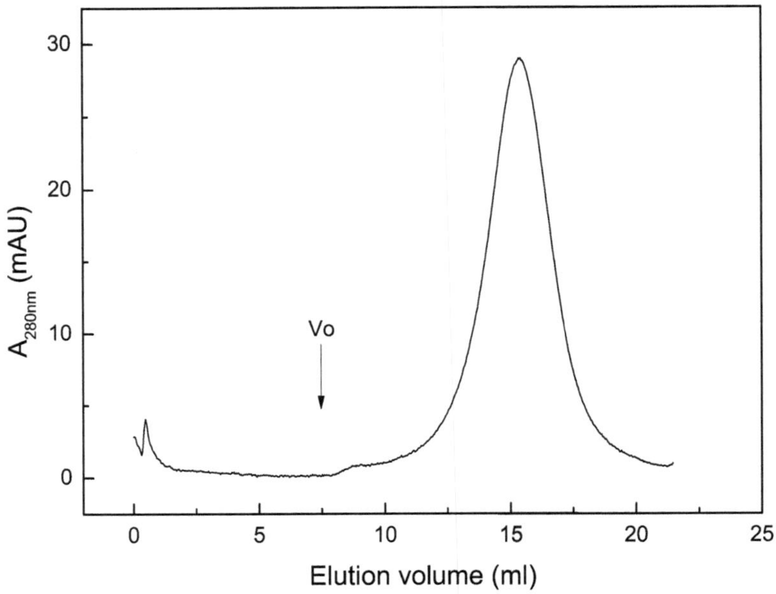

Fig. 4 Size exclusion chromatography of the purified bestrophin homologue MPSIL0347 on a Superose 6 10/300 column. The *arrow* indicates the void volume of the column

to size exclusion chromatography on a Superose 6 10/300 column in SAC dialysis buffer (see Note 28), at a flow rate of 1 mL per min, monitoring A_{280nm} to detect protein and collecting 0.5 mL fractions. Typical results are shown in Fig. 4.

13. Assess the functional state of the purified protein using a suitable assay (see Note 26).
14. To regenerate the column, wash it three times with 5 CV of SAC regeneration buffer. The color change from yellow to red indicates the progress of the regeneration process, which is complete when the intensity of the red color is uniform throughout the column. The column can then be stored at 4°C in the same buffer, which should be removed by washing the column two times with 4 CV of SAC wash buffer before the next purification run.

4 Notes

1. *E. coli* BL21 strains are used for protein expression because they lack the proteases OmpT and Lon, so minimizing protein degradation. (DE3) strains are employed because they carry a chromosomal copy of the T7 RNA polymerase gene under the control of the lacUV5 promoter. They are thus intended for expression of proteins under the control of a T7 promoter,

e.g., in pET vectors (Novagen), but are also suitable for expression under the control of other promoters, such as the *tac* promoter found in plasmid pTTQ18 (6). Strain BL21 Star™ (DE3) bears a truncated RNAse E gene to enhance mRNA stability (20), while strain C43(DE3) bears a mutation in the lacUV5 promoter, decreasing T7 polymerase levels and in consequence sometimes yielding improved levels of correctly folded membrane proteins (16). The pRARE2 plasmid encodes tRNAs for seven codons rarely used in *E. coli* and its presence can thus enhance translation efficiency of genes with codon usage different from endogenous *E. coli* ones (7).

2. In our laboratory we primarily utilize pTTQ18 (6) derivatives as expression vectors for membrane proteins. However, other inducible vectors, such as the pET series (Novagen), are also widely employed for membrane protein production.
3. Dissolving the lactose may take a considerable time, and can be speeded up by gentle warming.
4. Carbenicillin is used to select for pTTQ18 and its derivatives, which harbor a β-lactamase gene that confers ampicillin resistance. It is used because it is more stable than ampicillin. Other antibiotics, such as kanamycin, may be required to select for other expression vectors, such as some members of the pET series. Chloramphenicol is used to select for the pRARE2 plasmid, which harbors a chloramphenicol resistance gene.
5. Instead of making up autoinduction media from their individual components, they can also be purchased from a number of suppliers, including ForMedium™, although media from the latter do not contain glycerol.
6. If crystallization proves to be a problem the guanidine solution can be stored at 37°C.
7. An estimate of the amount of expressed protein per unit volume of culture can be obtained by comparison of the intensity of the spots seen with those for known amounts of a purified, water-soluble protein bearing the same affinity tag. The amount expressed per mg total protein can then be calculated using the results of the BCA assay performed in step 7. Alternatively, the latter can be used in order to load equal amounts of protein rather than equal volumes of culture onto the blot.
8. In our experience, although autoinduction usually gives better yields of expressed protein, for some membrane proteins IPTG induction is better. The temperature at which induction is performed can also be explored, lower temperatures (e.g., 16 or 25°C) sometimes yielding greater levels of natively folded, membrane-inserted protein that can be solubilized with mild nonionic detergents, and less protein in the form of inclusion bodies.

9. An IPTG-induced culture will typically have a D_{600nm} of ~1. For autoinduced cultures, which can reach much higher cell densities, measure the D_{600nm} and prepare membranes from a volume of cells equivalent to 50 mL of an IPTG-induced culture.
10. Formation of spheroplasts (spherical cells lacking an intact cell wall) can be confirmed by examination using phase-contrast microscopy at a magnification of 800×.
11. Heating membrane proteins in SDS-containing sample buffer for SDS-PAGE typically promotes their aggregation. Samples in SDS should therefore be prepared either at room temperature or by heating at 37°C for 20 min. Aggregation is also minimized by using a preparation of SDS containing a proportion of 14C and 16C alkyl sulfates in addition to dodecyl sulfate (21).
12. Following optimization, expression levels for the membrane protein of interest of up to ~25% of total cytoplasmic membrane protein are routinely achieved in our laboratory. However, because cytoplasmic membrane proteins represent only about 6% of the total protein in *E. coli*, typical yields of purified membrane protein per liter of culture are in the 1–10 mg range, necessitating larger cultures than for water-soluble protein expression.
13. We typically find that 0.5 mM IPTG gives good induction whilst minimizing cost and toxicity. Although some laboratories recommend testing a range of IPTG concentrations, the concentrative uptake of the inducer via the lactose transporter LacY means that there is no simple relationship between the extracellular and intracellular concentrations. If necessary, induction of expression by IPTG can however be titrated in *E. coli* BL21 Tuner™ strains (Novagen), which are *lacZY* deletion mutants. In these strains IPTG enters the cells passively and so induction can be achieved in a concentration-dependent fashion.
14. The susceptibility of individual types of membrane proteins to proteolytic degradation varies and so the requirement for inclusion of protease inhibitors during membrane preparation and protein purification must be tested empirically. If IMAC is to be used for purification, an inhibitor mix lacking EDTA should ideally be employed to avoid loss of metal ions from the column. However, a number of manufacturers, such as Roche, have recently introduced new affinity resins (e.g., complete His-Tag Purification Resin) which are reportedly stable in the presence of both EDTA and thiols.
15. If a cell disruptor is not available, cells can be lysed by sonication, although this is more laborious and typically not as efficient.

16. Snap-frozen samples should be stored at −80°C rather than at −20°C to prevent protein degradation.
17. Buffers used for solubilization and chromatography of membrane proteins typically contain up to 10% glycerol because of its stabilizing effect on membrane proteins (22).
18. In the case of $PepT_{So}$ and MPSIL0347, extensive solubilization of the protein can be achieved using 1% detergent (10 mg/mL) and a membrane protein concentration of 5 mg/mL, i.e., a detergent to protein weight ratio of 2:1. However, the optimal ratio of detergent to protein must be determined empirically and for other proteins a larger ratio is required. A good rule of thumb is to start with a protein concentration of 2 mg/mL. Typically solubilization can be achieved by incubation with detergent for 1 h but for some proteins, such as MPSIL0347, overnight incubation results in more complete solubilization.
19. Although both Ni-charged and Co-charged resins can be used for purification of His-tagged proteins, we find that the latter typically are more selective for the protein of interest in comparison to endogenous His-rich proteins, and so yield a purer protein preparation. The binding capacity of HisPur™ Co resin for His-tagged proteins is ~360 nmol/mL and so 0.6 mL resin should bind ~12 mg His_6-tagged $PepT_{So}$ (M_r 57962.5). The amount of resin required should be established empirically and the minimum quantity employed, in order to minimize nonspecific binding of endogenous His-rich proteins. The latter is also minimized by the inclusion of 7.5 mM imidazole in the IMAC solubilization buffer.
20. Binding of detergent-solubilized membrane proteins to IMAC resins can be slow, possibly because the detergent micelle sterically hinders access to the binding sites on the resin. Whatever the cause, we have found that binding in batch mode as described here is typically superior to loading the protein onto the resin in a column, even if recirculation is employed. Dissociation of the bound protein can also be slow, hence the inclusion of an extended incubation of the resin with elution buffer, as detailed in step 8.
21. The tightness with which membrane proteins bind to the IMAC resin can differ between different proteins, despite their bearing identical tags. The concentration of imidazole in the wash buffers required to elute weakly bound, endogenous His-rich proteins without elution of the tagged protein must therefore be determined empirically, and may differ from that described here. All buffers should contain a detergent concentration above the CMC to maintain protein solubility: the CMC for DDM in 0.2 M NaCl is ~0.12 mM (0.006%, w/v).

22. This fraction should contain just the IMAC wash buffer 2 displaced by the elution buffer from the column, i.e., it should not contain eluted protein, but it should be kept just in case!
23. Most batches of imidazole from commercial sources absorb light at 280 nm. Thus, the first fraction will have an anomalously low absorption, because it will contain a mixture of wash and elution buffer, but the subsequent fractions will contain only elution buffer and so the latter will represent an appropriate blank. Imidazole with a low A_{280nm} can be purchased, although it is of course more expensive.
24. In our experience, slow removal of imidazole by dialysis is preferable to its rapid removal by passage down a desalting column, because the latter can result in precipitation of the protein. A low concentration of DDM, at or above the CMC, is included in the dialysis buffer to prevent loss of detergent and thus precipitation of the membrane protein.
25. DDM forms roughly spherical micelles of reported size ranging from about 40–75 kDa. To prevent concentration of free detergent, a concentrator with a molecular weight cutoff (MWCO) of 100 kDa must therefore be used, and even so some increase in free detergent concentration is to be expected. Because the protein-detergent micelle is much larger than the protein alone, there is usually little loss of proteins with molecular masses of ≥ 50 kDa.
26. The methods used to assess whether the purified protein remains in its native folded state and retains function will of course depend on the identity of the particular protein under investigation. In some cases, function can be assessed directly using the detergent-solubilized protein, an example being the use of scintillation proximity assays to measure substrate binding to transporters (23). Alternatively, function can be assessed following reconstitution of the protein into a proteoliposome and measurement, for example, of the influx of radiolabelled substrate in the case of a transporter (24). The latter type of assay confirmed that $PepT_{So}$ purified by the method described retains its expected peptide transport activity. In the case of ion channels, electrophysiological assays are possible if the protein is reconstituted into giant unilamellar vesicles (25) or planar lipid bilayers (see Chapter 8 of this book).
27. The binding capacity of Strep-Tactin® Superflow® resin for Strep-tagged proteins is 50–100 nmol per mL, so 0.5 mL resin should bind between 1 and 2 mg Strep-tagged MPSIL0347 (M_r 38543).
28. In the case of MPSIL0347, the SAC dialysis buffer was determined to be the optimum for maintaining protein stability, using the assay described by Postis et al. (26). Once such an optimum buffer has been identified by this assay, it can also be

employed to good effect throughout the purification itself. The lack of any aggregated material eluting in the column void volume during size exclusion chromatography and the symmetrical appearance of the absorption peak (Fig. 4) indicates that the protein is monodisperse and suitable for further structural and functional investigations.

Acknowledgements

This work was supported by the U.K. BBSRC (grant BBS/B/14418; Membrane Protein Structure Initiative) and by the Wellcome Trust (ref. 019322/7/10/Z). Additional support from the University of Leeds is acknowledged. We are grateful to Michael McPherson, Peter Henderson, Sarah Deacon, Gareth Wright, Gerard Huysmans, Peter Roach, Jean Ingram, and David Sharples for their many contributions to the development of membrane protein expression and purification technologies in our laboratory.

References

1. Wallin E, von Heijne G (1998) Genome-wide analysis of integral membrane proteins from eubacterial, archaean, and eukaryotic organisms. Protein Sci 7:1029–1038
2. Rask-Andersen M, Almen MS, Schioth HB (2011) Trends in the exploitation of novel drug targets. Nat Rev Drug Discov 10: 579–590
3. Rahman M, Ismat F, McPherson MJ, Baldwin SA (2007) Topology-informed strategies for the overexpression and purification of membrane proteins. Mol Membr Biol 24: 407–418
4. Low C, Jegerschold C, Kovermann M, Moberg P, Nordlund P (2012) Optimisation of over-expression in *E. coli* and biophysical characterisation of human membrane protein synaptogyrin 1. PLoS One 7:e38244
5. Bill RM, Henderson PJ, Iwata S, Kunji ER, Michel H, Neutze R, Newstead S, Poolman B, Tate CG, Vogel H (2011) Overcoming barriers to membrane protein structure determination. Nat Biotechnol 29:335–340
6. Stark MJR (1987) Multicopy expression vectors carrying the lac repressor gene for regulated high-level expression of genes in *Escherichia coli*. Gene 51:255–267
7. Burgess-Brown NA, Sharma S, Sobott F, Loenarz C, Oppermann U, Gileadi O (2008) Codon optimization can improve expression of human genes in *Escherichia coli*: a multi-gene study. Protein Expr Purif 59:94–102
8. Norholm MH, Light S, Virkki MT, Elofsson A, von Heijne G, Daley DO (2012) Manipulating the genetic code for membrane protein production: what have we learnt so far? Biochim Biophys Acta 1818:1091–1096
9. Bernsel A, Viklund H, Hennerdal A, Elofsson A (2009) TOPCONS: consensus prediction of membrane protein topology. Nucleic Acids Res 37:W465–W468
10. von Heijne G (2006) Membrane-protein topology. Nat Rev Mol Cell Biol 7:909–918
11. Glover CA, Postis VL, Charalambous K, Tzokov SB, Booth WI, Deacon SE, Wallace BA, Baldwin SA, Bullough PA (2011) AcrB contamination in 2-D crystallization of membrane proteins: lessons from a sodium channel and a putative monovalent cation/proton antiporter. J Struct Biol 176:419–424
12. Huysmans GH, Chan N, Baldwin JM, Postis VL, Tzokov SB, Deacon SE, Yao SY, Young JD, McPherson MJ, Bullough PA, Baldwin SA (2012) A urea channel from *Bacillus cereus* reveals a novel hexameric structure. Biochem J 445:157–166
13. Hilf RJ, Dutzler R (2008) X-ray structure of a prokaryotic pentameric ligand-gated ion channel. Nature 452:375–379
14. Deacon SE, Roach PCJ, Postis VLG, Wright GS, Xia X, Phillips SE, Knox JP, Henderson PJ, McPherson MJ, Baldwin SA (2008) Reliable scale-up of membrane protein over-expression by bacterial auto-induction: from microwell

plates to pilot scale fermentations. Mol Membr Biol 25:588–598
15. Studier FW (2005) Protein production by auto-induction in high density shaking cultures. Protein Expr Purif 41:207–234
16. Wagner S, Klepsch MM, Schlegel S, Appel A, Draheim R, Tarry M, Högbom M, van Wijk KJ, Slotboom DJ, Persson JO, de Gier JW (2008) Tuning *Escherichia coli* for membrane protein overexpression. Proc Natl Acad Sci USA 105:14371–14376
17. Smith PK, Krohn RI, Hermanson GT, Mallia AK, Gartner FH, Provenzano MD, Fujimoto EK, Goeke NM, Olson BJ, Klenk DC (1985) Measurement of protein using bicinchoninic acid. Anal Biochem 150:76–85
18. Newstead S, Drew D, Cameron AD, Postis VL, Xia X, Fowler PW, Ingram JC, Carpenter EP, Sansom MS, McPherson MJ, Baldwin SA, Iwata S (2011) Crystal structure of a prokaryotic homologue of the mammalian oligopeptide-proton symporters, PepT1 and PepT2. EMBO J 30:417–426
19. Hagen AR, Barabote RD, Saier MH (2005) The bestrophin family of anion channels: identification of prokaryotic homologues. Mol Membr Biol 22:291–302
20. Lopez PJ, Marchand I, Joyce SA, Dreyfus M (1999) The C-terminal half of RNase E, which organizes the *Escherichia coli* degradosome, participates in mRNA degradation but not rRNA processing in vivo. Mol Microbiol 33: 188–199
21. Baldwin SA, Baldwin JM, Lienhard GE (1982) Monosaccharide transporter of the human erythrocyte. Characterization of an improved preparation. Biochemistry 21:3836–3842
22. Engel CK, Chen L, Prive GG (2002) Stability of the lactose permease in detergent solutions. Biochim Biophys Acta 1564:47–56
23. Harder D, Fotiadis D (2012) Measuring substrate binding and affinity of purified membrane transport proteins using the scintillation proximity assay. Nat Protoc 7:1569–1578
24. Xie H, Patching SG, Gallagher MP, Litherland GJ, Brough AR, Venter H, Yao SY, Ng AM, Young JD, Herbert RB, Henderson PJ, Baldwin SA (2004) Purification and properties of the *Escherichia coli* nucleoside transporter NupG, a paradigm for a major facilitator transporter sub-family. Mol Membr Biol 21: 323–336
25. Varnier A, Kermarrec F, Blesneac I, Moreau C, Liguori L, Lenormand JL, Picollet-D'hahan N (2010) A simple method for the reconstitution of membrane proteins into giant unilamellar vesicles. J Membr Biol 233:85–92
26. Postis VL, Deacon SE, Roach PC, Wright GS, Xia X, Ingram JC, Hadden JM, Henderson PJ, Phillips SE, McPherson MJ, Baldwin SA (2008) A high-throughput assay of membrane protein stability. Mol Membr Biol 25:617–624

Chapter 4

Transient Overexpression of Genes in Neurons Using Nucleofection

Hannah M. Kirton, Louisa Pettinger, and Nikita Gamper

Abstract

Nucleofection is a transfection method used to introduce substrates such as cDNA plasmids into primary cells or other cell lines. The method can be successfully applied to cells that are considered difficult to transfect or suffer from low transfection efficiency as seen with traditional transfection techniques. Neurons in primary cultures retain many properties of their in vivo state and therefore, in many instances, are considered better experimental systems than immortalized cell lines, thus becoming increasingly desirable cell types for biomedical research. However, being post-mitotic, primary neuronal cultures are particularly difficult to transfect using routine transfection reagents. There is therefore a growing need for the efficient delivery of expression vectors into such neuronal cultures. In this chapter we will discuss the application of nucleofection for the heterologous expression of genes in primary neuronal cultures. We also discuss the advantage of this technique relative to other conventional methods, and describe a reliable method for transfection of cultured rat dorsal root ganglion (DRG) and trigeminal (TG) neurons.

Key words Nucleofection, Neurons, Transfection efficiency and optimization

1 Introduction

Efficient delivery of an expression vector is an important tool in the study of neuronal biology. However, the expression of exogenous genes into nondividing cells such as neurons has become somewhat of a challenge due to the necessity to deliver genetic material directly into the cell nucleus (thus it has to cross both plasma and nuclear membranes). In contrast, in dividing cells, such as immortalized cell lines, successful transfection can be achieved by the delivery of constructs into the cytosol from where it can translocate into the nuclei during mitosis. Conventional

Nikita Gamper (ed.), *Ion Channels: Methods and Protocols*, Methods in Molecular Biology, vol. 998,
DOI 10.1007/978-1-62703-351-0_4,

nonviral transfection methods such as lipophilic transfection reagents have proven difficult and failed to deliver reasonable transfection efficiency with most primary neuronal cultures. While the viral-mediated gene transfer methods produce higher transduction efficiencies, these proved time consuming and require high levels of safety. Other transfection techniques used to transfect primary cultures include “physical transfection” methods such as direct nuclear injection of plasmid DNA (1) and the biolistic “gene gun technique” (2). The biolistic particle delivery system for heterologous expression of genes into primary neurons bombards neurons at high velocity with DNA-coated gold particles. Although it is reliable, it is time consuming and causes physical damage to the neurons. In addition, transfection efficiency of biolistic transfection is usually no greater than 10%.

The recent development of nucleofector technology is the first highly efficient nonviral gene transfer method that has vastly enhanced the number of cell types that are amenable to transfection. Nucleofection has the ability to transfect cDNA plasmids directly into the cell nucleus of nondividing cells by delivery of a defined set of electrical pulses to cells in suspension (3, 4). The newer modification of the nucleofector device also enables the transfection of adherent cultures. This technique is fast, reliable, and reproducible with a transfection success rate reaching 80–90% in some preparations (although transfection of neuronal cultures usually yields lower transfection rates).

Primary cultures of murine peripheral somatosensory neurons, such as neurons of the dorsal root or trigeminal (DRG and TG, respectively) ganglia, provide an invaluable in vitro system for studying the molecular processes that underlie somatosensory signalling and pain. Yet, as with most primary neuronal cultures, transfection of these neurons proves a considerable challenge. In this chapter we describe the nucleofection technique for the delivery of cDNA vectors into DRG and TG neurons that we have successfully applied to investigate the regulation of ion channels by G protein-coupled receptors in these neurons (5–8). This technique was also used for siRNA gene knock-down in sensory neuron cultures (8). The development of transfection methods, such as nucleofection, enabled the efficient transfection of ion channels, neuropeptides, G protein-coupled receptors as well as optical reporters of various intracellular signalling cascades which, in turn, greatly enhanced the “toolkit” available for the interrogation of neuronal signalling. Here we describe how to perform efficient nucleofection of cDNA in dissociated DRG and TG neurons from rats.

2 Materials

2.1 Dissociation of Rat DRG (TG) Neurons

1. 7-day-old male Wistar rat (see Note 1).
2. Cell culture medium: 50 mL DMEM (Invitrogen) pre-supplemented with GlutaMAX (see Note 2), 10% fetal bovine serum and 5% penicillin (50 U/mL), streptomycin (50 μg/mL) pre-warmed to 37°C.
3. 2 mL of dissection solution: dispase (10 mg/mL) and collagenase type 1A (1 mg/mL) dissolved in Ca^{2+}- and Mg^{2+}-free Hanks' balanced salt solution (HBSS).
4. Syringe filter (Millipore, 0.22 μM).
5. Ca^{2+}- and Mg^{2+}-free HBSS.
6. 2 mL of 100% Isoflurane (Merial Animal Health Ltd).
7. Sterile polypropylene falcon tubes, 15 mL containing ice-cold DMEM medium (supplemented with serum and antibiotics).
8. Sterile U-bottomed falcon tube, 15 mL.
9. Dissection tools (see Note 3):

   Microdissection forceps (Dumont #5 and #55 stainless steel forceps).

   DeBakey forceps with atraumatic teeth grip to prevent damage to tissue.

   Economy tweezers style 4.

   Dissecting scissors.

   Scalpel blade.

   Vannas scissors.
10. Box of ice.
11. Styrofoam pad (i.e., icebox cover), sprayed with 70% alcohol, four hypodermic needles.
12. Greiner sterile cell culture dishes (35 × 10 and 55 × 15 mm) filled HBSS.

2.2 Nucleofection of Rat DRG Neurons

1. Lonza 2B or 4D Nucleofector® device (formerly known as Amaxa Nucleofector).
2. Lonza rat neuron Nucleofector® kit: Certified cuvette, plastic pipette, and supplemented nucleofector buffer solution at room temperature.
3. Plasmid cDNA.

4. Tissue culture plate (Nunclon) with pre-coated poly-D-lysine and laminin glass coverslides stored at 37°C until required (see Note 4).
5. Hemocytometer (optional).
6. Sterile microcentrifuge tubes, 1.5 mL.

3 Methods

All procedures are carried out under sterile conditions using a class II inflow hood (unless indicated otherwise).

3.1 Dissociation of Rat DRG (TG) Neurons

1. Thoroughly spray the procedure table and all the instruments with 70% alcohol.
2. Dissection solution is prepared by dissolving dispase (10 mg/mL) and collagenase type 1A (1 mg/mL) in HBSS. The solution is then sterilized by filtering through a 0.22 μM syringe filter and stored at 37°C (tissue culture incubator) until required.
3. Rat is sacrificed using isoflurane overdose (or similar approved method).
4. The spine is removed and cut in two halves in the transverse plane. Each part is then cut in the sagittal plane using fine scissors and placed into the small cell culture dish filled with Ca^{2+}- and Mg^{2+}-free HBSS and kept on ice (see Note 5 *for TG dissociation*).
5. Pin one of the four spine sections to the Styrofoam pad using hypodermic needles with spinal canal facing up. Remove spinal cord with fine (Dumont #55) forceps.
6. Pool DRG from all levels with fine forceps and place into a small cell culture dish filled with the ice-cold Ca^{2+}- and Mg^{2+}-free HBSS. Repeat the procedure for the remaining sections of the spine. Refer to Chapter 25 of this book for further advice on DRG extraction.
7. Under sterile conditions DRG (or TG) are transferred into the U-bottomed tube containing the dissection solution, and incubated for 10–30 min (see Note 6) in a humidified incubator (37°C, air supplemented with 5% CO_2).
8. After 10–30 min gently triturate the cell suspension with a sterile 1 mL Gilson pipette to dissociate the ganglia (avoid excessive trituration as this may damage the neurons due to mechanical stress) (see Note 7).

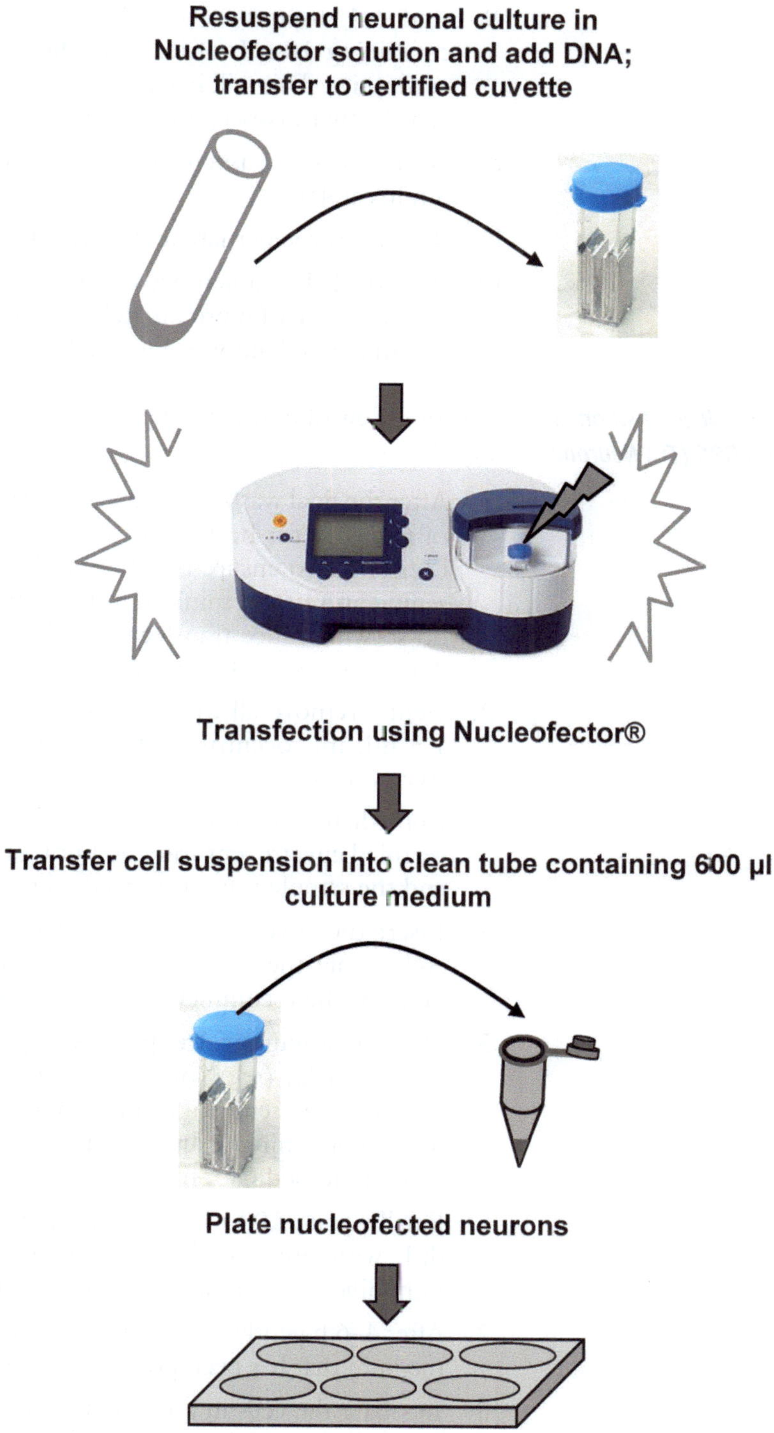

Fig. 1 Schematic representation for the nucleofection of cultured neurons using the Lonza Nucleofector® technique

9. Add cell suspension to a 15 mL polypropylene falcon tube containing 10 mL of ice-cold DMEM containing serum and antibiotics. This will inactivate the digestion enzymes and prevent further dissociation of the cell suspension (see Note 8).
10. Wash the cell suspension by centrifugation at ~800 × *g* for 5 mins at 4°C.
11. Remove the supernatant and repeat washing steps 8 and 9.
12. (Optional) Use a hemocytometer to check for appropriate cell culture density (approximately 2×10^6 cells per transfection is a recommended amount but see Note 9).

3.2 Nucleofection of Rat DRG (TG) Neurons

The overview of transfection procedure is schematically depicted in Fig. 1.

1. After the final wash (see Note 10) resuspend the cells in 100 μL of nucleofector buffer (pre-warmed to room temperature—it is essential to ensure that the buffer is warm; keep it at room temperature for a minimum of 30 min prior to use; failure to do so will reduce transfection efficiency). Do not remove the debris (see Note 8), and gently triturate 1–3 times.
2. Gently remove all the cell suspension and transfer to a clean 1.5 mL microcentrifuge tube containing 2 μg of total DNA (see Note 11).
3. Immediately transfer all the cell/DNA suspension into a certified cuvette, ensuring no bubbles are present in the sample and the sample is inserted to the bottom of the cuvette.
4. Insert the cuvette into the Nucleofector device and immediately apply the transfection program. For rat DRG and TG neurons the recommended program is O-03 using Lonza 2B.
5. Using a Pasteur pipette (provided by the Lonza kit), gently but without delay (see Note 12), transfer the sample into a sterile 1.5mL microcentrifuge tube containing 600 μL of DMEM media containing serum and antibiotics (pre-warmed to 37°C in a humidified CO_2 incubator).
6. Finally, plate 100 μL of the sample into the prepared culture dish with pre-coated coverslips, and incubate the cells in a humidified incubator (37°C, air supplemented with 5% CO_2).
7. After 4–6 h gently flood the culture dish wells with fresh culture medium that has been pre-warmed to 37°C (see Note 13).
8. Incubate the cells in a humidified incubator (37°C, air supplemented with 5% CO_2) for 24–48 h until analysis (see Note 13).

Examples of successfully transfected DRG and TG neurons are given in Fig. 2.

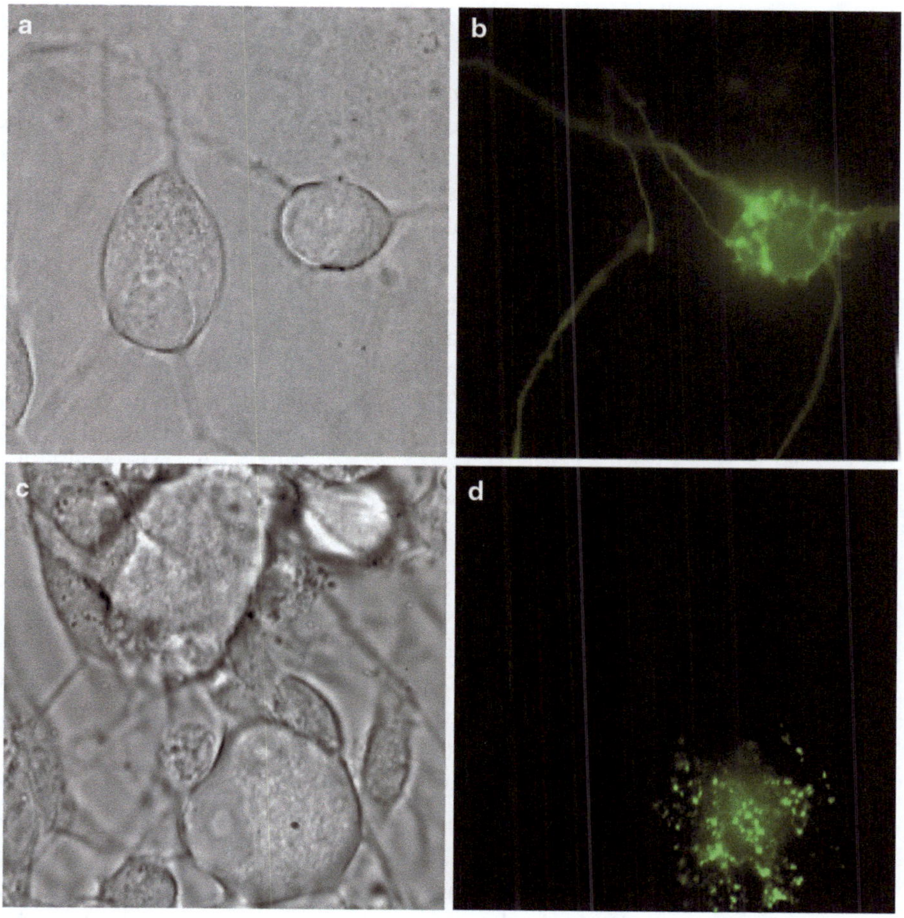

Fig. 2 Transfection of primary sensory neurons using nucleofection. (**a**, **b**) TG neurons transfected with GFP-tagged delta-opioid receptors (DOR-GFP). Bright-field image (**a**) shows two neurons one of which overexpresses DOR-GFP visible in epi-fluorescence illumination (**b**). (**c**, **d**) DRG neurons transfected with Venus-tagged neuropeptide Y (Venus-NPY). Bright-field image (**c**) shows two neurons and several satellite glial cells; one neuron displays vesicularly localized Venus-NPY fluorescence as imaged in total internal reflection fluorescence (TIRF) mode. Imaging was performed using Nikon TE200E swept-field confocal/TIRF microscope

4 Notes

1. This technique may also be applied to younger and older animals. Generally, the viability of neurons decreases with age; viable neuronal cultures can however still be achieved using this protocol from adult rat DRG neurons.
2. It is best to aliquot a desired amount of DMEM medium into a 50 mL polypropylene falcon tube and refrigerate to avoid metabolization of the GlutaMAX to glutamate which is known to be neurotoxic.
3. This dissection instrument list has been carefully chosen to ensure minimal damage to the tissue during (1) removal of the

spinal column, (2) cutting of the column into two halves in the transverse plane, and (3) cutting of the column in the sagittal plane. Ganglia can be pulled out using the forceps or can be cut out using the forceps and fine vannas scissors (the latter becomes necessary if older animals are used). Dorsal roots can be cut out if necessary (especially for older animals). We do not remove dorsal roots for preparation from 7-day-old rats.

4. Coated sterile cell culture coverslips need to be prepared before dissociation, preferably on the previous day. Sterile coverslips of desired size are placed into each well of a 6- or 24-well plate and subsequently coated with solution of laminin (50 μg/μL) and poly-D-lysine (0.001%) in HBSS. Plates are incubated in a humidified incubator for 6 h, and washed thereafter three times with sterile culture-grade water or sterile HBSS and allowed to thoroughly dry before storing at 4°C. On the day of the dissociation plates are pre-warmed to 37°C in a humidified incubator.
5. In the case of TG dissociation, the animal is decapitated (after isoflurane overdose). The cranium is opened at the back of the head and the brain is removed. Trigeminal ganglia are excised using vannas scissors. The rest of the procedure is similar for both TG and DRG. Each trigeminal ganglion hosts over 50,000 neuronal cell bodies (9) while each DRG neuron contains ~6,000 neuronal cell bodies (10). Therefore, one pair of TG provides an equivalent quantity of neurons comparable to that of 17 DRG.
6. This step is crucial for a good-quality culture. The timing of incubation depends on (1) age of the animal (the older the animal, the longer incubation time is required) and (2) the strength of the collagenase (this will decline with time of storage). For a 7-day-old Wistar rat and a fresh collagenase the recommended time of incubation is 10–12 min. After incubation ganglia are gently triturated using a 1 mL Gilson pipette (for even gentler trituration a disposable pipette tip can be cut at the very end of the tip to enlarge the opening) 1–3 times. If ganglia are still compact and there is little sign of dissociation, a further 5 min incubation is recommended (dissociation is characterized by a "cloudy" appearance to the media and few large clumps of cells). Alternatively it is suggested to gently "flick" the bottom of the tube to encourage the cells to de-aggregate. When dissociation of the ganglia becomes apparent, ganglia are triturated ten times in a similar manner as previously described. *Important:* there will be visible threads of undigested tissue—do not try to achieve a completely homogenized suspension as this will lead to the overexposure of neurons to the digestion enzymes and, as a result, lead to significant cell death. Presence of undigested tissue will not impair quality of culture.
7. Many DRG and TG neurons are mechanosensitive (that is, they express mechanosensitive ion channels that excite neurons upon mechanical stimulation). Excessive trituration may therefore

overexcite and kill mechanosensitive neurons. In cases where the aim of the experiment is to enrich the culture with polymodal nociceptors (which are not low-threshold mechanosensors), stronger trituration can be used to kill mechanosensitive neurons. On the contrary, if the mechanosensitive neurons need to be preserved, an extreme precaution should be taken when triturating the suspension. See Chapter 12 of this book for further advice on isolation of mechanosensitive DRG neurons.

8. Addition of ice-cold DMEM pre-supplemented with serum inactivates the digestion enzymes. Trituration dissociates the neurons from the ganglia, which increasingly exposes the neurons to the enzymes. The overexposure of the neurons to the enzyme increases cell damage. It is therefore important that the time between the dispersion of ganglia and addition of the ice-cold DMEM is kept to a minimum.
9. The high cell count required for transfection is needed for reducing the resistance of the solution in the transfection cuvette. If the resistance is too high (too few cells) the voltage applied by the device, to reach standard parameters of transfection, will increase, thus increasing the likelihood of cell death. Tissue debris, in addition to DRG neurons and glia, also lowers the resistance of the solution (therefore do not try to remove undigested tissue from the suspension as the presence of the debris can make up for the low cell count!). Empirically we have estimated that extraction of DRG from all spinal levels of one rat allows isolation of up to 600,000 cells (neurons and glia together); however with the undissociated tissue threads and debris this is enough for successful transfection. In the case of TG it is recommended to use two 7-day-old rats for one transfection.
10. After the final wash drain the pellet via gentle aspiration, followed by a further drain using a 1 mL disposable pipette tip. It is important to thoroughly drain the pellet prior to resuspending the pellet in nucleofector buffer since the DMEM medium can reduce efficiency of the transfection.
11. Transfection efficiency depends on the concentration of DNA in the transfection media. However, different constructs or their combination will have different optimal concentrations. It is suggested to start with 2 μg of total DNA. If optimization is required, this amount can be increased or decreased slightly (do not exceed total volume of 15 μL for DNA solution). The properties of the vector in which the cDNA is inserted can also affect the efficiency of transfection. It has been reported by Lonza that IRES plasmids generally give lower efficiencies as compared to other popular vectors.
12. The nucleofector buffer is mildly toxic and it is therefore highly recommended that the cell suspension should not be kept in the nucleofector buffer for more than 15 min as this may reduce cell viability. Reducing the time the cells spend in this buffer is

beneficial to cells. In addition, take care not to triturate cells when transferring and plating them immediately following transfection, as the membranes are weakened by the process.

13. Although it is common practice to use DMEM medium during the culturing of neurons our group have also used Neurobasal medium to reduce the production of satellite glial cells (SGC). Suppression of SGC proliferation can also be achieved by supplementing the culture media with the antimitotic drug, cytosine arabinoside (ara-C), at a concentration of 10 μM, for 24 h post-dissociation (8). For some studies it is necessary to include nerve growth factor (NGF) in the culture media as it promotes neuronal differentiation and neurite outgrowth (11). NGF (at 25 ng/mL) can be added at step 7 of the transfection protocol. We do not routinely use NGF in DRG/TG cultures since NGF is a potent inflammatory mediator (12) and our studies are often focused on the inflammatory mechanisms in sensory neurons.

Acknowledgements

This work was supported by the MRC, BBSRC and Wellcome Trust.

References

1. Abogadie FC, Vallis Y, Buckley NJ, Caulfield MP (1997) Use of antisense-generating plasmids to probe the function of signal transduction proteins in primary neurons. Methods Mol Biol 83:217–225
2. Gamper N, Shapiro MS (2006) Exogenous expression of proteins in neurons using the biolistic particle delivery system. Methods Mol Biol 337:27–38
3. Gresch O, Engel FB, Nesic D, Tran TT, England HM, Hickman ES, Körner I, Gan L, Chen S, Castro-Obregon S, Hammermann R, Wolf J, Müller-Hartmann H, Nix M, Siebenkotten G, Kraus G, Lun K (2004) New non-viral method for gene transfer into primary cells. Methods 33:151–163
4. Gartner A, Collin L, Lalli G (2006) Nucleofection of primary neurons. Methods Enzymol 406:374–388
5. Linley JE et al (2012) Reactive oxygen species are second messengers of neurokinin signaling in peripheral sensory neurons. Proc Natl Acad Sci U S A 109(24):E1578–E1586
6. Linley JE, Pettinger L, Huang D, Gamper N (2012) M channel enhancers and physiological M channel block. J Physiol 590:793–807
7. Linley JE, Rose K, Patil M, Robertson B, Akopian AN, Gamper N (2008) Inhibition of M current in sensory neurons by exogenous proteases: a signaling pathway mediating inflammatory nociception. J Neurosci 28: 11240–11249
8. Liu B, Linley JE, Du X, Zhang X, Ooi L, Zhang H, Gamper N (2010) The acute nociceptive signals induced by bradykinin in rat sensory neurons are mediated by inhibition of M-type K^+ channels and activation of Ca^{2+}-activated Cl^- channels. J Clin Invest 120: 1240–1252
9. Forbes DJ, Welt C (1981) Neurogenesis in the trigeminal ganglion of the albino rat: a quantitative autoradiographic study. J Comp Neurol 199:133–147
10. Arvidsson J, Ygge J, Grant G (1986) Cell loss in lumbar dorsal root ganglia and transganglionic degeneration after sciatic nerve resection in the rat. Brain Res 373:15–21
11. Ernsberger U (2009) Role of neurotrophin signalling in the differentiation of neurons from dorsal root ganglia and sympathetic ganglia. Cell Tissue Res 336:349–384
12. McMahon SB, Bennet DLH, Bevan S (2006) Inflammatory mediators and modulators. In: McMahon SB, Koltzenburg M (eds) Wall and Melzack's textbook of pain. Elsevier Churchill Livingstone, Edinburgh, pp 49–72

Chapter 5

Viral Gene Delivery: Optimized Protocol for Production of High Titer Lentiviral Vectors

James Hewinson, Julian F.R. Paton, and Sergey Kasparov

Abstract

HIV-derived lentiviral vectors (LVV) are among the most commonly used gene delivery vehicles. Their production in high quantities, which enables concentration of viral particles to high titers, is important for their successful application in both biomedical research and gene therapy. LVV are produced by co-transfection of three or more plasmids into a packaging cell line followed by several purification and concentration steps. Protocols currently in circulation differ from each other but the direct comparison of their efficacy based on the published information is extremely difficult because more than one variable may be changed and essential information may be omitted. We systematically evaluated three protocols and found that one single modification described here, using FuGene® 6 in the co-transfection step, increase LVV output almost 20 times as compared to the most commonly used calcium phosphate ($CaPO_4$) transfection technique. Unexpectedly FuGene® 6 was also much more efficient than another widely used reagent, Superfect. Dependent on requirements, this permits a dramatic downscaling of the packaging stage of viral production, and/or super-concentration of LVV to achieve stronger expression. For example we were able to prepare ~25 μL of high titer LVV suitable for injections into rodent brain using a single T75 cm^2 cell culture flask of packaging cells. The same output would require up to 20 times more packaging cells and reagents following conventional protocols. We illustrate the potential of our approach using transfection of primary neuronal cultures with LVV expressing an optogenetic actuator channelrhodopsin-2. Our observations should help to achieve reproducible production of high titer LVV for experimental and potential therapeutic applications.

Key words Lentivirus, High titer, FuGene, Transfection, Gene delivery, Transduction, Neuron, Channelrhodopsin

1 Introduction

Lentiviral vectors (LVV) are widely used for gene delivery in vitro and in vivo. They are non-immunogenic, fast to produce and offer a significant packaging capacity for complex expression cassettes (1). The most commonly used LVV are derived from HIV, and pseudotyped with the VSV-G protein from vesicular stomatitis virus. In the brain LVV can help to achieve high levels of transgene expression

Nikita Gamper (ed.), *Ion Channels: Methods and Protocols*, Methods in Molecular Biology, vol. 998,
DOI 10.1007/978-1-62703-351-0_5, © Springer Science+Business Media, LLC 2013

making them suitable for experiments where this feature is critical. Recent introduction of optogenetic actuator proteins, such as Channelrhodopsin-2 (Chr2), which allow control of neuronal activity and signalling with light (2–6), has made this feature particularly important because these proteins need to be present in cells at fairly high levels. The titer of the vector has a direct impact on the level of expression and therefore the protocol used for LVV production is decisive for the success of many applications.

Presently, LVV are prepared by co-transfection of three or four separate plasmids into a packaging cell line (7–9). In combination the plasmids encode the proteins sufficient to generate replication-deficient, self-inactivating LVV particles (individually the plasmids do not encode the necessary proteins to generate a functional virus). Functional LVV are capable of transducing mammalian cells delivering genetic material for the expression of genes of interest. The packaging cells are induced to produce viral vector particles for a period of time, followed by several steps of vector concentration. Concentration in most cases is achieved using ultracentrifugation although affinity purification methods are also available. If the initial production step is not efficient the only way to compensate for that is to start with large quantities of packaging cells and media and then to concentrate the diluted LVV using complex purification protocols (7, 8). Approaches adopted by different laboratories for production as well as titration of LVV vary significantly. Since multiple parameters are usually different and not all relevant information is presented, it is difficult to directly compare the efficiency of published protocols. We sought to optimize LVV production, focusing on the stage of transfection of the packaging cell line. The absolute majority of studies currently use calcium phosphate ($CaPO_4$) precipitation (7, 8) to introduce plasmids into packaging cell lines. Here we evaluate other transfection reagents and present the optimized protocol. In order to compare various methods and directly titer the purified LVV we used a construct, LVV EF1α-PLAP, which encodes placental alkaline phosphatase (PLAP), under control of the human elongation factor-1α promoter (EF1α) as a model (9). When PLAP is expressed in LVV-transduced cells its enzymatic activity can be visualized by a simple and highly sensitive reaction providing a direct and quantitative measure of viable transducing LVV particles.

2 Materials

2.1 Components for Lentiviral Vector Preparation

1. Lenti-X™293 T Cell Line (Clontech) (see Notes 1 and 2).
2. Full Media: Remove 55 mL Dulbecco's Modified Eagle Medium, high glucose, with GlutaMAX™ (D-MEM; Invitrogen,) from a 500 mL bottle. Add 50 mL heat inactivated fetal bovine serum (FBS; Invitrogen) and 5 mL penicillin–streptomycin (10,000

units of penicillin and 10,000 μg of streptomycin per mL; Invitrogen). Store at 4°C.

3. Trypsin, 0.05% (1×) with EDTA 4Na salt (Invitrogen).
4. Lentivirus Plasmids: pNHP, pHEF-VSVG, pCEP4-tat, and pTYF-EF1α-PLAP (available from AddGene or contact corresponding author).
5. FuGene® 6 Transfection Reagent (Promega).
6. Prepare 20% sucrose (Fisher Scientific) in phosphate buffered saline (Fisher Scientific) and sterilize using a 0.22 μm pore size syringe filter (Fisher Scientific). Store at 4°C.
7. Corning® 75 cm^2 rectangular canted neck cell culture flask with vent cap (Corning).
8. Filter unit with polyethersulfone membrane, 0.45 μm pore size (e.g., from Fisher Scientific).
9. Ultracentrifuge and rotor that can carry 28 mL centrifuge tubes and is capable of achieving 74,000 × *g*. [for example we use: Sorvall Discovery 90SE Ultracentrifuge with Sorvall AH-629 Swinging Bucket Aluminum Rotor (ThermoFisher)].
10. Ultracone thin wall ultracentrifuge tubes (e.g., from Seton Scientific).

2.2 Components for Lentiviral Vector Titration Using Placental Alkaline Phosphatase Staining

1. TE671 cell line (ATCC) (*see* **Note 3**).
2. Full Media, as described in Subheading 2.1.
3. Formaldehyde 37–41% (Fisher Scientific): dilute to 4% working solution in PBS when required.
4. Hexadimethrine bromide (Sigma) stock: 0.8 mg/mL in PBS, store at −20°C.
5. Levamisol (Sigma) stock: 50 mM in PBS, stored at −20°C.
6. 5-Bromo-4-chloro-3-indolyl phosphate disodium salt (BCIP; Sigma) solution: 10 mg/mL in dimethylformamide, stored at −20°C.
7. Nitro Blue Tetrazolium (NBT; Sigma) stock: 10 mg/mL in PBS, stored at −20°C.
8. Reaction Buffer: 100 mM Trizma-base (Sigma), 100 mM NaCl, 50 mM $MgCl_2$ (pH adjusted to 9.5 with NaOH); prepared as a 500 mL volume (see Note 4).
9. Levamisol working solution: Add 130 μL levamisol stock to 12.87 mL Reaction Buffer.
10. Reaction Solution: Add 130 μL NBT stock, 65 μL levamisol stock, and 65 μL BCIP stock to 6.24 mL Reaction buffer (see Note 5) 11. 12-well tissue culture treated plates (e.g., Greiner Bio-one).
11. Oven capable of maintaining 75°C.

12. Transmitted light microscope, we use Leitz Fluovert FU microscope with 10× objective (Wild Leitz, type. 090–128.017).

3 Methods

The following method details the optimization of the classical LVV generation protocol that was carried out by us. Batches of LVV EF1α-PLAP were produced by transient co-transfection of pNHP, pHEF-VSVG, pCEP4-tat, and pTYF-EF1α-PLAP plasmids (available from Addgene) into the Lenti-X™293 T Cell Line. Transfection was carried out using $CaPO_4$ (8), Superfect (9), or FuGene® 6. In each experiment at least two different protocols were run in parallel. LVV were purified using sucrose gradient and titered by recording the number of purple–blue diformazan precipitate-positive TE671 cells transduced with LVV EF1α-PLAP (see Fig. 1). Figure 1 demonstrates that the FuGene® 6—based protocol is approximately 20 times more efficient that $CaPO_4$ and also significantly outperforms the Superfect-based method (titers increase approximately seven times). Detailed protocols are described below.

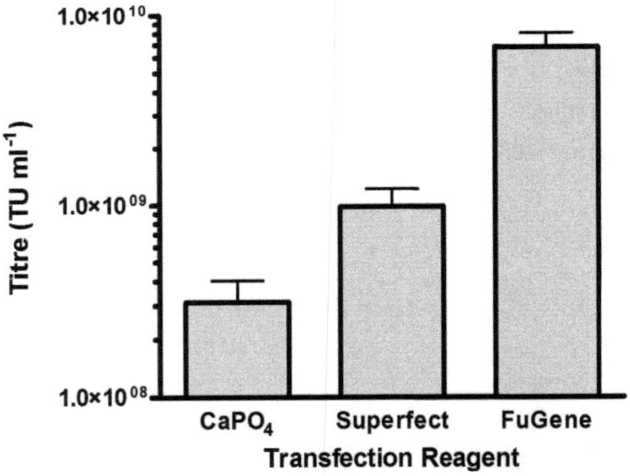

Fig. 1 Comparison of LVV production using FuGene® 6 with commonly used transfection reagents. Multiple preparations of LVV EF1α-PLAP were made and titered directly using the PLAP enzymatic assay in TE671 cells. Using the protocol described here, we found that transfection with FuGene® 6 resulted in an LVV titer approximately sevenfold higher compared to that using Superfect, and about a 20 times higher compared with $CaPO_4$. $n=7$, 14, and 15 for LVV EF1α-PLAP preps using $CaPO_4$, Superfect, and FuGene® 6, respectively. Error bars are SEM. Unpaired t-tests: $CaPO_4$ versus Superfect, $p=0.06$, Superfect vs FuGene® 6, $p=0.0002$, $CaPO_4$ versus FuGene® 6, $p=0.0025$. Out of 15 preparations with FuGene® 6 protocol, the lowest titer obtained was 1.2×10^9, while the highest titer was 1.5×10^{10}

It is important to highlight that while FuGene® 6 is a comparatively expensive transfection reagent, however, if all the costs are taken into account the updated protocol may overall be more economical than $CaPO_4$ use. First, it requires much less media and highly purified plasmid DNA. Second, it saves on the time required to handle large numbers of cell culture flasks. Finally it is also more eco-friendly since the requirement of disposable materials and laboratory plastics is greatly reduced.

To compare "classical" and optimized methods, batches of LVV encoding humanized Channelrhodopsin-2 (hChR2) fused to yellow fluorescent protein (YFP) under the control of the calcium/calmodulin-dependent kinase II promoter (LVV CamKII-hChR2:YFP) were prepared using FuGene® 6 and $CaPO_4$. Following transduction of 1-week-old primary neuronal cultures the percentage of transduced cells and fluorescence intensity of hChR2:YFP was compared. The percentage of cells transduced and fluorescence intensity within transduced cells significantly increased using LVV generated following the optimized protocol presented below compared to the traditional $CaPO_4$-based protocol (Fig. 2).

3.1 Optimized Lentiviral Vector Preparation Using FuGene®6 Transfection Reagent

Passage Cells (Day 1, pm)

1. Dissociate one confluent T75 flask of Lenti-X 293 T cells using 3 mL trypsin-EDTA solution.
2. Use 25% of the resulting cell solution to inoculate one T75 flask and bring volume to 10 mL using full media.

   Transfection of Cells (Day 2, am)

   The optimized transfection protocol is described here. For comparative purposes established transfection methods for LVV production were run in parallel, i.e., using Calcium Phosphate or Superfect. Descriptions of how these transfection methods were performed can be found in Subheading 3.1.1.

3. Approximately 17 h postinoculation cells should be approximately 60% confluent.
4. To prepare transfection mix, in a 1.5 mL centrifuge tube, add 20 μL FuGene® 6 directly to 400 μL serum- and antibiotic-free D-MEM; do not pipette FuGene® 6 onto the side of the 1.5 mL tube. Vortex by three 1 s pulses and allow to stand for 5 min at room temperature with no agitation.
5. Add plasmids pNHP (7.5 μg), pHEF-VSVG (3.1 μg), pCEP4-tat (0.7 μg), and pTYF-EF1α-PLAP (3.9 μg) to D-MEM-FuGene® 6, mix and vortex by three 1 s pulses. Allow to stand for 15 min at room temperature with no agitation.
6. Transfer transfection mix into 5.5 mL full media and gently mix.

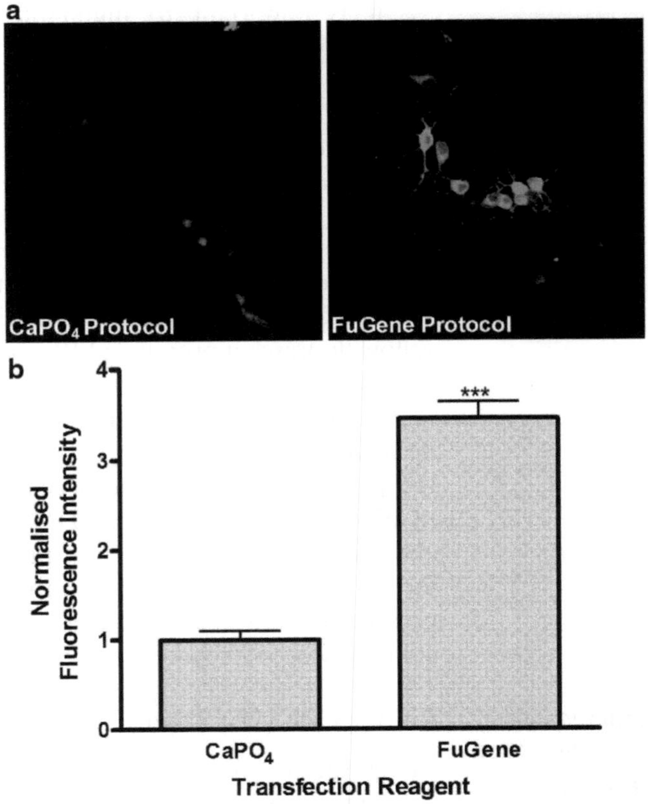

Fig. 2 Comparison of LVV transduction of primary neuronal cultures. Batches of LVV CamKII-hChR2:YFP were produced using the $CaPO_4$ and FuGene® 6-based protocols (three virus preparations for each protocol) and 1-week-old cortical neuronal cultures were infected with equal volumes of the LVV generated. Following 72 h incubation, cultures were analyzed under transmitted light for cell number, and for YFP fluorescence using a Leica Confocal TCS SP2 microscope (Leica Microsystems, Heidelberg, Germany). Six fields of view were randomly chosen for each virus preparation. The total number of cells analyzed were; 430 cells from cultures infected with $CaPO_4$—produced viruses and 380 cells from cultures infected with FuGene® 6—produced viruses, of which 20.7 and 52.6% cells were positive for YFP fluorescence, respectively. (**a**) Representative images of cortical neurones infected with LVV CamKII-hChR2:YFP prepared using the $CaPO_4$ and FuGene® 6—based protocols. (**b**) Average fluorescence intensity within individual YFP-positive cells was assessed using Leica Confocal Software (Version 2.61, Build 1537) and data were normalized to cultures infected with LVV produced using the $CaPO_4$-based protocol. Fluorescence intensity was significantly higher in cells infected with virus generated using the FuGene® 6-based protocol, (unpaired, two-tailed *t*-test, error bars represent SEM, $^{***}p < 0.0001$). Higher numbers of integrated viral copies were observed when FuGene® 6-based protocol was utilized

7. Aspirate media from Lenti-X 293 T cells and add transfection mix to the flask. Incubate cells under standard cell culture conditions overnight.

First Harvest (Day 3, pm)

8. Approximately 30 h following the application of the transfection mix collect media above cells (first harvest) and store at 4°C (see Note 6).
9. Add 7 mL fresh full media to each flask.
10. Place ultracentrifuge buckets, rotor, and filter unit at 4°C.

Second Harvest and Purification (Day 4, am)

11. Approximately 18 h following first harvest collect media above cells and pool with first harvest.
12. Filter the pooled lentivirus suspension through a 0.45 μm PES vacuum filter incubated on ice.
13. In an ultracentrifuge tube, overlay filtered lentivirus suspension over 0.5 mL 20% sterile sucrose in PBS, keep tubes on ice (see Notes 7 and 8).
14. Centrifuge at 74,000 × *g* for 2 h.
15. Aspirate medium from above the virus pellet and wipe excess medium from the side of tube using tissue paper.
16. Add 25 μL sterile PBS to the virus pellet, cover the tube, and incubate at 4°C overnight.

    Virus Resuspension (Day 5, am)
17. Resuspend lentivirus pellet by trituration (3–5 passes through the pipette are sufficient to resuspend the lentivirus pellet).
18. Aliquot the lentiviral suspension and store at −80°C.

3.1.1 *Alternative Transfection Protocols (Day 2)*

For comparative purposes the optimized protocol described above was run in parallel with established methods, i.e., using calcium phosphate or Superfect-based transfection of the packaging cell line (outlined below), producing LVV EF1α-PLAP or LVV CamKII-hChR2:eYFP.

Calcium Phosphate: Based Transfection (Adapted from (**8**))

Procedures for days 1 and 3–5 are identical to the protocol for FuGene® 6 described above (steps 1, 2, 8–18 of Subheading 3.1). The volumes used in the calcium phosphate protocol described in (8) were scaled down for use in a single T75 flask containing packaging cells.

Transfection of Cells (Day 2):

1. Approximately 17 h postinoculation cells should be 60–70% confluent.

2. To prepare transfection mix, in a 1.5 mL centrifuge tube, add lentiviral plasmids (see FuGene 6 procedure) to 57 μL 2 M $CaCl_2$, mix gently, and bring the volume to 238 μL with DNase free water.
3. Add 238 μL 2× HBS [50 mM HEPES, 1.5 mM $NaHPO_4$, 280 mM NaCl (Sigma)], mix quickly then transfer to 5.5 mL full media.
4. Aspirate media from Lenti-X 293 T cells and add transfection mix to the flask.

Superfect-Based Transfection (Adapted from Coleman et al. (9))

Procedures for days 1 and 3–5 are identical to the protocol for FuGene® 6 described above (steps 1, 2, 8–18 of Subheading 3.1).

Transfection of Cells (Day 2):

1. Approximately 17 h postinoculation cells should be 60–70% confluent.
2. To prepare transfection mix, in a 1.5 mL centrifuge tube, add lentiviral plasmids (see FuGene 6 procedure) to 400 μL serum- and antibiotic-free D-MEM.
3. Add 30 μL Superfect (Qiagen) and vortex for three-1 s pulses to mix. Allow to stand for 15 min at room temperature with no agitation.
4. Transfer transfection mix into 5.5 mL full media and gently mix.
5. Aspirate media from Lenti-X 293 T cells and add transfection mix to the flask.
6. Incubate cells for 6 h.
7. Aspirate transfection mix and replace with 6 mL full media.

3.2 Lentiviral Vector Titration: Placental Alkaline Phosphatase Staining

LVV EF1α-PLAP can be prepared in parallel with your virus of interest and be used in the titration procedure describe below, therefore giving an indication of the titer of your virus of interest. Alternatively your virus of interest can be titered using commercially available quantitative real-time PCR or viral capsid p24 ELISA kits (for example from Clontech).

Seed Cells (Day 1, pm)

1. Dissociate TE671 cells (see Note 3) using trypsin-EDTA and seed into a 12-well cell culture plate at a density of 175,000 cells/well in a final well volume of 1 mL full media.

Infect Cells with Lentiviral Vector (Day 2, am)

2. Prepare a hexadimethrine bromide working solution by adding 100 μL hexadimethrine bromide stock to 9,900 μL full media.

3. Remove media from abcve TE671 cells and add 500 μL hexadimethrine bromide working solution to each well (Caution: To maintain an even distribution of TE671 cells care should be taken not to dislodge cells during media addition).
4. Dilute the lentivirus in hexadimethrine bromide working solution as follows:
   (a) 100× dilution: 2 μL lentivirus + 198 μL hexadimethrine bromide working solution.
   (b) 1,000× dilution: 20 μL 100× dilution + 180 μL hexadimethrine bromide working solution.
   (c) 100,000× dilution: 5 μL 1,000× dilution + 495 μL hexadimethrine bromide working solution.
5. Gently apply diluted lentivirus to individual wells of TE671 cells as follows (each dilution should applied in duplicate):
   (a) 10 μL 100× dilution.
   (b) 10 μL 1,000× dilution.
   (c) 100 μL 100,000× dilution.
   (d) 10 μL 100,000× dilution.
   (e) No virus.
6. Gently tilt the plate to distribute the lentivirus.

Media Change (Day 3, am)

7. 24 h following the addition of lentivirus, remove media above cells and replace with 1 mL full media.

Placental Alkaline Phosphatase Stain (Day 4, am)

8. Warm 25 mL PBS in oven set at 75°C.
9. 48 h following the addition of lentivirus remove media above cells and wash cells in 1 mL PBS per well.
10. Remove PBS, add 1 mL 4% formaldehyde to each well, and incubate at room temperature for 10 min.
11. Remove formaldehyde and wash cells three times with 1 mL PBS.
12. Add 2 mL warmed PBS to each well and place plate at 75°C for 90 min.
13. Remove PBS, add 1 mL levamisol solution to each well, and incubate for 30 min at room temperature.
14. Remove levamisol, add 600 μL Reaction Solution to each well, and incubate at room temperature overnight to allow the purple-blue diformazan precipitate to form (see Note 5).

Cell Count (Day 5, am)

15. The number of positively stained cells should be recorded under six fields of view per well (see Note 9).

16. The lentiviral titer, expressed as Transducing Units per mL (TU/mL[1]), can be derived from the equation:

$$TU/mL = (A \times B)/(C \times D)$$

where,

A = Average positive cell count per field of view

B = Total fields of view per well (for the microscope and objective lens described there are 157 fields of view per well of the 12-well plates used, therefore $B = 157$. The total fields of view per well will need to be calculated if a different microscope and objective lens are used.)

C = Dilution factor expressed as a decimal fraction (i.e., 100,000 times dilution = 0.00,001)

D = The volume of virus added in mL

4 Notes

1. Lenti-X 293 T cells should be routinely cultured in D-MEM containing 10% FBS and 100 U/mL Penicillin and 100 μg/mL streptomycin (full media), and incubated in a humidified atmosphere containing 5% CO_2 at 37°C. To passage, when Lenti-X 293 T approach confluence cells should be dissociated using trypsin-EDTA and 10% of the cell suspension should be used to inoculate a new cell culture flask of equal surface area. For lentivirus preparation, Lenti-X 293 T cells should be used at low passage number (<p20).
2. The described protocol uses Lenti-X™ 293 T cells; however, similar results can be obtained with a more standard 293FT cell line (data not shown).
3. TE671 cells are cultured under the same conditions as described for Lenti-X 293 cells, see Note 1.
4. Store reaction buffer at 4°C to prevent/slow precipitation.
5. NBT is light sensitive, cover plate in foil during incubations with Reaction Solution.
6. All LVV-contaminated waste should be disposed of following local Waste Disposal Rules.
7. The simplest way to overlay the viral suspension over sucrose is to add the viral suspension to the centrifuge tube then carefully pipette 0.5 mL sucrose into the bottom of the tube.
8. A longer and more complicated iodixanol ultracentrifugation protocol (9) offered no obvious benefits over the sucrose gradient used here (data not shown).

9. To get an accurate cell count, only wells containing viral dilutions resulting in approximately 5–50 positive cells are used in the lentiviral titration calculation.

Acknowledgments

Funded by the British Heart Foundation RG07/006 and PG/08/009/24411.

References

1. Benzekhroufa K, Liu BH, Teschemacher AG, Kasparov S (2009) Targeting central serotonergic neurons with lentiviral vectors based on a transcriptional amplification strategy. Gene Ther 16:681–688
2. Tye KM, Prakash R, Kim SY, Fenno LE, Grosenick L, Zarabi H, Thompson KR, Gradinaru V, Ramakrishnan C, Deisseroth K (2011) Amygdala circuitry mediating reversible and bidirectional control of anxiety. Nature 471:358–362
3. Busskamp V, Duebel J, Balya D, Fradot M, Viney TJ, Siegert S, Groner AC, Cabuy E, Forster V, Seeliger M, Biel M, Humphries P, Paques M, Mohand-Said S, Trono D, Deisseroth K, Sahel JA, Picaud S, Roska B (2010) Genetic reactivation of cone photoreceptors restores visual responses in retinitis pigmentosa. Science 329:413–417
4. Gunaydin LA, Yizhar O, Berndt A, Sohal VS, Deisseroth K, Hegemann P (2010) Ultrafast optogenetic control. Nat Neurosci 13:387–392
5. Kleinlogel S, Feldbaur K, Dempski RE, Fotis H, Wood PG, Bamann C, Bamberg E (2011) Ultra light-sensitive and fast neuronal activation with the Ca^{2+}-permeable channelrhodopsin CatCh. Nat Neurosci 14:513–518
6. Gourine AV, Kasymov V, Marina N, Tang F, Figueiredo MF, Lane S, Teschemacher AG, Spyer KM, Deisseroth K, Kasparov S (2010) Astrocytes control breathing through pH-dependent release of ATP. Science 329:571–575
7. Naldini L, Blomer U, Gallay P, Ory D, Mulligan R, Gage FH, Verma IM, Trono D (1996) In vivo gene delivery and stable transduction of nondividing cells by a lentiviral vector. Science 272:263–267
8. Zhang F, Gradinaru V, Adamantidis AR, Durand R, Airan RD, de Lecea L, Deisseroth K (2010) Optogenetic interrogation of neural circuits: technology for probing mammalian brain structures. Nat Protoc 5:439–456
9. Coleman JE, Huentelman MJ, Kasparov S, Metcalfe BL, Paton JF, Katovich MJ, Semple-Rowland SL, Raizada MK (2003) Efficient large-scale production and concentration of HIV-1-based lentiviral vectors for use *in vivo*. Physiol Genomics 12:221–228

Part II

Electrophysiological Methods to Study Ion Channel Functions

Chapter 6

Two-Electrode Voltage Clamp

Bingcai Guan, Xingjuan Chen, and Hailin Zhang

Abstract

Two-electrode voltage clamp (TEVC) is a conventional electrophysiological technique used to artificially control the membrane potential (V_m) of large cells to study the properties of electrogenic membrane proteins, especially ion channels. It makes use of two intracellular electrodes—a voltage electrode as V_m sensor and a current electrode for current injection to adjust the V_m, thus setting the membrane potential at desired values and recording the membrane current to analyze ion channel activities. Here we describe the use of TEVC in combination with exogenous mRNA expression in *Xenopus* oocytes for ion channel recording.

Key words Voltage clamp, Membrane potential, Voltage electrode, Current electrode, *Xenopus* oocytes, Ion channel

1 Introduction

Ion channels of cell membranes may be gated by the membrane potential (V_m) and/or by specific chemicals. Those belonging to the superfamily of voltage-dependent ion channels are gated by the V_m directly, whereas others, classified as ligand-gated ion channels, are gated by their binding with specific chemicals called ligands. On the other hand, voltage-dependent ion channels can be subject to the modulation by chemical factors such as neurotransmitters, hormones, intracellular messengers or exogenous drugs; and some ligand-gated ion channels (such as NMDA receptors) are also affected by the change in V_m. Thus, to study the voltage dependent characteristics of ion channels readily, or to distinguish between the effects of V_m versus chemicals, an experimental procedure is needed to control the V_m (i.e., to change V_m in a desired pattern or set it at a desired level). This procedure, known as voltage clamp, was first designed by Cole and Marmont and improved by Hodgkin, Huxley and Katz for application to giant axons of squids as two-electrode voltage clamp (TEVC) in late 1940s (1). It utilizes two intracellular electrodes, one to monitor V_m and the other to inject

Nikita Gamper (ed.), *Ion Channels: Methods and Protocols*, Methods in Molecular Biology, vol. 998,
DOI 10.1007/978-1-62703-351-0_6, © Springer Science+Business Media, LLC 2013

a current to adjust V_m to desired values (the injected current being equal to the membrane current). Although the use of TEVC is limited to giant axons or large cells such as skeletal muscle cells, much of our understanding of the basic biophysical properties of ion channels has been acquired by using this approach. It was based on this method in combination with experiments using extracellular electrodes that Neher and Sakmann developed the currently widely used single electrode patch clamp technique (2). Nowadays popular though the latter is, it's only suitable for clamping relatively small cells, but unfit for large cells; for the large currents would cause a significant voltage drop across the recording electrode that cannot be compensated to an acceptable extent. Thus, TEVC is still irreplaceable in voltage clamping large cells, especially *Xenopus* oocytes which are often used for exogenous expression of ion channels or receptors. The *Xenopus* oocyte is a convenient expression system widely employed to study the structure and function of ion channels and receptors using TEVC in combination with various molecular biological approaches, since its membrane has low expression of endogenous channels and receptors.

The basic principle of TEVC recording from an oocyte is schematically depicted in Fig. 1. V_m is monitored by connecting the voltage electrode (electrode 1) to the input of a voltage follower (A_1), which has very high input impedance and draws negligible

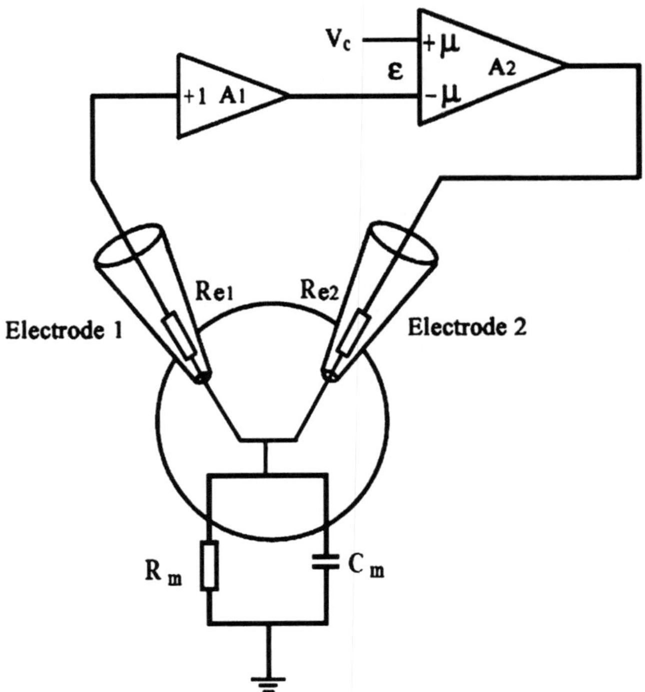

Fig. 1 Conventional two-electrode voltage clamp (TEVC) on an oocyte (see text for detail)

current from the cell. The output of A_1, the potential at which is approximately equal to V_m, is connected to one input terminal of a clamping amplifier (A_2), which is a high gain differential amplifier and compares V_m with the voltage command signal (V_c) applied to the other input terminal. The resulting output voltage of A_2 forces a current proportional to the difference (ε) between V_m and V_c, usually by means of a voltage-controlled current source mechanism, to flow through the current electrode (electrode 2) into the cell. The direction of the current is decided by the polarity of the voltage at the output of A_2, always reducing ε. As long as the gain (μ) of A_2 is sufficiently high ($\geq 10^3$), V_m can approach V_c very closely and thus achieves V_m clamp. Meanwhile, the current that passes through electrode 2 is measured (either upstream of electrode 2 or at the virtual grounding circuit) as the membrane current, since it counterbalances the V_m deviation the membrane current would have caused (3, 4).

Below we exemplify the TEVC procedures with the recording of Kv7.2/7.3 voltage-gated potassium channel currents.

2 Materials

All solutions are prepared using analytical grade reagents and ultrapure water (resistivity 18 MΩ·cm). Stock solutions and working solutions are stored at 4°C, and are used within 1 month. Longer time of storage is not advised. All waste materials, especially the discarded glass electrodes that may cause injury, should be securely disposed of in accordance with local laboratory regulations.

2.1 Equipment and Materials

1. TEVC amplifier (e.g., GeneClamp 500B, Molecular Devices).
2. Digitizer (e.g., Digidata 1440A, Molecular Devices).
3. TEVC software (e.g., Clampex 10, Molecular Devices).
4. Micromanipulators (e.g., M3301, World Precision Instruments).
5. Faraday cage (custom-made).
6. Vibroisolating table (e.g., Newport).
7. Stereomicroscope (e.g., PZMIII-BS, World Precision Instruments).
8. Perfusion system (e.g., ALA-VM8, ALA Scientific Instruments).
9. Perfusion chamber (custom-made).
10. Aspiration system (e.g., vacuum pump, ZT-I, Shaoxing Medical Apparatus).
11. Micropipette puller (e.g., P-97 Micropipette Puller, Sutter Instruments).
12. Glass capillaries (e.g., B10024F, VitalScan Scientific Instruments).

13. Laboratory microwave oven.
14. Water bath.
15. Microinjector (e.g., Nanoliter Injector 2000, World Precision Instruments).

2.2 Preparation of External Solutions

1. Oocyte Ringer 2 solution (OR-2): 82.5 mM NaCl, 2 mM KCl, 1 mM $MgCl_2$, 5 mM HEPES, pH7.4. Weigh 4.82 g of NaCl and 1.19 g of HEPES and transfer them to a 1 L glass beaker, add 800 mL of ultrapure water, add 2 mL of 2 M KCl and 1 mL of 1 M $MgCl_2$ stock solutions, mix and dissolve them completely with the aid of a magnetic stirrer. Adjust the pH of OR2 to 7.4 with NaOH. We usually prepare 1,000 mL of this solution using a volumetric flask (see Notes 1 and 2).
2. ND96: 96 mM NaCl, 1 mM KCl, 1.8 mM $CaCl_2$, 1 mM $MgCl_2$, 5 mM HEPES, pH 7.4. First prepare 1 L of ND96 stock (5×): weigh 28.05 g of NaCl, 372.75 mg of KCl, 998.91 mg of $CaCl_2$, 1,016.5 mg of $MgCl_2$, 5,957.5 mg of HEPES, respectively, and put them into a 1 L beaker. Add water and stir to dissolve them completely, and then bring the volume to 1,000 mL with water. The stock solution is stored at 4°C. Prepare the final solution before use by dilution; adjust the pH to 7.4 with NaOH (5) (see Note 3).

2.3 Preparation of Electrodes

1. Preparation of Ag/AgCl wires. Immerse silver wires (4–5 cm long and 0.25 mm thick) in chlorine bleach with only a short end (less than 1 cm) of each wire left in the air, until the immersed part becomes uniformly blackened (see Note 4).

   For making Ag/AgCl wires to be used in the voltage electrode and the current electrode, straight silver wires are used; for Ag/AgCl wires used for grounding the bath solution, it is better to coil part of the silver wire before coating, with narrow intervals left between adjacent turns of the coil. An alternative for grounding the bath solution is to use Ag/AgCl pellets (see Note 5).
2. Preparation of glass electrodes. The instructions below refer to the popular P-97 puller from Sutter Instruments; if using a different puller, refer to the instruction manual. Install a borosilicate glass pipette onto a P-97 puller, and run RAMP TEST to determine the heat value (RAMP value) for melting the glass (see Note 6). Repeat the determination and get an average RAMP value. Install a glass pipette of the same type and lot number, set the parameters of the puller as follows: HEAT = RAMP + 15, PULL = 150, VELOCITY = 100, TIME = 150, PRESSURE = 500 (see the P-97 puller manual for details) (6), and pull the pipette into two symmetric glass electrodes. Trim the tips of the glass electrodes using a pair a

fine scissors if needed (see Note 7), and store them in an electrode container.

3. Prepare 100 mL of 3 M KCl in a beaker, add 1 g of agarose, and microwave the mixture until the agarose is dissolved. Place the beaker containing agarose–KCl solution into a 80°C water bath to avoid agarose gelation. Warm up the syringe and its needle, with which the agarose–KCl solution is then injected into the glass electrode (see Note 8). Store the filled glass electrodes in a container (see Note 9).
4. Install a filled glass electrode onto the electrode holder (see Note 10), which, with Ag/AgCl wire pre-installed, has been secured to the headstage, the latter having been installed on a micromanipulator. Adjust the position of the tip of the glass electrode using the micromanipulator.
5. Measure the resistance of the electrode after its tip has been lowered into the bath solution with the micromanipulator. The electrode resistance should be in the range of 0.5–1 MΩ (see Note 11). If the electrode resistance is not within this range, adjust the parameters of the puller or modify the length of the tip for truncation.

2.4 Preparation of Xenopus Oocytes with Channels or Receptors Expressed

1. Preparation of *Xenopus* oocytes

   Anesthetize the clawed toad by freezing it with ice until it is immobilized. Make a small longitudinal incision (0.5 cm) about 0.5 cm apart from the midline at the lower abdomen, and cut several ovarian lobes off. Stitch up the incision with 5–0 gauge suture. Position the toad in a basin until it regains consciousness and then place it back to water.

   Cut the ovarian lobes into cubes of 3–4 mm in size, and digest them with OR2 containing collagenase (2 mg/mL) in a shaker at room temperature for 1.5–2 h. Stop digestion when most oocytes have been released and there are no connective tissues and capillaries left on the surface of the oocytes. Wash the oocytes repeatedly with OR2, and then wash them with ND96 three times. Select the large, healthy oocytes under a dissecting microscope, and keep them in ND96 at 18°C for a few hours before RNA injection.

2. Injection of mRNA into oocytes

   Fill a glass needle (which has a long tip truncated to 20–30 μm in OD at the tip opening) with corn oil, and install it onto the injector (Nanoliter Injector 2000, WPI). Drop 1–3 μL of mRNA solution onto a piece of clean parafilm, and draw it into the glass needle with the injector under the microscope. Make sure that the mRNA is free from RNAse contamination and no air has been drawn into the needle.

   Transfer the oocytes together with ND96 solution onto a Petri dish with meshes at its bottom with a sucking glass pipette (its opening fire-polished). Position the tip of the glass needle close to the surface of the oocyte and then impale it by using a micromanipulator. If the oocyte is in a good state, a resistance can be felt when the tip of the needle touches the oocyte membrane, and a "dimple" can be seen at the oocyte membrane before the needle can be inserted into the oocyte (in contrast, with unhealthy oocytes the impalement is much easier). Inject 46 nL of KCNQ2 or KCNQ3 mRNA solution (i.e., 3 ng mRNA) into each oocyte (the amount of injection of mRNA for other channels or receptors varies with the degree of difficulty in their expression and the status of the oocytes). After injection of an oocyte, wait for 1–2 s and then withdraw the needle carefully. Before turning to inject another oocyte, do an out-of-oocyte injection with the needle tip in the medium to make sure of the accuracy of next oocyte injection. After finishing the injection of all the oocytes, transfer them into ND96 solution containing gentamicin (50 mg/L) and sodium pyruvate (2.5 mM) and store at 18°C in an incubator. After 1–2 days TEVC experiments can be performed.

3 Methods

(All the following operations are carried out at room temperature).

3.1 Checkup of the Recording System

1. Check the perfusion system making sure that there is no blockage within it. Load solutions for the experiment, adjust the perfusion rate to about 2 mL/min (or to the value required for the specific experiment to be performed).
2. Check the aspiration system and drain tubing to make sure that they work properly so that the surface of the bath solution is stable.
3. Check the arrangement of the voltage headstage and current headstage installed on the micromanipulators, and make sure that they are positioned at a wide angle of over 90° (see Note 12).
4. Check the shielding and grounding of the apparatus around the input of the headstages to avoid introduction of external interferences, especially the power line frequency pickup (see Note 13).
5. Adjust the dissecting microscope so that the bath chamber to accommodate the oocyte is well within the visual field and clearly seen.
6. Turn on the system: turn the power of the digitizer on, and then turn on the power of the voltage clamp amplifier.
7. Configure the acquisition software and edit the voltage command protocol. For users of Clampex: open the software, click

"File"→"Set Data File Names" to designate the file name format and the folder to store the data to be recorded. Click "Configure"→"Digitizer" to change the configuration status from "Demo" to the digitizer being used. Create/edit the voltage command protocol for stimulation of an oocyte.

3.2 Load the Oocyte

Transfer a *Xenopus* oocyte expressing exogenous ion channel(s) and/or receptor(s) of interest gently to the bath chamber using a sucking pipette.

3.3 Observation and Recording of Currents

Steps described below are specific to Geneclamp 500B amplifier and the Clampex software; however, other TEVC hardware and software packages will have analogous steps/commands.

1. Lower the tips of the voltage electrode and current electrode into the bath using the micromanipulators, and position the electrode tips near (but not in contact with) the surface of the oocyte.
2. Zero offset potentials: press the buttons "ZERO V1" and "ZERO V2"on the front panel of the amplifier, and the "DC METERS" will read 0.
3. Measure electrode resistances: Press buttons "R1" and "R2" respectively, and the resistance of the electrodes will be displayed. The most appropriate values are within 0.5–1 MΩ (see Note 11).
4. Impale the oocyte: Under the dissecting microscope, press the tips of the electrodes against the surface of the oocyte gently using the micromanipulator, and impale the oocyte by further advancing the electrode tip slowly until it pops in (first do cell impalement with the voltage electrode and then with the current electrode) (see Note 14) (Fig. 2) (7).

Fig. 2 The visual field of the stereomicroscope showing the *Xenopus* oocyte in the bath chamber and the two glass electrodes

"DC METERS" will display the resting membrane potential (RMP) of the oocyte; for example, the RMP of oocytes with overexpressed exogenous Kv7.2/7.3 heteromeric voltage-gated K^+ channels is around −50 mV.

5. Position a grounded metal shield between the voltage and current electrodes if necessary (see Note 15).
6. Switch to voltage clamp mode: preset the holding potential at the RMP value (see Note 16). Then press button "VOLTAGE CLAMP."
7. Gain and stability adjustment: Apply a series of test voltage pulses of −10 mV in amplitude and observe the current response flowing through the current electrode. Increase the GAIN of the clamping amplifier and the transients at the onset and the finish of the test pulses will sharpen up and even begin to oscillate. Then increase phase lag with the STABILITY control, and the oscillation will disappear. Repeat the operation, that is, increase the GAIN and the phase lag iteratively until the clamping amplifier operates at the minimum phase lag that permits the use of the maximum gain (see Note 17) (8). Then turn off the test pulses.
8. Observation of currents before formal recording: click button "View Only" to run the voltage command protocol and observe Kv7.2/7.3 mediated potassium currents (Fig. 3). If the leak currents are not small enough to be negligible compared with the active Kv7.2/7.3 currents, use "LEAK SUBTRACT" to remove the leak currents for better observation of the active currents.
9. Recording of currents: when the currents have become stable, use "RECORD" command to record the currents, and apply drugs to test their effects according to the experimental design. Full wash-out of one drug with the normal external solution such as ND96, which takes at least 2 min, is needed before testing another drug. When the recording of an oocyte is finished and another oocyte is to be recorded, press "STOP"

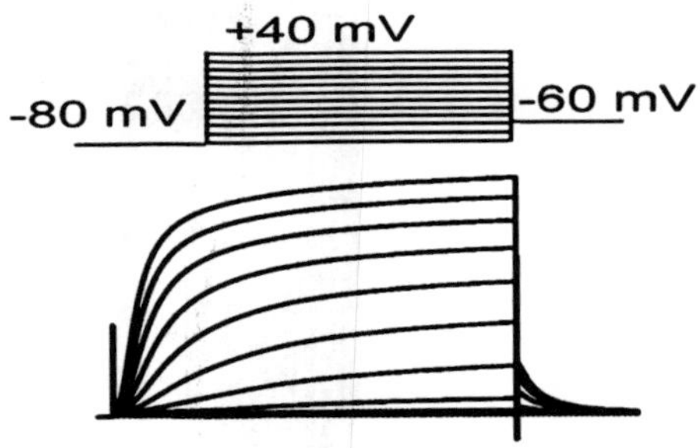

Fig. 3 Exemplary Kv7.2/7.3 mediated currents and the voltage clamp protocol used to induce the currents

to stop recording. Then press "SETUP" to stop voltage clamp and switch to setup mode, and then slowly withdraw the electrodes (see Notes 18 and 19). Repeat the aforementioned procedure to record currents from another oocyte.

3.4 Turning Off the Recording System and Cleaning the Setup

First close the software, then turn off the power of the digitizer and the amplifier respectively. Empty the remaining liquid in the perfusion system, and wash the system thoroughly with water in case its tubing(s) may be obstructed. Empty the waste liquid collected from the outlet of the vacuum pump.

4 Notes

1. OR2 is used to dissolve collagenase II for digestion of the layer of follicular cells surrounding the oocyte. It does not contain Ca^{2+} so as to avoid calcium overload in the oocyte during the digestion.
2. Only after all reagents are completely dissolved forming a clear solution by stirring the mixture should the pH adjustment be done.
3. The osmolality of normal external solutions for oocytes is ~189 mOsm/kg.
4. Silver wires can be well coated with AgCl layer after immersion in newly purchased or well sealed chlorine bleach for 20–30 min. Using outdated or poorly preserved bleach may take longer time or even produce no visible AgCl layer.
5. For recording large currents, a virtual ground circuit is generally used. Its voltage sensing electrode should be placed in the bath near the cell surface, so that the bath potential at the outside of the cell membrane can be clamped at zero without being affected by the large current flowing through the bath solution.
6. The result of the first RAMP test should be discarded because it often significantly deviates from those of the following tests due to some destabilizing factors.

   The RAMP value should be determined for three times to get a mean value as reference for HEAT value setting.
7. This is because the opening of the electrode tips should not be too thin. Otherwise the electrodes would have a large resistance after being filled (see Note 11).
8. For filling the electrodes, the agarose–KCl solution should not occupy more than half the total electrode length in case the solution may spill out of the electrode and spoil the holder. Care should be taken not to have air bubbles left in the agarose between the electrode tip and the AgCl coated silver wire.

9. The filled glass electrodes can be used repeatedly. With a small amount of water at the bottom of the electrode container, the filled electrodes can be stored for a few weeks without the agarose inside the electrode drying up.
10. If the amount of agarose filled into the electrode is more than needed, the excess can be squeezed out of the electrode using a silver wire before installing the electrode onto the holder. Note that the upper end of the silver wire (uncoated with AgCl) should be well in contact with the input of the headstage.
11. If the resistance of the electrode is too large, the electrode tip can be truncated using a pair of fine scissors such as ophthalmic scissors. A current electrode of too large resistance will affect the passage of currents, while a voltage electrode of too large resistance will affect its frequency response and distort the V_m unless electrode capacitance neutralization (which may cause oscillation) is used.
12. Enlargement of the angle between of the two headstages (i.e., between the two electrodes to be installed) reduces their coupling capacitance, thus reducing the possibility of current oscillation.
13. Take care not to ground the case of the headstage (such as the HS-2A-×1 LU for GeneClamp 500B amplifier) that uses a driven shield.
14. The tips of the two electrodes that have been inserted into the oocyte should be kept apart by some distance so as to reduce the error in V_m pick-up introduced by coupling between the two electrodes.
15. The grounded metal shield is used for reducing the coupling capacitance between the two electrodes, because this coupling capacitance is a destabilizing factor for TEVC. However, this shield should not contact the bath solution.
16. Presetting the holding potential at the RMP value can avoid the large current transient at the switch from SETUP mode to VOLTAGE CLAMP mode and large DC current after the switch. These currents may affect the cell becoming stable for recording.
17. The time constant (τ) for the change in V_m after a step voltage command is:

$$\tau = R_{e2} \times C_m / \mu$$

in which R_{e2} is the resistance of the current electrode, C_m is the membrane capacitance, and μ is the gain of the clamping amplifier (Fig. 1). It can be seen that increasing the gain of the clamping amplifier can make the V_m more close to V_c, thus improving its fidelity and speed of response. Whereas, increasing phase lag decreases the high-frequency gain, and the clamping speed will be compromised when V_c jumps.

18. Generally, each glass electrode that has been filled and installed onto the headstage can be used repeatedly, i.e., on a few oocytes in succession. However, the electrode should not be used too many times for its resistance would become larger or unstable due to the blockade by intracellular components. Check the resistance of the electrode before reusing it on another oocyte. If its resistance value appears unstable or is out of the acceptable range, replace it with another electrode.
19. After recording large currents or after long time of use, the AgCl coated silver wire will become white, indicating the exhaustion of the AgCl coat. This may occur either at the grounding silver wire or at that connected to the headstage input, depending on the current direction. Take the silver wire off, rinse it, clean it carefully with tissue paper and then with fine sandpaper, and re-chloride it.

References

1. Hodgkin AL, Huxley AF, Katz B (1952) Measurement of current–voltage relations in the membrane of the giant axon of Loligo. J Physiol 116:424–488
2. Neher E, Sakmann B, Steinbach JH (1978) The extracellular patch clamp: a method for resolving currents through individual open channels in biological membranes. Pflugers Arch 375:219–228
3. Halliwell JV, Plant TD, Robbins J, Standen NB (1994) Voltage clamp techniques. In: Ogden D (ed) Microelectrode techniques: the Plymouth workshop handbook, 2nd edn. Company of Biologists, Cambridge, pp 17–35
4. Sherman-Gold R (ed) (2008) The axon guide for electrophysiology and biophysics laboratory techniques, 3rd edn. Axon Instruments, Inc., Foster City, CA
5. Du X, Zhang H, Lopes C, Mirshahi T, Rohacs T, Logothetis DE (2004) Characteristic interactions with phosphatidylinositol 4,5-bisphosphate determine regulation of Kir channels by diverse modulators. J Biol Chem 279:37271–37281
6. Operation manual P-97 Flaming/Brown micropipette puller. (2009). Sutter Instrument Company, Novato
7. Bierwirtz A, Schwarz W (2007) Two-electrode voltage-clamp (TEVC). http://www.biophys.uni-frankfurt.de/~wille/prakt/anleitungen/03_elektrophys.pdf. Accessed 20 May 2012
8. GeneClamp 500B theory and operation (2002) Axon Instruments, Inc, Union City

Chapter 7

Conventional Micropipette-Based Patch Clamp Techniques

Jonathan D. Lippiat and David C. Wrighton

Abstract

The patch clamp technique revolutionized the study of ion channels and is considered the gold standard of measuring ion channel activity, from the academic laboratory to industrial-scale drug screening. This technique enables the study of ion channels, from single molecules up to the whole-cell ion channel population, and in their native environment. Whilst the study of single protein molecular behavior is the ultimate goal of biophysicists from all fields, this is a routine ability for ion channel specialists. This chapter is aimed at helping the beginner to design patch clamp experiments and to obtain the fundamental micropipette configurations with mammalian cells: cell-attached patch, whole cell, inside-out patch, and outside-out patch.

Key words Patch clamp, Electrophysiology, Ion channel, Whole-cell, Single channel

1 Introduction

The patch clamp technique was originally developed in order to record ionic currents from a tiny patch of membrane (1). A micrometer-aperture glass micropipette was pressed against a cell membrane and currents were recorded using a high-quality electronic amplifier. This revealed unitary currents in the picoampere range that fluctuated between discrete and quantal levels that were interpreted as the opening and closing of single ion-permeable pores in the membrane. We now know these pores or permeation pathways as the integral membrane proteins that we call ion channels. The two defining ion channel properties are permeation, which describes the types of ion that pass through the open pore and the speed at which this occurs, and gating, which describes the stimuli that cause the ion channel to spend more or less time with the pore open. Sometimes ion channels are further classified according to the ligands or drugs that activate or inhibit them. Patch clamp techniques enable us to determine all of these properties from cell membranes and much more. Even with automated chip-based or planar patch clamp equipment now available with

Nikita Gamper (ed.), *Ion Channels: Methods and Protocols*, Methods in Molecular Biology, vol. 998,
DOI 10.1007/978-1-62703-351-0_7, © Springer Science+Business Media, LLC 2013

ever-increasing throughput, there has been no noticeable decrease in the amount of manual or conventional micropipette-based patch clamp techniques being used, particularly in academic laboratories. It is this latter, conventional form of the techniques that is the subject of this chapter.

The headstage is at the heart of the patch clamp apparatus and it is to this that the micropipette holder and electrode wire are connected. This is where the current measurement and, importantly, where the voltage-clamping take place. The cell membrane where the ion channels reside is located at the tapered end of the micropipette, attached by a high-resistance (gigaohm) seal. The net charge movement between the pipette and bath electrodes, which are separated by the membrane and its complement of ion channels, is the recorded current. In the whole-cell configuration they will also be separated by the access resistance that lies in series with the membrane, which affect the reliability of macroscopic current recording. Unlike other voltage-clamp methods, in particular continuous two-electrode or discontinuous single-electrode (switch) voltage clamp, there is no feedback from the cell membrane of the actual transmembrane voltage. This always begs the question *does the transmembrane voltage faithfully follow the command voltage clamped at the headstage*? In whole-cell patch clamp recording, which is the most commonly used variation of the technique, the answer to this question is invariably negative to extents that may or may not be significant to the experiment in progress. It is important to understand the nature of this disparity and other issues associated with this technique, and to determine whether any adjustments to the experiment need to be made.

A comprehensive description of the various forms and applications of the patch clamp technique is beyond the scope of this chapter. Here, the basic configurations for current measurement, *cell-attached patch*, *whole cell*, *inside-out patch*, and *outside-out patch* (2), will be considered. Further methods can be found in this and previous volumes of *Methods in Molecular Biology*, e.g., perforated patch whole-cell recording (3), macropatch recording from *Xenopus* oocytes (4), and planar lipid bilayer recording of reconstituted channels (5). It is assumed that the reader has a sound theoretical understanding of ion channel biophysics and electrophysiological techniques. The reader is also advised to seek the support of coworkers, supervisors, or attend a practical workshop. These techniques are both theoretically and technically demanding to the beginner, there will be variations in the apparatus and biological material used, and these are skills that develop over a long time, if not a whole career. For the purpose of this chapter the recording of endogenous currents from HEK293 cells is described. A consideration of the range of experimental protocols and analysis methods used throughout the discipline is also beyond the scope of this chapter and so the reader is directed to research papers in their field.

2 Materials

2.1 Cell Culture

1. HEK293 cells cultured in 25 cm^2 flasks.
2. Culture medium: DMEM, supplemented with 10% fetal bovine serum (FBS) and antibiotics if required.
3. Phosphate-buffered saline (PBS) without Ca^{2+} and Mg^{2+}.
4. Trypsin-EDTA cell dissociation solution.
5. 35 mm plastic Petri dishes (see Note 1) and other sterile generic tissue culture and laboratory plasticware.

2.2 Micropipette Fabrication

1. Thin-walled 1.5 mm diameter borosilicate glass capillaries (Harvard Apparatus) for low-resistance pipettes for whole-cell recording.
2. Thick-walled 1.5 mm diameter borosilicate glass capillaries (Harvard Apparatus) for higher resistance and lower noise micropipettes for excised and cell-attached patch recording.
3. Micropipette puller (e.g., PP-830, Narishige).
4. Microforge (e.g., MF-830, Narishige).

2.3 Recording Solutions

These solutions approximate to physiological ionic conditions, but should be adjusted according to the channels and conditions under investigation (see Note 2). Highest quality purified water and chemicals should be used.

1. Extracellular solution: 137 mM NaCl, 5.6 mM KCl, 2.6 mM $CaCl_2$, 1.2 mM $MgCl_2$, 10 mM HEPES, pH 7.4 with NaOH (see Note 3). Solutions will have osmolarity typically in the 300–320 mOsm range. This solution is stable at 4°C for several months and can be sterile-filtered. Check the pH and clarity of the solution before use for evidence of contamination.
2. Intracellular solution: 140 mM KCl, 10 mM EGTA, 1 mM $MgCl_2$, 1 mM $CaCl_2$, 10 mM HEPES, pH 7.2 with KOH (see Note 3). If conducting whole-cell recording, adjust the osmolarity to 10 mOsm below that of the extracellular solution using an inert substance, such as mannitol (see Note 4). This solution is stable at 4°C for several months and can be sterile-filtered, but first check the pH and clarity of the solution before use for evidence of contamination. If small amounts of intracellular solution are used at a time as pipette solutions (e.g., for whole-cell experiments) then this solution can be frozen in aliquots, particularly if unstable substances such as ATP are included.
3. 5 mL syringe.
4. Syringe filter disc, 20 μm holes.
5. Microloader pipette tips (Eppendorf).
6. 200 μL pipette tips (generic type).

2.4 Patch Clamp Apparatus

The manufacturers of the following items are given as examples and are those presently in use in our laboratory. The whole system should be assembled and tested thoroughly with the model cell and then with micropipette and bath. Background noise and 50 Hz electrical noise should be eliminated to below 1 pA (see Note 5).

1. Antivibration (air) table (e.g., Linos, Thorlabs) and compressed air/gas supply.
2. Faraday cage. This could be supplied with the antivibration table or custom-made in-house if workshop facilities are available.
3. Inverted microscope (e.g., Nikon TE-2000) equipped with 4×, 10×, and 40× objective lenses (see Note 6).
4. Micromanipulators (e.g., Luigs-Neumann; microscope-specific mount and fixings can usually be provided by the manufacturer or fabricated in-house).
5. Patch clamp amplifier with headstage and micropipette holder (e.g., HEKA EPC 10).
6. Model cell (provided with amplifier).
7. Reference electrode: Ag/AgCl pellet (e.g., World Precision Instruments) soldered to thin wire and headstage pin connector.
8. Benchtop or mini-tower computer, fitted with interface if supplied with patch clamp amplifier, and installed with acquisition (e.g., Patchmaster, HEKA) and data analysis (e.g., Fitmaster, HEKA) software. If amplifier does not directly connect to the computer then digital-analogue interface hardware and software is required to be installed (e.g., Axon Digidata, Cambridge Electronic Design, National Instruments, etc.) to allow two-way communication between computer and amplifier. The computer-based acquisition system also acts as an oscilloscope.
9. Perfusion system: gravity-fed system fabricated in-house from syringe reservoirs, silicone tubing, two- or three-way stopcocks, and manifolds, or specialist system purchased from supplier (e.g., Automate Scientific, Biologic, etc.).
10. Recording chamber or mount for 35 mm Petri dish (fabricated in-house or supplied by World Precision Instruments, Warner Instruments, ALA Scientific Instruments, etc.).
11. 1 mL syringe, three-way syringe stopcock, and silicone tubing to attach to micropipette holder side-port to provide suction/pressure to the micropipette.

3 Methods

3.1 Cell Culture and Preparation

All procedures should be carried out using aseptic technique and in a sterile vertical flow cabinet suitable for mammalian cell culture. Procedures should also conform to local and national rules for disposal of contaminated microbiological or genetically modified material waste.

1. Aspirate culture medium from the culture flask.
2. Wash cells twice with PBS and harvest by adding trypsin-EDTA dissociation solution. 200 μL of trypsin-EDTA solution in one 25 cm^2 flask should suffice.
3. Rock the flask so that the dissociation solution covers the entire surface area and observe cells on an inverted microscope.
4. Once cells appear rounded and become loosely adherent, add 5 mL of culture medium and transfer cell suspension into a sterile plastic centrifuge tube (e.g., 15 mL).
5. Centrifuge cell suspension for 1 min at 200 × *g*.
6. Aspirate supernatant and resuspend in culture medium, diluting if necessary.
7. Seed cells into a fresh 25 cm^2 flask in a final volume of 6 mL for further propagation of the cell line and so that the cells have been diluted five- to tenfold.
8. Add 2 mL of cell suspension to each of the required number of 35 mm Petri dishes and incubate overnight (see Note 1).

3.2 Micropipette Fabrication and Filling

1. If required, cut capillary glass into shorter lengths (e.g., 150 mm capillaries should be cut into two approximately equal lengths). The capillary ends can be gently fire polished over a Bunsen flame to prevent damage to the electrode wire.
2. Optimize pipette puller settings (heat, pulling distance, force) for the specific device and for the type of glass capillary. For micropipettes fabricated from thin-walled glass and intended for whole-cell recording a final resistance as measured during the experiment should typically be 2–3 MΩ, whilst patch (cell-attached or excised) micropipettes are typically higher at 5–8 MΩ and are pulled from thick-walled glass. This resistance is proportional to the cross-sectional area of the micropipette tip and should be optimized to the cell type or size and channel density (see Note 7).
3. Polish the tip of the micropipette using the microforge. This enables fine control over the size of the aperture and makes a clean, smooth, tip surface that enhances seal formation.
4. Fill the 5 mL syringe with the solution to be loaded into the micropipette and attach a filter disc. For most experiments this

will be the intracellular solution, but variations will be described below. Attach the microloader pipette tip to the filter disc, making use of a regular 200 μL pipette tip cut to size as an adapter, if required. Label the syringe.

3.3 Cell-Attached Configuration

This is the starting point for each of the different patch clamp configurations described in this chapter. It can also be used to monitor currents across the enclosed patch of membrane in the intact cell. This is a particularly useful technique when determining whether particular ion channels are regulated by cytosolic second messengers, such as Ca^{2+} or cAMP, since the channels enclosed by the patch pipette would be isolated from membrane-impermeant receptor agonists applied by bath perfusion.

1. Check that the electrode wire in the micropipette holder has an adequate coating of silver chloride otherwise the offset voltage will drift (see Note 8).
2. Replace tissue culture medium with extracellular solution and mount Petri dish with cells onto microscope stage (or place coverslip with adherent cells into recording chamber).
3. Position reference electrode directly into the dish or connect by means of an agar bridge (see Note 9).
4. Fill a polished patch microelectrode with pipette solution (see Subheading 3.2) and mount into the micropipette holder, ensuring an airtight seal.
5. Position the micropipette above the bath solution and in the field of view on the microscope.
6. Using the 1 mL syringe and tubing connected to the side-port of the microelectrode holder, apply positive pressure. As an example, depress the syringe plunger from 0.2 to 0.1 mL. This prevents flow of debris from the bath into the micropipette that might prevent seal formation.
7. Lower the micropipette into the bath solution and position the tip in the proximity of the cells (see Note 10).
8. Zero the patch clamp amplifier using the control software ("LJ," in Patchmaster), or DC-offset if a manual amplifier (see Note 11).
9. Apply a 1 mV pulse and observe or calculate the pipette resistance (Fig. 1a).
10. Under high power magnification (40× objective) position the micropipette so that the tip is touching the cell membrane.
11. Using the 1 mL syringe and/or stopcock, release the positive pressure and apply negative pressure to the micropipette (e.g., retract syringe plunger by 0.1 mL). This should increase the resistance of the seal between the micropipette and the cell membrane, which will be reported by the control software

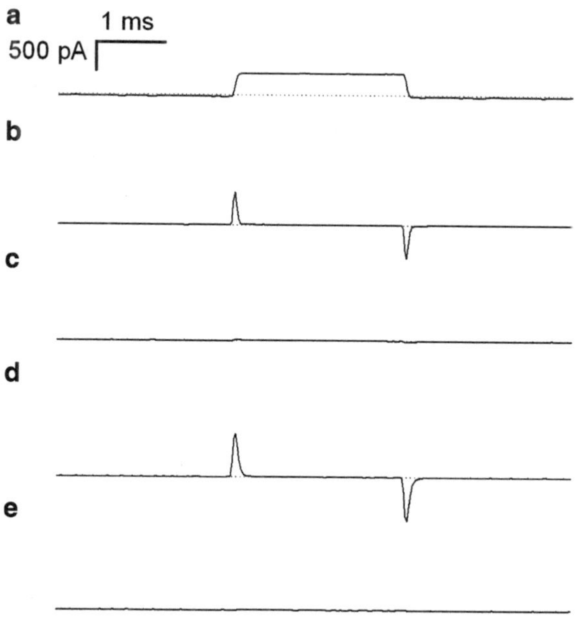

Fig. 1 Establishing the cell-attached (**a**–**c**) and whole-cell (**d**–**e**) configurations. (**a**) Current recorded with 0 mV holding potential and a 5 ms pulse to 1 mV. To "zero" the amplifier, the DC offset voltage is set so that current at 0 mV is zero either manually or automatically, depending on the amplifier. The 400 pA current at 1 mV gives, using Ohm's law, the pipette resistance as 2.5 MΩ. (**b**) Currents recorded following seal formation with the holding potential at −80 mV and a 5 ms pulse to 10 mV. The transient currents are caused primarily by the "fast" capacitance provided by the glass micropipette. (**c**) The transient currents can be cancelled using the fast capacitance cancellation function of the patch clamp amplifier (here, capacitance = 7.3 pF; $\tau = 0.9$ μs). (**d**) Appearance of whole-cell or "slow" capacitance after application of negative pressure and rupture of the membrane patch. (**e**) These transient currents can also be cancelled using the whole-cell (or "slow") capacitance cancellation function of the patch clamp amplifier (here, capacitance = 4.8 pF; series resistance = 7.2 MΩ)

and/or observed as a decrease in the current evoked by the 1 mV pulse (see Note 12). A gigaohm seal will be evident by a flattening of the current deflection on the computer monitor and also the appearance of capacitance currents (Fig. 1b, see also Note 13).

12. Increase the voltage pulse to 10 mV and cancel the fast capacitance currents (see Note 13) using the amplifier controls (Fig. 1c).
13. Spontaneous unitary currents can be observed by increasing the amplifier gain (>100 mV/pA) whilst holding the pipette voltage at 0 mV. A depolarizing holding potential or voltage pulses may need to be applied to evoke voltage-dependent channels.

14. In this configuration, the micropipette acts as an extracellular electrode; therefore negative voltages correspond to depolarizing potentials and positive/upward current deflections represent the movement of positive charge from the micropipette into the cell. Selecting the "on-cell" configuration in the Patchmaster software automatically reverses pipette voltage and current polarities so that these conventions of stating the voltage inside the cell with respect to the outside (usually where the reference electrode is placed) and efflux of positive charge from the cell displayed as an upward or positive current are followed.

3.4 Whole-Cell Configuration

This configuration permits the recording of membrane currents from the entire cell. The contents of the pipette dialyze and become continuous with the inside of the cell.

1. Fabricate low-resistance micropipettes and select the intracellular solution as the pipette solution (see Subheading 3.2).
2. Follow the procedure of obtaining the cell-attached configuration (see Subheading 3.3).
3. Set the holding potential to the intended whole-cell holding potential (e.g., –80 mV).
4. Using the syringe and tubing attached to the micropipette holder apply negative pressure by withdrawing the syringe plunger (e.g., 0.1 mL/s) until whole-cell capacitance currents appear in response to the 10 mV pulses (Fig. 1d).
5. Cancel the whole-cell (or "slow") capacitance using the control software, or adjust capacitance (Cm) and series resistance (Rs) settings manually. The time course of the whole-cell capacitance provides an indication of the charging time of the cell membrane when voltage steps are applied (Fig. 1d, see also Note 14).
6. Observe current amplitudes when conducting the experiment. The series resistance error, in mV, can be quickly estimated by multiplying the measured current in nA by the series resistance in MΩ. The voltage across the cell membrane will be the command or pipette potential minus the voltage drop across the series resistance (see Note 15).
7. If the voltage error is unacceptable then employ the series resistance compensation functionality of the patch clamp amplifier to a value >70%. Also consider working at voltages that give small currents or reduce the series resistance by using lower-resistance micropipettes.

3.5 Outside-Out Patch Configuration

This configuration enables the recording of current from a small patch of membrane, isolated from the cell, with the extracellular domains of the channel exposed to the bath. This permits the study of unitary or small numbers of channels whilst enabling the application of extracellular reagents to the bath, e.g., neurotransmitters or agonists in the case of ligand-gated ion channels.

1. Fabricate higher-resistance micropipettes from thick-walled capillary glass and select the intracellular solution as the pipette solution (see Subheading 3.2).
2. Obtain the cell-attached configuration (see Subheading 3.3).
3. Set the pipette potential to the intended holding potential (e.g., −80 mV).
4. Apply negative pressure to the micropipette in order to rupture the enclosed membrane patch (as explained in Subheading 3.4). This will be indicated by the appearance of slow membrane capacitance.
5. Using the micromanipulator controls, slowly withdraw the micropipette from the cell in both lateral and vertical directions. The successful excision and formation of the patch are indicated by the disappearance of the whole-cell capacitance (see Note 1) if excision is prevented by the loss of cell adherence).
6. The patch can be positioned at a shallow depth in the solution in order to reduce noise and/or positioned in the outflow of the perfusion system. The fast capacitance cancellation may need to be readjusted as these maneuvers will reduce the capacitance.
7. Apply test potentials or ligands to determine the presence of channels of interest.

3.6 Inside-Out Patch Configuration

This configuration enables currents to be recorded from a small patch of membrane, isolated from the cell, but with the intracellular-facing domains of the channel exposed to the bath. Here, unitary or small numbers of channels can be studied whilst enabling the application of intracellular reagents to the bath, e.g., nucleotides or Ca^{2+}, when studying ATP- or Ca^{2+}-sensitive potassium channels, respectively.

1. Fabricate higher-resistance micropipettes from thick-walled capillary glass and select the extracellular solution as the pipette solution (see Subheading 3.2).
2. With the cells bathed in intracellular solution, obtain the cell-attached configuration (see Subheading 3.3). If this is detrimental to the experiment it is possible to first bathe the cells in extracellular solution, but start perfusing intracellular solution once a gigaohm seal is formed and prior to patch excision.
3. Set the pipette potential to the intended holding potential (e.g., −70 mV). The polarity may need to be reversed (i.e., +70 mV) if this has not been done when establishing the

cell-attached configuration, e.g., by using the "on-cell" mode of Patchmaster software.

4. Using the micromanipulator controls, slowly withdraw the micropipette from the cell in both lateral and vertical directions (see Note 1 if excision is prevented by the loss of cell adherence).
5. The patch can be positioned at a shallow depth in the solution in order to reduce noise and/or positioned in the outflow of the perfusion system. The fast capacitance cancellation may need to be readjusted as these maneuvers will reduce the capacitance.
6. Apply test potentials or ligands to determine the presence of channels of interest or if a vesicle has unintentionally been formed during the process of excising the patch (see Note 16).
7. Because the pipette electrode is on the extracellular side, the voltage polarity and direction of charge flow will be the opposite of the usual convention. The "inside-out" setting in Patchmaster can account automatically for this, similar to the "on-cell" mode.

3.7 Conducting the Experiment

The choice of electrophysiological protocol, along with pipette and bath solutions, deserves consideration long before the experiment and should not be an afterthought once a successful gigaohm seal has been obtained. The protocol should be appropriate for the maximal interrogation of the ion channel under investigation, to the detriment of any contaminating currents. Consult published papers for recording conditions and protocols used successfully for the channel of interest. Some basic properties that will direct the structure of the protocol are outlined here. The example provided in this chapter describes currents recorded in the whole-cell configuration from a HEK293 cell, and the protocol is designed to isolate voltage-gated currents (Fig. 2).

3.7.1 Voltage-Gated Channels (Fig. 2)

1. Select a holding potential that is below the activation threshold of the channel of interest, e.g., −80 mV.
2. Use square-wave voltage pulse protocols.
3. The duration of the voltage pulse should be sufficient for the channel to reach steady state (e.g., 50 ms is sufficient for most Kv channels).
4. If the channel undergoes inactivation, consider the use of hyperpolarizing pre-pulses.
5. If the channel undergoes inactivation, ensure that the inter-pulse interval is long enough to allow full recovery from inactivation, unless you intend measuring time- and voltage-dependent recovery from inactivation.

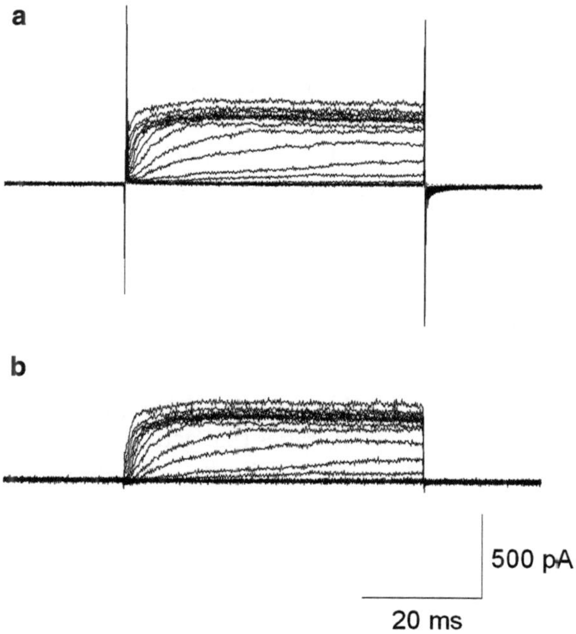

Fig. 2 Representative recordings from a HEK293 cell in the whole-cell configuration. From a −80 mV holding potential, depolarizing pulses of 50 ms duration were applied at 3 s intervals to a variety of potentials up to 100 mV, with 80% series resistance compensation. (**a**) Transient currents that occur where the pulse starts and ends are often recorded when series resistance compensation is applied, even though fast and slow capacitance have been cancelled. Complex cell morphologies may also provide additional capacitance components that cannot be fully cancelled. (**b**) Currents from the same cell, but employing a P/4 leak subtraction protocol, which subtracts residual leak and capacitance currents by extrapolating linear current components recorded at voltages below the activation threshold of the voltage-dependent channel. Note the removal of the residual capacitance currents, but at the expense of lowering the signal to noise ratio

6. To measure voltage-dependent parameters (e.g., half-maximal activation voltage) apply pulses across the voltage range that spans maximal and minimal channel activation.
7. When measuring the effects of drugs/modulators on channel activation, choose a depolarizing voltage pulse at which the channel has intermediate open probability, so that both increases and decreases in activity can be determined.
8. To isolate voltage-dependent currents from currents through voltage-independent channels, residual leak, and capacitance currents use a P/N leak-subtraction voltage protocol. This is usually an option in the protocol sections of the acquisition software where voltage pulses to 1/N of the test pulse amplitude are applied to subthreshold voltages, usually N times, and evoked currents are summed or averaged and scaled (by N) to

predict the passive linear currents, which can be subtracted from the current evoked by the test pulse. An example of leak-subtracted traces is provided in Fig. 2b. Ensure that leak currents are obtained at voltages below the activation threshold of the voltage-gated channel of interest or apply pulses to hyperpolarizing potentials (P/-N).

3.7.2 Voltage-Independent or Weakly Rectifying Channels

1. Select a holding potential close to the channel reversal potential (i.e., no net current ion flow).
2. Do not use leak subtraction; voltage-independent currents are considered by these protocols and amplifier functions as leak.
3. Use either voltage pulses or voltage ramp protocols and avoid voltages that activate voltage-gated channels that might also be present in the cell membrane.
4. There is little benefit in using a wide range (i.e., >50 mV) of test voltages; measure channel activity as conductance from a narrow voltage range through the reversal potential.

3.7.3 Ligand-Gated Ion Channels

1. Holding potentials close to the native resting membrane potential (e.g., −60 mV) are often selected in order to determine whether the ligand-gated channel activation is excitatory or inhibitory, whereby the ligand evokes either an inward or outward current, respectively.
2. Research the ligand/agonist concentration range and select a concentration that will stimulate the channel of interest, but not affect other types of channel that may be present in the cell.
3. Take into consideration a possibility that the ligand or agonist causes desensitization of the channel, this will determine for how long the ligand is applied.
4. Low-bandwidth continuous recordings are often used to study macroscopic currents.
5. Ligand-gated channels are usually voltage-independent so do not use automated leak-subtraction methods.
6. Determine the reversal potential by applying voltage pulses or a voltage ramp whilst the channels are exposed to ligand/agonist.

3.7.4 Single-Channel Recording

1. The apparatus should be sufficiently grounded and shielded to resolve single-channel (pA range) currents from the background noise. If it does not, improve grounding and shielding (see Note 5), and micropipette fabrication and coating (see Note 14 for micropipette coating).
2. With the exception of the negligible series resistance errors due to the small currents, the same considerations and types of protocol for studying macroscopic currents (see Subheadings 3.7.1–3.7.3) generally apply to single-channel recording. However,

continuous recording for longer duration (minutes) or many pulse repetitions are often used in order to collect a sufficient number of channel opening and closing events. Make sure that the same type of stimulus that evokes macroscopic channels (voltage steps, ligand, etc.) also affects single channels.

3. Is there just one single channel that is active in the patch? Multiple channels will give multiple open levels if they are open at the same time. Use conditions that raise the channel open probability so that multiple channel openings are likely in order to determine the number of channels in the patch.
4. Select low-pass filter frequencies and digitization rates that enable brief opening and closing events to be recorded (6).

4 Notes

1. 35 mm Petri dishes are convenient for culture of the cells and can be placed directly onto an inverted microscope for electrophysiological study. Alternatively, cells can be seeded and cultured on glass coverslips, which can be transferred into the electrophysiological recording chamber using forceps. Depending on the cell type, the cell adherence can be improved, if required, by purchasing pretreated coverslips or by preparing substrates coated with poly-d-lysine, laminin, collagen, etc., depending on the cell type. To enable patches to be excised successfully for inside-out and outside-out patch recording cell adherence may need to be optimized. Alternately, cells could be grown into clusters by seeding at higher density and culturing for >2 days so that adherence is improved via a greater surface area contact between the substrate and cell monolayer. This, however, is not desirable for whole-cell recording from cells that form electrical synapses with neighboring cells via gap junctions, and these include HEK293 cells.
2. Some ion channels require additional substances in order to be significantly active, e.g., nucleotides or Ca^{2+}. If the current of interest is overwhelmed by other ionic conductances then the conditions could be optimized to enhance the current of interest or by reducing currents through other channels by including impermeant ions or drugs.
3. The pH of the solutions should be adjusted using acid or base that will add cations/anions that are already in high concentration, i.e., NaOH for Na^+-rich extracellular solutions, and KOH for K^+-rich intracellular solution. Likewise, HCl is used if the solution is required to be adjusted to more acidic pH. If the final concentrations of certain ions are of particular concern, e.g., [K^+] in intracellular solution, then either the pH could be

adjusted with a measured volume of KOH and then by adding the required amount of KCl to give the final [K^+], or the solution could be deliberately made alkaline using 40 mM KOH + 100 mM KCl and adjusting to the final pH with HCl.

4. This prevents cell shrinking during whole-cell recording where water leaves the cell by osmosis to the solution with the higher osmolarity. Likewise, a hypertonic intracellular solution will cause the cell to swell over the course of the experiment. Both may have adverse effects on the stability of the recording and the activation of certain types of ion channel if they have mechanosensitive properties.
5. A single grounding point, e.g., a metal bar fixed to the Faraday cage, should be connected to the ground of the patch clamp amplifier. Whilst observing current noise on the computer screen or oscilloscope, a multimeter and/or a test wire can be used to determine which metallic components inside the Faraday cage do not have a low-resistance connection to the grounding point and if they require grounding or shielding. Check the following: all sides of the Faraday cage, air table breadboard, micromanipulator components, and lamp housings. Check also if any nearby nonessential electrical items and lighting in the laboratory could be turned off to reduce the electrical noise.
6. For isolated cell preparations or cultured cells the use of an inverted microscope provides adequate working space above the stage for positioning of the recording micropipette, perfusion outlet, reference electrode, etc. When working with samples of intact tissue, e.g., brain slice, the convenient cells from which to record will be towards the uppermost part of the tissue, which will not be easily resolved on an inverted microscope, and the micropipette would need to pass through most of the slice to reach cells closest to the objective. Therefore upright microscopes are frequently used for slice recording, with the micropipette positioned between the tissue and the objective.
7. For whole-cell recording, errors associated with series (access) resistance are reduced by lowering the resistance between the cell membrane and pipette electrode wire. This is influenced by the initial resistance of the micropipette, which sets the theoretical lower limit for the series resistance. In practice, series resistance is always higher than the micropipette resistance. Therefore, as low resistance micropipettes as possible should be fabricated for this purpose. However, the size of the cell then becomes a limiting factor as a large aperture micropipette has an increased chance of sucking up the cell when negative pressure is applied. For higher resistance micropipettes intended for single-channel recording, the micropipette fabrication

should be optimized according to the channel density. Expect to make several attempts to obtain a single-channel recording, if this is desired, since there will be comparable probabilities of obtaining zero, one, or more channels in each membrane patch.

8. A sufficient coating of silver chloride over the silver wire is required to complete the Ag/AgCl half cell. The silver wire can be coated with silver chloride either by electroplating (connecting the wire into an electrical circuit with 3 M KCl and a 9 V battery), immersing in bleach for at least an hour, or by dipping the wire into molten silver chloride. The immersion in bleach is usually the most convenient and practicable approach in most laboratories.

9. There are a couple of reasons why the bath may need to be separated from the Ag/AgCl reference electrode by an agar bridge. Firstly, some ion channels may be sensitive to contaminating levels of silver ions in the bath solution that may originate from the electrode wire. Secondly, a junction potential exists between the pipette and bath solutions via the two half cells and the headstage circuit, which can be quantified as a chloride electrochemical potential using the Nernst equation. Should the chloride concentration of the bath solution need to be changed during the course of the experiment then this electrochemical potential will change the voltage offset, which is usually cancelled on a single occasion prior to seal formation. If the bath [Cl^-] is to be changed during the experiment then the reference electrode, usually a Ag/AgCl pellet, should be kept in constant [Cl^-], which could be the intracellular, extracellular solution, or a solution high in KCl, and connected to the bath solution via an agar bridge. Agar bridges can be made from glass capillaries that are gently molten into a "C" shape and filled with recording solution or 1–3 M KCl containing 1–3% agar. Heat the solution to dissolve the agar and load into the capillary as it starts to cool. Agar bridges should be stored in the same, but agar-free, solution at 4°C.

10. Bringing the tip of the micropipette near to the cell is one of the more difficult technical steps for the novice and takes some practice with each apparatus. Monitoring the position of the tip in the *X–Y* dimensions using the microscope field of view is relatively straightforward, but judging the depth (*Z*-axis), or how far the micropipette tip is above the cell, is more difficult. A good technique is to use the microscope focusing controls to judge the vertical distance between the cell and micropipette tip. The following procedure can be followed: focus the microscope on the micropipette tip, focus down to the cell, focus back up to a plane in between the two, lower the micropipette until it comes into focus, and repeat until the focal planes of the micropipette tip and cell become close. During this process

the objective lens will likely be changed to give increased magnification, until the highest is reached.

11. This DC offset will also cancel the liquid junction potential that exists between the bath and pipette solutions at the micropipette tip aperture. This junction disappears upon formation of the gigaohm seal but the DC offset remains. The junction potential can be calculated from the ion concentrations and relative ion motilities (e.g., using the calculator in the Axon Clampex software) or can be measured by the patch clamp apparatus in current clamp mode. This is done with the micropipette placed in the bath, first with extracellular solution in both bath and micropipette, zeroing the amplifier, and then measuring the voltage recorded when a micropipette containing intracellular solution is mounted and placed in the bath solution. Either way, the junction potential can be accounted prior to the experiment, as part of the zeroing procedure, or by adjusting voltages after the experiment.

12. The process of seal formation depends on the health and type of cell, the geometry of the micropipette tip, and the lack of debris between the two. Seal formation can be rapid (<1 s) or occur over a longer duration (>1 min). Applying negative pipette potentials can often promote seal formation, as can removing and reapplying gentle suction. As a general rule, if the seal resistance is increasing then leave it to improve, but change pipette pressure or voltage if it stalls before achieving >1 GΩ.

13. Fast capacitance currents arise in the cell-attached configuration because the micropipette glass, which is very thin towards its tip and with conducting solutions on either side, can act as a capacitor. The amplitude of the capacitance currents will change if the bath solution levels increase and decrease over the course of the experiment because the fluid level changes the external area of the micropipette in contact with solution. The size of the capacitance currents can be reduced by minimizing the bath volume or by coating the micropipette with a hydrophobic substance. An easy way to do the latter is to dip the filled micropipette tip into Sigmacote (Sigma) immediately prior to mounting onto the pipette holder. Sylgard or dental wax can also be used, but these both need to be coated and cured prior to micropipette polishing and filling.

14. The time constant of the decay of the whole-cell capacitance current, $\tau = Cm \times Rs$, and so membranes with larger surface area or higher series resistance have longer charging times. This may affect the study of channels with rapid voltage-dependent kinetics. Rs should be kept as low as possible and the use of series-resistance compensation functionality of the amplifier should be considered in these cases. Series resistance issues can also be minimized by studying channels in excised macro-patches that contain many channels.

15. The resistance of the cell membrane and the resistance between the micropipette and the cell membrane are in series and form a voltage divider. The current flow is the same throughout the circuit, but the voltage (the pipette potential) across the two is shared between each resistance according to its proportion of the total resistance. The series resistance (Rs, a few MΩ) is usually much lower than the membrane resistance when very few channels are active (Rm, a few 100 MΩ) and so the voltage across the membrane (Vm) is very close to the pipette voltage (Vp):

$$Vm = Vp \times (Rm/(Rm + Rs)).$$

If Rm decreases, i.e., through the opening of many ion channels then the contribution of Rm to the total resistance (Rm + Rs) decreases along with Vm. The application of Ohm's law, $Vs = I \times Rs$, is a quick way of obtaining the voltage drop across the series resistance, Vs, from the measured current I and Rs; then $Vm = Vp - Vs$.

16. The formation of an excised vesicle is a common problem during attempts to form an inside-out patch and occurs when the excised membrane edges close up, in a manner akin to the formation of outside-out patches, and results in a "cell-attached" vesicle. A bathing solution that is very low in Ca^{2+} and Mg^{2+} ions can help prevent this. The outward-facing part of the vesicle can be ruptured to leave the inside-out patch by passing the tip of the micropipette quickly out and then back into the bath solution, through the air-solution interface. This does however have limited success and a complete loss of the patch occurs with at least equal frequency. For more tips on dealing with such vesicles refer to Chapter 9 of this volume.

References

1. Neher E, Sakmann B (1976) Single-channel currents recorded from membrane of denervated frog muscle fibres. Nature 260:799–802
2. Hamill OP, Marty A, Neher E, Sakmann B, Sigworth FJ (1981) Improved patch-clamp techniques for high-resolution current recording from cells and cell-free membrane patches. Pflugers Arch 391:85–100
3. Lippiat JD (2008) Whole-cell recording using the perforated patch clamp technique. Methods Mol Biol 491:141–149
4. Tammaro P, Shimomura K, Proks P (2008) Xenopus oocytes as a heterologous expression system for studying ion channels with the patch-clamp technique. Methods Mol Biol 491: 127–139
5. de Wet H, Lippiat JD, Allen M (2008) Analysing steroid modulation of BK(Ca) channels reconstituted into planar lipid bilayers. Methods Mol Biol 491:177–186
6. Colquoun D, Sigworth FJ (1995) Fitting and statistical analysis of single-channel records. In: Sakmann B, Neher E (eds) Single channel recording, 2nd edn. Plenum, New York, pp 483–587

Chapter 8

Recording of Ion Channel Activity in Planar Lipid Bilayer Experiments

Eleonora Zakharian

Abstract

Planar lipid bilayer is an electrophysiological technique that enables study of functional activities of ion channels, porins, and other pore-forming molecular complexes. The main purpose of this method is to monitor ion channels' behavior at the single molecule level in the artificial membranes. Here, I describe the details of this technique that will underline formation of the lipid bilayers and incorporation and activation of the ion channel protein.

Key words Planar lipid bilayer, Black lipid membranes, Ion channel, Ionic current, Conductance

1 Introduction

Electrophysiological approaches are designed to evaluate physical properties and characteristics of ion channels. Planar lipid bilayer is one of the unique electrophysiological techniques that is intended to study specific channel properties of the purified complexes in a well-controlled artificial environment (1). The effectiveness of this method reveals the possibility to study and characterize ion channel behavior at the single molecule level. Furthermore, this technique is very useful for investigation of the direct effects of chemicals and enzymes on the channel while excluding possible indirect effects, which might be caused by the presence of regulatory proteins of the native membranes.

Planar lipid bilayer experiments enable monitoring the transport rates of ions across membranes through incorporated ion channels. The experiments are performed in the chamber with *cis*- and *trans*-compartments that are connected through a small aperture with a diameter ranging from 50 to 250 μm. A lipid solution is applied to the aperture with subsequent formation of a planar bilayer membrane on the hole. This is followed by reconstitution of ion channel proteins. Ion channels can be inserted into bilayer lipid

Nikita Gamper (ed.), *Ion Channels: Methods and Protocols*, Methods in Molecular Biology, vol. 998,
DOI 10.1007/978-1-62703-351-0_8, © Springer Science+Business Media, LLC 2013

membrane (BLM) directly from a micellar solution or fused with liposomes. After the channel is incorporated into the bilayer, the ionic current can be induced by applying a driving force. This force represents an electrochemical potential that has two components—electrical, which is called membrane potential (V_m or $\Delta\psi$) derived from charges or voltages applied across the membrane, and chemical component that derives from a chemical gradient with asymmetric ionic solutions. The ionic current through typical ion channel is measured in pico-amperes (pA), and potential or voltages are measured in millivolts (mV). Conductance of ions at a single-channel level is usually measured in a range of 10–1,000 pico-siemens (pS).

Some ion channels such as ligand-gated channels would also require the presence of specific molecules-activators (ligands) in order to induce channel openings. In case of many mammalian channels this regulation can be quite complex and would require simultaneous presence of a number of molecular components and/or various physical factors (temperature, pressure) in order to stimulate channel openings. The advantage of planar lipid bilayer is that it increases likelihood to precisely identify all the chemical and physical factors that are required for, or supporting, the channel activity. The other benefits of this method include a possibility to alternate lipid composition or chemical compounds regulating channel activity that can be applied to either side of the membrane.

Like any technique, planar lipid bilayer also has its disadvantages. Among the disadvantages of lipid bilayers are large capacitances due to the large sizes of the aperture, comparing it to the diameters used in excised patches. The larger capacitances cause for a slow voltage response time. Another disadvantage of this technique, relative to the native membrane patch-clamp, is the generation of high-amplitude noise due to the large area of bilayers in traditional systems. In order to reduce the peak-to-peak amplitude of the noise in the system, intensive filtering may be applied. However low-pass filtering of the single-channel signal produces a significant loss of time resolution and at some point will make fast-gating events undetectable. This issue has been addressed by recently developed alternative BLM systems that allow performing low-noise and higher bandwidth recordings (2, 3).

Overall, the complexity of planar lipid bilayer technique at first glance seems to arise from the intricacy of the steps in formation of the artificial membranes that require thorough knowledge of physical chemistry of BLM. Nevertheless, the theoretical insights that describe physicochemical properties of lipid bilayer as well as methodological details are well established (for more references see refs. 1, 4–6), along with the practical side of the method (including advanced technology for the electrophysiological setup, voltage-clamp recordings, and availability of high quality and purity of synthetic lipids). All these factors aid in the formation of artificial lipid bilayer suitable for channel study and make it a straightforward and reproducible method. The main challenge of this technique is

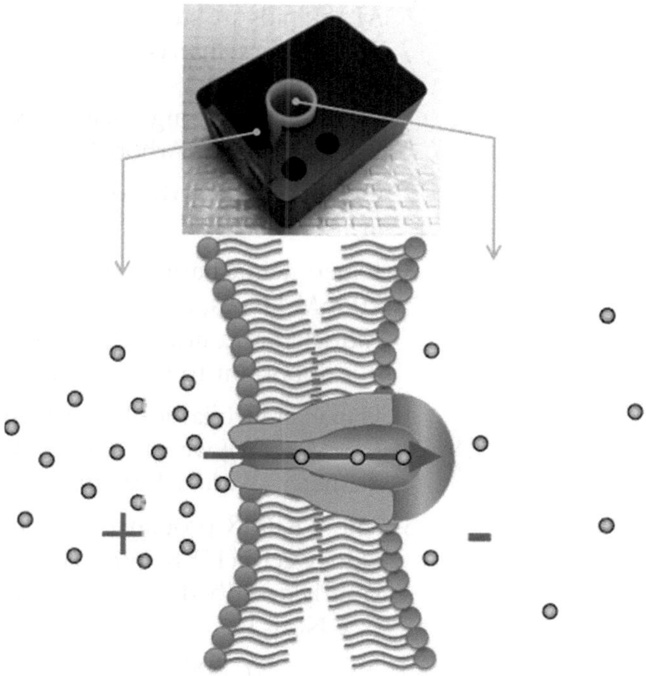

Fig. 1 Planar lipid bilayer chamber with inserted cuvette, lipid bilayers are painted on the aperture in the middle of the cuvette. Silver-silver chloride electrodes are connected to each compartment (*cis* and *trans*), those allow to obtain the ionic current that is amplified using a voltage-clamp amplifier. Cups and chambers are filled with equal volumes in the *cis*- and *trans*-sides, which results in a balanced solution height, thus minimizing mechanical gradients across the bilayer membrane. *Below* is a graphical representation for formation of planar lipid bilayers and incorporated ion channel that conducts ions across the membrane

rather hindered in the capricious nature of the membrane proteins, which ion channels represent, and therefore the success of the experiment often heavily relies on the protein part, which includes protein isolation, purification, folding, and incorporation. The reason for this is that each protein is unique and ways to handling it may vary greatly. Here, as an example, I describe planar lipid bilayer method (Fig. 1) of incorporation and activation of the cold and menthol receptor—TRPM8, which we have successfully applied in our studies (7, 8). The method is comprised of (1) formation of the artificial membranes, (2) incorporation of the ion channel, and (3) activation of the channel protein by its agonists and other molecules that are involved in the channel activity.

2 Materials

2.1 Solutions and Reagents

1. Lipids include synthetic 1-palmitoyl-2-oleoyl-glycero-3-phosphocoline (POPC) and 1-palmitoyl-2-oleoyl-glycero-3-phosphoethanolamine (POPE, Avanti Polar Lipids, Birmingham,

AL). Lipids are stored in septum glass vials and screw caps with Teflon discs inserts that are from Thermo Scientific Pierce (Rockford, IL).

Short- and long-chain phosphoinositides are from Cayman Chemicals (Ann Arbor, MI).

2. Organic solvent to dissolve the lipids—*n*-decane (Sigma-Aldrich, St. Louis, MO).
3. Experimental bathing solution: 150 mM KCl, 0.2 mM $MgCl_2$, 20 mM Hepes, pH 7.4. All salts are ultrapure (>99%) (Sigma-Aldrich, St. Louis, MO). To increase the purity of the solutions it is important to filter all the buffers with 0.22 μm filters.
4. Planar lipid bilayer chamber and cuvette are made of Delrin (acetyl resin); we use cuvette with an aperture of ~150 μm in diameter (Warner Instruments, Hamden, CT).
5. The TRPM8 protein is purified from the human embryonic kidney cells (HEK-293), stably expressing the channel as previously described (7), for updates in the purification procedure (see Note 1).
6. NCB buffer: 500 mM NaCl, 50 mM NaH_2PO_4, 20 mM Hepes, 2 mM Na-orthovanadate, 10% glycerol, pH 7.5.
7. Cell homogenization buffer: NCB buffer with addition of 1 mM of protease inhibitor PMSF, 5 mM β-mercaptoethanol.
8. Protein isolation buffer: NCB buffer with addition of one tablet of protease inhibitor cocktail, 1 mM PMSF, 20 μg/mL DNase, 20 μg/mL RNase, 0.1% Nonidet P40 (Roche, Indianapolis, IN), and 0.5% dodecyl-maltoside (DDM) (CalBiochem, Darmstadt, Germany).
9. Phosphate buffer saline (PBS).

2.2 Equipment and Materials

1. Electrophysiological setup: Axopatch 200B amplifier; CV-203BU headstage; Baseplate; Electrode holder; Series Resistance Dither Box; Patch-1U Model Cell (Molecular Devices, Sunnyvale, CA).
2. DD1440A Digidata 1440A data acquisition system and pClamp-10 Electrophysiology software (Molecular Devices, Sunnyvale, CA).
3. 8-Pole Bessel filter—950 TAF (Frequency Devices, Ottawa, IL).
4. The entire setup is fixed in the Faraday cage to block sound and electrical noise (AutoMate Scientific, Inc, Berkeley, CA).
5. Bilayer chamber and cuvette (Warner Instruments, Hamden, CT).
6. For temperature experiments we use bilayer chamber made of a thermally conductive plastic; the pyroelectric heating/cooling stage connected to temperature controller CL-100 (Warner Instruments, Hamden, CT).

7. Stereo microscope (Olympus, Center Valley, PA).
8. Bath sonicator (Branson, Hatfield, PA).
9. Glass capillary tubes (Fisher Scientific).
10. Bunsen burner (Fisher Scientific).
11. Molecular sieves (Sigma).
12. Speed-Vac (VWR).
13. Nitrogen gas tank (GTS Welco, Inc., Newark, NJ).

3 Methods

3.1 Preparing Working Equipment for Planar Lipid Bilayer: Chambers, Cuvettes, and Glass Capillaries

Preparation of the working equipment such as bilayer chambers and cuvettes is the very first and important step and should be carried before each experiment. To optimize the rate for a successful BLM experiment and eliminate false positive results due to possible contaminants, the bilayer chamber and cuvette have to be cleaned thoroughly.

1. Wash with dishwashing detergent in warm water; rinse well several times with warm running water to remove detergent; rinse three times with milli-Q water.
2. Wash with organic solvents—first with ethanol and then with methanol; dry with nitrogen gas. It is useful to press on top of cuvette when filled with solvent, the intense solvent flow through the aperture helps to remove residual lipids and other content from the aperture. When switching from one target-protein to another, or in case of difficulties to remove contaminants, all washing steps can be done along with bath sonication (see Note 2).
3. Delrin cuvettes can also be cleaned with chloroform, when necessary to remove highly hydrophobic content from the aperture. Note that not all materials are suitable for a chloroform use; be sure that your cuvette will not be damaged during this cleaning procedure (see Note 3).
4. Prepare a set of air-bubble glass capillaries. Holding to the end of capillary, flame the other end, simultaneously pooling the edge with forceps until the thin prolonged glass tube is formed. Remove the forceps and continue to flame the thin tube in order to make the bubble-like structure.

3.2 Preparation of the Lipids

1. Open the ampoules with lipids in chloroform solution and transfer the content into new and prerinsed with organic solvent glass vials, add the molecular sieves (4 Å) and cover with Teflon-side screw cap. For storage conditions see Note 4.
2. Prepare the mixture of POPC/POPE in 3:1 ratio and dry the lipids by lyophilization in Speed-Vac at least for 2 h to

completely remove the chloroform content. Next, resuspend the lipidic mixture in an organic solvent such as *n*-decane to a final concentration of 25–30 mM. Mix the lipids under the stream of nitrogen gas.

3.3 Formation of Planar Lipid Bilayer

1. Fill the chamber in the following order: first add 1 mL of bathing solution to the *trans*-compartment (cuvette); press slightly on the top of cuvette until a tiny droplet will appear in the area of aperture (see Note 5); add 1 mL of bathing solution to the *cis*-compartment.
2. Immerse a pair of matched Ag-AgCl electrodes in the *cis*- and *trans*-compartments. Adjust the background-leak current near 0 values, this can be done by changing the voltage in a range of ±1 mV and adjusting manually the offset appropriately so that at positive voltage baseline would go upwards and at negative it would shift downwards (see Note 6).
3. Dip the air-bubble glass capillary in the lipid/decane solution and carefully apply the lipids to the aperture in the cuvette by "painting." After several gentle movements of the air-bubble in the area of aperture, the film will form. We prefer not to pre-apply the lipidic solution beforehand; instead, we distribute the lipids by painting with the air-bubble glass capillary when the aqueous solvents are present (see Note 7).
4. Monitor the thinning of the lipid film by either observing it in the microscope—planar lipid bilayers reflect no light and appear black after formation, or alternatively estimate the capacitance of the membrane—as lipid bilayers will form the capacitance will gradually increase, indicating that the film is thinning. The capacitance of the bilayer can reach 100–200 pF, depending on the size of aperture used (see Note 8).

3.4 Incorporation and Activation of Ion Channel

1. After the bilayers are formed, add 0.2 μL of the TRPM8 micellar solution (0.002 μg/mL) to the *cis*-compartment with gentle stirring. Normally it will take some time for the channel to incorporate in the membrane. Fusion of the proteo-liposomes or micelles can be observed by fusion spike appearance. After the incorporation stop the stirring.
2. To activate TRPM8 in planar lipid bilayer, add its agonist menthol (500 μM) to the *cis*- and/or *trans*-compartment. TRPM8 inserts into lipid bilayer without preferences and may be oriented either way. To induce TRPM8 channel openings, apart from the ligand, it also requires the presence of its gating factor phophoinositol 4,5-biphosphate (PIP_2) (Fig. 2). The short acyl-chain dioctanoyl diC_8PIP_2 (2.5 μM) can be added to either side of the membrane, but it activates the channel from the intracellular side (see Note 9). Channel current recordings can be then obtained with different voltages. Orientation of TRPM8 can be determined according to the outward rectification

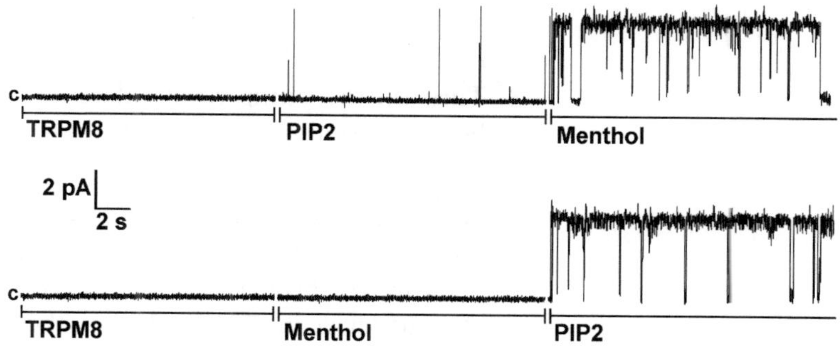

Fig. 2 Activation of TRPM8 channels in planar lipid bilayers by menthol and PIP_2. Representative single-channel current recordings of TRPM8 channels incorporated in planar lipid bilayers formed from POPC/POPE (3:1) in *n*-decane, between symmetric bathing solutions of 150 mM KCl, 0.2 mM $MgCl_2$ in 20 mM Hepes buffer, pH 7.4 at 22°C. 0.2–0.5 μL of 0.2 μg/mL TRPM8 protein was incorporated in POPC/POPE micelles, which were added to the *cis*-compartment. Clamping potential was +60 mV. Upper and lower traces consist of three segments with subsequent additions of components as indicated in the figure: 2 μM of diC_8 PIP_2 and 500 μM of menthol were added to both compartments (the figure is reproduced from ref. 8 with permission)

and/or by an intracellular block with polylysine (polyK can be added at the end of the experiment).

3. Outward currents of TRPM8 exhibit mean slope conductance values of ~72 pS, and P_o of ~0.89 at 100 mV, and inward currents may be observed in two conductance states with main conductance level of ~42 pS and P_o of ~0.4 (at −100 mV) and rarely detected burst openings of a subconductance state with mean conductance of ~30 pS ($P_o \leq 0.001$), which would step to the fully open magnitude (72 pS) of the channels. In case if inconsistency in channel behavior is observed see Note 10.
4. Unitary currents are recorded with an integrating patch clamp amplifier. The *trans*-solution (voltage command side) is connected to the preamplifier headstage input, and the *cis*-solution is held at virtual ground via a pair of matched Ag-AgCl electrodes. Currents through the voltage-clamped bilayers (background conductance <3 pS) are filtered at the amplifier output (low pass, −3 dB at 10 kHz, 8-pole Bessel response). Data are secondarily filtered at 100 Hz and digitized at 1 kHz.
5. Single-channel conductance events, all points' histograms, open probabilities, and other parameters are identified and analyzed using the appropriate software.

3.5 Temperature Studies

Planar lipid bilayer is a convenient system to study temperature sensitivity of the cold receptor TRPM8.

1. For temperature studies, position a Delrin cuvette in the bilayer chamber made of a thermally conductive plastic.

2. The chamber is fitted on a conductive stage containing a pyroelectric heater/cooler.
3. Turn on the pump to circulate deionized water through the heating/cooling stage: the water is pumped into the system to remove the generated heat.
4. Set the temperature of interest on the temperature controller connected to the pyroelectric heating/cooling stage. The temperature of the bath is monitored constantly with a thermoelectric device in the *cis*-side, i.e., the groundside of the cuvette. Although there will be a temperature gradient build between the bath solution and conductive stage, the temperature within the bath could be reliably controlled within ±0.5°C.

4 Notes

1. HEK-293 cells stably expressing Myc-tagged TRPM8 are grown to ~80% confluence, washed and collected with PBS. Cells are harvested and resuspended in cell homogenization buffer. Then the cells are lysed by freeze-thawing method and centrifuged at low speed to remove cell debris and DNA. The supernatant is further centrifuged at 40,000 ×*g* for 2.5 h., and the pellet is resuspended in protein isolation buffer. The suspension is incubated overnight at 4°C on a shaker with gentle agitation and then centrifuged for 1 h at 40,000 ×*g*. The supernatant is collected and incubated with the beads, where the TRPM8 protein is purified by immunoprecipitation with anti-Myc-IgG conjugated to A/G protein magnetic beads (Pierce, Thermo Scientific), following the procedure provided by the manufacturer. All steps of purification are performed at 4°C. For the planar lipid bilayer experiments the protein is eluted with Myc-peptide (50 μg/mL).
2. Place chamber and cuvette into a beaker and perform all the washing steps (warm water with detergent; rinse; organic media) while using bath sonicator at low intensity, do not exceed 5 minutes for each washing step. Longer and more frequent sonication may affect the quality of the chamber.
3. Polystyrene cuvettes have poor resistance to organic solvents, which can lead to degradation of the aperture; therefore these cannot be cleaned with strong organic solvents. We also do not recommend using chloroform-cleaning step for the bilayer chamber, although it is made of Delrin, but the observation window is made of a different plastic polymer that can be entirely destroyed with chloroform.
4. Stock lipids/chloroform solutions are degassed with nitrogen stream and stored in the presence of molecular sieves 4 Å (Sigma) at −20°C. The molecular sieves are used for dehydration

and absorb molecules with 4 Å effective diameter (H_2O capacity ~20% by weight). Vials with lipids should be tightened with a Teflon-insert cap and covered with parafilm.

5. This step is important and helps to remove air from the aperture and create nice contact between the *cis-* and *trans-*compartments. Air entrapped in the aperture may cause a problematic formation of the bilayers.
6. At certain experimental conditions, such as with gradient solutions that introduce different ionic strengths across the membrane and cause formation of junction potentials, it is preferable to connect the electrodes with bath solutions using agar bridges. The agar bridges are made of glass capillaries and filled with 3 M KCl-agar solution.
7. Although it slightly increases the time for forming the bilayer, overall avoiding pre-application of lipids helps to minimize introduction of bulky amounts of lipids on the aperture. This in turn is useful in formation of good/thin bilayer with reduced torus of amorphous lipid that surrounds the membrane.
8. Planar lipid bilayer represents a capacitor that is capable of storing charge in electric field. The bilayer capacity is directly proportional to its surface area and dielectric constant (ε), and is inversely proportional to the thickness of the bilayer (d): Capacitance (C) of the bilayer formed across circular aperture can be calculated using the following formula $C = \varepsilon \cdot \pi r^2 / d$ (where r is the radius of the bilayer). Note that r is the radius of the actual bilayer, not the aperture, and the value of r generally reflects the quality of the bilayer. A phospholipid membrane has a capacitance in the range of 1 μF/cm². A steady-state charge on a capacitor is expressed as $q = CV$, where q is the charge on the capacitor (in Coulombs) and V is the potential between the capacitor plates (in Volts). The time-dependent derivative for the charge can be obtained from the following expression: $I = \mathrm{d}q/\mathrm{d}t = C \cdot dV/\mathrm{d}t$. This expression is useful to estimate the capacitance of the bilayer, when the current measured under the conditions of the linear voltage ramp (constant $\mathrm{d}V/\mathrm{d}t$) is directly proportional to the bilayer capacitance.
9. It may take several (5–10) minutes for diC_8PIP_2 to incorporate in the membrane. Stirring the bath solution is helpful and can accelerate the incorporation of the phosphoinositide. Alternatively, PIP_2 can be incorporated in the bilayers in a form of long dipalmitoyl chain—$diC_{16}PIP_2$ by adding it to the mixture of phospholipids beforehand at a final concentration of 1%.
10. In case if the channel behaves irregularly, do not rush to discard the solution and restart the experiment, sometimes the channel's final folding may occur after incorporation in the lipid

bilayer. In other words, channel may at first be misfolded; however it is possible that the protein will finally obtain the proper conformation state in the bilayer. This can be evidenced by stabilization of the channel currents that would change from irregular activity to nice strait openings and closures. Likewise, the channel may continue to misbehave. This would mean that most likely the protein purification encountered some problems and the protein is unstable. Alter the incorporation method and insert the channel with the liposomes. If this would not improve the result, the most logical would be to repeat the purification procedure. Note that presence of Mg^{2+} is important for the stability of the TRPM8 protein. We found that Mg^{2+} is important in supporting the native supramolecular complex of TRPM8 with polyhydroxybutyrate/polyphosphate (PHB/polyP). Affecting the molecular integrity of this supramolecular complex of the channel has an immense effect on its activity (8).

References

1. Miller C (1986) Ion channel reconstitution. Plenum, New York
2. White RJ, Ervin EN, Yang T, Chen X, Daniel S, Cremer PS, White HS (2007) Single ion-channel recordings using glass nanopore membranes. J Am Chem Soc 129:11766–11775
3. Zhang B, Galusha J, Shiozawa PG, Wang G, Bergren AJ, Jones RM, White RJ, Ervin EN, Cauley CC, White HS (2007) Bench-top method for fabricating glass-sealed nanodisk electrodes, glass nanopore electrodes, and glass nanopore membranes of controlled size. Anal Chem 79:4778–4787
4. White SH (1986) The physical nature of planar lipid bilayer membranes. In: Miller C (ed) Ion channel reconstitution. Plenum, New York, pp 3–35
5. Montal M, Mueller P (1972) Formation of bimolecular membranes from lipid monolayers and a study of their electrical properties. Proc Natl Acad Sci U S A 69:3561–3566
6. Williams AJ (1994) An introduction to the methods available for ion channel reconstitution. In: Ogden D (ed) Microelectrode techniques: the plymouth workshop handbook, 2nd edn. Company of Biologists, Cambridge, pp 79–99
7. Zakharian E, Cao C, Rohacs T (2010) Gating of transient receptor potential melastatin 8 (TRPM8) channels activated by cold and chemical agonists in planar lipid bilayers. J Neurosci 30:12526–125348
8. Zakharian E, Thyagarajan B, French RJ, Pavlov E, Rohacs T (2009) Inorganic polyphosphate modulates TRPM8 channels. PLoS One 4:e5404

Chapter 9

Recording Macroscopic Currents in Large Patches from *Xenopus* Oocytes

Tibor Rohacs

Abstract

The excised inside-out patch clamp technique gives rapid access to the intracellular surface of the plasma membrane while measuring channel activity. This way the effects of intracellular regulators of ion channels or transporters can be studied in isolation, in the absence of most of the cellular machinery. This chapter summarizes our experience with this technique using large patches to study various ion channels expressed in *Xenopus* oocytes.

Key words *Xenopus* oocytes, Excised patch, Macropatch, TRP channels, Patch clamp

1 Introduction

The giant patch or "macropatch" technique was originally developed to measure currents of electrogenic transporters or pumps, which are far smaller than those carried by ion channels (1, 2). Large excised patches are also very useful for the study of ion channels. Generally this technique gives similar information to that obtained by using single-channel measurements in excised patches. It does not give the detailed information that single-channel measurements give on gating kinetics, conductance, etc., but it is a lot more robust, and simpler to analyze, as it gives composite current of hundreds or even thousands of channels. The excised inside-out configuration allows the perfusion of various compounds to the intracellular surface of the patch membrane, thus the study of the effects of intracellular regulators in isolation from the cellular environment. This chapter is based on our experience with this technique with inwardly rectifying K^+ (Kir) channels and various transient receptor potential (TRP) channels (3–8).

Nikita Gamper (ed.), *Ion Channels: Methods and Protocols*, Methods in Molecular Biology, vol. 998,
DOI 10.1007/978-1-62703-351-0_9,

Xenopus laevis oocytes have been used for quite a long time as an expression system to study recombinant ion channels (9). There are many articles that provide excellent description of oocyte preparation and RNA injection (10); we will provide a brief description here based on our experience.

2 Materials

2.1 Oocyte Preparation and Injection

1. *OR2 solution for oocyte preparation and storage.* The composition of OR2 is 82.5 mM NaCl, 2 mM KCl, 1 mM $MgCl_2$, 5 mM HEPES, pH = 7.4. We designate this Ca^{2+}-free solution as OR2(−). Since large volumes are used because of the repetitive wash steps during oocyte isolation, it is practical to prepare a 20× concentrate, and dilute several liters to working concentration to keep in a large bottle. After dilution the pH needs to be adjusted, and the solution can be kept at 4°C for several weeks; it is not necessary to sterile filter the OR2(−) solution. After digestion, the eggs are kept in OR2 solution with an additional 1.8 mM Ca^{2+} and penicillin-streptomycin (designated as OR2(+) solution); we recommend that this solution is sterile filtered.
2. *Collagenase solution.* We use 0.1–0.2 mg/mL of type IA collagenase from Sigma. We dissolve it freshly in OR2(−) just before starting the digestion process (see Note 1).
3. *MS222 solution.* 0.25% Ethyl 3-aminobenzoate methanesulfonate (tricaine, MS222) dissolved in H_2O, pH adjusted to ~7.4. We prepare 1 L, which can be reused several times, when kept in a cold room.
4. Scissors, forceps.
5. Sutures (Chromic gut).
6. Oocyte incubator (18°C) preferably with an electrical outlet inside to power the rotating platform.
7. Dissecting microscope and light source.
8. Injector (Drummond Nanoject).
9. Nuclease-free water and mMessage Machine kit (Ambion).

2.2 Patch Clamp Measurements

We recommend that all solutions are sterile filtered to avoid clogging of the patch clamp pipette and the perfusion system.

1. *Bath solution.* The oocyte is placed in the bath solution during the experiment, and the same solution is used to fill up the perfusion system. The composition is 97 mM KCl, 5 mM HEPES, and 5 mM EGTA, pH = 7.4. The purpose of including EGTA in the bath solution is to avoid the activation of the Ca^{2+}-activated Cl^- current endogenous in the oocytes (see Notes 2 and 3).

2. *Pipette solution*. For most ion channels, a standard extracellular NaCl-based solution can be used: 97 mM NaCl, 2 mM KCl, 1 mM $MgCl_2$, 5 mM HEPES, pH = 7.4; we use this solution to record TRPV1 and TRPM8 (see Note 4).
3. Some channels require modified pipette solutions, for example, TRPV6 is blocked by extracellular $Mg^{2+}$and Ca^{2+}; thus, a divalent-free solution is used, 99 mM $LiCl_2$, 1 mM EGTA, and 5 mM HEPES, pH = 7.4. The use of lithium as a charge carrier minimizes endogenous currents that may become activated upon removal of extracellular divalent cations.
4. *Cl^--free bath solution*. 93 mM K-gluconate, 5 mM HEDTA, 5 mM HEPES, pH = 7.4. This solution is used when intracellular Ca^{2+} is applied. For Ca^{2+} concentration up to 10 μM, we use HEDTA to buffer Ca^{2+}. Above this concentration, Ca^{2+} cannot be reliably buffered with HEDTA, so we leave this chelator out of the solution, but the weak buffering of Ca^{2+} by gluconate should be taken into consideration (11). The free Ca^{2+} concentrations can be calculated with the Max Chelator program (http://maxchelator.stanford.edu/). The Ca^{2+} free HEDTA-containing solution can be used for pipette solution, to eliminate the Ca^{2+} induced Cl^-current (see Note 3).
5. *Patch pipettes*. We use 4 in. long 1.5 mm OD patch clamp pipettes (WPI). Other types of pipettes may also work well (see Notes 5–7 and Fig. 1).
6. A micropipette puller (see Note 6).
7. *A pair of very sharp forceps to devitellinize (peel) the oocyte*. We use Dumont 55 forceps (Fine Science Tools). The forceps can easily be damaged; they can be sharpened with a sharpening

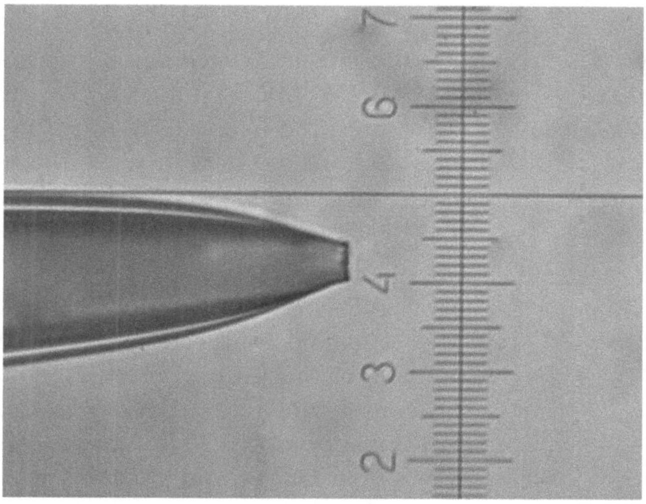

Fig. 1 A typical patch pipette for macroscopic current measurements in excised inside-out patches. The small divisions correspond to 2.5 μm

stone, also available from the same company. This is best done under the dissecting microscope used for oocyte injection.

8. Patch clamp setup, equipped with a stereo microscope (see Chapter 7 of this book for more details).
9. *Micromanipulator.* We use a three-axis manual manipulator with one motorized axis (Siskiyou). The manual manipulators can be used to approach the oocyte, and the fine final movement is done with the motorized axis.
10. *Perfusion system.* Various gravity-based perfusion systems with solenoid or pinch valves are available; we use models from ALA Scientific and Automate Scientific. We use our custom-made perfusion pencils, since we need very low perfusion rates, as we often use expensive compounds that require keeping fluid flow and dead space low. A perfusion pencil can be made by gluing seven thin plastic tubes into a gel-loading tip using two-component epoxy glue, available at any hardware store. Generally it is possible to use less than 1 mL, even as low as 0.5 mL of perfusion solution with a handmade system. As an alternative, commercial perfusion pencils can be used for local perfusion of the patch pipette, also available from ALA and Automate. To achieve low volumes, thin tubes need to be used. It is important to use as few connections as possible, since every connection is a source for air bubbles. We use 1 mL plastic syringes as reservoirs, and instead of connecting them with a luer connector, we melt the bottom part, using a gas burner, and pull it to form a thin tube, which we directly connect to the silicone tubing used for the pinch valves.

3 Methods

3.1 Oocyte Preparation and RNA Injection

1. A female *Xenopus laevis* frog is anesthetized by placing into a 0.25% MS222 solution. This usually takes a couple of minutes; the frog can be monitored by gently knocking on the side of the bucket. Once the frog is unresponsive, it can be lifted up and put on its back. If this can be done without resistance and movement from the frog, anesthesia is deep enough (see Note 8).
2. An ~1–1.5 cm incision using a scalpel is made on the left or right side of the abdomen through the skin and the muscular layer. Small scissors can be used to lengthen the incision. Using a pair of forceps, the oocyte sacs can be lifted out of the abdominal cavity, cut free with scissors, and put into OR2(−) solution. Once enough oocytes are obtained, the abdomen can be closed (see below), if the frog is to be reused. One oocyte sac, that can be visually identified, contains about 200 oocytes. We

usually remove far more oocytes than we need since many are lost or damaged during the digestion, and in a poor quality prep, sometimes a large number of oocytes need to be screened to find enough good-quality ones (see Note 9).

3. Using dissolvable chromic gut sutures, the abdominal layer and then the skin is sutured back with 3–4 stitches.
4. The frog is placed back to a recovery tank. They recover from the anesthesia usually within 15–20 min. It is important not to over-anesthetize them, since they can drown, if the recovery time is excessive.
5. The oocyte sacs are either cut to smaller pieces with a pair of scissors or torn to smaller pieces using two pairs of forceps. After several washes with OR2(−), until the solution is not turbid from debris, the oocytes are placed into a 0.1–0.2 mg/mL collagenase solution, and the tube is rotated very gently at 18°C overnight. We use a 50 mL Falcon tube filled with 40 mL collagenase solution, and place it in a rotating device inside an incubator, which is kept at 18°C. The temperature is critical. The plane of rotation is parallel to the axis of the falcon tube, which creates a very gentle mechanical agitation (see Note 10).
6. The next morning, the oocytes are washed several times with OR2. This step is critical, since collagenase needs to be completely removed before the oocytes are placed into a Ca^{2+}-containing solution. Ca^{2+} greatly enhances the activity of collagenase; thus even trace amounts of the enzyme can damage the oocytes. Washing the oocytes is simple, they sediment within seconds in a 50 mL tube, then the solution can be gently poured off, and replaced with fresh, repeating several times.
7. Next the oocytes are shaken for an additional hour in OR2(−), still Ca^{2+}-free. This step removes the follicular layer by mechanical agitation from most of those eggs that still have it (see Note 11). We usually do this step with the rotation being perpendicular to the axis of the 50 mL tube with lower solution levels (usually 35 mL) to create stronger mechanical agitation.
8. Next the oocytes are placed in OR2(+) solution that contains Ca^{2+} and antibiotics. Older protocols also included nutrients such as pyruvate, this is unnecessary, since the yolk of the oocytes provides enough nutrients, and additional nutrients make infection of the solution more likely.
9. Select oocytes for injection, usually it is good to select at least 2× the required number to have non-injected controls, and extra eggs for unforeseen purposes, they usually last a lot longer than the ones that are not selected out from the debris and damaged oocytes.

10. *Maintenance of oocytes.* We keep the selected oocytes in 35 or 60 mm tissue culture dishes. It is important to check the oocytes at least once every day, and dispose the damaged ones, and change solution to clean the debris. Even one disintegrated oocyte can damage the rest of eggs in the dish.

3.2 RNA Preparation and Injection

1. RNA is prepared from linearized DNA using the mMessage Machine kit from Ambion (see Note 12 for oocyte vectors). Gloves, RNAase-free solutions, pipette tips, and other tools are necessary. Usually we store all tubes, pipette tips, and other materials used for RNA work separately, and use them for this purpose only.
2. A Drummond Nanoject injector is used; we usually inject 50 nL, even though the volume can be adjusted. After pulling, the tip of the needle needs to be broken to a diameter of ~20–25 μm. This can be achieved either by pushing the needle through a Kimwipe paper or by carefully "cutting" the tip with a scalpel. It is advisable to check the size of the tip after breaking it. Injecting pipettes with too small diameters get clogged easily; too large ones damage the oocytes (see Notes 13–15).
3. The purpose of the digestion is to remove the oocytes from the follicular sac, and to remove the follicular layer. Oocytes without the follicular are easy to inject, since the sharp needle easily penetrates the vitelline layer and the plasma membrane. Oocytes with the follicular layer on are very difficult to inject and most of the time injection causes significant damage to the oocytes, reducing viability. The follicular layer can easily be noticed by the thin blood vessels. If the follicular layer lacks visible blood vessels it is a lot harder to notice; often it is only recognized during injection by the difficulty of penetrating with the injection needle. If oocytes are under-digested, and a significant number of them have the follicular layer, it is worth manually defolliculating them with a fine pair of fine forceps before injection. This is a lot easier than devitellinization before patching, see later, and generally pays back by better oocyte quality and quantity.

3.3 Patch Clamp

1. The vitelline layer of the oocyte needs to be removed using two pairs of sharp Dumont 55 forceps (see Note 16).
2. After the pipette is filled up with the pipette solution, the pipette is lowered into the bath solution, and pipette potential is zeroed. Gentle positive pressure is applied and the pipette is moved close to the oocyte using the manual axes on the manipulator (see Note 17).
3. Upon touching the oocyte, using the motorized axis, pipette resistance increases slightly, then the positive pressure can be relieved, which should further increase pipette resistance.

In some cases, this procedure is enough to form a gigaseal, after some waiting period. If the seal is not formed spontaneously, gentle negative pressure may be applied. Keep in mind that, according to Laplace's law, the tension in the membrane will be higher with large pipettes than with small pipettes for the same pressure applied, so be gentle, especially with larger pipettes. Generally it takes longer to form a seal with a large pipette than with standard patch pipettes, used for single-channel or whole-cell patch clamp measurements. With larger pipettes it is not always possible to reach gigaohm resistance, which is not a problem as long as the leak is substantially less than the measured current. Seal formation can take anywhere between 30 s and 5 min. The required negative pressure is quite variable from patch to patch. Since the oocyte is a quite large cell, the pipette can be pushed deeper in the oocyte than into a mammalian cell for seal formation. Most channels we study have significant activity at 0 mV, where the seal test is performed, and if channel activity is high, this reduces the apparent seal resistance. Thus we often switch to a standard ramp protocol during seal formation. Since most channels we study are either inward or outward rectifiers, it is usually easy to tell from the shape of the current (i.e., by the presence of appropriate rectification) if we are close to a seal or not. For constitutively active channels without significant rectification, it is often not trivial to tell if we have a good seal or not, i.e., if the current to be studied is large enough compared to the leak. In these cases a nonsymmetrical solutions can be used to "induce" apparent rectification.

4. Inside-out configuration is established by pulling out the patch using the motorized manipulator. After this is done, we usually quickly try to move air bubbles out of the perfusion system, by applying pressure with a syringe plunger to make sure all tubes are flowing (see Note 18), then place the perfusion system right on top of the patch pipette almost touching it.
5. Measurements and analysis proceed with standard electrophysiological procedures (see Note 19).

4 Notes

1. *Collagenase.* The individual lots of collagenase can differ in activity. When a new lot is tried, it is worth doing parallel digestions with two or three different concentrations and comparing the results. Once conditions are established, we usually order another batch from the same lot, to avoid frequent changes.
2. The oocyte membranes have endogenous currents, the two most notable are the Ca^{2+}-activated Cl^- current (see Note 3)

and a stretch-activated current. The latter can be inhibited by gadolinium. Gadolinium (100 μM) can be used in the pipette solution when recording Kir channel activity, but it also inhibits TRP channels; thus, it should be avoided in those experiments. There are numerous other smaller currents; the best way to avoid them is obtain good expression levels of the channel in study that makes the small endogenous currents negligible.

3. The oocytes contain a large endogenous calcium-activated Cl^- current. To avoid this activity our internal solutions contain 5 mM EGTA. If one wants to study the effects of intracellular Ca^{2+}, this current needs to be taken into consideration. We use two different methods. First, Cl^- can be replaced in both the pipette and in the bath solution with gluconate. In this case, the silver chloride electrode needs to be connected to the bath solution via an agar bridge. In the pipette, only the first 1 cm should be filled up with Cl^--free solution, and the back of the pipette with a solution containing Cl^-, so that the silver wire electrode can contact Cl^- (11). As an alternative, regular Cl^--containing solution can be used for the pipette solution, and Cl^- free for the bath solution. Under those conditions, measuring at around −100 mV, Cl^- currents are negligible, whereas at positive potentials the Cl^- current can be monitored. Since gluconate also permeates the Cl^- channel to some extent, it is best to titrate out the negative exact voltage where Cl^- currents reverse, in non-injected oocytes applying micromolar Ca^{2+} concentrations to activate the Cl^- current; for us −103 mV worked well (7). This protocol works well with inwardly rectifying channels, such as TRPV6 (7).

4. *Solutions.* Our solutions are isotonic with frog body fluids (~96 mM cations), but mammalian solutions (~140 mM cations) are also used by many people, and since oocytes do not express aquaporins, the osmotic difference usually does not cause significant problems.

5. We usually use pipette openings between 5 and 15 μm the typical being around 8–12 μm. These are considerably smaller pipettes that those described for macropatch measurements with transporters (2). The pipette size matters, but it is usually not critical; generally, the smaller ones are easier to seal with; the larger ones have more current. The smaller pipettes form vesicles more often after excision (see Note 7). We fire-polish the two ends before pulling the pipettes to avoid damage to the pipette holder. We find that it is generally better to use freshly pulled pipettes. Pipettes pulled the day before the experiment still usually work well, but ones pulled several days before usually do not. We do not use a microforge to fire-polish the tip of the pipettes after pulling.

6. We use a P-97 Flaming Brown puller from Sutter instruments. The website of Sutter contains protocols to generate stable pipettes and to make larger pipettes. Generally, it takes a lot longer to pull a macropatch pipette than a regular patch pipette. This usually makes the puller block to heat up, and the pipette tips become smaller and smaller upon sequential pullings. This can be overcome either by waiting a couple of minutes between pulling individual pipettes or, for the less patient ones, to gradually decrease the heat level of the last pulling step to keep the pipette size within the required range. See Fig. 1 for a typical pipette.
7. *Vesicle formation* (see Fig. 2). After establishing the inside-out configuration, sometimes the patch reseals, and forms a vesicle on the tip of the pipette. This can be noticed by the immediate disappearance of current activity. The vesicle can be broken either by gently "brushing" the pipette tip on the surface of

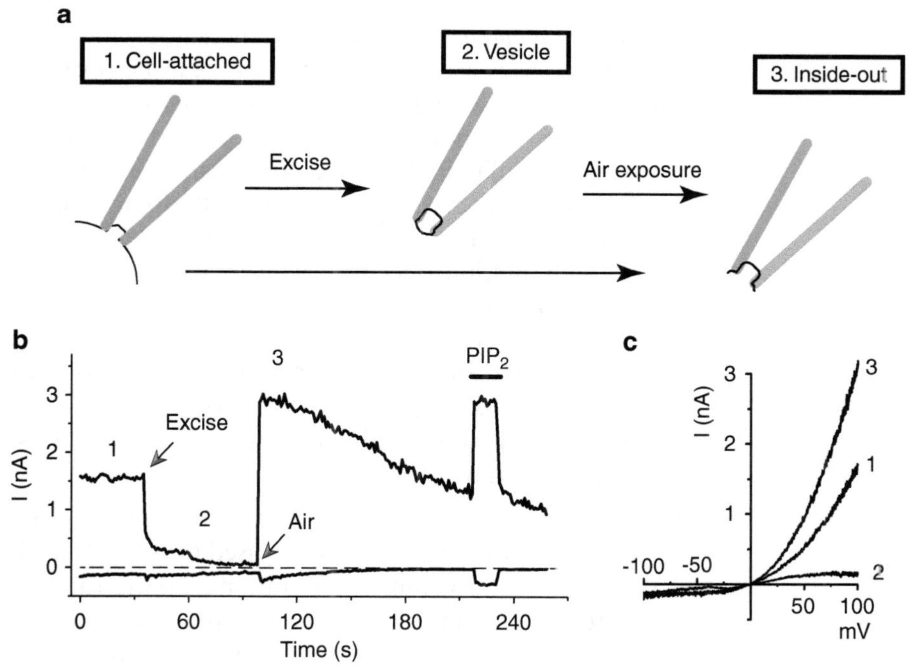

Fig. 2 (**a**) Cartoon depicting the patch pipette (*thick grey lines*), and the cell-attached configuration (1), formation of a vesicle after patch excision (2), and formation of an inside-out patch after a brief air exposure (3). Note that most patches do not form a vesicle, but rather go straight into the inside-out configuration; see *long arrow*. (**b**) Time course of a measurement with TRPM8 channels with 500 μM menthol in the patch pipette, current traces are shown at −100 and +100 mV from a voltage ramp from −100 to +100 mV. After excision, the low current levels correspond to a vesicle, which was broken by a brief exposure to an air bubble, generated with a P20 pipette tip next to the patch pipette. The slow current decrease after the inside-out configuration is due to the loss of phosphatidylinositol 4,5-bisphophate (PIP_2) from the patch membrane, which is required for TRPM8 activity. The application of 50 μM diC_8-PIP_2 is shown by the *horizontal line*

the oocyte or by briefly applying air to the vesicle. We find that pulling the pipette out then lowering it back to the solution usually destroys the seal. In our hands the best way is to use a P2 or P20 pipettor, move the end of the pipette tip very close to the tip of the patch clamp electrode, and pipette out a small air bubble touching the electrode tip with the air bubble for as briefly as possible. If the vesicle is destroyed, but the seal remained stable, current activity similar to that observed in the cell-attached configuration will reappear. Sometimes this procedure destroys the seal. We found that the tendency to form a vesicle depends on the batch of oocytes, on most days it never happens, and on some days most attempt to form an inside-out patch results in a vesicle. In the latter case it is worth trying a larger pipette.

8. The frog is placed on its back, on ice; lack of movement signifies deep enough anesthesia for operation. Low temperature itself is usually enough to anesthetize a frog, but today it is considered unacceptable to use solely cold for this purpose. Nevertheless keeping the frog on ice helps maintain anesthesia.
9. *Number of operations on a frog*. Generally, if only a small number of oocytes are removed, the frog can be operated almost infinite times on alternating sides. One month is generally considered to be enough recovery time for the next operation, which is usually done on the other side. For a frog that gives good-quality oocytes, it might be worth doing repetitive operations many times. In practice however, oocyte quality deteriorates with multiple operations in most cases, especially if a large number of oocytes are taken. In practice we cannot recommend more than four operations, and local animal ethics bodies are usually against more than four operations. It is reasonable alternative to sacrifice the frog every time we collect eggs, which increases the number of frogs used, but reduces the suffering of the individual frogs.
10. *Overnight vs. acute digestion*. Higher concentrations of collagenase (1–2 mg/mL) can also be used for ~1 h. Usually we use somewhat lower volumes, ~35 mL in a 50 mL tube and rotate the tube perpendicular to its axis to evoke more vigorous mechanical agitation. This makes the procedure a lot faster, finished on the same day, but in our experience it is easy to over-digest the eggs, and it is usually necessary to monitor the digestion every 5–10 min after the first 30 min. This makes the overall hands-on time longer and the results, in our experience, more variable.
11. The additional shaking step in OR2(−) after collagenase digestion is not always necessary. If the oocytes are slightly under-digested, it helps. If they are thoroughly digested it is not necessary, if over-digested it can even be harmful.

12. It is advisable to use a vector designed to obtain optimal expression in oocytes. To our knowledge none of these vectors are commercially available, but many of them are in circulation by different labs. Vectors include the pGEMHE (12), and its derivative pGEMSH, pBF (13), pEXO (14), pGH-19 (15), and pBSTA (16). Essentially all of them have the 5′ and 3′ untranslated regions of the *Xenopus* globin gene that gives stability to the RNA, and a long poly A tail or polyadenylation signal. What usually varies between the different vectors is the multiple cloning site, and the linearization site, which is after the poly A tail. It often helps remove the original 5′ and 3′ untranslated regions from the coding sequence of the ion channel cDNA before subcloning it into the oocyte vector. Most oocyte vectors use the T7 polymerase but there are some other variants to that use either T3 or the SP6 to make RNA. Kits are available from Ambion for all three variants. Using the wrong primer is a frequent cause for not getting any RNA.
13. The expression of a given ion channel in the membrane depends on the distance from the site of injection, the further away the injection site is, the less channel activity is expected in the membrane patch. To identify the injection site, we generally inject in the middle of the black hemisphere, and try to patch close to the injection site.
14. *Variability*. Expression levels between individual oocytes can be highly variable. This is often frustrating when seal after seal the patch does not have enough current do an experiment. Efficiency can be greatly enhanced by prescreening the oocytes with good expression levels using a two-electrode voltage clamp setup (see Chapter 6 of this book); this can be done the day before the experiment. Most oocytes survive well till the next day despite the small damage created by the electrodes.
15. The amount of RNA required to obtain good expression of a given ion channel is highly variable among different ion channels, just like the time required to have good expression. Kir2.1 channels for example require less than one ng of RNA, and express well the next day. Many TRP channels require 2–3 days to express, and 10–50 of ng-s of RNA. On the extreme end, the mechanosensitive piezo1 channels require at least 25 ng of RNA and at least 3–4 days to express. Generally, the larger the protein the more time it takes to express it.
16. The vitelline layer is a glycoprotein matrix that gives some structural rigidity to the oocyte. It is permeable for small molecules, and easily penetrated by sharp electrodes used in two-electrode voltage clamp experiments; thus it does not have to be removed for those measurements. For patch clamp measurements, however, the vitelline layer needs to be removed

from the oocyte, to gain access to the plasma membrane. We do this usually by poking a hole in the vitelline layer with a very sharp forceps, and then grabbing the vitelline layer surrounding the hole with two forceps and gently pulling it apart. With a couple of hours of practice this can be achieved within less than a minute, and we find this procedure to be more efficient on the long run than shrinking the oocyte with a hyperosmotic solution to separate the vitelline layer from the plasma membrane.

17. *Room temperature.* It is advisable to keep the room temperature at 20°C or below for patch clamp experiments, 18–19°C being optimal. Room temperatures above 22–23°C usually make patching very difficult, often impossible. This often requires installing extra air-conditioning units, especially in the summer months.
18. Air bubbles in the perfusion system are a common problem. The most important thing to minimize them is to keep the perfusion solution at room temperature overnight before the experiment. Solutions straight from the cold room will have a lot more bubbles after warming up than solutions equilibrated with room temperature longer. If expensive compounds are used, low dead space is critical, which can be achieved using thin tubes, which get clogged very easily by small air bubbles. We usually apply pressure to all tubes for a second or two to make sure they all are flowing, right after establishing the excised patch configuration, but before moving the perfusion pencil close to the patch pipette. Doing this in the cell-attached mode is not advisable, since the movement of the table might destroy the seal.
19. The oocyte is a large cell, and for two-electrode voltage clamp, the equivalent of whole-cell configuration, we do not use vibration isolation tables. The vibration isolation table however is absolutely critical for seal formation in patch clamp measurements, until the excision. If seals are unstable and break easily, it is worth checking if the floating table is working properly, i.e., none of the corners are deflated and touch the bottom.

References

1. Hilgemann DW (1990) Regulation and deregulation of cardiac Na^{+}/Ca^{2+} exchange in giant excised sarcolemmal membrane patches. Nature 344:242–245
2. Hilgemann DW (1995) The giant membrane patch. In: Sakmann B, Neher E (eds) Single channel recording, 2nd edn. Kluwer Academic, pp 307–327
3. Lukacs V, Thyagarajan B, Varnai P, Balla A, Balla T, Rohacs T (2007) Dual regulation of TRPV1 by phosphoinositides. J Neurosci 27:7070–7080
4. Rohacs T, Lopes C, Mirshahi T, Jin T, Zhang H, Logothetis DE (2002) Assaying phosphatidylinositol bisphosphate regulation of potassium channels. Methods Enzymol 345:71–92

5. Rohacs T, Lopes CM, Jin T, Ramdya PP, Molnár Z, Logothetis DE (2003) Specificity of activation by phosphoinositides determines lipid regulation of Kir channels. Proc Natl Acad Sci U S A 100:745–750
6. Rohacs T, Lopes CM, Michailidis I, Logothetis DE (2005) PI(4,5)P_2 regulates the activation and desensitization of TRPM8 channels through the TRP domain. Nat Neurosci 8:626–634
7. Thyagarajan B, Lukacs V, Rohacs T (2008) Hydrolysis of phosphatidylinositol 4,5-bisphosphate mediates calcium-induced inactivation of TRPV6 channels. J Biol Chem 283:14980–14987
8. Zakharian E, Cao C, Rohacs T (2011) Intracellular ATP supports TRPV6 activity via lipid kinases and the generation of PtdIns(4,5)P_2. FASEB J 25:3915–3928
9. Barnard EA, Miledi R, Sumikawa K (1982) Translation of exogenous messenger RNA coding for nicotinic acetylcholine receptors produces functional receptors in Xenopus oocytes. Proc R Soc Lond B Biol Sci 215:241–246
10. Smith LD, Xu WL, Varnold RL (1991) Oogenesis and oocyte isolation. Methods Cell Biol 36:45–60
11. Csanady L, Torocsik B (2009) Four Ca^{2+} ions activate TRPM2 channels by binding in deep crevices near the pore but intracellularly of the gate. J Gen Physiol 133:189203
12. Liman ER, Tytgat J, Hess P (1992) Subunit stoichiometry of a mammalian K^+ channel determined by construction of multimeric cDNAs. Neuron 9:861–871
13. Baukrowitz T, Tucker SJ, Schulte U, Benndorf K, Ruppersberg JP, Fakler B (1999) Inward rectification in KATP channels: a pH switch in the pore. EMBO J 18:847–853
14. Fink M, Lesage F, Duprat F, Heurteaux C, Reyes R, Fosset M, Lazdunski M (1998) A neuronal two P domain K^+ channel stimulated by arachidonic acid and polyunsaturated fatty acids. EMBO J 17:3297–3308
15. Venkatachalan SP, Bushman JD, Mercado JL, Sancar F, Christopherson KR, Boileau AJ (2007) Optimized expression vector for ion channel studies in Xenopus oocytes and mammalian cells using alfalfa mosaic virus. Pflugers Arch 454:155–163
16. Shih TM, Smith RD, Toro L, Goldin AL (1998) High-level expression and detection of ion channels in Xenopus oocytes Methods Enzymol 293:529–556

Chapter 10

Combined Single-Channel and Macroscopic Recording Techniques to Analyze Gating Mechanisms of the Large Conductance Ca^{2+} and Voltage Activated (BK) Potassium Channel

Nguyen V. Nguyen, Aleksandra Gruslova, Wojciech A. Kosiba, and Bin Wang

Abstract

Ion channels are integral membrane proteins that regulate membrane potentials and signaling of cells in response to various stimuli. The patch-clamp technique enables the study of single channels or a population of channels. The macroscopic recording approaches are powerful in revealing population-averaged behaviors of channels both under basal conditions and in response to various stimuli, modulators and drugs. On their own, however, these approaches can be insufficient for determinations of channel gating mechanisms as they do not accurately report channel open probabilities below 10^{-2} to 10^{-3}. This obstacle can be overcome with the use of single-channel recording techniques. Single-channel recording techniques can be applied to one or a few channels to estimate P_o over a larger range than macroscopic recordings. The combination of heterologous overexpression of ion channels with macroscopic and single-channel recordings can be applied to hundreds of channels to estimate P_o between 1 and 10^{-8}. Here, we describe practical approaches of single-channel recordings that our laboratory utilizes. We also provide examples where the combined macroscopic and single channel approach can be employed to study gating mechanisms of the BK type, large conductance, Ca^{2+} and voltage activated potassium channel in a mammalian expression system. The techniques presented should be generally applicable to the studies of ion channels in heterologous expression systems.

Key words Patch clamp, Single channel recording, Heterologous expression, Ion channel, BK, Mammalian expression system, Transient transfection

1 Introduction

Ion channels are integral membrane proteins that open or close transmembrane ion conduction pathways in response to various stimuli such as changes of temperature, pH, membrane potential, or ligand binding. By increasing or reducing membrane permeabilities to select ions, ion channels regulate membrane potentials and affect

Nikita Gamper (ed.), *Ion Channels: Methods and Protocols*, Methods in Molecular Biology, vol. 998,
DOI 10.1007/978-1-62703-351-0_10, © Springer Science+Business Media, LLC 2013

signaling of cells. The important role of ion channels in many physiological processes has made ion channel recording techniques an important staple of many laboratories.

Among the techniques that record ion channel activity, various configurations of the so called "patch-clamp technique" allow the recording of either populations of channel openings (macroscopic currents) or single channel openings. Both approaches offer advantages. Recordings of macroscopic currents offer insight into the cellular and average behavior of ion channels that reveal the magnitude of ionic currents and population response properties. Single-channel recordings directly reveal individual ion channel behavior that underlies the macroscopic ion currents. Properties such as single-channel conductance, number of conformational states and rate constants, as well as heterogeneity of channels in the patch (both in terms of types and subunit compositions) can be directly evaluated using single-channel recordings (1). A disadvantage of single-channel recording is that analysis is more time consuming.

One application of the single-channel recording technique is to provide accurate estimations of low steady-state open probabilities (P_o < 0.01). Obtaining low P_o data under conditions where certain aspects of gating can be investigated independently is very useful for understanding changes in gating mechanisms. This is in contrast to data from macroscopic recordings such as conductance–voltage (*G–V*) relationships of voltage-gated channels, where they describe channels' steady-state behavior well, but is often not sufficient to achieve mechanistic understanding of channel gating.

Single-channel recordings have been widely used to extend P_o measurements in biophysical analysis of the BK type, large-conductance, Ca^{2+} and voltage-activated potassium channel (2–24). The BK channel gate can open at low probabilities in the absence of voltage-sensor activation and Ca^{2+} binding (intrinsic gating) (3, 25). Voltage-sensor activation and Ca^{2+} binding increase channel opening by allosteric coupling between the sensors and the gate (3, 25). Effects of modulators on these three aspects of BK channel gating: intrinsic-, voltage-, or Ca^{2+}-dependent gating, are indistinguishable using *G–V* analysis alone. Simulations using a BK channel gating model (3) demonstrate this point (Fig. 1a). A tenfold increase in intrinsic gating and a −34 mV shift in voltage-sensor activation cause almost identical changes in the *G–V* relation. However, these differences can be distinguished by recording under conditions where gating transitions are studied in relative isolation (Fig. 1b). For example, intrinsic gating of BK channels can be studied by combining macroscopic and single-channel recording in the absence of Ca^{2+} (channels are ligand-unbound) and at very negative voltages (where voltage sensors are forced into resting states) (3). Similarly, Ca^{2+}-dependent gating can be studied independent of voltage-sensor activation at very negative voltages over a wide range of Ca^{2+} (3). Under these conditions, P_o values are often too

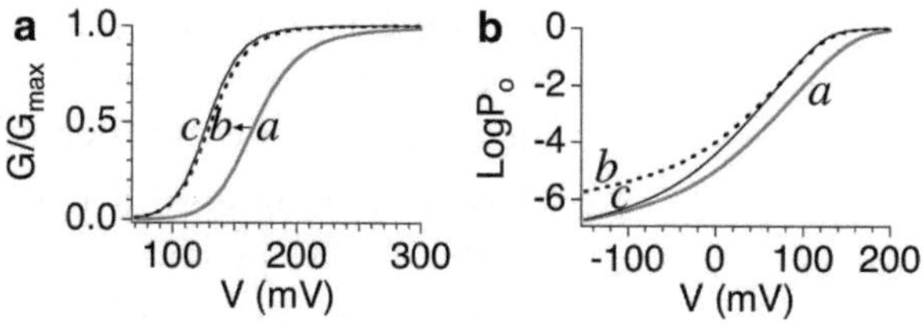

Fig. 1 Macroscopic and G–V relation alone do not distinguish unique gating changes. (**a**) Simulated 0 Ca^{2+} *G–V* relations based on three sets of gating parameters *a*, *b*, and *c*. Relative to *a* (*grey curve*), *b* (*dashed curve*) represents a tenfold increase of intrinsic gating and *c* (*solid curve*) represents a −34 mV shift of the half-activation voltage of Q_0 (Vh_0 (3). (**b**) Simulated 0 Ca^{2+} Log P_0–*V* relations based on the same sets of gating parameters. Log P_0–*V* but not *G–V* relations distinguish *b* and *c*

low to be accurately estimated using macroscopic recordings alone and thus require single channel recordings.

Here we describe an approach that combines single-channel and macroscopic recording techniques to extend measurement of BK channel P_0 over a wide range of Ca^{2+} and voltages. Heterologous expression of up to thousands of channels in a patch is used to extend the range of P_0 measurements. Macroscopic recordings are used to estimate (a) P_0 between 0.01 and 1 and (b) numbers of channel in a patch (*N*). Single channel recordings are used to estimate P_0 below 0.01. Using this method, one can estimate P_0 between 10^{-8} and 1. The application of this approach is not limited to BK channels. For example, similar approaches have been used to characterize shaker potassium channel and sodium channels in the past (26, 27).

2 Materials

2.1 Transient-Expression of BK Channels in HEK-293 Cells

1. HEK-293T cells (HEK 293T/17, American Tissue Culture Collection, Manassas, VA).
2. Dulbecco's modified Eagle's medium (DMEM).
3. Fetal bovine serum.
4. Penicillin–streptomycin.
5. L-Glutamine.
6. 0.01% Trypsin/0.1 mM EDTA solution.
7. T25 (25 cm^2) cell culture flasks with filter screw.
8. 35 mm tissue culture dishes.
9. Phosphate buffered saline (PBS) without $CaCl_2$.
10. Lipofectamine (Life Technologies Corporation, Grand Island, NY, USA).

11. pEGFP-C1 (Clontech, Mountain View, CA, USA) or any other fluorescent protein-expressing plasmid to enable the identification of cotransfected cells.
12. Plasmids containing BK Channel subunit cDNAs downstream of the CMV promoter. Mouse BK α subunit (GenBank: U09383.1) was subcloned in pCDNA3.1Hygro(+) (Life Technologies Corporation, Grand Island, NY 14072, USA) (15). Mouse BK β1 subunit (GenBank: EDL23733.1) was subcloned in pIRES2-EGFP (Clontech, Mountain View, CA, USA).
13. 25 mm diameter, #0 glass coverslip (e.g., PLATINUM LINE manufactured by Knittel Glaeser, Germany).

2.2 Analysis of Recombinant BK Channel Using the Patch-Clamp Method

1. Patch clamp amplifier: We use an EPC8 amplifier (HEKA Elektronik, Pfalz, Germany), a 16-bit A/D converter (ITC18/USB-18, HEKA Elektronik, Pfalz, Germany), a Macintosh-based computer system in combination with the Patchmaster acquisition software (HEKA Elektronik, Pfalz, Germany). Alternative patch clamp equipment and acquisition software can be used as appropriate (see Chapter 6 of this book).
2. Micromanipulator: We use a MHW-3 three-axis water hydraulic micromanipulator (Narishige, Japan).
3. Micro-vibration isolation table with Faraday cage floating on outlet air pressure (~150 psi).
4. Inverted microscope: We use a TE300 microscope (Nikon, Melville, NY, USA) fitted with excitation and emission filters for EGFP (FITC/RSGFP LP emission: #21012, Chroma Tech. Corp., Rockingham, VT, USA).
5. Micropipette puller: We use a Model P-87 flaming/brown micropipette puller (Sutter Instr. Co., Novato, CA, USA).
6. Homemade setup for dipping pipettes into melted wax (Fig. 2a) (Sticky Wax, Kerr Inc, Orange, CA, USA).
7. Microforge: We use a homemade device assembled with a Labophot-2 microscope with 40× objective (Nikon, Japan), a small micromanipulator (Narishige, Japan), a micropipette slide (WPI, Sarasota, FL, USA), and a medium gauge heating filament (WPI, Sarasota, FL, USA) (Fig. 2b).
8. Borosilicate glass capillaries (Sutter Instrument, Novato, CA, USA) pulled and forged to 1–3 mΩ for excised patch recording.
9. Fast exchange recording chamber (e.g., Model RC-22, Warner Instruments, Hamden, CT, USA).
10. Perfusion system: We use a six-channel perfusion valve control system (Catalog #VC-6, Warner Instruments, Hamden, CT, USA).

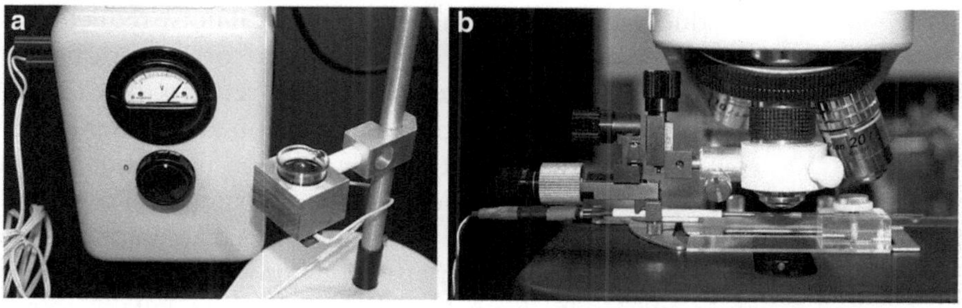

Fig. 2 Homemade sticky wax coating device and microforge. (**a**) A power transformer is connected to an aluminum block with 30 Ω 25 W resistor (Dale) as a heating element. The aluminum block has been machined to fit a small pyrex beaker for melting of wax. (**b**) A micromanipulator is mounted onto a 40× Nikon objective. The micromanipulator places the heating filament within the focal plane of the objective. A glass electrode holder is mounted to the stage in order to position the glass pipette tip in the proximity of the heating filament and within the objective focal plane

11. Diamond-tipped scribe (Bioindustrial Products, San Antonio, TX).
12. Internal Recording solution: a pH 7.2 solution of 20 mM HEPES, 140 mM $KMeSO_3$, 2 mM KCl, and buffered with appropriate calcium chelators. Appropriate ratios of chelators and calcium can be calculated using the Maxchelator program (http://www.stanford.edu/~cpatton/webmaxcS.htm (Stanford University)). Ba^{2+} was chelated with 40 μM (+)-18-crown-6-tetracarboxylic acid (28).
13. External Recording solution: 20 mM HEPES, 140 mM $KMeSO_3$, 2 mM KCl, 2 mM $MgCl_2$, pH 7.2.

3 Methods

Expression and analysis of BK channels in human embryonic kidney (HEK293T) cells is a three-step process. First, cDNAs encoding channel subunit(s) are transiently transfected into HEK293T cells to express BK channels. Second, channel activities are assayed using standard patch-clamp electrophysiology techniques that include both macroscopic and single-channel recordings. Third, macroscopic and single-channel records are analyzed offline.

To obtain high quality single-channel recordings, it is essential to carefully and consistently maintain the cultured cells. Healthy, rapidly dividing cells have the highest transfection efficiency. Moreover, high resistance and stable recordings at "extreme conditions" (e.g., 0 Ca^{2+}, very negative voltages) require healthy membranes. Finally, analysis is facilitated by stable, low noise recordings.

The techniques for inside-out patch-clamp recording of BK channel are standard. One aspect that is important for single-channel recordings, particularly of channels with small single-channel conductance and fast kinetics, is electrical noise reduction. Readers interested in details of denoising and other various aspects of patch-clamp recording techniques can refer to the excellent book, "Single-channel recording," by Sakmann and Neher (1).

An important aspect of P_o estimation based on NP_o data is to correctly estimate N, the number of channels in a patch. A relatively large N can be estimated by macroscopic recordings: $N = \frac{I}{I_s \times P_o}$ (P_o, I, and I_s are the open probability, tail current amplitude, and single-channel amplitude at a given voltage, respectively). For BK channels, we use a voltage that evokes maximum opening ($I = I_{max}$). At I_{max}, P_o is close to 1 for BK channels (BK/α channels at almost all Ca^{2+} (28), and BK/αβ1 and BK/αβ4 channels at high Ca^{2+} (12, 13)) and the above equation can be simplified to $N = \frac{I_{max}}{I_s}$.

The power of single-channel recording techniques extends beyond applications described here. Readers interested in using single-channel recording techniques to study channels' conformations and rate constants are strongly encouraged to read Chapters 18–19 of the "Single-channel recording" book (1).

3.1 Expression of BK Channels in HEK-293T Cells (See Note 1)

1. HEK293T cells are thawed and passaged using sterile techniques in a tissue culture hood with sterile solutions and supplies.
2. HEK-293T cells are maintained in T25 tissue culture flasks in a humidified tissue culture incubator at 37 °C and 5% CO_2. Each flask contains 5 ml of DMEM complete media (supplemented with 10% FBS, 2 mM L-glutamine and 5% streptomycin/penicillin).
3. Cell cultures are passaged every 2–3 days to maintain a confluence between 10 and 85% (see Note 2).
4. Passage of HEK cells: Remove media and wash cells twice with PBS (without Ca^{2+} or Mg^{2+}). Add 0.5 ml of trypsin/EDTA (the minimum volume to cover the entire bottom of the flask) to release the cells from the flask. Within a minute, tap the dish against the palm of your hand to dislodge the cells for about 30 s. Inactivate trypsin with 6 ml of tissue culture media and gently pipette the cell/media suspension to obtain a largely single-cell suspension. Redistribute a fraction of cells into a new T25 culture flask to maintain the stock (a 1:4 to 1:6 split will generally achieve a 15–20% confluency). For transfection, cells are placed into a 35 mm tissue culture dish at approximately 40% confluence (see Note 3). The goal is to obtain ~80–90% confluence the following day for transfection.
5. Prepare DNA mix for transfection: For BK channels composed of only the pore-forming α subunit, 0.25 μg plasmid cDNA encoding the α subunit and 0.25 μg of the pEGFP-C1 plasmid encoding the EGFP marker are used. For BK channels composed of pore-forming α subunit and the auxiliary β1 subunit, 0.25 μg plasmid cDNA encoding the α subunit and 1.5 μg of

the pIRES2-EGFP- β1 plasmid encoding both the β1 subunit and the EGFP marker are used. The excess β1 to α ratio ensures that transfected cells uniformly express the BK channels with saturating β1 subunits.

6. Prepare DNA and lipofectamine mix: For each transfection, prepare 5 ml of transfection media (DMEM with 10 mM HEPES, see Note 4). In a 1.5 ml microfuge tube, dilute DNA into 100 μl transfection media, mix by pipetting 2–3 times. In another 1.5 ml microfuge tube, dilute 10 μl lipofectamine into 100 μl transfection media, mix by pipetting 2–3 times. Combine diluted lipofectamine and DNA, and mix by gentle tapping of the tube. Incubate at room temperature for ~30 min.
7. Transfection: Cells are rinsed once with ~2 ml transfection media (DMEM/HEPES) and covered with 0.8 ml DMEM/HEPES. 0.2 ml DNA–lipofectamine mixture (see Note 5) is dropped onto cells for a final volume of 1 ml. The solution is mixed with a slow swirling motion so as not to dislodge attached cells.
8. Transfer cells to glass coverslips: Two to four hours after transfection (see Note 6) the transfection reagent is removed, cells are rinsed with PBS, and trypsinized using 0.1 ml 0.01% Trypsin at room temperature. Dishes are gently tapped to facilitate release and dislodge of cells. 3 ml culture media are then added followed by gentle pipetting to achieve single cell suspension. Place a piece of 25 mm autoclaved cover glass (see Note 7) each into two 35 mm tissue culture dishes. One dish receives 2 ml of transfected cells for analysis 24 h post-transfection. The other dish receives 1 ml of transfected cells and 1 ml of culture media for analysis 48 h post-transfection.

3.2 Analysis of Recombinant BK Channels

1. Analysis of BK channels expressed in HEK cells employs standard inside-out patch-clamp methods (1).
2. Pipettes are pulled to the tip resistance of 1–3 mΩ using a predetermined program (see Note 8).
3. Coating pipettes: To reduce the capacitance and noise associated with pipettes, we coat the pipettes with sticky wax (see Note 9). Pipettes are dipped into melted wax to coat the outer wall (Fig. 2a). To prevent wax from filling the tip of the pipette, air is blown out of the tip via a compressed air tubing attached to the back of the pipette. Pipettes are fire polished on a homemade microforge (Fig. 2b) (see Note 10). Small amounts of wax at the tip of pipettes will melt and recede from the tip during the polishing process.
4. A culture dish containing transfected cells is removed from the tissue culture incubator. A small chip of the cover glass containing transfected HEK cells is cut out with a diamond-tipped scribe.

Fine forceps are convenient for handling of the glass chip. Before placed into the perfusion chamber containing the internal solution, the glass chip is dipped into a dish containing the same internal solution to wash off residual culture media (see Note 11).

5. Transfected cells are targeted for patch clamp recording based on EGFP fluorescence and health of cells. Healthy HEK cells are those with smooth, shiny edges.
6. A wax coated and polished glass pipette is used for patch clamp recording. After compensating offsets, the patch pipette is lowered onto the EGFP positive cell to form a high resistance cell-attached seal (see Notes 12 and 13). Pipette fast capacitance is compensated. For excised, inside-out patches, the pipette in the on-cell configuration is quickly withdrawn from the cell allowing the membrane to be excised.
7. After establishing the inside-out seal, a voltage ramp protocol (−150 mV to +150 mV over a 200 ms period) is used to roughly estimate the number of channels in a patch (see below; Fig. $3a_1$). Depending on the number of channels in a patch, macroscopic or single-channel recordings will be performed. Data are sampled at 10 μs intervals and low-pass filtered at 8.4 kHz using the HEKA EPC8 four-pole Bessel filter. Small single-channel currents (e.g., between −20 and +20 mV) are filtered at 1–3 kHz.
8. When there are greater than five channels in a patch, N is estimated by macroscopic recordings. BK currents are elicited by a series of depolarizing test potentials (Fig. $3a_2$). Instantaneous tail current amplitudes (200 μs after repolarization to −80 mV) are plotted and fitted by the Boltzmann function to estimate the maximum tail current amplitude (I_{max}) (Fig. $3a_3$). N is estimated by dividing I_{max} by the single-channel amplitude (I_s) at −80 mV (18 pA) (Fig. $3a_3$).
9. When there are only a few channels in the patch ($N \leq 5$), N is estimated using all-point amplitude histogram of single-channel records over a range of voltages (Fig. 3b). The maximum amplitude is tentatively estimated at depolarizing voltages, where P_o is near 1 (e.g., +40 mV, Fig. $3b_1$). The size of leak current is estimated recording at negative voltages (e.g., −80 mV, Fig. $3b_2$). Together, these data suggest $N = 5$ channels in the patch shown in Fig. 3b. Finally, inspection of single-channel records alone can be sufficient to determine N when N is sufficiently small (Fig. 4a).
10. Once N is estimated, single-channel recordings are carried out under conditions where single opening events can be resolved. The sweep durations vary depending on channels' open probabilities at a particular voltage and Ca^{2+} concentration (see Note 14).

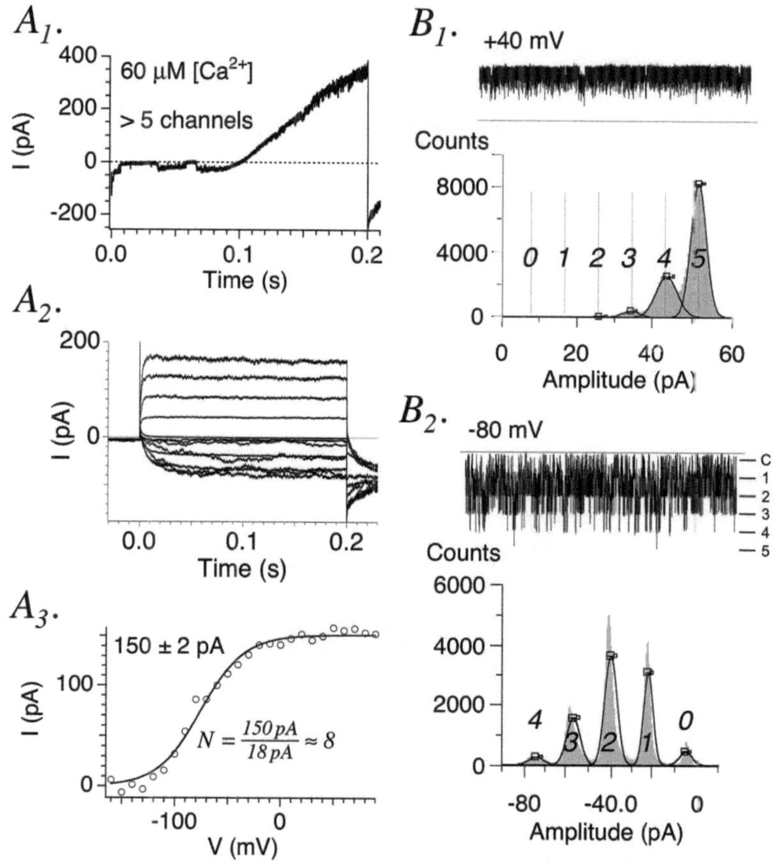

Fig. 3 Determining channel number (*N*) using single-channel or macroscopic recordings. (**a**) Counting *N* of a patch using macroscopic recordings. (**a1**) A BK current elicited with the voltage ramp protocol suggests multiple channels in the patch. (**a2**) BK currents elicited with a voltage step protocol. (**a3**) Fitting the *I*–*V* relation to the Boltzmann function shows that at I_{max}, the maximum tail current amplitude (obtained at −80 mV tail current voltage) is 150 pA. Dividing the I_{max} by the single-channel current amplitude at tail current voltage (18 pA) provides an estimation of *N*. (**b**) Counting *N* in a patch using single-channel recordings. (**b1**) A single-channel recording trace at +40 mV (*top*) and analysis with all-point histogram (*bottom*). The analysis suggests a maximum of five channels in the patch, if leak current is relatively small. (**b2**) Same patch as in (**b1**), but the reduced P_o at a negative command voltage (−80 mV) makes the individual openings easier to discern. The analysis suggests that the leak current is relatively small, and therefore there are indeed five channels in the patch

11. We use TAC and TACFit programs (V4.1, Bruxton Corporations, Seattle, WA, USA) for data analysis (see Note 15). Pulse data (HEKA Elektronik, Pfalz, Germany) can be opened by TAC directly and Patchmaster data will need to be exported as PULSE v8.6. We use two methods: all-point amplitude histogram and event detection to estimate NP_o.

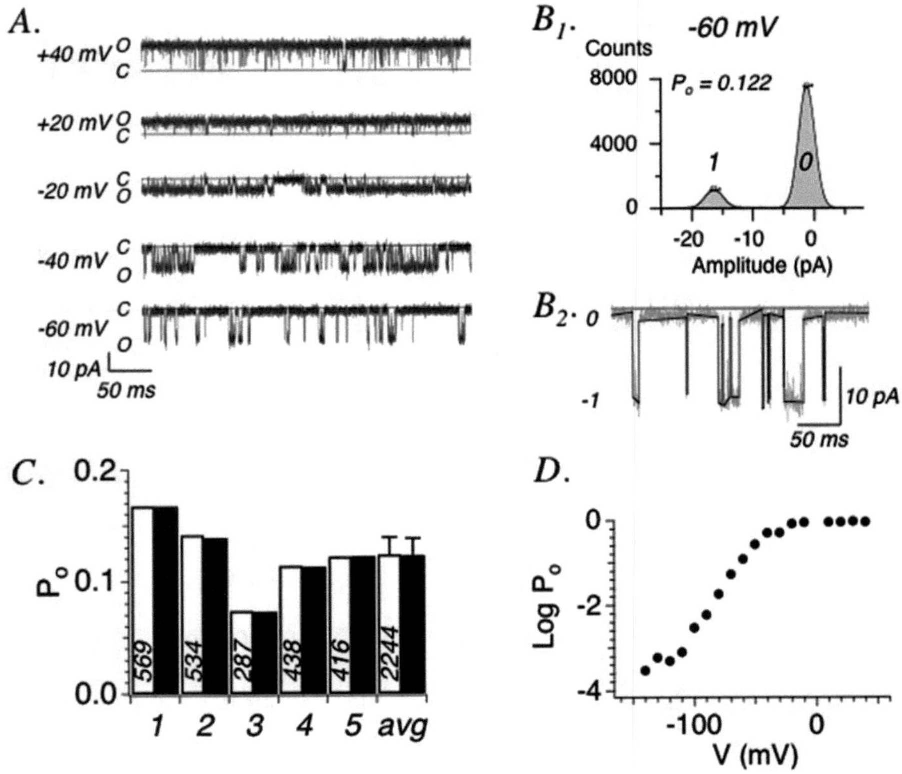

Fig. 4 Analysis of NP_0 using all-point amplitude histogram or event detection. (**a**) Selected single channel records of a patch at indicated voltages. There is one channel in the patch and currents between −60 and −140 mV are not shown. (**b1**) Analysis of a −60 mV sweep using the all-point amplitude histogram. (**b2**) Analysis of the same −60 mV sweep using event detection (only a fraction of the analysis is shown). Raw data are shown in *grey*, idealized events are shown in *black*. (**c**) Estimated P_0 values (at −60 mV) using event detection (*empty bars*) or all-point amplitude histogram (*solid bars*) from five sweeps. Event numbers are indicated in the *empty bars*. Averaged P_0 and SEM are shown on the *right*. (**d**) LogP_0–V data obtained from the single-channel recording

12. All-point amplitude histogram analysis (1): In this type of analysis, distribution of digitized current amplitude values is plotted as a histogram. Peaks represent either the single shut level or open levels. The Gaussian fit of area under each peak provides an estimate of proportional time spent at that level. Below is the simple procedure for estimation of NP_0 by performing all-point amplitude histogram analysis in the program TAC.

    The data is opened in the TAC program and a sweep is selected for analysis without further filtering. The TAC software generates an all-point amplitude histogram of the data when "sweep data histogram" is selected under "view" in TAC (Fig. 4b$_1$). Based on the number of discernable channel states, we set initial fit components manually by double clicking on the *X*-axis (which provides a manual fitting tool). We adjust the peak amplitude and width of each fit component to roughly fit each level. Following the adjustments, we have the TAC program provide a more precise fit of the data by selecting "Automatic Fit" under "Sweep Histogram" (see Note 16). We inspect the fits to make sure that they are reasonable. Otherwise, each fit component of the "Sweep Histogram" has to be readjusted and "Automatic Fit" has to be attempted again. Fit parameters are provided by the software (displayed in "Sweep Histogram Fit Parameters") that includes the individual current amplitudes and probability (weight). From these, we calculate NP_o based on the weight of each channel level. $NP_o = \sum_{i=1}^{N} i \times P_i$ (P_i is the probability that i channels open simultaneously). The advantage of all-point amplitude histogram is that it is simple and rapid.

13. Event detection (1): We use event detection if (a) channel openings are infrequent, (b) drift occurs in the baseline or (c) in rare occasions, there is an endogenous channel in the patch. Under either condition, all-point amplitude histogram does not provide accurate estimation of NP_o. The event detection approach detects current transitions between two adjacent levels and measures the time spent between all transitions. In most cases, we detect events automatically, followed by visual inspections to remove false transitions. Below is the simple procedure for estimation of NP_o by performing event detection analysis in the program "TAC."

    (a) TAC provides optional criteria for detecting events: The window for setting detection criteria is in "Data," "Setting." We always select the "Track events" option which idealizes events using actual amplitudes. (We do not select "Fixed" option which sets a fixed amplitude to idealized events.) We set "Search" and "Noise limit" to be "3" (1). Other parameters are set based on the data. We set "Threshold" to be approximately 50% of the single-channel amplitude for the software to use as basis for detecting transitions. Based on the data, we also set "Rise time," "noise estimate" and "Sweep starting level" to maximize accurate event detection (1). For example, for the −60 mV BK channel recording shown in Fig. 14a, we set "Threshold" to be 7 pA, "Rise time" to be 0.1 ms, "Noise estimate" to be 4 pA, and "Sweep starting level" to be 0. "Sweep starting level" is the first event detected in a sweep. For example, if the sweep begins with a single channel inward current, then a "Sweep starting level" of −1 is used. For a sweep starting with no openings, then 0 is given, etc.

    (b) Inspection and correction of idealized data: Under "View," we open "Events" to identify errors in event detection. We mostly inspect "level" to identify false events. For example, for a single-channel recording at a negative voltage with a single channel in the patch, a level of −2 (indicating two channels) or +1 (indicating an outward current despite an inward driving force) is impossible and indicates an error. We identify the first detected false event in "Data" by selecting it in "Events." We then select the false event in the "Data" (indicated by idealized data trace in a distinct color). Under "Data," "Events," we select "Erase Sweep Remaining." We will then use manual event detection to obtain a reasonable idealized data trace to bypass segments "confusing" to the program before resuming automatic event detection again. We repeat these steps until false events are deleted (see Note 17).

14. After analyzing all relevant sweeps, the events are saved under "File," "Save Events As." The event file is opened in the TACFIT program (Bruxton Corporations). If every sweep of interest includes opening events, under "Settings," "Index," select "Select All." Click "Statistics" under "View." Under "Statistics," "Setting," choose "levels." If some sweeps lack open events, under "Settings," "Index," select "Include" and the number of relevant sweeps. Click "Statistics" under "View." Under "Statistics," "Setting," choose "Relevant Segment."
15. We calculate NP_o based on "level" and "P": $NP_o = \sum_{i=1}^{N} i \times P_i$.
16. When NP_o is relatively high, the all-point amplitude histogram is much less time consuming and relatively accurate. In the example shown in Fig. 10.4 estimated NP_o (at −60 mV) using all-point amplitude histogram and event detection are 0.122 ± 0.02 and 0.124 ± 0.02, respectively (Fig. 10.4c).

4 Notes

1. We choose HEK293T cells over Xenopus oocyte because of relative simplicity of cell culture and transfection. One potential shortcoming of HEK293T cells is that the number of channels in a patch can vary dramatically. However, the EGFP reporter allows some selection of expression levels based on EGFP fluorescence intensity. To better control channel expression having an inducible promoter (such as tetracycline-regulated expression system) in a stable cell line may be advantageous (29). Alternatively, Xenopus oocyte can be used where expression levels are more easily controlled by the amount of RNA injected (29).

   For studying BK channel Ca^{2+} dependent gating, variations in expression levels can be useful. For example, P_o of BK channel α and αβ (e.g., αβ4) channels at high Ca^{2+} (>60 μM) and very negative membrane potentials (where voltage sensors are forced into resting states) is ≈10^{-3}. This can be easily analyzed with one to tens of channels. At nominally 0 Ca^{2+} and very negative membrane potentials, $P_o \leq 10^{-8}$ (3, 13). This needs to be analyzed with hundreds to thousands of channels. One can obtain useful data by adjusting recording conditions (intracellular Ca^{2+} and voltage) based on the number of channels in a patch.

2. It is critical to consistently maintain healthy, proliferating cells for transfection (30). One should make sure that cells are in log phase growth the day of transfection (30). From our experience, high quality single-channel recordings are more readily obtained from healthy cells.

3. The goal here is to maintain most cells in a monolayer, log phase growth. One should avoid overgrowth. In addition, under- and over-trypsinization both lead to clustering of cells that will not grow as a monolayer.

4. We found that HEPES stabilizes the pH of the transfection media during lipofectamine incubation, which largely reduces cell death.

5. In addition to Lipofectamine, Polyfect and Superfect reagents (Qiagen, Valencia, CA) (31), FuGen reagent (Roche Applied Science, Indianapolis, IN, USA) (32) and biolistic particle delivery system (gene-gun) (33) have been successfully used by others to introduce exogenous cDNA into mammalian cells. See also Chapter 2 of this book for more information on transfection techniques.

6. The length of transfection can influence the average number of channels in a patch such that the longer the transfection, the higher the channel expression. However, reducing the length of transfection tends to improve cell health and quality of recordings. The number of channels per patch can also be adjusted by varying the amount of DNA used in transfection and pipette diameters.

7. We use PLATINUM LINE coverslips manufactured by Knittel Glaeser, Germany as HEK cells adhere to this specific brand of coverslips without poly-lysine treatment.

8. See "pipette cookbook" for tips on pipette pulling using Sutter Instruments pipette pullers: http://www.sutter.com/contact/faqs/pipette_cookbook.pdf.

9. Pipettes can also be coated with Sylgard® (Sylgard® 184 silicone elastomer kit, Dowcorning, USA) (1) although we find that the wax-dipping approach is less time consuming.

10. Microforges can be purchased from the Narishige Group (Japan) and World Precision Instrument (WPI, Sarasota, FL, USA).

11. We prefer to excise BK channel patches at relatively high Ca^{2+} and subsequently perfuse other Ca^{2+} solutions, as necessary. In our experience, seals are easier to form at higher Ca^{2+}. At high Ca^{2+}, it is also easier to verify that P_o at I_{max} is ~1. Thereby *N*, the number of channels in a patch, can be accurately estimated.
12. Although >1 GΩ is generally sufficient for macroscopic recordings, single-channel recordings generally require >3 GΩ seals, which is not difficult to achieve with healthy cells.
13. We generally start forming seal at 0 mV. However, depolarizing voltages or hyperpolarizing voltages sometimes speed seal formation. Once a seal is formed, we hold the cell at −80 mV. Occasionally, intracellular Ca^{2+} can be high enough to turn on BK channels at 0 mV, which can be mistaken as leaky seals. This is particularly true in the presence of the β1 subunit, which negatively shifts GV at greater than 1 μM Ca^{2+}.
14. Ideally, we like to estimate channel opening with greater than 1,000 events. For α channels, this is achievable under almost all conditions. However, for αβ1 channels, this is not achievable at low Ca^{2+} and negative membrane potentials. In this case, it is necessary to extend the length of recordings and use data with fewer opening events.
15. TAC V4.1 software can be used with Mac OS9. For PC users, there is a newer version of TAC (V4.3.3). Alternative single-channel analysis programs include Fitmaster (HEKA Elektronik, Pfalz, Germany) and Clampfit (Molecular Devices, Sunnyvale, CA, USA).
16. TAC limits the maximum number of levels to be 10.
17. If these procedures do not detect and eliminate all false events, manual event detections should be performed.

References

1. Sakmann B, Neher E (1983) Single-channel recording. Plenum Press, New York, NY
2. Horrigan FT, Cui J, Aldrich RW (1999) Allosteric voltage gating of potassium channels I. Mslo ionic currents in the absence of Ca^{2+}. J Gen Physiol 114:277–304
3. Horrigan FT, Aldrich RW (2002) Coupling between voltage sensor activation, Ca^{2+} binding and channel opening in large conductance (BK) potassium channels. J Gen Physiol 120:267–305
4. Horrigan FT, Heinemann SH, Hoshi T (2005) Heme regulates allosteric activation of the Slo1 BK channel. J Gen Physiol 126:7–21
5. Zhang G, Horrigan FT (2005) Cysteine modification alters voltage- and Ca^{2+}-dependent gating of large conductance (BK) potassium channels. J Gen Physiol 125:213–236
6. Ma Z, Lou XJ, Horrigan FT (2006) Role of charged residues in the S1-S4 voltage sensor of BK channels. J Gen Physiol 127:309–328
7. Horrigan FT, Ma Z (2008) Mg^{2+} enhances voltage sensor/gate coupling in BK channels. J Gen Physiol 131:13–32
8. Ma Z, Wong KY, Horrigan FT (2008) An extracellular Cu^{2+} binding site in the voltage sensor of BK and Shaker potassium channels. J Gen Physiol 131:483–502
9. Bao L, Cox DH (2005) Gating and ionic currents reveal how the BKCa channel's Ca^{2+} sensitivity is enhanced by its (1)1 subunit. J Gen Physiol 126:393–412

10. Sweet TB, Cox DH (2008) Measurements of the BKCa channel's high-affinity Ca^{2+} binding constants: effects of membrane voltage. J Gen Physiol 132:491–505
11. Sweet TB, Cox DH (2009) Measuring the influence of the BKCa (2)1 subunit on Ca^{2+} binding to the BKCa channel. J Gen Physiol 133:139–150
12. Wang B, Brenner R (2006) An S6 mutation in BK channels reveals beta1 subunit effects on intrinsic and voltage-dependent gating. J Gen Physiol 128:731–744
13. Wang B, Rothberg BS, Brenner R (2006) Mechanism of beta4 subunit modulation of BK channels. J Gen Physiol 127:449–465
14. Wang B, Rothberg BS, Brenner R (2009) Mechanism of increased BK channel activation from a channel mutation that causes epilepsy. J Gen Physiol 133:283–294
15. Gruslova A, Semenov I, Wang B (2012) An extracellular domain of the accessory beta1 subunit is required for modulating BK channel voltage sensor and gate. J Gen Physiol 139: 57–67
16. Yang H, Hu L, Shi J, Delaloye K, Horrigan FT, Cui J (2007) Mg^{2+} mediates interaction between the voltage sensor and cytosolic domain to activate BK channels. Proc Natl Acad Sci U S A 104:18270–18275
17. Yang J, Krishnamoorthy G, Saxena A, Zhang G, Shi J, Yang H, Delaloye K, Sept D, Cui J (2010) An epilepsy/dyskinesia-associated mutation enhances BK channel activation by potentiating Ca^{2+} sensing. Neuron 66: 871–883
18. Koval OM, Fan Y, Rothberg BS (2007) A role for the S0 transmembrane segment in voltage-dependent gating of BK channels. J Gen Physiol 129:209–220
19. Hou S, Horrigan FT, Xu R, Heinemann SH, Hoshi T (2009) Comparative effects of H^+ and Ca^{2+} on large-conductance Ca^{2+}- and voltage-gated Slo1 K^+ channels. Channels 3:249–258
20. Hou S, Xu R, Heinemann SH, Hoshi T (2008) The RCK1 high-affinity Ca^{2+} sensor confers carbon monoxide sensitivity to Slo1 BK channels. Proc Natl Acad Sci U S A 105:4039–4043
21. Chen RS, Geng Y, Magleby KL (2011) Mg^{2+} binding to open and closed states can activate BK channels provided that the voltage sensors are elevated. J Gen Physiol 138:593–607
22. Orio P, Latorre R (2005) Differential effects of beta 1 and beta 2 subunits on BK channel activity. J Gen Physiol 125:395–411
23. Santarelli LC, Wassef R, Heinemann SH, Hoshi T (2006) Three methionine residues located within the regulator of conductance for K^- (RCK) domains confer oxidative sensitivity to large-conductance Ca^{2+}-activated K^+ channels. J Physiol 571:329–348
24. Cui J, Aldrich RW (2000) Allosteric linkage between voltage and Ca(2+)-dependent activation of BK-type mslo1 K^+ channels. Biochemistry 39:15612–15619
25. Rothberg BS, Magleby KL (2000) Voltage and Ca^{2+} activation of single large-conductance Ca^{2+}-activated K^+ channels described by a two-tiered allosteric gating mechanism. J Gen Physiol 116:75–99
26. Zagotta WN, Hoshi T, Dittman J, Aldrich RW (1994) Shaker potassium channel gating. II: Transitions in the activation pathway. J Gen Physiol 103:279–319
27. Hirschberg B, Rovner A, Lieberman M, Patlak J (1995) Transfer of twelve charges is needed to open skeletal muscle Na^+ channels. J Gen Physiol 106:1053–1068
28. Cox DH, Cui J, Aldrich RW (1997) Separation of gating properties from permeation and block in mslo large conductance Ca-activated K^+ channels. J Gen Physiol 109:633–646
29. Trapani JG, Korn SJ (2003) Control of ion channel expression for patch clamp recordings using an inducible expression system in mammalian cell lines. BMC Neurosci 4:15
30. Hawley-Nelson P, Ciccarone V, Moore, ML (2008) Transfection of cultured eukaryotic cells using cationic lipid reagents. Curr Protoc Mol Biol Chapter 9: Unit 9.4
31. Staruschenko A, Booth RE, Pochynyuk O, Stockand JD, Tong Q (2006) Functional reconstitution of the human epithelial Na^+ channel in a mammalian expression system. Methods Mol Biol 337:3–13
32. Magnus CJ, Lee PH, Atasoy D, Su HH, Looger LL, Sternson SM (2011) Chemical and genetic engineering of selective ion channel-ligand interactions. Science 333: 1292–1296
33. Gamper N, Shapiro MS (2006) Exogenous expression of proteins in neurons using the biolistic particle delivery system. Methods Mol Biol 337:27–38

Chapter 11

Perforated Whole-Cell Patch-Clamp Recording

John E. Linley

Abstract

Perforated whole-cell patch-clamp is a variant of the patch-clamp technique used to measure the sum activity of ion channels in the plasma membrane of a single cell. Its defining feature is that electrical access to the cell is obtained through inclusion of a pore-forming antibiotic in the patch pipette which perforates the sealed patch of membrane in contact with the patch pipette. The antibiotic pores allow equilibration of small monovalent ions between the patch pipette and the cytosol whilst maintaining endogenous levels of divalent ions such as Ca^{2+} and signalling molecules such as cAMP. Therefore, the perforated patch-clamp technique is ideal for studying ion channels whilst maintaining the integrity of second messenger signalling cascades. Other benefits of using perforated patch-clamp over conventional patch-clamp include reduced current rundown and stable whole-cell recording lasting >1 h. In this chapter, the application of the perforated patch-clamp technique for the study of heterologously expressed Kv7 potassium channels will be discussed in detail including benefits and limitations of the technique.

Key words Patch-clamp, Perforated-patch, Amphotericin B, Nystatin, Gramicidin, Run-down, Electrophysiology, Whole-cell

1 Introduction

Conventional whole-cell patch-clamp electrophysiology involves sealing a glass micropipette (patch pipette) onto the surface of a single cell and subsequent rupture of the seal by sharp suction to obtain electrical access to the cell. The sum activity of all ion channels in the cell membrane can then be measured by voltage clamp. Using this technique, the experimenter has complete control over the cytosolic composition due to equilibration of the patch pipette solution with the cytosol. This allows Ca^{2+} to be buffered to a fixed value using EGTA or BAPTA and membrane impermeant drugs to be applied to the cytosol. However, conventional whole-cell patch-clamp also results in washout of key endogenous cytosolic signalling molecules such as cAMP and Ca^{2+} resulting in the potential disruption of receptor-mediated signalling events. In 1988, Horn and Marty reported a new, less invasive method of obtaining electrical

Nikita Gamper (ed.), *Ion Channels: Methods and Protocols*, Methods in Molecular Biology, vol. 998,
DOI 10.1007/978-1-62703-351-0_11, © Springer Science+Business Media, LLC 2013

access to a cell called the perforated-patch technique (1). In this technique, the sealed patch of membrane is perforated rather than ruptured by inclusion of a pore-forming antibiotic in the patch pipette. In this case, the equilibration between the pipette solution and cytosol is determined by the permeability properties of the antibiotic. Nystatin and amphotericin B are both commonly used antibiotics for perforated patch and have similar permeability properties, being permeable to small monovalent cations and anions such as Na^+, K^+, and Cl^- and impermeable to large ions such as tetraethylammonium and cytosolic signalling molecules. The pore radius of nystatin is calculated to be ~0.4 nm (2), whereas amphotericin B forms a larger pore which is ~0.8 nm in radius (3), thereby excluding nonelectrolytes with a molecular weight >200. The cation selectivity sequence for both nystatin and amphotericin is $Rb^+ > K^+ > Na^+ > Li^+$, and is directly related to the size of the hydrated ions. Divalent cations such as Ca^{2+} and Mg^{2+} are impermeant. Chloride permeability of nystatin and amphotericin B is less than cation permeability ($P_{anion}/P_{cation} = 0.1$) although Cl^- still readily equilibrates in the perforated patch configuration. For electrophysiological recordings in which the endogenous intracellular Cl^- concentration is desired to be maintained (for example when recording GABA responses in primary neurons), then the antibiotic gramicidin can be used. Gramicidin pores are permeable to small monovalent cations but impermeable to anions (4). However, gramicidin perforation is very slow compared to amphotericin B, typically taking ~30 min to stabilize, thereby impacting on the number of experiments which can be conducted in a day. Note that perforation is dependent on sterol composition of the lipid membrane (5).

1.1 Advantages of Perforated Patch

1. Minimal disruption of cytosolic components making it ideal for studying regulation of ion channels by intracellular signalling cascades.
2. Stable recording for >1 h.
3. Maintenance of endogenous Ca^{2+} concentration and buffering.
4. Maintenance of endogenous Cl^- concentration (Gramicidin only).
5. Reduced rundown of current.

1.2 Disadvantages of Perforated Patch

1. Typically higher series resistance than that in conventional whole-cell patch-clamp (5–20 MΩ or 2–5× pipette resistance).
2. Takes longer to achieve whole-cell configuration than conventional whole-cell patch-clamp due to time spent waiting for perforation. Time for perforation is cell type dependent but typically takes 15 min for amphotericin B. Nystatin and gramicidin can take considerably longer.

3. It is impossible to apply drugs through the patch pipette.
4. No control over cytosolic Ca^{2+}.

Due to the higher series resistance associated with perforated patch recording, this technique is not typically used for studying large-amplitude fast-gating ion channels such as voltage-gated sodium channels (see Note 1).

The following chapter will describe practical methodology for conducting perforated patch-clamp recording on the voltage-gated potassium channel Kv7.2/3 heterologously expressed in CHO cells using amphotericin B as the pore-forming antibiotic.

2 Materials

2.1 Cell Culture

1. CHO-K1 cells (ECACC, Sigma).
2. Ham's-F12 media (Gibco, Invitrogen).
3. Fetal bovine serum (Sigma).
4. Trypsin-EDTA (0.25%).
5. Phosphate buffered saline (PBS), Ca^{2+} and Mg^{2+} free.
6. Glass coverslips (10 mm) or plastic 35 mm culture dishes.

2.2 Amphotericin Stock

1. Amphotericin B powder (~80% purity), 100 mg vial.
2. Aluminum foil.
3. Vortex.
4. Eppendorf tubes.
5. Ice bucket.
6. Dimethylsulfoxide (DMSO).

2.3 Solutions

1. Extracellular solution: 140 mM NaCl, 2 mM KCl, 2 mM $CaCl_2$, 1 mM $MgCl_2$, 10 mM HEPES. pH is titrated to 7.4 using NaOH. Test osmolarity and adjust to 300 mOsm with sucrose (1 mM sucrose = 1 mOsm). On the day of experiments add 0.9 g/L d-glucose (5 mM) to the solution.
2. Intracellular solution: 120 mM KCl, 10 mM NaCl, 5 mM $MgCl_2$, 10 mM HEPES. pH is titrated to pH 7.2 with KOH. Test osmolarity and adjust to 280 mOsm with sucrose (see Note 2). Store solution at 4°C.

General rules for solutions used for electrophysiology are as follows:

1. Use purified deionized water (18 MΩ) for all solutions.
2. Filter solutions using a 0.2 μm filter attached to a syringe.

3. Store solutions at 4°C (unless otherwise stated) and discard if any precipitation is observed or after 1 month.
4. For solutions which contain glucose, add on the day of experiments to avoid bacterial growth in solution.
5. A microbalance with an accuracy of 0.001 g is required for accurate weighing of chemicals.
6. Osmolarity of solutions should be tested using an osmometer (e.g., Osmomat 30, Gonotec).
7. Use the purest salts available.

2.4 Electrophysiology

1. Inverted microscope (e.g., Nikon TS-100).
2. Micromanipulator (e.g., Sutter MP-285).
3. Faraday cage.
4. Patch-clamp amplifier and digitizer (e.g., Axon multiclamp 700B; Axon digidata 1440).
5. Acquisition and analysis software, e.g., pCLAMP.
6. Pipette puller (e.g., Sutter P-97 or Narishige PC-10).
7. Pipette borosilicate glass, thin walled, filamented. For example outer diameter 1.2 mm, inner diameter 0.69 mm, wall thickness 0.51 mm (GC-120F, Clarke).
8. Pipette-filling needle (e.g., Microfil, 28 gauge 250 μm ID 350 μm OD, World precision instruments).
9. Microforge (e.g., Narishige).

3 Methods

3.1 Cell Culture

CHO cells are grown as an adherent culture in Ham's F-12 supplemented with 10% fetal calf serum, penicillin, and streptomycin (0.1%) in an incubator at 37°C, 5% CO_2. Note that CHO cells have an absolute requirement for proline in the media. To passage cells, they are first washed in PBS then detached by trypsinization. For electrophysiological recording from transfected CHO cells, cells are plated onto 35 mm Nunclon culture dishes at a density of ~50% confluency and transfected 24 h later using a lipofection reagent (e.g., Fugene HD) in accordance with the manufacturer's instructions. For transfection of Kv7.2/Kv7.3 channels we use 1 μg Kv7.2, 1 μg Kv7.3, 0.2 μg GFP plasmid cDNA and 3 μL Fugene HD per transfection. 24 h post transfection cells are trypsinized and replated onto sterile 10 mm glass coverslips precoated with poly-l-lysine (0.001% poly-l-lysine solution for 20 min followed by triple wash) in 24-well plates in a volume of 1 ml culture media. Cells are recorded from 48 to 72 h post transfection. Plasmids encoding human Kv7.2 and human Kv7.3 (GenBank accession numbers AF110020 and AAC96101, respectively) were given to us by David McKinnon (State University of New York at Stony Brook, NY, USA) and Thomas Jentsch (Zentrum fuer Molekulare Neurobiologie, Hamburg, Germany) and subcloned into pcDNA3.1 (Invitrogen, Grand Island, NY, USA). For electrophysiological recording, best results are often obtained when recording from single cells which are not in direct contact with neighboring cells.

3.2 Amphotericin B Pipette Solution

1. Amphotericin B powder should be stored at 4°C wrapped in foil as it is light sensitive.
2. On the day of experiment, a stock solution is made from powder by adding 4 mg amphotericin B to 100 μL DMSO (see Note 3) in a 1.5 mL Eppendorf tube.
3. Vortex for 1–2 min to form a clear orange/yellow solution. Store this stock in an ice-filled box with lid or covered with foil (see Note 4). Final stock concentration is 40 mg/mL (see Note 5).
4. In a new 1.5 mL Eppendorf add 1 mL of filtered intracellular solution. To this add 10 μL of amphotericin B stock solution (40 mg/mL) to give a final working concentration of 400 μg/mL.
5. Cover in foil and vortex for 1–2 min resulting in a hazy yellow solution which is supersaturated (see Note 6).
6. This solution should be drawn up in a 1 mL plastic syringe and a fine needle suitable for patch pipette filling inserted onto the end (e.g., Microfil).
7. The pipette solution should be kept at 4°C in an ice box with lid/covered with foil.
8. The amphotericin B pipette solution will fall out of solution over a period of 2–3 h and therefore must be made up regularly from the stock solution throughout the day. Failure to do so will reduce perforation success.

3.3 Perforated Patch-Clamp: Setup

1. Pull patch pipettes using a micropipette puller and heat polish to a resistance of 2–4 MΩ using a microforge.
2. Place patch pipettes in a covered holder or in a lidded petri dish containing blu-tack in order to keep them clean. Note that patch pipettes should be pulled fresh each day.
3. Remove glass coverslip from 24-well plate using fine forceps, place in 35 mm dish containing extracellular solution, and rock gently to wash cells (serum in media inhibits seal formation).
4. Place in perfusion chamber (break coverglass to obtain multiple coverglass "chips" if required) on microscope stage.
5. Identify positively transfected cells by green fluorescence.

6. Dip the tip of a patch pipette in intracellular solution which does not contain amphotericin B for 0.5–5 s (tip dipping) (see Note 7).
7. Backfill the patch pipette with intracellular solution containing amphotericin B so that it is ¾ full.
8. Place the patch pipette on the electrode holder and manipulate the patch pipette through the air liquid interface whilst applying light positive pressure through the electrode holder.
9. Maneuver the pipette so that it is a few microns directly above a single green fluorescent cell.
10. Block or turn off all microscope light sources.

3.4 Perforated Patch-Clamp: Achieving Seal and Whole-Cell

1. Apply a 5 mV voltage test pulse and zero off any offset potential.
2. Make a note of the open circuit resistance as this is indicative of the patch pipette resistance.
3. Slowly micromanipulate the pipette in the *z*-axis towards the cell whilst watching for a reduction in amplitude of the test pulse (an increase in pipette resistance).
4. As soon as this occurs, apply light suction to the cell to obtain a giga ohm seal. This can be trickier than in conventional whole-cell as amphotericin B inhibits seal formation. If no seal is obtained then increase the tip-dipping time (see Note 7).
5. Compensate for fast capacitance generated by the glass pipette bath solution interface.
6. Change the membrane holding voltage to −70 mV. A slow capacitance transient should begin to develop (see Fig. 1) over a period of 5–15 min. This is caused by the gradual inclusion of amphotericin pores in the patch of membrane forming the seal allowing measurement of the cell capacitance. Initially this will appear as a small (~20 pA) capacitance spike with a slow decay kinetic (Fig. 1a). This will increase in magnitude and the decay

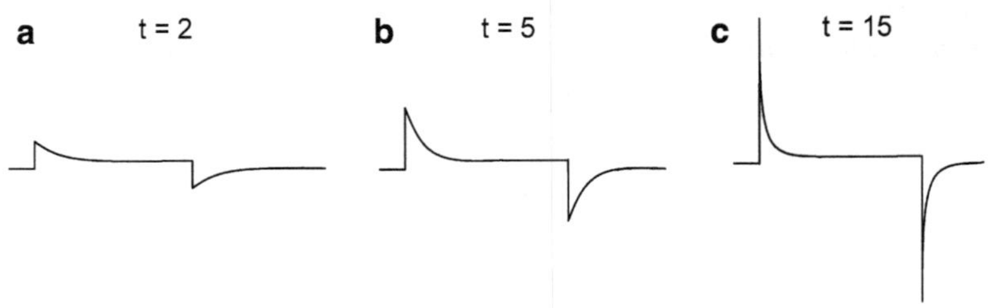

Fig. 1 Development of whole-cell capacitance in perforated patch. Current in response to a 5 mV test pulse (10 ms duration) is shown at various points after seal formation. *t*=time in minutes. Amplitude of capacitance current at *t*=15=200–400 pA

kinetic will speed up as electrical access to the cell is achieved. Note that during perforation, a slow decrease in seal resistance can occur as electrical access to the cell is obtained and seal resistance becomes whole-cell resistance.

7. Monitor series resistance and cell capacitance continuously until it stabilizes. Aim for 2–3 times the pipette resistance or 5–20 MΩ. Compensate this transient using the amplifier circuitry.
8. Apply series resistance compensation using the amplifier circuitry of 50–70% (see Note 8).
9. Care must be taken to ensure that the cell does not enter the conventional whole-cell configuration by rupture of the membrane seal. This is normally observed as a rapid change in series resistance and can occur at any time during sealing and during the experiment. During an experiment this can be seen as a sharp capacitance spike appearing on current traces during voltage pulses often accompanied by a change in the size of the measured current and usually followed by the development of a "leak"-like current as the amphotericin perforates the whole cell. If this occurs then the experiment must be terminated.
10. Once series resistance has stabilized, voltage clamp the cell at the desired holding potential which in the case of the example given here is -70 mV. Apply a voltage protocol designed to investigate the channel of interest; in the case of the Kv7.2/Kv7.3 channels, we often use a depolarizing voltage pulse to 0 mV applied every 5 s to monitor the magnitude of the current. Once this has stabilized the experiment can begin. Figure 2 shows an example of Kv7.2/Kv7.3 current measured using perforated patch. The voltage dependence of activation is plotted in Fig. 2b.

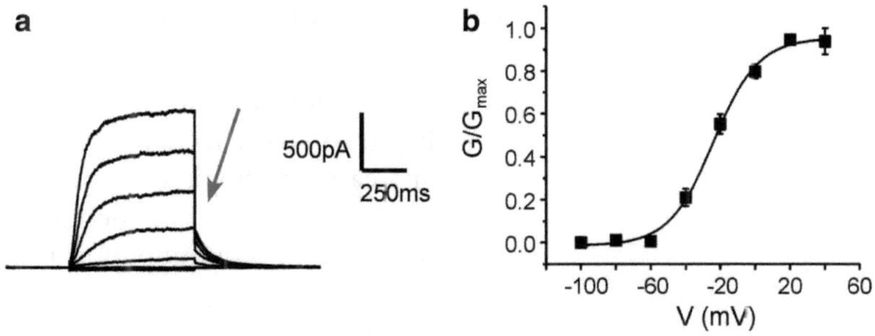

Fig. 2 Example recording from CHO cell expressing Kv7.2 and Kv7.3. (**a**) Current trace in response to depolarizing voltage steps from a holding potential of −70 mV in 20 mV increments from −100 to +60 mV. (**b**) Voltage dependence of activation calculated from the magnitude of the deactivating tail current when stepping from the test potential to −100 mV as indicated by the *arrow* in (**a**). Data are normalized to the maximum tail current and presented as mean ± SEM from ten cells

4 Notes

1. Large-amplitude fast-gating ion channels such as voltage-gated sodium channels are more susceptible to errors associated with series/access resistance. The resistance between the pipette and the cell (series resistance) causes a voltage drop at the pipette resulting in the voltage which the cell sees being less than the amplifier command voltage. The magnitude of this voltage drop can be calculated using the following equation $V_m = V_p - IR_s$ where V_m = membrane voltage, V_p = pipette voltage, I = whole-cell current, and R_s = series resistance. Therefore large currents (>1 nA) combined with high R_s result in a large voltage error. High R_s also reduces the speed with which the membrane potential is changed thereby introducing further errors for fast-gating ion channels which activate and inactivate within a few milliseconds. In this case it is often desirable to use conventional whole-cell.
2. Intracellular solution osmolarity is typically made ~20 mOsm hypotonic compared with the bath solution. This is to avoid cell swelling due to the oncotic effect of cellular proteins.
3. DMSO is hygroscopic and therefore will become "wet" with time. Ensure the cap is not left off and replace with new solution if solubility of compounds appears to decrease.
4. Some labs use sonication of the stock and/or final pipette solutions for 5–10 min at this stage; however I find that vortexing is sufficient.
5. The amphotericin B stock solution should be discarded at the end of the day and a new stock made at the beginning of every days experiments.
6. Amphotericin B has poor water solubility therefore will visibly fall out of solution. The amphotericin B containing intracellular solution should not be filtered.
7. Tip-dipping time affects the time it takes for the amphotericin B to begin to take effect. If you find that the cell will not seal then increase tip-dipping time. If a seal is obtained but no series resistance is obtained then decrease tip-dipping time. For some cell types, no tip dipping is required.
8. Series resistance compensation can cause "ringing" of the feedback amplifier. This can be seen as a large oscillating current which often results in loss of whole-cell configuration. If this occurs regularly then reduce the % R_s compensation used.

References

1. Horn R, Marty A (1988) Muscarinic activation of ionic currents measured by a new whole-cell recording method. J Gen Physiol 92:145–159
2. Hladky SB, Haydon DA (1970) Discreteness of conductance change in bimolecular lipid membranes in the presence of certain antibiotics. Nature 225:451–453
3. de Kruijff B, Gerritsen WJ, Oerlemans A, Demel RA, van Deenen LL (1974) Polyene antibiotic-sterol interactions in membranes of Acholeplasma laidlawii cells and lecithin liposomes I. Specificity of the membrane permeability changes induced by the polyene antibiotics. Biochim Biophys Acta 339:30–43
4. Rhee JS, Ebihara S, Akaike N (1994) Gramicidin perforated patch-clamp technique reveals glycine-gated outward chloride current in dissociated nucleus solitarii neurons of the rat. J Neurophysiol 72:1103–1108
5. Andreoli TE, Monahan M (1968) The interaction of polyene antibiotics with thin lipid membranes. J Gen Physiol 52:300–325

Chapter 12

Piezo-Electrically Driven Mechanical Stimulation of Sensory Neurons

Jizhe Hao, Jérôme Ruel, Bertrand Coste, Yann Roudaut, Marcel Crest, and Patrick Delmas

Abstract

Mechanotransduction, the conversion of a mechanical stimulus into a biological response, constitutes the basis of a variety of physiological functions such as the senses of touch, balance, proprioception, blood pressure, and hearing. In vertebrates, mechanosensation is mediated by mechanosensory neurons, whose cell bodies are located in trigeminal and dorsal root ganglia. Here, we describe an in vitro model of mechanotransduction that provides an opportunity to explore the properties of mechanosensitive channels in mammalian sensory neurons. The mechano-clamp method allows applying local force on plasma membrane of whole-cell patch-clamped sensory neurons. This technique uses a mechanical probe driven by a computer-assisted piezoelectric microstage to repeatedly stimulate sensory neurons with accurate control of stimulus strength, duration, and speed.

Key words Mechanosensation, Pain, Sensory neuron, Mechanoreceptor, Skin, DRG, Primary culture, Mechanical probe, Piezoelectric microstage, Patch clamp, Ion channels

1 Introduction

Somatosensory neurons detect a wide variety of mechanical stimuli. Some are specialized to detect external mechanical stimuli (e.g., mechanoreceptors, mechanonociceptors), while others inform the nervous system about self-generated stimuli (e.g., stretch receptors). The ability of these mechanoreceptors to detect mechanical cues relies on the presence on the specialized sensory endings of mechanotransducer channels that rapidly transform mechanical forces into electrical signals and depolarize the sensory ending (1–3). This local depolarization, called receptor potential, eventually leads to the generation of action potentials that propagate toward the central nervous system.

Progress has been made in establishing the functional properties, specificity, and perceptual functions of mechanoreceptors.

Nikita Gamper (ed.), *Ion Channels: Methods and Protocols*, Methods in Molecular Biology, vol. 998,
DOI 10.1007/978-1-62703-351-0_12,

This progression led to the recognition that mechanoreceptors serve as selective peripheral encoding devices able to extract information about the various parameters of the mechanical stimulus and to supply the central nervous system with a neural "image" of the peripheral situation. However, molecular mechanisms of mechanotransduction remain poorly understood. Compared with other types of ion channels, including voltage-gated and ligand-activated channels, that have been substantially elucidated, we are largely ignorant of the properties of the force transducers that contribute to our perception of mechanical cues. The technique described here ("mechano-clamp") adds a new dimension to the study of mechanosensation (Fig. 1). It has opened up new pathways for the investigation of molecular mechanisms of mechanosensation. It can be exploited in the effort to bridge the gap between the properties of mechanotransducer channels in vitro and the characteristics of mechanoreceptors in vivo. This technique is applicable to both primary neuronal cultures as well as immortalized cell lines. The aim of this paper is to describe how mechano-clamp can be applied to cultured DRG neurons (4, 5) and to discuss potential applications in mechanotransduction studies with emphasis on the strengths and limitations of the technique.

2 Materials

2.1 Reagents for Recording DRG Neurons in Primary Culture

1. Experimental animals: rats (male Wistar, 120–130 g). All animal experiments have to be performed in accordance with the guidelines on the use of animals by the relevant authorities.
2. Halothane (Belamont, Nicholas Piramal). Harmful by inhalation, use adequate ventilation.
3. Collagenase IA (C9891, Sigma).
4. Bovine Serum Albumin (BSA, A9647, Sigma).
5. Cell culture media, DMEM (Invitrogen).
6. Hank's balanced salt solution, HBSS (Invitrogen).
7. Penicillin/Streptomycin (Invitrogen).
8. Glutamine (Invitrogen).
9. Nerve growth factor, NGF (Millipore).
10. Glial cell derived neurotrophic factor, GDNF (Invitrogen).
11. Laminin (Sigma).
12. Tetrodotoxin (Ascent Scientific).

2.1.1 Solution Setup for Patch Clamp Recording

Pipette (internal) solution for whole-cell recordings

1. For voltage-clamp experiments, prepare pipette solution as follows: 125 mM CsCl, 1 mM $MgCl_2$, 4.8 mM $CaCl_2$, 10 mM HEPES, 10 mM EGTA, 4 mM Mg-ATP, and 0.4 mM Na-GTP. Adjust pH 7.4 with CsOH. Adjust osmolarity to 300 mOsm with CsCl.

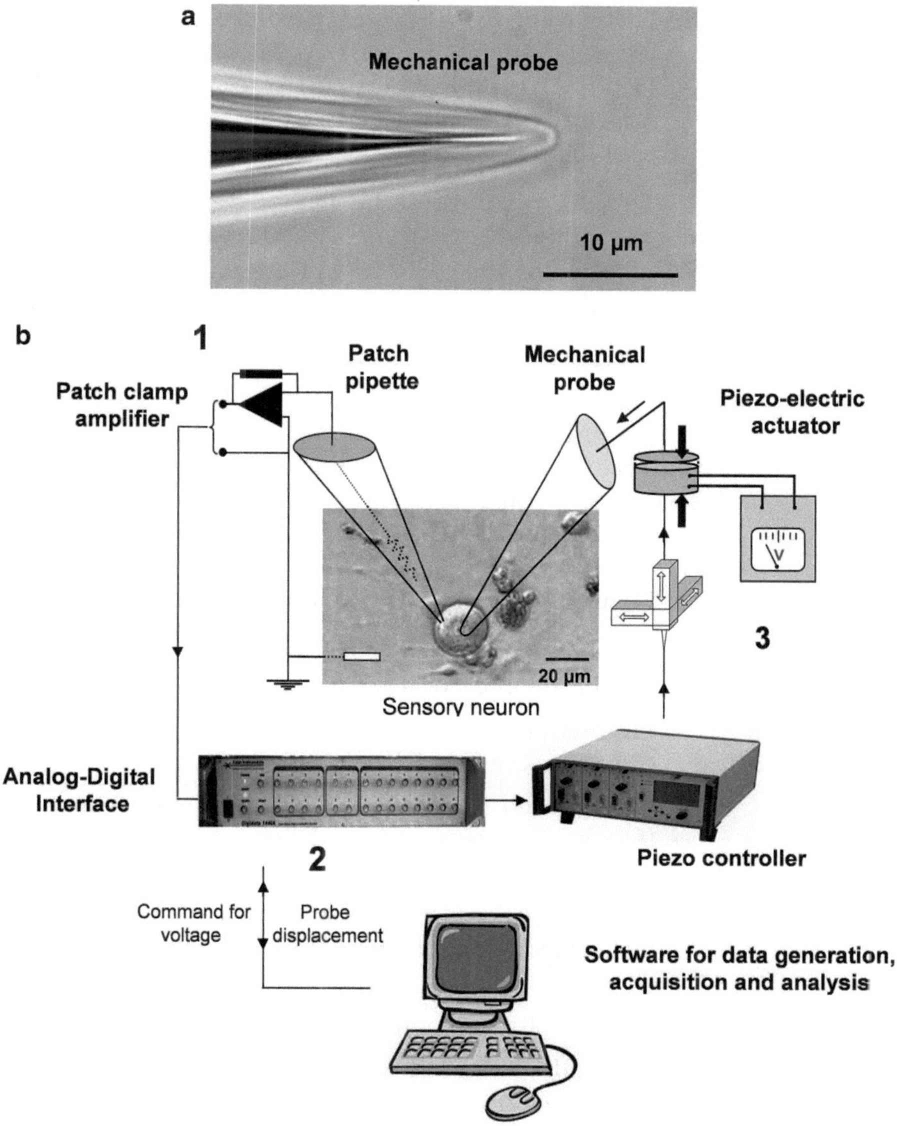

Fig. 1 Mechano-clamp setup. (**a**) Photograph of the mechanical probe. (**b**) Schematic representation of the mechano-clamp setup. Cultured sensory neurons are recorded using the patch clamp technique (1). A glass pipette is filled with a pipette solution and an Ag/AgCl wire connects the cell to the patch clamp headstage, which is a sensitive current-to-voltage converter. The patch clamp amplifier is connected to a computer through an analog-digital interface, allowing data generation, acquisition and analysis (2). The mechanical probe is connected to a piezoelectric ceramic actuator, which is ideal for applications requiring high-resolution movements for micro- and nanopositioning (3). Piezoceramic actuators make use of the deformation of the piezoelectric material when an electric field is applied. The piezoelectric actuator is connected to a linear amplifier-driver and stimulus parameters, such as duration and speed of stimuli, are user configurable via pCLAMP software

2. For current-clamp experiments, prepare pipette solution as follows: 134 mM KCl, 1 mM $MgCl_2$, 1 mM $CaCl_2$, 10 mM HEPES, 10 mM EGTA, 4 mM Mg-ATP, and 0.4 mM Na-GTP. Adjust pH 7.4 with KOH. Adjust osmolarity to 300 mOsm with KCl.

Internal solutions can be aliquoted and stored at −20°C for months without Na-GTP and Mg-ATP. Stock solutions of 40 mM Na-GTP and of 400 mM Mg-ATP are aliquoted, stored at −20°C, and added to the internal solution on the day of experiment.

Bath (extracellular) solution for all recordings

Prepare extracellular solution as follows: 132 mM NaCl, 3 mM KCl, 1 mM $MgCl_2$, 2.5 mM $CaCl_2$, 10 mM HEPES, and 10 mM Glucose. Adjust pH 7.4 with NaOH. Adjust osmolarity to 300 mOsm with NaCl. This solution is stored at 4°C for up to 1 week.

2.2 Equipment for Patch Clamp Recording and Piezo-Electric Stimulation

1. Double-headed peristaltic pump (i.e., Masterflex L/S).
2. Gravity-fed bath perfusion system.
3. Microforge (i.e., Beaudouin).
4. Piezo-electric actuator, e.g., Step Driver PZ-100 (Burleigh), or more recent model.
5. Thick-walled borosilicate glass tubing (inner diameter 0.86 mm, outer diameter 1.5 mm; Sutter). Patch pipettes are pulled every day.
6. Micropipette puller; we use Brown-Flaming P-97 programmable pipette puller (Sutter Instrument Company).
7. Patch pipette fillers (MicroFil; WPI) with solution filter (4 mm diameter, 0.2 μm pore).
8. Inverted microscope, e.g., Nikon with external power supply to the light source.
9. Micromanipulators and pipette holder for patch-clamping (e.g., Sutter Instruments,, Narishige, or Burleigh). For positioning local superfusion, micromanipulator can be of lower grade.
10. Noise eliminator, e.g., Humbug (Quest Scientific).
11. Patch-clamp amplifier, e.g., Axopatch 200B (Molecular Devices).
12. Software and data acquisition interface, e.g., pCLAMP Digidata 1322A (Molecular Devices).
13. Vibration isolation table and perimeter Faraday cage (i.e., TMC).

2.2.1 Setup for Patch Clamp Recording (Fig. 1)

There are many papers that illustrate the assembly of a functional patch-clamp setup and perfusion systems, i.e., (6, 7); see also Chapter 7 of this book.

2.2.2 Equipment Setup for Mechanical Stimulation (Fig. 1)

Fix the piezo-electric actuator on a three-axis micromanipulator. A pipette holder is mounted on the actuator. The probe is positioned at an angle of 45–65° from horizontal plane and moved downward toward the selected neuron. The timing and the displacement of the mechanical probe is monitored and controlled by pCLAMP program. The mechanical stimulation system must be firmly fixed, in particular the mechanical probe, in order to avoid mechanical artifacts caused by vibration of probe during recording.

3 Methods

3.1 DRG Neuron Preparation

1. Preparation of coated Petri dishes. Nunclon dishes (35 mm ∅) need to be prepared at least 1 h before seeding cells. With a Pipetman loaded with 600 μL of laminin (10 μg/mL), position the pipette at the center of the dish, then release the solution until it forms a large droplet in the middle of the dish. Avoid covering the entire surface of the dish to prevent cells moving at the edge of the dish. Incubate the dishes for at least 1 h at 37°C. Before dissociation of DRGs, wash the dishes twice with HBSS solution and once with culture medium. Allow culture dishes to remain in a safe environment.
2. Dissection of DRGs. Young rats are deeply anesthetized with halothane and humanly killed in accordance of local ethical guidance. DRGs are removed from their connective tissue sheaths and incubated in HBSS solution containing 2 mg/mL collagenase IA for 45 min at 37°C. Fire-polish two glass Pasteur pipettes so that relatively small tip diameters are achieved.
3. Remove the supernatant containing the enzymatic solution and wash seven or eight times DRGs with HBSS solution. Resuspend DRGs with 800 μL of HBSS.
4. Triturate DRGs with Pasteur pipettes having large then small tip diameters (1 and 0.5 mm) until the suspension becomes visually opaque. Trituration consists in filling and emptying the barrel at a rate of about 2.0 mL/s. The trituration should be gentle because mechanosensory neurons are damaged by shear force produced by backward and forward motion (see Note 1).
5. The resulting suspension is plated at the center of Nunclon dishes coated with laminin. Culture medium is DMEM supplemented with 10% heat-inactivated fetal calf serum, 2 mM L-glutamine, 100 U/mL penicillin–streptomycin, 25 ng/mL NGF, and 2 ng/mL GDNF. Neurons are maintained in a humidified atmosphere (5% CO_2, 37°C) for 12–16 h before recording.

3.2 Fabrication of Mechanical Probe

Fabricate mechanical probe from borosilicate glass tubing. Pull the pipette to have a final tip diameter of 1–2 μm on a Brown-Flaming P-97 pipette puller. Fire-polish micropipette using a microforge for 15 s. Examine the tip pipette under a microscope or a binocular loupe and ensure that the tip is blunt with diameter of 3–4 μm (Fig. 1a) (see Note 2).

3.3 Setting the Patch-Clamp Setup

1. Prepare patch pipettes with resistances of 3–4 MΩ. The tip diameter is determined by the pulling parameters of the pipette puller.
2. Make solutions as appropriate for your experiment (basic solutions are given in Subheading 2.1.1).
3. Transfer the Nunclon dish containing DRG neurons into a 500 μL-recording chamber in the inverted microscope for visualization and subsequent patch-clamping. Cells are superfused using a gravity-fed bath perfusion system at a laminar flow rate of ~4 mL/min under continuous suction. Position the inlet tubing at one end of the recording chamber opposite from the suction tubing. Keep the perfusion flowing at a constant rate in order to avoid shear stress perturbation during recording.
4. Select a neuron based on cell health. Choose an isolated neuron not bearing neurites (improved voltage clamp) if you aim at characterizing biophysical properties of mechanosensitive currents. Select a neuron-bearing neurites if you aim at investigating mechanical responses of nerve processes. The success rate of experiments is almost entirely based on cell health conditions. Unhealthy cells display granular cytosol and poorly refracting surface membrane (see Note 3).
5. Fill the patch pipette with the internal solution and place the pipette on the electrode holder. Use a syringe to apply a small positive pressure to the pipette solution.

3.4 Positioning Mechanical Probe and Getting Whole-Cell (See Note 4)

1. Bring both patch pipette and mechanical probe above the selected neuron. Place initially the mechanical probe at 1.4 μm to the edge of the cell by estimating the distance between the probe and the cell surface by using cycling mechanical sweeps of 0.7 or 1.4 μm.
2. Move patch pipette down into the bath solution. By focusing back and forth between the target cell and the pipette tip, move the pipette into position above the neuron. Readings of pipette resistance and pipette offset are taken at this time. Move downward the pipette until the pipette resistance increases by 0.1–0.2 MΩ. Once contact is made, release the pipette positive pressure. We usually observed a gradual rise in pipette resistance from the values taken before contacting the cell, often accompanied by the appearance of small capacitive transients in the recording. If seals do not form spontaneously apply a continuous suction by mouth until seal resistance reaches 100 MΩ. Apply negative potentials, ranging from −10 to −60 mV, via the voltage clamp command to help seal formation (≥1 GΩ).

3. To get whole-cell, apply a pulse of gentle suction by mouth. Strong and brief suctions are necessary to break the membrane. Hence, electrical contact is established between the electrode and the intracellular milieu. After obtaining the whole-cell configuration, wait for 2–3 min before beginning recording, so that the pipette solution diffuses into the cell and equilibrates with the intracellular medium.

4. Before recording, adjust the final position of mechanical probe. In voltage-clamp configuration, record neuron's responses in gap-free mode. Downward movement of probe is stopped once mechanical response artifact is detected on the current baseline and confirmed under visual control.

3.5 Performing Electrophysiological Recordings

To test whether the neuron of interest exhibits mechanosensitive currents, apply a series of mechanical stimulus in 0.7 μm increments till obtaining saturated currents (Fig. 2) (see Note 5). The best

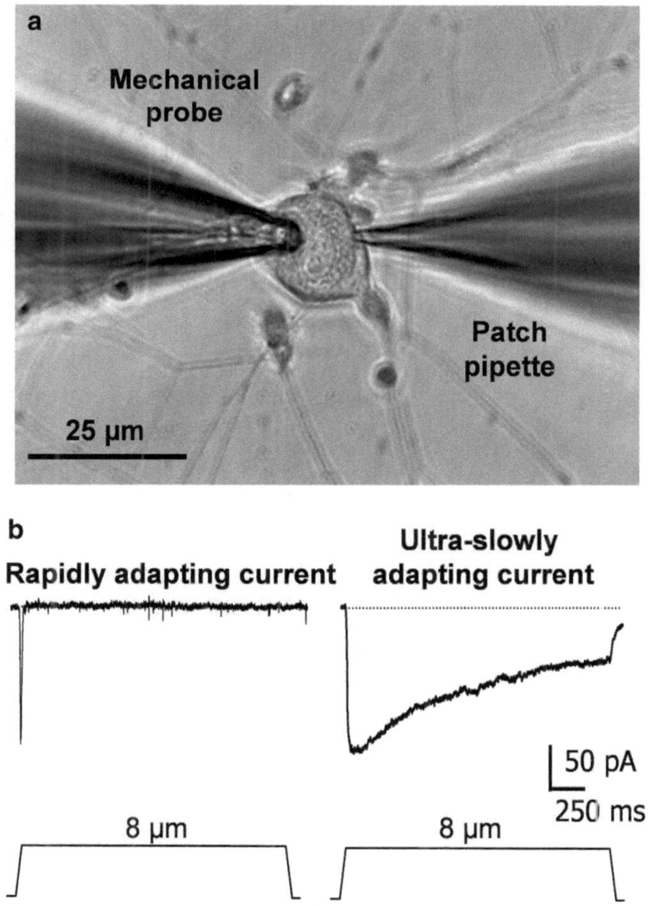

Fig. 2 Recording of mechanosensitive currents in sensory neurons. (**a**) Photograph showing mechanical stimulation of a patch-clamped DRG neuron. The neuron was subjected to a mechanical stimulus of 8 μm. (**b**) Recording of rapidly or ultraslowly adapting mechanosensitive currents in DRG neurons at membrane voltage of −60 mV

method for monitoring the quality of the recording while it is in progress is to note the amplitude of currents periodically (see Note 6). Typically, mechanosensitive current amplitude rose over the first few minutes of a recording. Current amplitude could be maintained for many minutes to several hours. If amplitude of currents decreased unexpectedly, then discard the cell (see Notes 7 and 8). To investigate the effects of mechanosensitive currents on neuronal excitability, recordings are made using a KCl-based pipette solution (see Subheading 2.1.1). Record changes in membrane potential and action potential firing upon mechanical stimulation. Assess access resistance throughout the experiment.

3.6 Basic Post-Hoc Analysis of Recordings

1. Kinetics of mechanosensitive currents is determined by using nonlinear least-squares regression analysis applied to the decay phase, also called relaxation, of the currents (8–12). Current traces can be fitted with two exponential components (I_1 and I_2) as follows: $I(t) = A_1 \cdot \exp(-t/\tau_1) + A_2 \cdot \exp(-t/\tau_2) + A_o$, where τ_1 and τ_2 represent the more rapid and slower exponential components, A_1 and A_2 represent the amplitude of each respective component, and A_o represents the baseline current. This adequately describes the decaying phase of mechanosensitive currents in 40% of cases. In other cases, a monoexponential function adequately fits the relaxation; fitting with functions with greater than two exponential components does not significantly enhance description of the relaxation, as judged by residual analysis.
2. Current–stimulus (*I–X*) relationships can be obtained by plotting peak current amplitude against mechanical probe displacement (9). *I/X* curves are fitted according to the Boltzmann function: $I/I_{max} = 1/\left(1 + \exp\left[\left(\mathrm{Stim}_{50} - \mathrm{Stim}\right)/k\right]\right)$, where I/I_{max} is the normalized current, Stim_{50} is the stimulus of half-maximum channel activation, and *k* is the steepness factor.
3. Availability relationships can be determined by plotting normalized peak currents against the conditioning probe displacement using Boltzmann equation, as above (9).
4. DRG neurons can be subjected to stimuli having different velocities to examine whether mechanotransducer current amplitude depends on the rate of stimulus application (9, 13) (Fig. 3). Correlation between the amplitude of mechanosensitive currents and the onset speed of the stimulus can be analyzed by plotting current amplitude as a function of velocity stimulus.
5. Current amplitude can be plotted against holding potential to determine the reversal potential (E_{rev}) of the current under investigation (9–15). E_{rev} is taken as the voltage at which the current is null. Permeability ratio (P_X/P_{Cs}) for mechanosensitive currents can be determined for each cation tested from the reversal potential of the current when that cation (*X*) is the

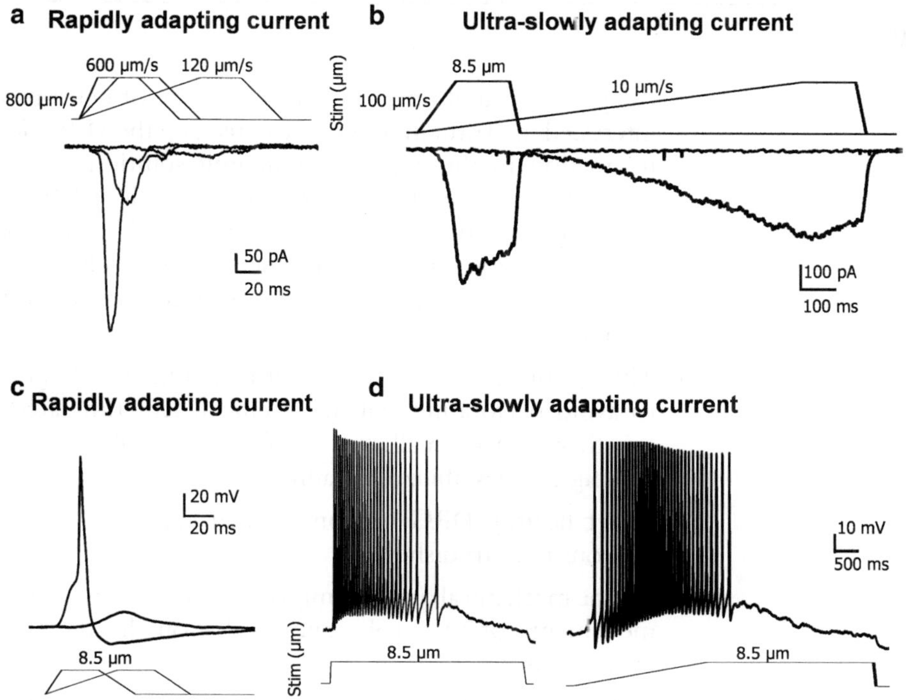

Fig. 3 Investigating velocity dependence of mechanosensitive currents. (**a**, **b**) Effects of varying the rate of onset mechanical stimulus on the amplitude of rapidly adapting (**a**) and ultraslowly adapting (**b**) mechanosensitive currents. (**c**, **d**) Current-clamp responses evoked by mechanical stimuli of varying velocities (*lower traces*) in DRG neurons expressing rapidly adapting (**c**) or ultraslowly adapting (**d**) mechanosensitive currents

major cation in the extracellular solution. For monovalent cations, the simplified Goldman-Hodgkin-Katz equation can be employed:

$$E_{\text{rev}} = \frac{RT}{zF}\ln\frac{P_{\text{X}}[X]_0}{P_{\text{Cs}}[\text{Cs}]_i}$$

where RT/zF has the value of 25.5 at 23°C. For divalent cations, the following equation can be used:

$$E_{\text{rev}} = \frac{RT}{F}\ln\left(\sqrt{\frac{4P_X[X]_0}{P_{\text{Cs}}[\text{Cs}]_i}+\frac{1}{4}}-\frac{1}{2}\right).$$

Most mechanosensitive currents in DRG neurons are non-selectively carried by cations, with reversal potential ranging from −4 to +8 mV (1).

6. Firing frequency adaptation in response to mechanical stimulation can be recorded (Fig. 3c, d). Firing frequency can be plotted against amplitude of the mechanical stimulus to obtain frequency–stimulus relationships (9). See Note 9 for the limitations of the mechano-clamp technique.

4 Notes

1. Adapt the trituration protocol to the type of neurons you are interested in. Trituration serves to break up the DRGs following incubation in the enzyme-containing solution. If done too vigorously, low-threshold mechanosensory cells will be damaged lowering viability; small nociceptive cells will be enriched. Too weak trituration and tissue fragments will be left intact lowering yield. You can best determine a suitable rate for your tissue through trial and error.
2. The tip surface must to be smooth and relatively large in order to avoid perforating cell membrane when stimulated. Probe tips that are too small are harmful to the cells; those that are too big can destabilize cell adhesion.
3. Select healthy DRG neurons, which appear as shiny, round neurons in Petri dishes.
4. Fix the mechanical probe firmly on its holder to avoid unwanted movements of the probe and oscillation during mechanical indentation.
5. Design mechano-clamp protocol carefully. Avoid too rapid onset speed for forward motion of the mechanical probe that may dislodge the recording patch pipette and generate oscillation. Avoid withdrawing the probe too rapidly to ensure stability of patch clamp recording. Avoid high frequency stimulation that causes cumulative desensitization of mechanosensitive currents in successive sweeps. Avoid too large stimulus amplitude that may alter cell integrity.
6. Record cells with small leak currents at negative holding potentials. Discard cells that display mechanosensitive currents with weird kinetics.
7. Make sure that voltage-clamp conditions are satisfactory, in order to avoid contamination of mechanosensitive currents by voltage-dependent sodium currents escaping voltage control.
8. Pay attention to several important parameters. Stimulus duration: examining the entire cohort of relaxation kinetics of mechanosensitive currents in DRG neurons requires using mechanical stimuli of various durations lasting from 100 ms to up to 4 s. Interstimulus interval: this interval allows mechanosensitive currents to fully recover between each stimulus; it depends on stimulus duration too. Typically, a 10 s interval is required for MS currents to fully recover. If the interstimulus interval is too short, a use-dependent decrease in current amplitude will manifest. Stimulus intensity: avoid using excessive stimulus amplitude (>10 μm probe displacement) as it may cause irreversible damage to the cell. Stimulus velocity: rapid onset rate of

mechanical stimulation (>800 μm/s) may sometimes damage the cell and cause oscillation of the probe. Conversely, low rates of onset (<200 μm/s) are inadequate to record rapidly adapting mechanosensitive currents that desensitize quickly.

9. As for many techniques the mechano-clamp has its own limitations. First, there are still uncertainties as to whether this experimental paradigm recapitulates physiological forces that activate mechanosensory nerve endings in vivo. Second, it is not known whether mechanosensitive currents in cultured sensory neurons recapitulate the properties of mechanotransducers at nerve terminals. This should be taken into account when interpreting the data. Last, the stimulus–current relationship of mechanosensitive currents determined using the mechano-clamp technique can be subjected to experimental bias, especially when dealing with mechanical threshold. The difficulty resides in the fact that the positioning of the mechanical probe at a standardized distance from the cell membrane is technically difficult. In addition, the distance travelled by the probe before impacting the cell does not directly reflect the force exerted by the probe on the cell surface, especially with neurons of different sizes.

Acknowledgments

This work was supported by the CNRS and by grants from the Agence Nationale de la Recherche, Fondation Schlumberger, ARCInca-2006, UPSA, IRME, and Fondation pour la Recherche Médicale.

References

1. Delmas P, Hao J, Rodat-Despoix L (2011) Molecular mechanisms of mechanotransduction in mammalian sensory neurons. Nat Rev Neurosci 12:139–153
2. Basbaum AI, Bautista DM, Scherrer G, Julius D (2009) Cellular and molecular mechanisms of pain. Cell 139:267–284
3. Lumpkin EA, Caterina MJ (2007) Mechanisms of sensory transduction in the skin. Nature 445:858–865
4. McCarter GC, Reichling DB, Levine JD (1999) Mechanical transduction by rat dorsal root ganglion neurons in vitro. Neurosci Lett 273:179–182
5. Hao J, Delmas P (2011) Recording of mechanosensitive currents using piezoelectrically driven mechanostimulator. Nat Protoc 6: 979–990
6. Cummins TR, Rush AM, Estacion M, Dib-Hajj SD, Waxman SG (2009) Voltage-clamp and current-clamp recordings from mammalian DRG neurons. Nat Protoc 4:1103–1112
7. Davie JT, Kole MH, Letzkus JJ, Rancz EA, Spruston N, Stuart GJ, Hausser M (2006) Dendritic patch-clamp recording. Nat Protoc 1:1235–1247
8. Drew LJ, Wood JN, Cesare P (2002) Distinct mechanosensitive properties of capsaicin-sensitive and -insensitive sensory neurons. J Neurosci 22:RC228
9. Hao J, Delmas P (2010) Multiple desensitization mechanisms of mechanotransducer channels shape firing of mechanosensory neurons. J Neurosci 30:13384–13395
10. Coste B, Crest M, Delmas P (2007) Pharmacological dissection and distribution of NaN/

Nav1.9, T-type Ca^{2+} currents, and mechanically activated cation currents in different populations of DRG neurons. J Gen Physiol 129:57–77
11. Coste B, Mathur J, Schmidt M, Earley TJ, Ranade S, Petrus MJ, Dubin AE, Patapoutian A (2010) Piezo1 and Piezo2 are essential components of distinct mechanically activated cation channels. Science 330:55–60
12. Coste B, Xiao B, Santos JS, Syeda R, Grandl J, Spencer KS, Kim SE, Schmidt M, Mathur J, Dubin AE, Montal M, Patapoutian A (2012) Piezo proteins are pore-forming subunits of mechanically activated channels. Nature 483: 176–181
13. Rugiero F, Drew LJ, Wood JN (2010) Kinetic properties of mechanically activated currents in spinal sensory neurons. J Physiol 588:301–314
14. Hu J, Lewin GR (2006) Mechanosensitive currents in the neurites of cultured mouse sensory neurones. J Physiol 577:815–828
15. McCarter GC, Levine JD (2006) Ionic basis of a mechanotransduction current in adult rat dorsal root ganglion neurons. Mol Pain 2:28

Chapter 13

Automated Planar Patch-Clamp

Carol J. Milligan and Clemens Möller

Abstract

Ion channels are integral membrane proteins that regulate the flow of ions across the plasma membrane and the membranes of intracellular organelles of both excitable and non-excitable cells. Ion channels are vital to a wide variety of biological processes and are prominent components of the nervous system and cardiovascular system, as well as controlling many metabolic functions. Furthermore, ion channels are known to be involved in many disease states and as such have become popular therapeutic targets. For many years now manual patch-clamping has been regarded as one of the best approaches for assaying ion channel function, through direct measurement of ion flow across these membrane proteins. Over the last decade there have been many remarkable breakthroughs in the development of technologies enabling the study of ion channels. One of these breakthroughs is the development of automated planar patch-clamp technology. Automated platforms have demonstrated the ability to generate high-quality data with high throughput capabilities, at great efficiency and reliability. Additional features such as simultaneous intracellular and extracellular perfusion of the cell membrane, current clamp operation, fast compound application, an increasing rate of parallelization, and more recently temperature control have been introduced. Furthermore, in addition to the well-established studies of over-expressed ion channel proteins in cell lines, new generations of planar patch-clamp systems have enabled successful studies of native and primary mammalian cells. This technology is becoming increasingly popular and extensively used both within areas of drug discovery as well as academic research. Many platforms have been developed including NPC-16 Patchliner® and SyncroPatch® 96 (Nanion Technologies GmbH, Munich), CytoPatch™ (Cytocentrics AG, Rostock), PatchXpress® 7000A, IonWorks® Quattro and IonWorks Barracuda™, (Molecular Devices, LLC); Dynaflow® HT (Cellectricon AB, Mölndal), QPatch HT (Sophion A/S, Copenhagen), IonFlux HT (Fluxion Bioscience Inc, USA), which have demonstrated the capability to generate recordings similar in quality to that of conventional patch clamping. Here we describe features of Nanion's NPC-16 Patchliner® and processes and protocols suited for this particularly flexible and successful high-throughput automated platform, which is based on planar patch-clamp technology. However, many of the protocols and notes given in this chapter can be applied to other automated patch-clamp platforms, similarly.

Key words Automated electrophysiology, Planar patch-clamp, Planar chip, Micro-fluidics, Intracellular perfusion, Voltage-gated ion channels, Ligand-gated ion channels, Voltage-clamp, Current-clamp, Temperature control

Nikita Gamper (ed.), *Ion Channels: Methods and Protocols*, Methods in Molecular Biology, vol. 998,
DOI 10.1007/978-1-62703-351-0_13, © Springer Science+Business Media, LLC 2013

1 Introduction

The well established patch-clamp technique developed by Erwin Neher and Bert Sakmann (1) involves the manual positioning of a glass microelectrode onto the surface of a single cell. A seal is formed, and with the application of a small amount of negative pressure (or a voltage pulse), the cell membrane at the tip of the pipette (the patch) ruptures allowing direct access to the interior of the cell, enabling control of the membrane voltage. Here we describe the automated planar patch-clamp approach using the platform Patchliner as an example (Fig. 1a). This approach requires the manual preparation of the cells in suspension, and because capturing cells is a random process it is vital that they are of extremely high quality with uniform homogeneity. The automated planar approach involves the robotic delivery of solutions, cells, and compounds onto the planar chip (Fig. 1b–d), which is a tiny borosilicate sheet of glass with a micron-sized aperture in the center.

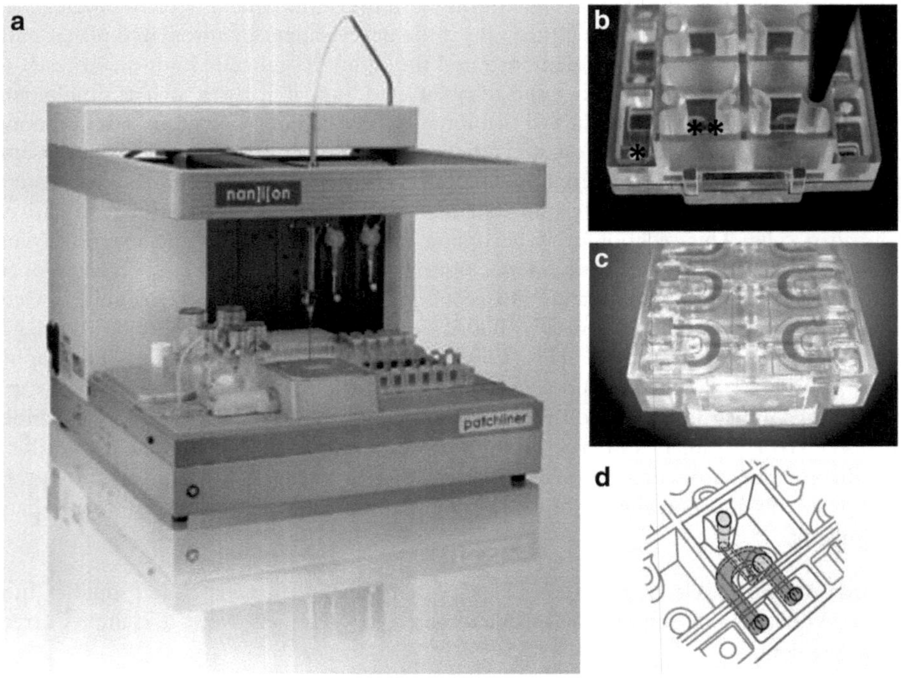

Fig. 1 (**a**) The automated planar patch-clamp platform consists of the NPC-16 Patchliner, amplifiers, and a computer. It has a much smaller foot print compared to a conventional rig. The Patchliner takes up to three NPC-16 chips per session. (**b**) Image of part of the NPC-16 multi-well microfluidic cartridge with a solution delivery pipette inserted into the extracellular inlet of one chamber. Waste chamber highlighted by ** and intracellular chamber highlighted by *. (**c**) Image of the underside of the cartridge in which the intracellular solution compartments are highlighted with dye. (**d**) Schematic of a single chamber showing the arrangement of extracellular, intracellular, and waste chambers as well as the position of the planar chip

Chips are embedded in microfluidic chambers within a chip cartridge (Fig. 1d). The aperture in the chip is the equivalent of the pipette tip in the conventional method. There are 16 individual chips within a single cartridge and 8 of these are used simultaneously for each experiment. Once the solutions and cells have been robotically delivered into the microfluidic chambers, negative pressure is applied from below each chamber independently, to attract a single cell onto each chips aperture. Further application of negative pressure ruptures the membrane sitting over the aperture, establishing the whole-cell configuration. The exchange of external solution for both manual and automated methods is relatively straightforward. However, the exchange of intracellular solution is significantly easier with Patchliner, where it is a routine feature. Apart from manual preparation of the cells, solutions and compounds, the whole process is controlled by preprogrammed computerized protocols in PatchControl^{HT}, resulting in routine reproducible comparable data output. Another versatile feature of automated platforms is that they enable non-electrophysiologists to perform patch-clamp experiments after a minimal learning time (counted in days, not in months as for manual patch-clamping).

Cell lines such as HEK 293, CHO, LTK, RBL, NIH 3T3, and Jurkat are routinely used on Patchliner. Neuroblastoma cell lines such as SHSY5Y, IMR 32, and NG108 have also been used with a high success rate (2–4), and primary cells such as human erythrocytes, human T-lymphocytes, human synoviocytes, human saphenous vein smooth muscle cells, human neutrophils, and DRG satellite glia cells have also been used successfully ((5–8), see Note 1).

Certain Patchliner voltage-clamp protocols for cell-lines as well as primary cells have already been described (7). This chapter describes in more detail the application of automated planar patch-clamp protocols involving the ligand gated ion channel α7 nicotinic acetylcholine receptor (α7nAChR), using stacked solution protocols combined with temperature control. The α7nAChR is widely distributed throughout the central nervous system and several lines of evidence indicate that pharmacological enhancement of α7nAChRs function by type II positive allosteric modulators (PAMs), e.g., PNU120596, could be a potential therapeutic route to alleviate disease-related cognitive deficits (9). α7nAChRs are rapidly activating and desensitizing, therefore, to enhance receptor activity the ligand (nicotine) is co-applied with PNU120596 in a stacked solution protocol (Fig. 2a). This stacking procedure allows for extremely rapid application and washout of compounds, attenuating desensitization and enabling subsequent additions to the same cell (Fig. 2b). Furthermore, the temperature sensitivity of this allosteric modulator is also demonstrated by preheating the stacked solutions before application, resulting in the attenuation of channel activity at near physiological temperatures ((10); Fig. 2b, c).

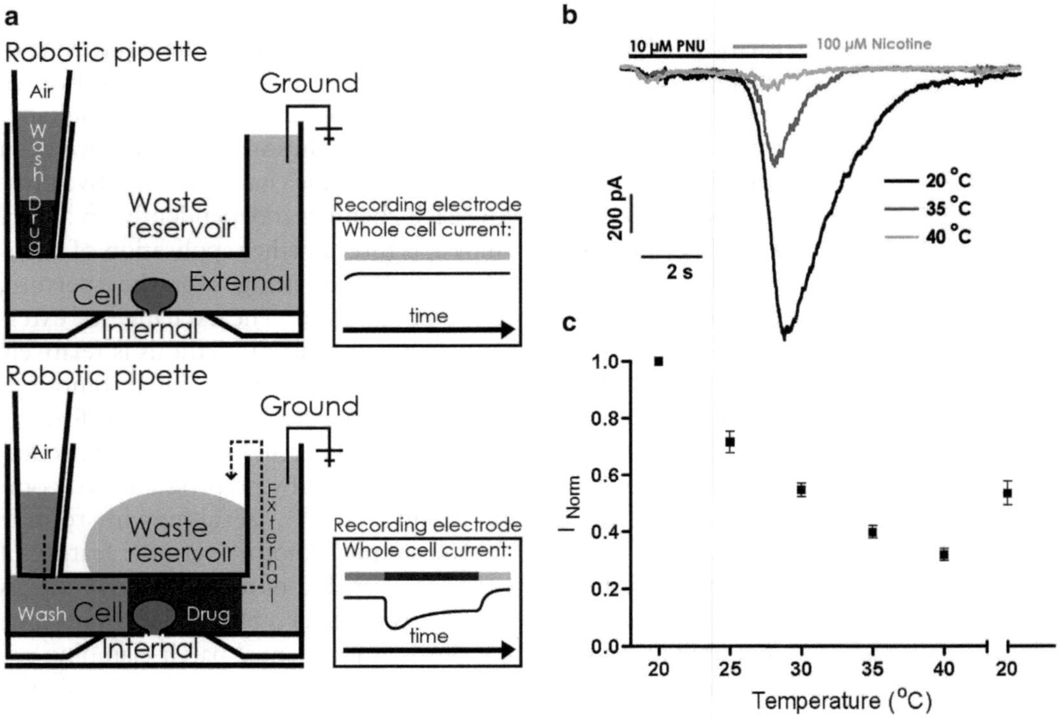

Fig. 2 (**a**) Schematic, illustrating the mechanism of double stacked solution application. Stacking solutions allow brief compound applications with timed intervals. The pipette aspirates buffer first, followed by compound. In this way, the cell is exposed to ligand first followed by rapid buffer washout. (**b**) Example responses to 100 μM nicotine in the maintained presence of 10 μM PNU120596 at three different temperatures using a triple stacked application protocol. 10 μM PNU120596 was applied to GH4C1 cells expressing the human α7nAChR, followed by a brief co-application of 100 μM nicotine, followed by wash buffer. PNU120596 shows strong temperature dependence. (**c**) Pooled data plotting the amplitude of response to 100 μM nicotine in the presence of 10 μM PNU120596 at various temperatures. Data are sampled at 1 kHz and filtered at 333 Hz. Data are normalized to the corresponding responses recorded at room temperature ($n = 10 \pm$ SEM)

While patch-clamp methods do provide a direct readout of ion channel function, it has been discussed that for some channels the data measured from patch-clamping using ion channels in heterologous expression systems might not resemble channel function in the physiological situation. This is particularly important where ion channels might be regulated by second messenger systems, auxiliary beta-subunits, or show interactions with other membrane proteins. Recent advances in the measurement of ion channel activity in primary cells, provides an approach to address some of these concerns. When measured in primary cells, ion channels reside within a more physiological environment with intact regulatory components. This becomes especially relevant when cardiac ion channels are addressed.

Measurements of heterologously expressed cardiac ion channels, especially the hERG potassium channel which is associated with cardiac arrhythmias, have become standard practice using different assays to evaluate cardiac safety in the early development of pharmaceutical compounds. The action potential (AP) of the human heart is however generated by an ensemble of ionic currents, and it has been shown that the effects of compounds on different cardiac ion channels can mitigate or aggravate potential proarrhythmic effects (11–13). In addition, different stimulation protocols and temperatures can impact the effects of compounds on ion channels (14–16).

Using planar systems to measure native channels in acutely dissociated mammalian cardiac myocytes has been challenging. However, the recent introduction of stem cell-derived cardiomyocytes to examine the electrophysiological properties of cardiac ion channels has provided a useful tool to overcome some of these hurdles. The pharmacology of a set of reference compounds on individual ionic currents as well as on the AP of these cells is in good agreement to expected values (16, 17).

It should however be stressed that the pharmacology of stem cell-derived cardiomyocytes requires further validation before it can be applied to evaluate the safety of cardiac compounds, because some modes of action of cardiac toxicity are not necessarily well detected in assays that are based on these cells (18) and they most likely represent a neonatal state which is not representative of the adult organism (19, 20). No in vitro assay is expected to provide a complete picture of potential cardiotoxicities; however, automated planar electrophysiology can provide useful relevant information to enhance our understanding and possibly improve cardiac safety of pharmacological compounds. Voltage-clamp and current-clamp modes for recordings of mouse embryonic stem cell (mESC)-derived cardiomyocytes are also discussed in this chapter.

2 Materials

2.1 Equipment and Reagents

1. NPC®-16 Patchliner Probe Selector/Quattro/Octo, PatchControl^{HT} software, NPC®-16 chips (single use, disposable). NPC®-16 electrode set (see Note 2) (Nanion Technologies GmbH).
2. Patch-clamp amplifiers (EPC-10 quadro, HEKA Instruments Inc.).
3. Computer with 24-in. TFT monitor (Dell).
4. Software for multichannel data acquisition (Patchmaster, HEKA Instruments Inc.) and analysis (GraphPad Prism 4.0; GraphPad Software Inc., La Jolla, USA or IGOR-Pro; WaveMetrics Inc., Lake Oswego, OR).
5. Cell culture flasks (T25 & T75, BD Falcon).

6. 2-, 10-, 200-, and 1,000-μL pipettes.
7. Fisherbrand AccuSpin 400 centrifuge (Fisher Scientific).
8. Dulbecco's modified Eagle's medium 4,500 mg/L glucose and GlutaMAX supplemented (DMEM, Cambrex), heat-inactivated fetal bovine serum (FBS, Biosera), geneticin (G418, Invitrogen).
9. Dulbecco's phosphate buffer saline with Ca^{2+} and Mg^{2+} (D-PBS) (Invitrogen).
10. Dulbecco's phosphate buffer saline Ca^{2+} and Mg^{2+} free (D-PBS) (Invitrogen).
11. Cell dissociation buffer Enzyme free PBS-based (GIBCO Invitrogen).
12. Trypsin/EDTA solution (Invitrogen).
13. Dimethyl sulfoxide (DMSO).
14. (−)-nicotine (Sigma Aldrich, UK), PNU120596 (Tocris Biosciences, UK).

2.2 Cell Culture

2.2.1 GH4C1-rα7 Cells

1. GH4C1 cells stably transfected with rat α7nAChRs (GH4C1-rα7 cells) are cultured and passaged in standard T75 tissue culture flasks in DMEM supplemented with 10 % FBS and 100 μL/mL geneticin at 37 °C in a humidified atmosphere composed of 95 % air and 5 % CO_2.
2. The cells should be passaged every 2–3 days using an enzyme free PBS-based cell dissociation buffer and kept to a confluence of 60–80 % (see Note 3).

2.2.2 mESC-Derived Cardiomyocytes

1. mESC-derived cardiomyocytes (Axiogenesis, Germany) are seeded at a density of 10^5 viable cells per square centimeter culture area in one T25 cell culture flask with 5 mL Cor.At Complete Culture Medium (see ref. 17 and Note 4).
2. Pre-culture mESC-derived cardiomyocytes for 2–4 days. 1 day after seeding, a continuous rhythmic contraction of the cells (beating) can already be observed (see Note 5).

2.3 Planar Patch-Clamp Recording Solutions

Prepare solutions in deionized water, filter (see Note 6), and measure the osmolarity (see Note 7) and the pH. Solutions can be stored at 4 °C for up to 5 days. Solutions should be warmed to room temperature (20–22 °C) before use.

1. Standard extracellular solution: 140 mM NaCl, 4 mM KCl, 1 mM $MgCl_2$, 2 mM $CaCl_2$, 5 mM d-Glucose monohydrate, 10 mM HEPES; adjust to pH 7.4 with NaOH. Osmolarity 298 mOsm.
2. Standard extracellular solution for enhancing seals and recording with GH4C1-rα7 cells: 80 mM NaCl, 3 mM KCl, 10 mM

$MgCl_2$, 35 mM $CaCl_2$, 10 mM HEPES (Na^+ salt); adjust pH 7.4 with HCl (see Note 8).

3. Standard intracellular solution for whole-cell recordings: 50 mM KCl, 10 mM NaCl, 60 mM KF, 20 mM EGTA, 10 mM HEPES; adjust to pH 7.2 with KOH Osmolarity: 285 mOsm (see Note 9 and ref. 21).

3 Methods

3.1 Harvesting Cells for Planar Patch-Clamp

The use of healthy cells is absolutely critical to the success of planar patch-clamp recordings because the capture of cells onto the planar chip is a random process. The health of the cell impacts on the quality of the seal formed between the cell membrane and the planar chip, which ultimately influences the quality of the recording. Unlike conventional patch-clamp recording where healthy cells can be visually selected, once cells are prepared in suspension, there is no opportunity to optically select the healthiest looking cells.

3.1.1 GH4C1-rα7 Cells

1. In sterile conditions, remove the medium from the cells and gently wash with D-PBS with Ca^{2+} and Mg^{2+} (see Note 10).
2. Pipette 2 mL cold cell dissociation buffer on to the cells and incubate at room temperature for 3 min.
3. Add 5 mL culture media and pipette up and down gently.
4. Transfer the cells in culture media from the flask to a 15 mL conical centrifuge tubes and centrifuge at $100 \times g$ for 2 min at room temperature and discard the supernatant by decanting.
5. Resuspend the cell pellet in 2 mL of extracellular solution, centrifuge at $100 \times g$ for 2 min at room temperature and discard the supernatant by decanting.
6. Resuspend the cell pellet in a mixture of extracellular solution for capturing cells and culture media (50:50 ratio, see Note 11) at a density of 1×10^6 to 5×10^7 cells/mL (see Note 12) (cells can be counted using a hemocytometer).
7. Transfer the cells to the cell hotel on the Patchliner where they are continuously pipetted up and down to maintain single cells and viability.

3.1.2 mESC-Derived Cardiomyocytes

1. Wash twice in 5 mL of D-PBS with Ca^{2+} and Mg^{2+} and incubate at 4 °C for 15 min.
2. Wash once with 5 mL of D-PBS without Ca^{2+} and Mg^{2+}.
3. Dissociate with 2 mL pre-warmed trypsin/EDTA solution for 4–5 min at 37 °C and 5 % CO_2 (see Note 13).
4. Transfer cell suspension to 6 mL Cor.At Complete Culture Medium in a 50 mL tube.

5. Centrifuge for 2 min at $100 \times g$.
6. Discard supernatant and resuspend cell pellet to a final density of 1×10^6 to 5×10^7 cells/mL in external recording solution and transfer into the cell hotel (see Note 14).

3.2 Automated Planar Patch-Clamp Using Patchliner

PatchControl^{HT} software is coupled to Patchmaster software and when this program is opened, Patchmaster also opens. This allows amplifier functions and data acquisition to be controlled through commands in the experimental protocol (PatchControl^{HT} *Tree*). Preprogrammed experimental protocols can be loaded and modified for optimization with different cell types/characteristics. Chip cartridge selection is based on the resistance of the hole in the chip (aperture) and the size of the cells (see Note 15). The motorized stage (chip wagon), which houses the chip cartridges, can accommodate three cartridges allowing for 48 recordings. When a *Tree* is activated, recording solutions are dispensed by the pipetting robot into the microfluidic chambers of the chip cartridge and the cartridge is moved into the measuring head, which contains the pneumatic and electric contacts and moves up and down to address the chip cartridges. The recording head can accommodate up to eight chip chambers. A slight positive pressure is applied to each chip chamber, independently, and the offsets are corrected. Once the cell suspension is added, a small suction of 50 mBar is applied to attract cells to the holes and this leads to a small increase in the seal resistance. The seal resistance is increased to a gigaohm seal when seal enhancing solution is applied to the cells along with suction pulses and the application of a negative voltage to the cells. The whole-cell access is achieved by short suction pulses. For some cells it is helpful to support this process by zapping to help rupture the patch of membrane (see Note 16). The software adapts the applied negative pressure according to parameters such as chip resistance, series resistance, and slow capacitance. In this way, the program can determine if a cell has been sealed to the chip aperture and whether the parameters correspond to the cell-attached or the whole-cell configuration. Cells that do not meet selected quality control parameters (e.g., seal resistance, series resistance) will be disabled at this stage. The user can adjust pressure, voltage, and condition settings according to the cell type/characteristics.

3.2.1 General Experimental Set-up

1. Load a preprogrammed *Tree* and select the edit mode to make modifications according to your cell type/characteristics. This enables you to modify amplifier and pump parameters and also allows access to the full range of commands, which can be inserted into a protocol via a generic drag-and-drop function (see Note 17).
2. Select the appropriate chip resistance for your cells and place three chip cartridges on the chip wagon.

3. Prepare compound solutions directly before each experiment (see Note 18).
4. Place recording solutions and compounds in position, according to those defined in the job list (see Note 19).
5. Place the cells into the cell hotel, where they will be aspirated every 30 s throughout the experiment to prevent clumping and sedimentation (see Note 20).
6. Select and activate the initialization folder to initialize the robot and wash the pipette. This also generates a new data file within the HEKA software and sets all amplifier and robot parameters to reasonable starting values. This folder only needs to be activated once at the beginning of each day of experiments.
7. The robot will start when the *Tree* is activated. At the end of each run, it will loop back to the start and continue this process until all chips on the chip wagon have been used.

3.2.2 Stacked Solution Application with GH4C1-rα7 Cells

As with most ligand gated ion channels, nAChRs exhibit receptor desensitization, which is a common phenomenon for ligand gated ion channels. The kinetics and level of desensitization of ligand gated ion channels are determined by ligand concentration and exposure time, or both. For rapidly desensitizing ion channels, it is important that compound application is fast, so that the entire ion channel population is exposed to maximum concentration before entering the desensitized state. Therefore rapid solution exchange combined with brief drug exposure times can minimize or correct for the deleterious effect caused by receptor desensitization. This is achieved by using a stacked solution application where two or three zones of solution are aspirated into the pipette before administration onto the cells (see Note 21).

1. Load a preprogrammed ligand *Tree* for triple stacked solution application. Volumes and speeds of applications can be adjusted (see Note 22).
2. Select high resistant single hole chips (5–6 MΩ, ~10 pF) for GH4C1-rα7 cell recordings and load three chips onto the chip wagon.
3. Within the *Tree*, adjust the holding potential to −75 mV and select a continuous recording protocol, with a 13 s sweep, from the pulse generator file in the job list to record the fast ligand activated currents.
4. Follow steps 3–6 in Subheading 3.2.1 (see Note 23).

In the example shown in Fig. 2b GH4C1-rα7 cells were exposed to the modulator PNU120596 (10 μM), directly followed by nicotine (100 μM) in the presence of PNU120596 and subsequently by wash buffer using a triple stacked solution application.

In the presence of PNU120596 at room temperature (20 °C) α7nAChR desensitization was largely not apparent.

3.2.3 Temperature Controlled Compound Application with GH4C1-rα7 Cells

The measuring head, chip wagon, and solution inside the pipette can be heated simultaneously to maintain physiological temperatures during the course of an experiment. Alternatively, the solution in the pipette can be heated independently, exposing the cells to transient temperature increases, for the study of heat activated channels.

1. Load a preprogrammed ligand *Tree* for triple stacked solution application combined with temperature control. Temperature settings can be adjusted (see Note 24).
2. Select high resistant single hole chips (5–6 MΩ, ~10 pF) for GH4C1-rα7 cell recordings and load three chips onto the chip wagon.
3. Within the *Tree*, adjust the holding potential to −75 mV and select a continuous recording protocol, with a 13 s sweep, from the pulse generator file in the job list to record the fast ligand activated currents.
4. Follow steps 3–6 in Subheading 3.2.1.

In the examples shown in Fig. 2c, GH4C1-rα7 cells were exposed to nicotine in the presence of the PNU120596 at different temperatures (20, 25, 30, 35, 40, and 20 °C). Responses are greatly attenuated when the temperature is raised to physiological levels and higher. Current is partially recovered when the temperature is lowered back to 20 °C. Activation of α7nAChRs in the absence of modulator is not significantly inhibited by elevated temperatures (9). Therefore, these data demonstrate the strong temperature dependent action of PNU120596 at α7nAChRs.

3.2.4 Current and Voltage Clamp Recordings with mESC-Derived Cardiomyocytes

mESC-derived cardiomyocytes can be measured on Patchliner in the voltage-clamp mode and current-clamp mode, to assess the electrophysiological properties of ionic currents and APs, respectively.

1. Load a preprogrammed *Tree* for cardiomyocytes. Potassium (I_K), sodium (I_{Na}), and L-type calcium ($I_{Ca,L}$) currents are recorded first in the voltage-clamp mode and then APs are subsequently recorded in the current-clamp mode (see Note 25).
2. Select low resistant single hole chips (1–2 MΩ, cell size ~30 pF) for mESC-derived cardiomyocyte recordings and load three chips onto the chip wagon.
3. Follow steps 3–6 in Subheading 3.2.1.

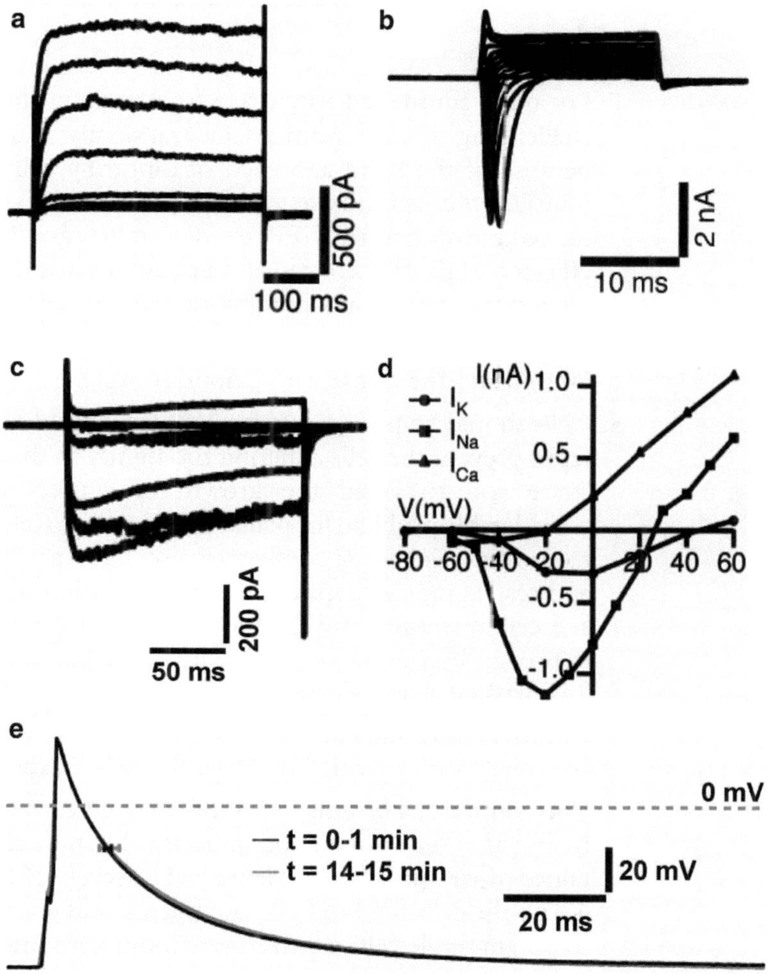

Fig. 3 (**a–c**) Parallel voltage clamp recordings of mESC-derived cardiomyocyte ion channel currents in the automated patch-clamp system Patchliner. I_K (**a**), I_{Na} (**b**), $I_{Ca,L}$ (**c**) current traces and corresponding current–voltage relationship curves (**d**). (**e**) APs in control solution are highly reproducible over a time period of 15 min. Representative traces are shown at the beginning (t=0–40 s) and after t=14 min 20 s to 15 min, traces represent the mean of four consecutively recorded APs. The error bars represent the mean at APD_{50} SEM. Sweep interval was 10 s

In the example shown in Fig. 3, to induce I_K and I_{Na} currents, cells were held at a holding potential of −80 and −100 mV, respectively and stepped in 20 mV increments to +60 mV. Cells were held at a holding potential of −80 mV and stepped in 10 mV increments to +60 mV to induce $I_{Ca,L}$ currents. Typical current recordings of I_K, I_{Na}, and $I_{Ca,L}$ in the voltage-clamp mode are shown in Fig. 3a–c. Representative traces demonstrate how highly reproducible APs are over time (Fig. 3e).

4 Notes

1. For most automated patch-clamp systems on the market it is challenging to use primary or transiently transfected cells because of the blind approach of capturing cells. Transiently transfected cells can have low expression rates so it is advisable to co-transfect with GFP to determine good transfection efficiency. Primary cells can also be problematic if the cell preparation is impure containing more than one cell type.
2. Place electrodes in bleach filled chloridation chambers for 30 min and then rinse with deionized water.
3. Cells should be passaged every 2–3 days with dissociation buffer to prevent the cells adhering too tightly to the support substrate and to avoid the growth of clusters which are a considerable problem for planar patch-clamp. Cells left to grow to greater than 80 % confluency also leads to cell aggregates and results in poor capture rates. Dissociation buffers encourage cell separation and isolation. Seal rates can sometimes be improved by optimizing cell culture methods, e.g., by using T75 flasks instead of smaller T25.
4. Cor.At culture medium is a complete medium that has been optimized for use with Cor.At mESC-derived cardiomyocytes.
5. The continuous beating of mESC-derived cardiomyocytes up to and after 4 days is a good indicator that the cells have maintained their quality and viability. However, after more than 4 days the cells tend to form aggregates, and it is more difficult to obtain single cell suspensions without harming them, which reduces the success rate.
6. All recording solutions should be filtered. The internal solution is particularly important and should be filtered with a 0.22 μm-pore diameter filter on the day of the experiment. The internal solution should not be older than a few days, and should be stored at 4 °C.
7. Recording solution osmolarity is measured directly using a freezing-point osmometer and adjusted if necessary with mannitol or sucrose, non-permeating molecules, to 285 mOsm/L for all internal solutions and between 290 and 310 mOsm/L for all external solutions. External osmolarity should always be higher than the internal osmolarity.
8. It is helpful to use seal enhancer solution to increase the chance of giga-seal formation. The extracellular seal enhancer solution is a high calcium containing solution, whereas the intracellular solution for capturing cells contains fluoride. These seal enhancing solutions are usually replaced once a giga-seal is achieved and before establishing whole-cell access. To avoid

calcium-dependant inactivation of voltage-gated calcium channels and increase calcium channel currents, calcium in the extracellular solution can be replaced by barium as charge carrier since most voltage-gated calcium channels also conduct barium ions (22, 23).

9. Inclusion of fluoride has long been known to improve patch-clamp sealing and stabilizes the cell membrane, resulting in longer, more stable patch-clamp recordings (21). The mechanism of this effect is unknown.
10. Enzyme free PBS-based cell dissociation buffer should be stored at 4 °C and used at this temperature in sterile conditions. The buffer is designed for gentle dissociation of mammalian cells from support substrates and each other. It is recommended for use in studies requiring intact cell surface proteins such as ligand binding. This buffer is not recommended for routine passage of strongly adherent cell types.
11. Cells in suspension remain viable for up to 4 h, stored at room temperature, when resuspended in a mixture of recording solution and culture media (50:50).
12. Cell densities of 1×10^6 to 5×10^7 cells/mL for use on the Patchliner appear to work well for most cells. For some cell types (especially more challenging primary cells) it might however be difficult to obtain this number of cells. For a number of cells we have had good success rates with much lower cell densities, down to below 1,000 cells/mL.
13. Several different detachment agents can be used for harvesting the cells, e.g., accutase or detachin, although in our experience, the standard trypsin/EDTA solution works very well. Trypsin alone yields good results for robust cell types and channels, but can have degrading effects on cells and/or channels, and EDTA alone does not always sufficiently detach the cells from the flask, or results in insufficient separation of aggregated cells. Increased temperature (e.g., 37 °C) will increase the speed of the dissociation process, while reduced temperature (e.g., 4 °C) will slow it down.
14. When resuspending the mESC cardiomyocytes pellet, pipette up and down twice very slowly and gently. When you pipette the cells into the cell hotel this will be the third time. Let the cells rest for 5 min before activating the cell hotel. If the cells are not handled with great care at this stage the cell membranes will rupture and seals will not be formed. Following the recovery period the cells need to be pipetted more frequently in the cell hotel and this feature is already incorporated into the preprogrammed *Tree* for these cells.
15. Selection of the appropriate NPC-16 chips is based on cell size and chip resistance (aperture size). NPC-16 chip resistance can

be manufactured within a range from 1 to 8 MΩ with an accuracy of ±0.5 MΩ for resistances between 1 and 4 MΩ and ±1 MΩ for resistances between 5 and 8 MΩ, making them suitable for a variety of cell types of different sizes. Chips are generally manufactured in the following ranges: low resistance (1–2 MΩ), medium resistance (2–4 MΩ) and high resistance (5–6 MΩ), but can be custom made according to specific requirements. Ensemble chips are also available, with four or eight holes per chip with low, medium and high resistance and, because the currents are summated, these are ideal for enhancing current size when recording from cells which express very small currents.

16. If it is difficult to establish whole-cell then try harvesting the cells 1 day after plating. In addition, adjust the size of the high voltage pulses to 600–800 mV ("zap"), which can be applied to help rupture the patch of membrane, thus establishing the whole-cell configuration.
17. A full range of preprogrammed *Trees* for specific cell and channel types with various experimental protocols can be accessed in PatchControl^{HT}, e.g., sodium channel current–voltage relationship (NaIV), potassium pharmacology (Kvpharm). There are various aspects of a *Tree* that can be optimized, including suction parameters to capture cells, form giga-seals and establish whole-cell, as well as voltage paradigms. For more information, refer to the PatchControl^{HT} application notes (http://www.nanion.de). Furthermore, once a *Tree* has been optimized it is not usually necessary to modify it further for future use with the same cell type.
18. Observe the solubility of your compounds in solution, as some compounds precipitate over time. Store all compound solutions, where possible, in glassware as some compounds will adhere to other substrates. Adhesion or precipitation can lead to an underestimation of concentration.
19. Compound position, volume, and speed of application are defined in the job list within the *Tree*. It is also here that the user can select the appropriate Patchmaster pulse generator file (pgf).
20. Regular aspiration of the cell suspension in the cell hotel, prevents clumping and sedimentation and at the same time improves viability. We find that cells retain good viability for at least 4 h after they have been prepared in suspension, although some deterioration in success rates has been observed after 3 h.
21. For a double stack with two zones of solution, the wash buffer is aspirated first, directly followed by the ligand resulting in the ligand zone being applied to the cells first followed immediately by the wash buffer (Fig. 2a). To examine the effects of a

ligand modulator, a triple stack with three solution zones should be used. The wash buffer is aspirated first, followed directly by solution containing both ligand and modulator, followed immediately by modulator.

22. The volumes of the different solution zones can be adjusted, as well as the speed of application, enabling exposure times of as little as 100 ms. Typical volumes for a triple stacked application are 150 μL wash buffer zone, 10–50 μL ligand/modulator zone and 125 μL modulator zone, but these can be optimized for different ligands and modulators accordingly. The maximum volume that the pipette can aspirate at one time is 350 μL. The zone containing the ligand should be applied more rapidly (e.g., 171 μL/s; speed 12) than the other zones (e.g., 24 μL/s; speed 20).

23. PNU120596 and nicotine are dissolved in DMSO at a concentration of 10 mM and then diluted with standard extracellular solution for enhancing seals, which in these experiments is used for making the recordings.

24. The temperature can be set between room temperature and 80°C for the chip wagon, the measuring head and the pipette, independently. The temperature can be raised, but not lowered. The temperature controls for the chip wagon and the measuring head are used to keep the temperature constant for measurements at physiological temperatures. For temperature regulated channels, like TRPV's, it is possible to apply short pulses at elevated temperatures by heating the pipette. The cells survive temperatures up to 70 °C, only when applied for a very short time.

25. After evaluation of ionic currents in the voltage-clamp mode, the *Tree* switches the amplifiers into current-clamp mode and proceeds on to AP measurements. To induce AP responses in mESC-derived cardiomyocytes, cells are held at a constant membrane potential, i.e., between −80 and −100 mV (but this is cell-type dependent). The program then finds the stimulus threshold for each individual cell by applying a 1 ms depolarizing pulse. The values for the thresholds are then set in the pgf in Patchmaster. PatchControlHT then loads another pgf and the Cclamp protocol uses these set values so that each individual cell is stimulated accordingly.

Acknowledgments

Thanks to Dr Sonja Stoelzle and Dr Niels Fertig (Nanion Technologies GmbH) for cardiomyocyte data.

References

1. Neher E, Sakmann B (1976) Single channel currents recorded from membrane of denervated frog muscle fibres. Nature 260: 799–802
2. Farre C, Stoelzle S, Haarman C, George M, Brueggemann A, Fertig N (2007) Automated ion channel screening: patch clamping made easy. Expert Opin Ther Targets 11:557–565
3. Brüggemann A, Farre C, Haarmann C, Haythornthwaite A, Kreir M, Stoelzle S, George M, Fertig N (2008) Planar patch clamp: advances in electrophysiology. Methods Mol Biol 491:165–176
4. Majeed Y, Bahnasi Y, Seymour VAL, Wilson LA, Milligan CJ, Agarwal AK, Sukumar P, Naylor J, Beech DJ (2010) Rapid and contrasting effects of rosiglitazone on transient receptor potential TRPM3 and TRPC5 channels. Mol Pharmacol 79:1023–1030
5. Xu SZ, Sukumar P, Zeng F, Li J, Jairaman A, English A, Naylor J, Ciurtin C, Majeed Y, Milligan CJ, Bahnasi YM, Al-Shawaf E, Porter KE, Jiang LH, Emery P, Sivaprasadarao A, Beech DJ (2008) TRPC channel activation by extracellular thioredoxin. Nature 451:69–72
6. Li J, Sukumar P, Milligan CJ, Kumar B, Ma ZY, Munsch CM, Jiang LH, Porter KE, Beech DJ (2008) Interactions, functions, and independence of plasma membrane STIM1 and TRPC1 in vascular smooth muscle cells. Circ Res 103:e97–104
7. Milligan CJ, Li J, Sukumar P, Majeed Y, Dallas ML, English A, Emery P, Porter KE, Smith AM, McFadzean I, Beccano-Kelly D, Bahnasi Y, Cheong A, Naylor J, Zeng F, Liu X, Gamper N, Jiang LH, Pearson HA, Peers C, Robertson B, Beech DJ (2009) Robotic multiwell planar patch-clamp for native and primary mammalian cells. Nat Protoc 4:244–255
8. Naylor J, Li J, Milligan CJ, Zeng F, Sukumar P, Hou B, Sedo A, Yuldasheva N, Majeed Y, Beri D, Jiang S, Seymour VAL, McKeown L, Kumar B, Harteneck C, O'Regan D, Wheatcroft SB, Kearney MT, Jones C, Porter KE, Beech DJ (2010) Pregnenolone Sulphate- and Cholesterol-Regulated TRPM3 Channels Coupled to Vascular Smooth Muscle Secretion and Contraction. Circ Res 106:1507–15
9. Hurst RS, Hajos MM, Raggenbass M, Wall TM, Higdon NR, Lawson JH, Rutherford-Root KL, Mitchell B, Berkenpas MB, Hoffmann WE, Piotrowski DW, Groppi VE, Allaman G, Ogier R, Bertrand S, Bertrand D, Arneric SP (2005) A novel positive allosteric modulator of the α7 neuronal nicotinic acetylcholine receptor: in vitro and in vivo characterization. J Neurosci 25:4396–4405
10. Sitzia F, Brown JT, Randall AD, Dunlop J (2011) Voltage and temperature-dependent allosteric modulation of α7 nicotinic receptors by PNU120596. Front Pharmacol 2:1–9
11. Witchel HJ, Hancox JC, Nutt DJ (2003) Psychotropic drugs, cardiac arrhythmia and sudden death. J Clin Psychopharmacol 23: 58–77
12. Fenichel RR, Malik M, Antzelevitch C, Sanguinetti M, Roden DM, Priori SG, Ruskin JN, Lipicky RJ, Cantilena LR (2004) Drug-induced torsades de pointes and implications for drug development. J Cardiovasc Electrophysiol 15:475–495
13. Möller C, Witchel H (2011) Automated electrophysiology makes the pace for cardiac ion channel safety screening. Front Pharmacol 2:1–7
14. Zhou Z, Gong Q, Ye B, Fan Z, Makielski JC, Robertson GA, January CT (1998) Properties of HERG channels stably expressed in HEK 293 cells studied at physiological temperature. Biophys J 74:230–241
15. Polonchuk L (2012) Toward a new gold standard for early safety: automated temperature-controlled hERG test on the Patchliner®. Front Pharmacol 3:1–7
16. Stoelzle S, Obergrussberger A, Brüggemann A, Haarmann C, George M, Kettenhofen R, Fertig N (2011) State-of-the art automated patch clamp devices: heat activation, action potentials, and high throughput in ion channel screening. Front Pharmacol 2:1–11
17. Stoelzle S, Haythornthwaite A, Kettenhofen R, Kolossov E, Bohlen H, George M, Brüggemann A, Fertig N (2011) Automated Patch Clamp on mESC-Derived Cardio myocytes for Cardiotoxicity Prediction. J Biomol Screen 16:910–6
18. Jonsson MK, Vos MA, Mirams GR, Duker G, Sartipy P, de Boer TP, van Veen TA (2012) Application of human stem cell-derived cardiomyocytes in safety pharmacology requires caution beyond hERG. J Mol Cell Cardiol 52:998–1008
19. Binah O, Dolnikov K, Sadan O, Shilkrut M, Zeevi-Levin N, Amit M, Danon A, Itskovitz-Eldor J (2007) Functional and developmental properties of human embryonic stem cells-derived cardiomyocytes. J Electrocardiol 40: S192–S196

20. Xi J, Khalil M, Shishechian N, Hannes T, Pfannkuche K, Liang H, Fatima A, Haustein M, Suhr F, Bloch W, Reppel M, Saric T, Wernig M, Janisch R, Brockmeier K, Hescheler J, Pillekamp F (2010) Comparison of contractile behavior of native murine ventricular tissue and cardiomyocytes derived from embryonic or induced pluripotent stem cells. FASEB J 24:2739–2751
21. Kostyuk PG, Krishtal OA, Pidoplichko VI (1975) Effect of internal fluoride and phosphate on membrane currents during intracellular dialysis of nerve cells. Nature 23: 691–693
22. Veselovskii NS, Fedulova SA (1986) Effects of substituting barium for calcium ions during research into inward currents in mammalian neurons. Neurophysiology 18:227–231
23. Ferreira G, Yi J, Rios E, Shirokov R (1997) Ion-dependent inactivation of barium current through L-type calcium channels. J Gen Physiol 109:449–461

Chapter 14

Recording Single-Channel Currents Using "Smart Patch-Clamp" Technique

Anamika Bhargava and Julia Gorelik

Abstract

Microdomains that form on the plasma membrane of cells are essential for signalling compartmentation within cells. The localization of ion channels in these surface microdomains is important in defining what signalling cascades will be generated. For example, in cardiomyocytes, similar to other excitable cells, action potential propagation depends essentially on the properties of ion channels that are functionally and spatially coupled. In this chapter we describe a novel advanced patch-clamp technique, "Smart patch-clamp," which enables the study of functional ion channels in the cell surface microdomains in a wide variety of biological cells and tissues. Smart patch-clamp combines conventional patch-clamp and Scanning Ion Conductance Microscopy (SICM). SICM uses a glass micropipette as scanning probe and generates a high-resolution topography image of the cell surface. Next, same micropipette is used as a patch-clamp pipette to record ion channel signals from specific spots on the cell surface as determined by SICM. In this chapter we focus on recording single channel L-type calcium channel currents from T-tubules of adult rat cardiomyocytes.

Key words Smart patch-clamp, Scanning ion conductance microscopy, Cardiomyocyte, L-type calcium channel, Single-channel recording

1 Introduction

The spatial distribution of ion channels on the cell surface has been investigated by microscopy based techniques such as immunochemistry (1, 2). However, these techniques provide little information on the functional characteristics of ion channels. By contrast, the patch-clamp technique provides information about ion channel currents and hence functionality of ion channels on the cell surface. However, the conventional patch-clamp technique does not allow the precise selection of a region of interest on the cell to be investigated and thus provides limited spatial resolution (3, 4).

Smart patch-clamp technique combines Scanning Ion Conductance Microscopy (SICM) and patch-clamp electrophysiology using a glass micropipette as a dual probe: scanning probe

Nikita Gamper (ed.), *Ion Channels: Methods and Protocols*, Methods in Molecular Biology, vol. 998,
DOI 10.1007/978-1-62703-351-0_14, © Springer Science+Business Media, LLC 2013

and patch-clamp pipette (5, 6). In this method, the glass micropipette first scans the cell surface using current feedback and a high-resolution topographic image of the cell surface is obtained. The same glass micropipette is then positioned over the region of interest on the cell surface as determined by SICM and single ion channel currents are recorded using patch-clamp technique. Therefore, smart patch-clamp can record ion channel activity from precise microdomains on the surface of live cells and thus can generate a spatial functional map of surface ion channels. In contrast, in conventional patch-clamp method the position of the pipette with respect to the cell topography cannot be controlled with a nanoscale precision. This technique, though recently developed, has already led to the determination of the spatial distribution of ion channels which may help to unravel the mechanisms of the localized control of cell signalling and function (6). In this chapter we describe the smart patch-clamp technique and its application to various cell types.

1.1 Principle of Operation of SICM

SICM is a microscopy, which generates high-resolution topography images of live cells. A glass micropipette used as the scanning probe, scans over the cell surface at the distance of one pipette tip radius without touching the surface. SICM is based on the phenomenon that the ion current through the solution-filled micropipette is partially occluded when the pipette approaches the surface of a cell (7). Therefore, the position of the tip of the pipette relative to the cell surface strongly influences the ion current through the pipette. This ion current is digitized and fed into the feedback and scan control system which provides the feedback signal to control the vertical position of the tip of pipette keeping a constant pipette-sample separation (Fig. 1). Distance-modulated feedback control system is designed to maintain the ion current through the pipette constant and this results in keeping pipette-sample separation constant (8). The ion current through the pipette depends on overall resistance of the tip, which is combination of the resistance of the micropipette itself (Rp) and the access resistance (Ra) of the micropipette opening. Access resistance is a complex function of the distance between the sample and the probe, and the geometry and electrochemical properties of the sample surface. The current (I) through the pipette, which is measured directly, is given by:

$$I = V / \left(\mathrm{Rp} + \mathrm{Ra}\,(d)\right)$$

where "V" is the voltage applied to the electrode and "d" is the distance between sample and the probe.

In the first implementation of SICM a continuous feedback mechanism kept the pipette always in the proximity of the sample surface by moving the pipette up and down while the pipette was raster scanning the sample (9, 10). However, because the tip of the

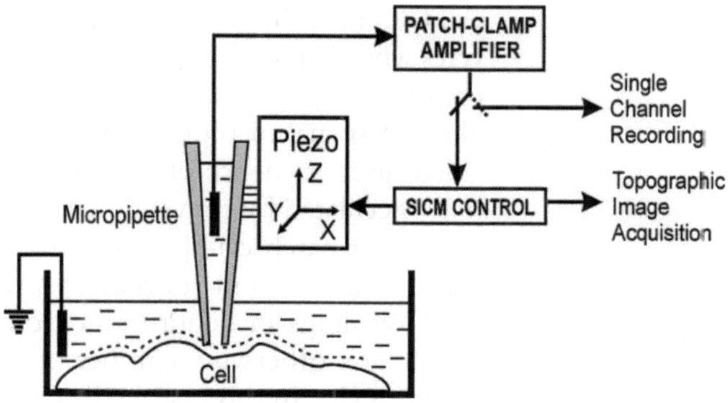

Fig. 1 Schematic of the smart patch-clamp setup. The micropipette is mounted on a three-axis piezo actuator controlled by a computer. The ion current that flows through the pipette is measured by a patch-clamp amplifier, and it is used for the feedback control to keep a constant distance between the micropipette and the sample during scanning. Upon completion of the scanning protocol, computer control is used to position the micropipette at a place of interest based on the topographic image acquired, and finally the same patch-clamp amplifier is used for electrophysiological recording. Modified from ref. 5

probe was very close to the sample, it had limited ability to scan cell surfaces with prominent tall structures such as microvilli. When the structures were taller than the pipette-sample separation the probe often broke itself or damaged the surface structures.

Therefore the technique was further developed and "hopping mode" SICM was introduced (11). In hopping mode SICM, at each imaging point, the pipette approaches the sample from a starting position that is well above any of the surface features. At this starting position the reference current is measured at an applied voltage. The pipette then approaches the cell surface until the current is reduced by a predefined set-point, usually 0.25–1 %. The position of *z*-dimension actuator when the current achieves this reduction is recorded as the height of the sample at this imaging point. The *z*-dimension actuator then withdraws the pipette away from the cell surface to the starting point and the sample is moved laterally to the next imaging point. The reference current is continuously updated while the pipette is away from the surface to account for any slow drifts in the pipette current. This achieves unprecedented accuracy in revealing the topography of various cell surfaces.

1.2 Applications of Smart Patch-Clamp

Since its development in 2002 by Yuri Korchev and colleagues (5, 6), smart patch-clamp technique has been applied in a wide variety of experimental settings. The most popular application of smart patch-clamp is recording ion currents from structures that are too small to be resolved conventionally or that cannot be detected by light microscopy. The best example of this kind is L-type calcium channel

(LTCC) recordings from T-tubules of isolated adult rat cardiomyoyctes where these channel are preferentially localized and are in close proximity to other proteins involved in excitation-contraction coupling. A spatial map of LTCC obtained by smart patch-clamp provided direct evidence of functional ion channel location which was only speculated by conventional microscopy techniques (6).

Smart patch-clamp can be used to record ion currents from very small cells and subcellular structures (e.g., sperm cells, fine focal swellings in dendritic neurons, synaptic neuronal boutons) which are otherwise cumbersome to patch with conventional methods (5). In this case feedback control is very useful as the glass micropipette will not touch the cell until it is desired and feedback control is switched off; therefore there is no danger of damage to the cell or pipette.

Another remarkable application of smart patch-clamp is ion channel recordings from precise location on non-transparent samples (e.g., aorta, brain slices). It has been otherwise impossible to get information about the position or type of cellular structures from which the recordings are being made.

1.3 Future of Smart Patch-Clamp

The full potential of smart patch-clamp is yet to be explored. For instance, smart patch-clamp can potentially be used to identify electrophysiological changes associated with morphological changes that are associated with cell differentiation from precursor cells. This approach may prove very useful in characterizing cells when differentiating from either embryonic stem-cells or induced pluripotent stem cells.

Another useful implementation would be to combine smart patch-clamp with studying cell signalling. Ion channels are membrane proteins involved in protein–protein interactions and are also target proteins for many effector molecules such as second messengers. Therefore smart patch-clamp can be applied to study the role of ion channel in cell signalling in precise microdomains of the cell. Modulation of ion channels by signalling molecules such as cAMP or cGMP can be studied together with the use of biosensors for these signalling molecules. Successful recent study of cAMP signalling in different membrane domains of cardiomyocytes using SICM and imaging techniques presented possible application of smart patch-clamp to cell signalling studies (13).

2 Materials

2.1 Cells

Smart patch-clamp can be applied to a wide variety of cells ranging from large adult rat cardiomyocytes to small cells like sea urchin spermatozoa and even to subcellular structures such as neuronal synaptic boutons (5, 6). Isolated live cells can be evaluated for a number of ion channels and receptors. In this chapter we describe the study of LTCCs on the surface of isolated adult rat cardiomyocytes. The procedure for isolation of adult rat cardiomyocytes from whole rat hearts has been described in details elsewhere (12).

2.2 Recording Solutions

Smart patch-clamp similarly to conventional patch-clamp requires specific ionic solutions to be used for each ion channel. For example for recording LTCC currents from adult cardiomyocytes in cell-attached voltage-clamp mode following solutions can be used:

1. Pipette solution: 90 mM $BaCl_2$, 10 mM Sucrose, 10 mM HEPES, pH 7.4 with TEA-OH; ~250 mOsm.
2. Bath solution: 120 mM k-gluconate, 25 mM KCl, 2 mM $MgCl_2$, 1 mM $CaCl_2$, 2 mM EGTA, 10 mM Glucose, 10 mM HEPES, pH 7.4 with NaOH; ~290 mOsm.

2.3 Micropipettes

Pipettes used in smart patch-clamp are essentially no different from those used in conventional patch-clamp. The best and most commonly used material is borosilicate glass. We use borosilicate glass capillaries of 1.00 mm outer and 0.5 mm inner diameters (Intracel LTD or Harvard apparatus). However, the size of pipette tip used in smart patch-clamp is much smaller than in conventional patch-clamp in order to obtain a high-resolution topography image. Once pulled using a laser based puller, there is no need for additional fabrication such as fire polishing or Sylgard coating for smart patch-clamp pipettes. We pull pipettes with inner tip diameter of about 100 nm and when filled with pipette solution that gives an average tip resistance of 100 MΩ.

2.4 Pipette Puller

Although any puller can be used, from our experience laser based puller (P-2000, Sutter Instrument Co., San Rafael, CA) is a reliable instrument for reproducible pulling of pipettes of same shape and size.

2.5 Instrumentation for SICM

The instrument uses an inverted microscope (e.g., Diaphot 200 or TE200; Nikon Corporation, Tokyo, Japan). On the microscope stage a computer controlled three axis piezotranslation system with 100×100 μm *x–y* piezo-stage for sample positioning and 25 μm *z*-axis piezo-actuator for pipette movement is mounted (ICnano sample scan system, Ionscope Ltd, UK). The glass micropipette is mounted on the *z*-piezo stage and connected to the headstage of an amplifier. A number of high quality patch-clamp amplifiers are now available from a number of suppliers. The most common suppliers are HEKA (Germany) and Molecular devices (Sunnyvale, CA, USA). Smart patch-clamp can use any of the commercial amplifiers with either a resistive or capacitive feedback headstage like Axopatch 200B or Multiclamp 700B (Molecular devices Sunnyvale, CA). Ag/AgCl electrodes, in the bath and pipette, provide an electrical connection in a conventional electrophysiological circuit.

2.6 Instrumentation for Patch-Clamp

Smart patch-clamp uses the electronic components of a conventional patch-clamp setup: a patch-clamp amplifier that is used by both SICM and patch-clamp recordings, a digitizer (*analog-to-digital converter*), and a computer with appropriate software. Digitizer and compatible software (pCLAMP) for data acquisition are available easily from molecular devices (Sunnyvale, CA); *see* also Chapter 7 of this book for further details.

3 Methods

1. Plate cells onto culture dishes or glass coverslips coated with laminin (see Note 1).
2. Once the cells adhere to the culture dish, wash once with external recording solution and mount on the microscope for smart patch-clamp bathed in the external recording solution.
3. Pull a glass pipette using borosilicate glass capillary, about 100 MΩ resistance when filled with internal recording solution (see Note 2), and mount on the piezo stage.
4. Lower the pipette into the external recording solution and adjust for any current offset using amplifier's offset protocol.
5. Apply a command holding voltage ranging from 100 to 200 mV to measure the ion current flowing through the pipette (see Notes 3 and 4).
6. Using the scanning software select a current feedback set-point where the Z piezo is hopping with regular movements.
7. Perform SICM of the desired cell surface in the hopping mode by computer controlled software.
8. Once a high-resolution topographic image is obtained, click on the coordinates of this image to precisely position the pipette onto the desired structure on the cell surface (exemplified in Fig. 2).
9. Turn feedback off through the scanning software and lower the pipette manually with 20–30 nm step movements.
10. Monitor the pipette resistance when lowering the pipette, and apply suction when an increase in the resistance is seen. A GΩ seal is usually obtained at this point.
11. Perform single ion channel recording by conventional patch-clamp technique using pCLAMP software with any configuration desired, e.g., cell-attached, inside-out or outside-out mode (exemplified in Fig. 3; see Notes 5–7). Some further details on single channel recording techniques can be found in Chapters 7 and 10 of this book.

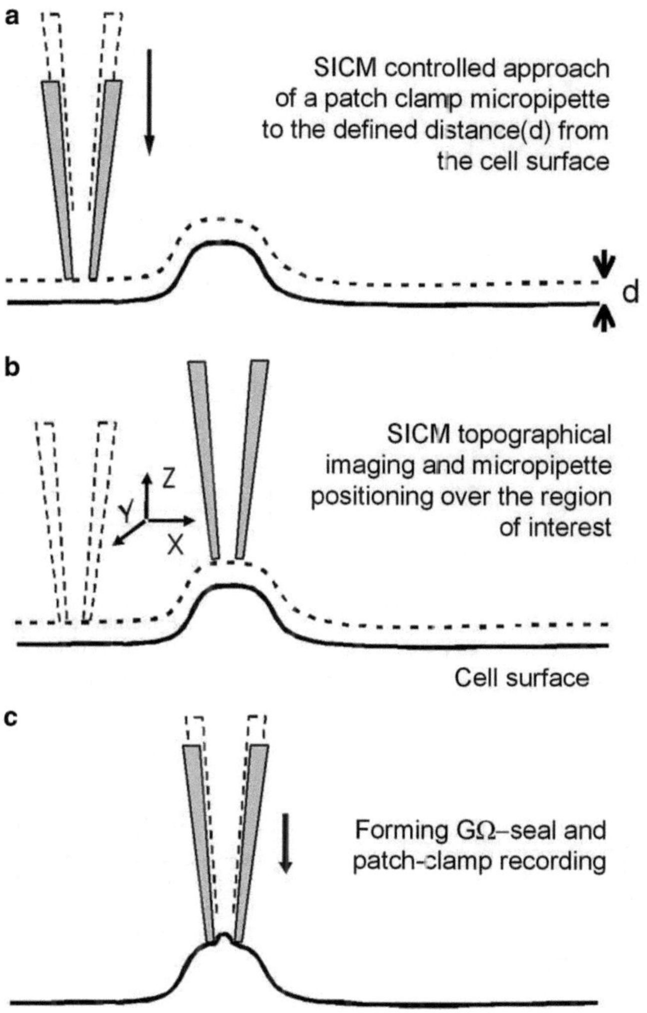

Fig. 2 Principle of operation of smart patch-clamp. (**a**) A micropipette approaches the cell surface and reaches a defined separation distance (*d*) whereupon the distance is kept constant by SICM feedback control. (**b**) The SICM software scans this micropipette over the cell surface and positions it at a place of interest for patch-clamp recording. (**c**) The micropipette is lowered to form a gigaohm seal for patch-clamp recording from the selected structure. Reproduced with permission from ref. 5

4 Notes

1. It is not necessary to coat commercial culture dishes with laminin if the cells can adhere without. Also poly-d-lysine can be used instead of laminin depending on the cell type.
2. A common problem with these high-resistance scanning pipettes is that they get blocked very easily if the solutions are not properly filtered. We use 0.02 μm filters from Whatman.
3. If there are bubbles in the pipette tip, the high tip resistance can be deceptive and you may not get a high-resolution image.

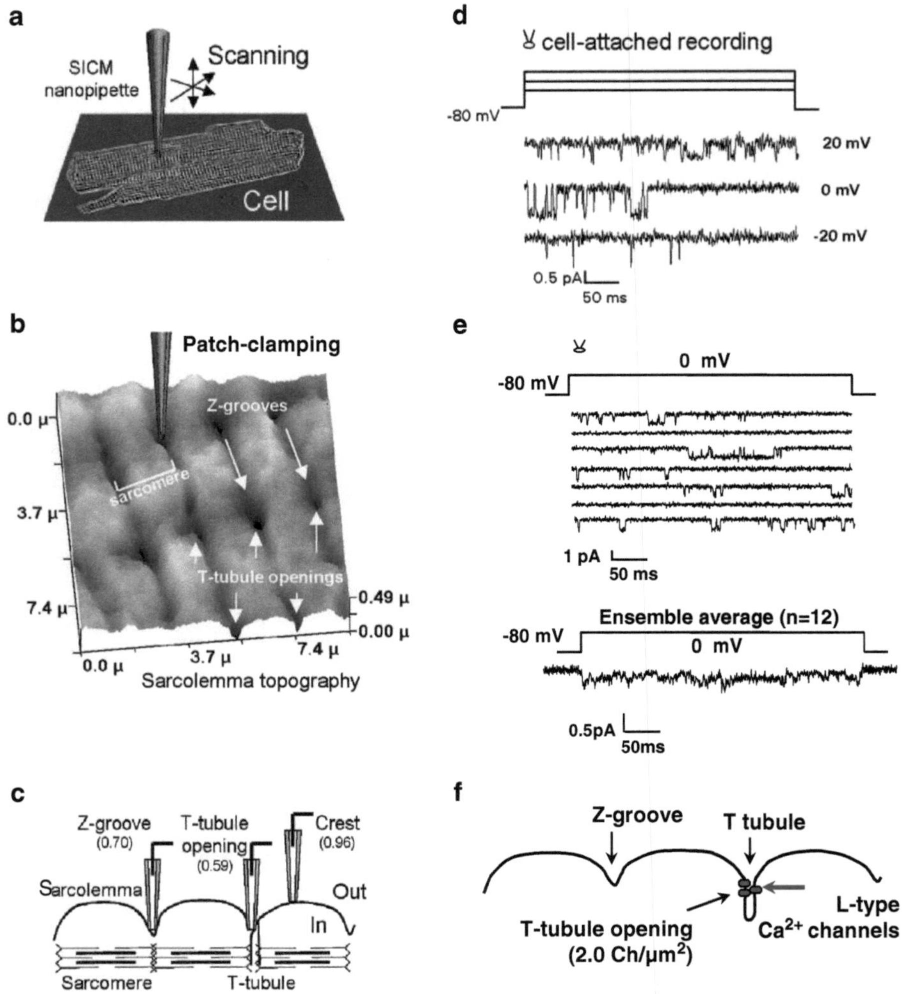

Fig. 3 LTCC channel distribution in the cardiomyocyte membrane: mapping of ion channels by the smart patch-clamp technique. (**a**) Adult rat cardiomyocyte and micropipette as seen optically by the microscope. (**b**) Experimental topographic image of a representative rat cardiomyocyte membrane. Z-grooves, T-tubule opening, and characteristic sarcomere units are marked. (**c**) Functional schematic of sarcomere units showing position of the probed region (Z-groove, T-tubule opening and scallop crest). Probabilities of forming a GΩ seal as a function of surface position are shown in parenthesis. (**d**) Cell-attached Ba^{2+} currents at voltages of +20, 0, −20 mV. (**e**) Several current traces elicited at 0 mV from one patch and ensemble average of 12 traces showing typical LTCC current kinetics. (**f**) Statistical distribution of LTCCs with the highest density near the T-tubule opening. Reproduced with permission from ref. 6

We use a microscope attached to a camera to check the pipette for any bubbles after filling with electrolyte every time we mount a pipette.

4. As with conventional patch-clamp, do not try to scan with blocked or dirty pipette. It will never obtain a good image.
5. Composition of external (bath) and internal (pipette) solutions also depend on the configuration being used. For example in

cell attached and inside out configurations external solution for cell is filled inside the pipette. Also, the membrane potential of cell-attached and inside out patch is negative of the command voltage set on the patch-clamp amplifier. Therefore the experimenter has to adjust the polarity of the applied voltage. This can be done easily be setting the command voltage as negative of the desired voltage. For example to hold the patch at −80 mV the experimenter has to apply 80 mV command voltage.

6. It is common to record calcium currents with barium as a charge carrier to better resolve the currents through voltage gated calcium channels.
7. It is also common to invert the current traces recorded in cell attached and inside out configurations so that inward currents are displayed as downward deflections to agree with the current conventions in electrophysiology.

References

1. Alonso G, Widmer H (1997) Clustering of KV4.2 potassium channels in postsynaptic membrane of rat supraoptic neurons: an ultrastructural study. Neuroscience 77:617–621
2. Angelides KJ (1986) Fluorescently labelled Na+ channels are localized and immobilized to synapses of innervated muscle fibres. Nature 321:63–66
3. Alkondon M, Pereira EF, Albuquerque EX (1996) Mapping the location of functional nicotinic and gamma-aminobutyric acidA receptors on hippocampal neurons. J Pharmacol Exp Ther 279:1491–1506
4. Frosch MP, Dichter M (1992) Non-uniform distribution of GABA activated chloride channels in cultured cortical neurons. Neurosci Lett 138:59–62
5. Gorelik J, Gu Y, Spohr HA, Shevchuk AI, Lab MJ, Harding SE, Edwards CR, Whitaker M, Moss GW, Benton DC, Sanchez D, Darszon A, Vodyanoy I, Klenerman D, Korchev YE (2002) Ion channels in small cells and subcellular structures can be studied with a smart patch-clamp system. Biophys J 83:3296–3303
6. Gu Y, Gorelik J, Spohr HA, Shevchuk A, Lab MJ, Harding SE, Vodyanoy I, Klenerman D, Korchev YE (2002) High-resolution scanning patch-clamp: new insights into cell function. FASEB J 16:748–750
7. Korchev YE, Bashford CL, Milovanovic M, Vodyanoy I, Lab MJ (1997) Scanning ion conductance microscopy of living cells. Biophys J 73:653–658
8. Shevchuk AI, Gorelik J, Harding SE, Lab MJ, Klenerman D, Korchev YE (2001) Simultaneous measurement of Ca^{2+} and cellular dynamics: combined scanning ion conductance and optical microscopy to study contracting cardiac myocytes. Biophys J 81: 1759–1764
9. Hansma PK, Drake B, Marti O, Gould SA, Prater CB (1989) The scanning ion-conductance microscope. Science 243:641–643
10. Korchev YE, Milovanovic M, Bashford CL, Bennett DC, Sviderskaya EV, Vodyanoy I, Lab MJ (1997) Specialized scanning ion-conductance microscope for imaging of living cells. J Microsc 188:17–23
11. Novak P, Li C, Shevchuk AI, Stepanyan R, Caldwell M, Hughes S, Smart TG, Gorelik J, Ostanin VP, Lab MJ, Moss GW, Frolenkov GI, Klenerman D, Korchev YE (2009) Nanoscale live-cell imaging using hopping probe ion conductance microscopy. Nat Methods 6: 279–281
12. Vescovo G, Jones SM, Harding SE, Poole-Wilson PA (1989) Isoproterenol sensitivity of isolated cardiac myocytes from rats with monocrotaline-induced right-sided hypertrophy and heart failure. J Mol Cell Cardiol 21:1047–1061
13. Nikolaev VO, Moshkov A, Lyon AR, Miragoli M, Novak P, Paur H, Lohse MJ, Korchev YE, Harding SE, Gorelik J (2010) Beta2-adrenergic receptor redistribution in heart failure changes cAMP compartmentation. Science 327:1653–1657

Part III

Imaging and Fluorescence Methods to Study Ion Channels

Chapter 15

Using Total Internal Reflection Fluorescence Microscopy to Observe Ion Channel Trafficking and Assembly

Sarah Schwarzer, Gregory I. Mashanov, Justin E. Molloy, and Andrew Tinker

Abstract

Ion channels are integral membrane proteins that allow the flow of ions across membranes down their electrochemical gradients and are a major determinant of cellular excitability. They play an important role in a variety of biological processes as diverse as insulin release from beta cells in the pancreas through to cardiac and smooth muscle contraction. We have used total internal reflection fluorescence (TIRF) microscopy to watch ion channels being transported in vesicles along microtubules within the cytoplasm of the cell. Furthermore, we can directly observe the fusion of these vesicles with the plasma membrane and the release and radial dispersion of single ion channels into the membrane. Finally, automated single-particle tracking of these objects allowed us to determine their diffusional behavior.

Key words Total internal reflection fluorescence microscopy (TIRFM), Single molecule detection, Potassium channel, Vesicle, Fluorescence, Plasma membrane

1 Introduction

Membranes consist of a lipid bilayer which makes them impermeable for most molecules including ions. In biological membranes, ion channels form pores which, with varying selectivity, allow ions to pass at a high rate through the membrane down their electrochemical gradient. Ion channel activity can be modulated directly by ligand binding, membrane potential, second messengers (such as Ca^{2+} or cyclic-AMP), protein–protein interactions, and phosphorylation (1). Potassium channels are the largest family of ion channels and they are important for maintenance of the membrane resting potential, hormone secretion, regulation of excitability, repolarization of the action potential, and its shape and frequency (2–4). Potassium channels are homo- or heteromeric proteins consisting of the pore-forming alpha subunit and in some cases additional beta subunits, which can be cytosolic or integral membrane proteins (5).

Nikita Gamper (ed.), *Ion Channels: Methods and Protocols*, Methods in Molecular Biology, vol. 998,
DOI 10.1007/978-1-62703-351-0_15, © Springer Science+Business Media, LLC 2013

The function of ion channels can be studied in exquisite detail using patch clamping to the point where opening of single channels can be observed with a millisecond time resolution. In addition to this measurement of electrical activity it would be revealing to directly observe single vesicles carrying ion channels towards the plasma membrane and the movement of single ion channels in the membrane of cells. Confocal microscopy allows the analysis of the ion channel subcellular localization and can identify single vesicles; however it is not easy to identify single molecules. In contrast, laser-based total internal reflection fluorescence microscopy (TIRFM) combined with use of highly sensitive camera systems has sufficient signal-to-noise ratio and sensitivity to identify single fluorophores. Furthermore, by tracking the motion of individual molecules moving at the plasma membrane, using automated algorithms, information about protein mobility, membrane structure, and protein–membrane and protein–protein interaction can be obtained. We have developed methodologies to study the trafficking, membrane fusion, and single molecule diffusion of potassium channels in the plasma membrane of living cells using a combination of recombinant gene technology, cell transfection, and novel imaging techniques (6). We first tag the protein of interest with a fluorescent protein such as enhanced green fluorescent protein (eGFP) or monomeric red fluorescent protein (mRFP) using standard cloning methodologies. Transfection techniques then allow the expression of a low level of the channel in HEK293 or the cardiac derived HL-1 cells (7). Using tagged fluorescent proteins is convenient, because we know they are bound to the protein of interest; the biogenesis and transport of the channel can be studied and adjusting the transfection method can regulate the expression level.

2 Materials

2.1 Cloning

1. pEGFP-N1 vector from Clonetech.
2. mRFP vector (8).

2.2 Cell Culture and Transfection

2.2.1 HEK293-Cells (Human Embryonic Kidney Cell Line)

1. Phosphate-buffered saline (PBS) solution, pH 7.4.
2. Minimal essential medium with Earle's Salts, l-glutamine supplemented with 10% fetal bovine serum, 1% penicillin/streptomycin.
3. Lipofectamine 2000 (Invitrogen) or TurboFect (Fermentas) transfection reagent.

2.2.2 HL-1 Cells (from Prof. W. C. Claycomb) (9)

1. PBS solution, pH 7.4.
2. Gelatin/fibronectin for culturing.

3. Claycomb medium (proprietary formulation SAFC, http://www.sigmaaldrich.com/etc./medialib/docs/Sigma/Product_Information_Sheet/p51800.Par.0001.File.tmp/p51800.pdf) (see Note 1) supplemented with 10% fetal bovine serum (Sigma), 2 mM l-glutamine, 100 units/mL of penicillin, 100 μg/mL of streptomycin, and 0.1 mM norepinephrine (Sigma).
4. FuGENE HD transfection reagent (Roche Applied Science).
5. 25 mm glass coverslips No. 1 (BDH).

2.3 Single Molecule Microscopy and Analysis

1. Sealed imaging chambers.
2. Hanks' balanced salts solution containing 10% fetal calf serum (FCS), 20 mM HEPES, pH 7.4 (all Sigma).
3. Inverted microscope, i.e., Zeiss Aviovert-100TV (Ziess, UK).
4. Objective lens, Alpha-plan 100×, NA 1.45 (i.e., from Ziess, UK).
5. Blue laser, 488 nm, 20 mW (we use Protera 488, Novalux, Sunnyvale, CA).
6. Green laser, 532 nm, 50 mW (we use Suwtech 532–50, SP3-Plus Tunbridge Wells, UK).
7. Laser grade beam expanders, lenses, and mirrors for TIRF illumination.
8. Microscope 37 °C incubator (i.e., Solent Scientific, Segensworth, UK).
9. EMCCD camera, i.e., iXon-897BV (Andor, UK).
10. GMimPro image analysis software (www.nimr.mrc.ac.uk/gmimpro/).

3 Methods

3.1 Generation of Fluorescent Constructs

1. Standard molecular cloning approaches can be used to fuse the relevant potassium channels or other genes of interest in frame with a fluorescent protein.
2. In our studies a fusion between human KCNQ1 and the enhanced variant of the green fluorescent protein (eGFP) was created using PCR. The long isoform of KCNQ1 without the termination codon was amplified by PCR using a high-fidelity polymerase with EcoRI/XbaI restriction sites and subcloned into pcDNA3. eGFP was amplified by PCR including 5′ and 3′ XbaI and ApaI restriction sites and cloned in frame to the C-terminus of KCNQ1 in pcDNA3. An alternative strategy is to use the eGFP vectors (see Note 2).

3.2 HL-1 Cell Preparation and Transfection

1. Seed HL-1 cells at very low density (1:20) on 25 mm glass coverslips 48 h before transfection.
2. Transfect the cells using FuGENE HD transfection reagent; the transfection complex consists of 97 μL Claycomb medium, 5 μL FUGENE HD, and the 1–2 μg of DNA. Add this mix to the cell culture and incubate for 24 h.
3. Perform experiments 24–48 h after transfection (see Note 3).

3.3 HEK293 Cell Preparation and Transfection

1. Seed HEK293 cells at a higher density (1:5) in 35-mm dishes 24 h before transfection.
2. Transfect the cells using Lipofectamine 2000 or TurboFECT and split the day after onto 25 mm glass coverslips at a very low density (1:15) (see Note 4).

3.4 Solution and Imaging Chamber

1. It is important to use Hank's buffered salt solution for TIRF microscopy to reduce background fluorescence; additionally it buffers the pH in our sealed imaging chamber.
2. For lengthy experiments it is advisable to use medium with the addition of 10% FCS. Our experience shows that the fully sealed, custom-built, imaging chamber works best.
3. Cells cultured on 25 mm glass coverslips are placed in the chamber topped by a silicone O-ring and covered by another coverslip to seal the chamber. This custom-built chamber allows one (via two ports on opposite sides of the chamber) to insert hypodermic syringe needles connected to syringes to fill the chamber with the medium.
4. The chamber is fixed securely to the stage of the microscope.
5. The temperature during the experiment is regulated by using a microscope 37 °C incubator and a custom-made airflow heater system.

3.5 Microscopy

1. Microscope. Our TIRFM imaging system was custom-built around an inverted microscope mounted on vibration isolation air table. Two laser beams (488 and 532 nm) were combined and made colinear using a dichroic mirror and were then expanded using a Galilean beam expander. Light was then focused using a 100 mm focal length achromatic doublet lens onto a small (3 mm diameter) silvered mirror located at one edge of the back focal plane of a high-numerical-aperture objective lens suitable for TIRFM (we use AlphaPlan, 100×, NA 1.45, Zeiss, Jena, Germany). The incident laser beam angle was adjusted to 64° so that it was totally internally reflected at the glass–water interface creating an evanescent field at the coverslip surface. A second mirror, positioned at the opposite edge of the objective lens back aperture, was used to couple the returning, totally internally reflected, laser beam out of

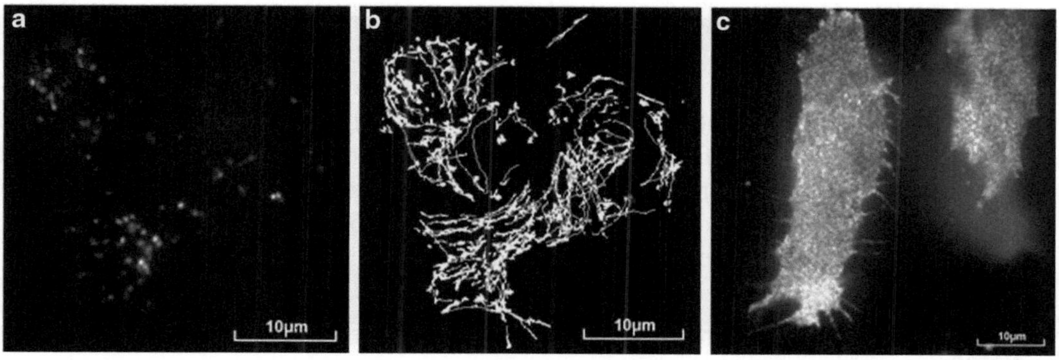

Fig. 1 TIRF microscope images of HEK293 cells transfected with KCNQ1-GFP. (**a**) Image of a video showing vesicles containing KCNQ1-GFP. (**b**) Tracking analysis of the video displaying the vesicle movements. (**c**) Single molecules (KCNQ1-GFP) in the plasma membrane

the microscope light path. This beam was then directed onto a position-sensitive photodiode to enable automatic focus control.

2. To visualize eGFP and mRFP in the same specimen, appropriate laser excitation (488 or 532 nm) and emission filter (ET525/50M for eGFP; HQ590/50M for mRFP) were selected.
3. Important criteria for the choice of camera are high sensitivity, low readout noise and fast acquisition rate (20–50 frames s-1). We used an electron multiplied CCD camera (EMCCD), which is currently the best choice for high-speed single fluorophore imaging within living cells because the signal generated at every CCD element is multiplied before the readout giving increased sensitivity and lower relative readout noise (see Note 5).

3.6 Image Analysis

1. Vesicles can be identified as bright punctate objects moving just below the plasma membrane in following straight-line trajectories (Fig. 1a, b). On occasion these vesicles would stall and fuse with the membrane and single molecules (Fig. 1c) would radially disperse from the site of fusion (see Note 6).
2. Video sequences are stored to computer hard disc and then analyzed using custom-written image processing software, which is available free of charge for academic users (http://www.nimr.mrc.ac.uk/gmimpro/) (10). The computer program automatically detects and tracks individual fluorescent spots which are diffraction limited in size, have an average intensity corresponding to that of a single fluorophore (measured under controlled conditions), and which show single-step photobleaching. The software allows simultaneous tracking of multiple objects (up to a thousand) within each record and produces an output file that contains the inte-

grated intensity (measured over a 5 × 5 pixel region) and the *x*, *y* coordinates (with sub-pixel resolution) for every spot on all video frames. This permits downstream analysis of the spatial and intensity trajectories which might include analysis of diffusional behavior from plots of mean squared displacement versus time interval and intensity changes by inspection of individual trajectories or by histogramming methods (see Note 7).

4 Notes

1. Contact details: Professor William C. Claycomb, PhD, Department of Biochemistry and Molecular Biology, 1901 Perdido St., Box P7-2, Medical Education Building Room 7238, New Orleans, LA 70112, wclayc@lsuhsc.edu.
2. Fluorescent proteins are convenient and widely used to track protein localization and biogenesis. However the protein tag can affect function and this has to be independently assessed. For instance, here we were able to check the function of KCNQ1-GFP channels by patch-clamping and comparing our results with those obtained from wild-type KCNQ channels (11). A drawback of fluorescent protein tags is that they are not as bright and photobleach more rapidly than small organic fluorophores (such as Cy3B). We have found that observation of single eGFP fluorophores within live cells is readily achievable but requires careful choice of equipment and optimization of imaging conditions (12).
3. For single molecule experiments with TIRFM very low cDNA concentrations are necessary. Systematic variations of the transfection conditions are important to optimize a low level of protein expression. For example it may be necessary to vary the amount of DNA, change the transfection reagent, the transfection time or cell density. For example, the transfected DNA concentration for KCNQ1 was 40 ng in our studies (6). In some circumstances we would transfect a range of concentrations and use the one that worked best.
4. If several plasmids need to be transfected together it can be helpful to sequentially transfect them one after the other to ensure comparable expression at the site of interest. This arises from differences in protein turnover rates of the various constructs. Similar titrations of transfection conditions might also be necessary with HEK293 cells.
5. We found that image intensified CCD cameras (ICCDs) have sufficient sensitivity and signal-to-noise ratio to image single fluorophores, but the intensifier system creates spatial and temporal noise which makes them inferior to EMCCD cameras.

A recent generation of CMOS cameras (made by various manufacturers) presents exciting new possibilities because they allow very fast readout rates. However, currently CMOS cameras appear to have slightly lower signal-to-noise than EMCCD architectures.

6. To obtain cells in which vesicular cargo transport can be observed, it is best to examine cells relatively early after transfection. Vesicle fusion events are relatively rare.
7. We found it difficult to identify multistep photobleaching of single ion channels as might be expected to occur for the known tetrameric structure of KCNQ. We think this is due to a combination of factors: photobleaching of GFP is relatively rapid so it is not always possible to observe the early bleaching events (expected for a multimer); since GFP undergoes fluorescence "blinking," oligomeric complexes can exhibit complicated intensity trajectories; finally because ion channels diffuse rapidly at the membrane they often come close (within the diffraction-limited spacing, ~350 nm) to one another, causing large intensity fluctuations, and it is difficult to disentangle the intensity contribution arising from each object. However, one approach is to analyze histogram of the intensity distribution of the whole population measured at each time frame. If the intensity of each spot arises from multiple fluorophores then, for statistical reasons, one expects the shape of the histogram to change in a systematic fashion over time (as the population shifts, on average from 4 > 3 > 2 > 1 fluorophores per object). This approach is discussed and presented in detail in (6).

Acknowledgments

The work in our laboratories is supported by the Medical Research Council, Wellcome Trust, and British Heart Foundation. This work forms part of the research themes contributing to the translational research portfolio of Barts Cardiovascular Biomedical Research Unit which is supported and funded by the National Institute for Health Research.

References

1. Cooper EC, Jan LY (1999) Ion channel genes and human neurological disease: recent progress, prospects, and challenges. Proc Natl Acad Sci U S A 96:4759–4766
2. Jan LY, Jan YN (1997) Cloned potassium channels from eukaryotes and prokaryotes. Annu Rev Neurosci 20:91–123
3. Chandy KG, Gutman GA (1995) Voltage-gated potassium channel genes. In: North RA (ed) Handbook of receptors and channels. Ligand and voltage-gated ion channels. CRC, Boca Raton, FL, pp 1–71
4. Isomoto S, Kondo C, Kurachi Y (1997) Inwardly rectifying potassium channels: their molecular heterogeneity and function. Jpn J Physiol 1997(47):11–39
5. Tinker A (2002) The assembly and targeting of potassium channels. In: Henley J, Moss SJ (eds) The assembly and targeting of ion channels. Oxford University Press, Oxford, pp 28–57

6. Mashanov GI, Nobles M, Harmer SC, Molloy JE, Tinker A (2010) Direct observation of individual KCNQ1 potassium channels reveals their distinctive diffusive behaviour. J Biol Chem 285:3664–3675
7. Claycomb WC, Lanson NA Jr, Stallworth BS, Egeland DB, Delcarpio JB, Bahinski A, Izzo NJ Jr (1998) HL-1 cells: a cardiac muscle cell line that contracts and retains phenotypic characteristics of the adult cardiomyocyte. Proc Natl Acad Sci U S A 95:2979–2984
8. Campbell RE, Tour O, Palmer AE, Steinbach PA, Baird GS, Zacharias DA, Tsien RY (2002) A monomeric red fluorescent protein. Proc Natl Acad Sci U S A 99:7877–7882
9. White SM, Constantin PE, Claycomb WC (2004) Cardiac physiology at the cellular level: use of cultured HL-1 cardiomyocytes for studies of cardiac muscle cell structure and function. Am J Physiol Heart Circ Physiol 286:H823–H829
10. Mashanov GI, Molloy JE (2007) Automatic detection of single fluorophores in live cells. Biophys J 92:2199–2211
11. Wilson AJ, Quinn KV, Graves FM, Bitner-Glindzicz M, Tinker A (2005) Abnormal KCNQ1 trafficking influences disease pathogenesis in hereditary long QT syndromes (LQT1). Cardiovasc Res 67: 476–486
12. Nenasheva TA, Mashanov GI, Peckham M, Molloy JE (2011) Imaging individual myosin molecules within living cells. Methods Mol Biol 778:123–142

Chapter 16

Förster Resonance Energy Transfer-Based Imaging at the Cell Surface of Live Cells

Sonya M. Bierbower and Mark S. Shapiro

Abstract

Understanding the molecular mechanisms of protein–protein interactions at the cell surface of living cells is fundamental to identifying the nature of cellular processes. Here, we discuss how fluorescence-based approaches have been successfully developed to visualize protein–protein interactions in living cells. Förster resonance energy transfer (FRET) is unique in generating fluorescence signals between proteins that are highly spatially sensitive. Furthermore, total internal reflectance fluorescence (TIRF) microscopy combined with FRET is a robust technique used to assay protein/protein interactions and the functionality of proteins assembled at the cell surface membrane.

Key words Protein–protein interactions, Cell surface, FRET, TIRF, Live-cell imaging

1 Introduction

Biological systems are composed of basic physiological and chemical processes in which molecular events among thousands can be feasibly studied through optical imaging techniques. Fluorescence microscopy is a powerful tool in the biomedical and biological sciences enabling visualization of the myriad proteins and processes of cellular physiology. Progress has been made in the last decade in developing methods in different modes of epifluorescence microscopy. One method of detecting molecular interactions involves Förster resonance energy transfer (FRET) in which energy is transferred from a "donor" fluorophore in an excited state whose emission spectrum overlaps the absorption spectrum of an "acceptor" fluorophore (1). The FRET efficiency depends on many factors, but the most critical is the donor–acceptor distance, with the efficiency maximal at ~50Å (r_0), a distance only expected if the donor and acceptor fluorophores are in intimate proximity. What makes FRET such good evidence of intermolecular interactions is that the FRET efficiency declines with the inverse sixth power of

Nikita Gamper (ed.), *Ion Channels: Methods and Protocols*, Methods in Molecular Biology, vol. 998,
DOI 10.1007/978-1-62703-351-0_16, © Springer Science+Business Media, LLC 2013

the donor–acceptor distance, more precisely $1/(R-r_o)^6$, where R is the distance between donor and acceptor (2). Thus, although the FRET efficiency is dependent on other factors, such as the degree of overlap of the donor emission and acceptor absorption spectra, the relative orientation of the donor absorption and acceptor transition moments, and the refractive index of the medium, the steep dependence on distance makes this approach powerful in establishing the physical interaction between fluorescent molecules (3).

Traditional FRET experiments on membrane proteins performed under confocal microscopy have often been hampered by contamination of events at the plasma membrane, which are most relevant, by those occurring in the cytoplasm. Thus, a major tool in the field is the use of total internal reflectance fluorescence (TIRF) microscopy, which exploits the behavior of light at the interface between two media of differing refractive index. At such an interface, at an angle greater than a critical angle, the main light beam does not penetrate the second medium, but rather is totally internally reflected back into the first medium. However, a component of the light energy, called the evanescent wave, does penetrate into the second medium at a perpendicular angle and, importantly, decays away exponentially with distance (4, 5). At a glass/water interface typical of living cells in solution, and using a fluorescent microscope objective with numerical aperture (N.A.) of typically 1.45, the evanescent wave, and thus the excitation of the cell, only penetrates to a depth of ~200 nm. Thus, if one is performing FRET experiments under TIRF illumination, only fluorophores located within or close to the plasma membrane are selectively detected within the evanescent field, whereas molecules in the cytoplasm will not be illuminated. This allows for a drastic reduction in the fluorescence background coming from irrelevant proteins in the cytoplasm, and a very high signal-to-background ratio in and around the membrane of living cells. TIRF combined with FRET is especially useful for studying the interactions among plasma membrane proteins such as ion channels, receptors, and their associated signaling molecules, and trafficking events in which the movement of proteins into or away from the membrane can be visualized in real time. Examples of the TIRF/FRET approach include interactions among G-proteins, enzymes, ion channels, scaffolding proteins, and other associated proteins. For example, GTPases were shown to regulate trafficking of $K_{ir}2.1$ channels (6), calmodulin, and A-kinase anchoring protein 79/150 binding to M-Type (KCNQ) K^+ channels (7, 8) and activation of GIRK K^+ channels by $G_{\beta\gamma}$ subunits (9).

The methods and equipment discussed in this chapter utilize enhanced cyan fluorescent protein (eCFP) and enhanced yellow fluorescent protein (eYFP), which is the most commonly used fluorescent protein FRET pair, and we will focus on this set of donor and acceptor fluorescent proteins (FPs). Given that TIRF works best at a glass/water interface and that we are usually most interested in dynamic events happening in living cells, our TIRF/FRET experience is solely using live cells, as opposed to fixed cells in which the typical mounting medium and flattened cell architecture are both not very suitable. Cultured cells are most convenient since multiple FP-tagged proteins can be transfected simultaneously and also these cells adhere well to a glass substrate. Furthermore, these cells can be feasibly imaged and the FRET efficiency quantified using a regular epifluorescence microscope. Such FRET efficiency is often measured by either quantifying the increase in emission of the acceptor fluorophore upon energy transfer from the donor (sensitized emission) or by comparing the emission of the donor before and after photobleaching of the acceptor (acceptor photobleaching, donor dequenching). The former method is best when dynamic FRET is sought, since neither fluorophore is destroyed during a sequence of images, but necessitates a careful series of control images. On the other hand, the latter method is most efficient when a static FRET measurement is the goal, and is also relatively simple to quantify accurately. This chapter describes how to perform TIRF/FRET measurements using the acceptor photobleaching method on heterologously transfected cultured cells. We describe the setup as currently sold by Nikon Instruments, the vendor who sold us our TIRF/FRET setup (still in use) some 10 years ago.

2 Materials

2.1 Equipment

1. Inverted Eclipse Ti Microscope with through-the-lens TIRF imaging (Nikon Instruments).
2. Vibration isolation system ("air table") to minimize drift and noise on which the microscope is situated (i.e., Technical Manufacturing Corp., Peabody, MA, USA).
3. Laser light excitation is controlled by a computer-controlled acoustic optical tunable filter (AOTF) that controls output by computer. The lasers should include at least solid-state diode lasers with 442 and 514 nm lines. Although it should be noted that it is also possible to use an Argon (40 mW) laser outputting 488 and 514 nm lines for excitation of GFP and YFP, respectively, and a 442 solid-state diode laser for excitation of CFP, the development of economical, highly durable solid-state lasers has largely eclipsed the use of gas lasers.
4. Microscope optics equipped with an Apo TIRF 60× oil immersion high-resolution (1.45 N.A.) objective.
5. It is most helpful to have the microscope contain the following filter cubes: wide-field CFP and YFP, TIRF CFP and YFP, and

a cube for using the dual-view chip-splitter containing only a dichroic mirror, as described below.

6. High-resolution charge-coupled device (EMCCD) camera (e.g., Photometrics Cool SNAP HQ2).
7. Dual-view chip splitter (Optical Insights, Photometrics, Tucson, AZ, USA).

2.2 Software

1. NIS-Elements for image acquisition and data analysis. The software controls laser light delivery, microscope, and the EMCCD camera (Nikon Instruments).
2. Microsoft Excel (used in conjunction with NIS-Elements for formula-based data analysis).

2.3 Epifluorescence Imaging

1. Chinese Hamster Ovary (CHO) or other suitable cell line expressing a membrane targeted tandem construct of eCFP and eYFP (Rho-pYC) to act as a positive control for strong FRET efficiency. It consists of the C-terminal prenylation site of Rho (RQKKRRGCLLL) appended to the C-terminus of a YFP-CFP fusion protein (10) (see Note 1).
2. CHO (or equivalent) cells expressing the CFP-tagged protein to be tested and eYFP-M (membrane bound). These act as a negative control for FRET efficiency, since they are known to be too far apart in the membrane to exhibit FRET, and make a good control for "spurious" or incidental FRET.
3. Tissue-culture cells expressing the eCFP-tagged and eYFP-membrane bound proteins that you wish to assay for molecular interactions.
4. Cell imaging solution: 160 mM NaCl, 5 mM KCl, 1 mM $MgCl_2$, 2 mM $CaCl_2$, 10 mM HEPES, pH 7.4 with NaOH. Other suitable saline solutions are equally suitable.
5. Glass-bottom culture dishes with No. 1.5 coverglass (0.16–0.19 mm) on the bottom, 35 mm (Mat-tek) (see Note 2).
6. Objective immersion oil.

In this chapter we are not describing protocols (and therefore not listing any materials required) for cell culture and transfection as these can be found elsewhere (i.e., Chapters 2, 4, and 5 of this book).

3 Methods

3.1 TIRF/FRET Experiments on Cultured Cells Using Epifluorescence Microscopy

It is most convenient to image eCFP and eYFP emission simultaneously, which can be easily performed using the dual-view chip splitter, which is put between the side-camera port and the CCD camera. This device is equipped with a filter cube containing a dichroic mirror, and emission filters of your choice. For concurrent eCFP/eYFP imaging, we use an HQ470 nm/30 m and

HQ550 nm/30 m emission filters for eCFP and eYFP and a 505 nm dichroic mirror for separation of emission wavelengths. In this configuration, the microscope cube can contain only a dual-band TIRF dichroic mirror to selectively send excitation light to the sample, and emission light to the detector. The approximate TIRF angle is adjusted by eye to give the signature TIRF illumination to the experimental chamber. Any fine TIRF angle adjustment can be made through the *NIS-Elements* software. Fluorescence images are collected and processed with a 16-bit, cooled charge-coupled device camera (e.g., Photometrics Cool SNAP HQ2) interfaced to the *NIS-Elements* software. This camera uses a front-illuminated EMCCD with on-chip multiplication gain. Images are collected (100–600 ms exposure time) immediately before and after photobleaching eYFP. Images are not binned or filtered, with pixel size corresponding to a square of 122 × 122 nm.

3.2 Imaging of Cells

1. Just prior to fluorescence imaging, replace the cell culture medium with an appropriate cell imaging solution (see Subheading 2.3, step 4).
2. Place the chamber on the stage on top of the oil immersion 60× objective (1.45 N.A.).
3. Using bright field transmitted light, focus on an isolated cell. Note that there will not be contrast optics in the light path, so focusing on the cells will be hard. We often "find" the cells under epifluorescence, without using transmitted light (see Note 3).
4. Using the 200 W metal-halide lamp (which gives similar spectral output to more traditional mercury vapor lamps) and alternating CFP and YFP filter cubes, pick a cell with robust expression of both fluorophores (see Note 4).
5. Under TIRF illumination and either the TIRF CFP or YFP filter cubes, the focal plane is adjusted immediately before each image acquisition to obtain a "sharp" TIRF image. The membrane proteins should appear punctate if the cell is focused correctly (see Note 5).
6. Take the "before" photobleaching images of the cell using the 442 and 514 nm laser lines (see Note 6).
7. Photobleach the eYFP fluorophores by using the metal-halide lamp and the wide-field YFP filter cube for a minimum of 7–10 min, which should be sufficient to achieve >80% photobleaching of the eYFP. Photobleaching is done through the wide-field metal-halide lamp since the molecules in live cells are diffusible in the membrane and are likely to move. Thus, wide-field illumination allows for the eYFP fluorophores in

the entire cell to be photobleached and eliminates the problem of limited area laser photobleaching under TIRF illumination, which would be insufficient for donor dequenching (see Note 7).

8. After photobleaching, take the "after" photobleaching images of the cell using the same 442 and 514 nm laser lines.
9. Using the CFP "before" image, use the *NIS-Elements* analysis package draw tool to trace the cell perimeter. Export the averaged intensity to *Excel* using the log data function.
10. Using the draw tool in *NIS-Elements*, draw a second circle in the blank region of the CFP "before" image to estimate the background level. Export the averaged intensity to *Excel* using the log data function.
11. Subtract the background averaged intensity from the averaged intensity of the cell measured above. When the CFP image is used, this yields the CFP intensity; when the YFP image is used, this yields the YFP intensity.
12. Repeat steps 10 and 11 for each of the four images (i.e., "before" and "after" photobleaching for both CFP and YFP emission images).

3.3 Quantification of FRET Efficiency in TIRF Illumination

FRET efficiency using the acceptor photobleaching (donor dequenching) paradigm is calculated as the percentage of increased CFP emission after YFP photobleaching. The % FRET efficiency is calculated by drawing the entire area of the cell in the CFP TIRF image and subtracting the background in a cell-free region for each image (see formula below). In addition, % photobleaching efficiency is calculated by drawing the entire perimeter of the cell in the YFP TIRF image and subtracting the background in a cell-free region for each image.

$$\%\text{FRET efficiency} = \left(\frac{(\text{eCFP}_{\text{post}} - \text{eCFP}_{\text{pre}})}{\text{eCFP}_{\text{pre}}}\right) \times 100,$$

where CFP_{pre} and $\text{eCFP}_{\text{post}}$ are the CFP emissions before and after YFP photobleaching, respectively.

Using the protocol detailed in this chapter, Fig. 1 shows representative images of cells transfected with CFP and YFP before and after YFP photobleach. Specifically, Chinese hamster ovary cells were transfected with CFP-tagged angiotensin II AT_1 receptors and YFP-tagged KCNQ3 K^+ channels. Here, FRET was measured under TIRF illumination by the donor dequenching method and CFP emission was significantly stronger after YFP photobleach, indicating robust FRET.

Positive controls are used as a measure of the highest percentage of FRET possible. Rho-pYC is a CFP and YFP tandem

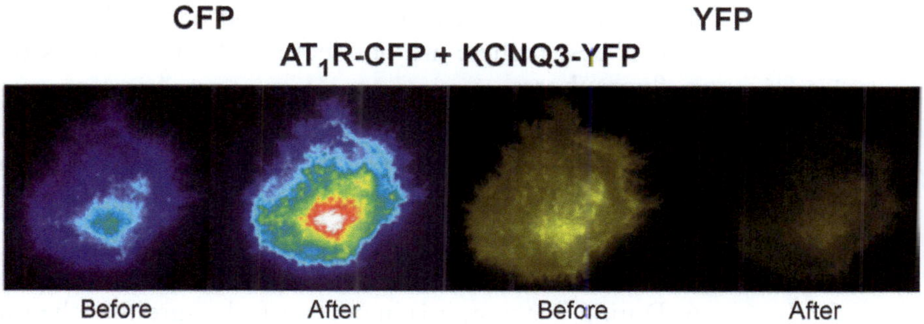

Fig. 1 Shown are images of Chinese hamster ovary cells transfected with CFP-tagged angiotensin II receptors (AT_1R) and YFP-tagged KCNQ3 K^+ channels under TIRF illumination, using 442 nm or 514 nm laser lines, respectively. Images of eCFP (*left*, in "*rainbow pseudocolor*") and eYFP (*right, in yellow pseudocolor*) emissions are shown before or after YFP photobleach, as labeled. Note the significantly brighter eCFP emission (*warmer colors*) after eYFP photobleach and the profoundly dimmer eYFP emission after 7 min photobleach under wide-field illumination with the YFP filter cube

protein that is anchored to the membrane and thus gives an estimate for the maximal energy transferable between two interacting proteins. Conversely, a good negative control pair should give an upper limit for "spurious FRET" between two proteins in the membrane that do not interact, but might be accidentally close. We commonly use a CFP-tagged membrane protein (such as an ion channel) paired with membrane-localized YFP as such a good negative control. Thus, the controls provide the range of levels of FRET for data comparison, both for estimating the strength of FRET for interacting proteins, compared to the maximal possible, and for distinguishing two truly interacting proteins from those that are incidentally close, but do not interact.

4 Notes

1. Cells can typically be kept in culture for imaging for 3 days post-transfection, but often appear to be unhealthy and do not provide reliable data on the fourth and subsequent days post-transfection.
2. The coverglass with bottom thickness No. 1.5 is preferable to accommodate the short working distance of most high-power objectives.
3. If one is comparing groups of cells, it is imperative to select cells at a random basis that express differing fluorescence emission levels, as long as both eCFP and eYFP fluorophores are robustly expressed. Choosing cells of mostly the same expression will lead to bias of the data.

4. For the donor-dequenching method of FRET quantification, it is best if there are not more CFP fluorophores than YFP fluorophores.

5. It is imperative to be sure that the cell is under TIRF illumination. A quick check can be done by moving the TIRF angle in and out of the critical angle. Non-TIRF illumination will show high fluorescence in the cytoplasmic area of the cell. The nucleus should not be visible under TIRF.

6. During imaging, it is important to be sure that the excitation and fluorescence detection parameters are set to include peak emission wavelengths and such that saturation levels are not reached for any pixel.

7. It is also very important to note that when searching for a cell, a long period of exposure using the metal-halide lamp light source has the possibility of damaging the cell as well as photobleaching the FPs.

References

1. Centonze VE, Sun M, Masuda A, Gerritsen H, Herman B (2003) Fluorescence resonance energy transfer imaging microscopy. Methods Enzymol 360:542–560
2. Stryer L (1978) Fluorescence energy transfer as a spectroscopic ruler. Annu Rev Biochem 47:819–846
3. Erijman EA, Jovin T (2003) FRET imaging. Nat Biotechnol 23:1387–1395
4. Axelrod D, Thompson NL, Burghardt TP (1983) Total internal reflection fluorescent microscopy. J Microsc 129:19–28
5. Funatsu TY, Harada M, Tokunaga K, Saito T, Yanagida T (1995) Imaging of single fluorescent molecules and individual ATP turnovers by single myosin molecules in aqueous solution. Nature 374:555–559
6. Boyer SB, Slesinger PA, Jones SV (2009) Regulation of Kir2.1 channels by the Rho-GTPase, Rac1. J Cell Physiol 218:385–393
7. Bal M, Zaika O, Shapiro MS (2008) Calmodulin binding to M-type K+channels assayed by TIRF/FRET in living cells. J Physiol 586:2307–2320
8. Bal M, Zhang J, Hernandez CC, Zaika O, Shapiro MS (2010) Ca2+/calmodulin disrupts AKAP79/150 interactions with KCNQ (M-Type) K+channels. J Neurosci 30: 2311–2323
9. Riven I, Iwanir S, Reuveny E (2006) GIRK channel activation involves a local rearrangement of a preformed G protein channel complex. Neuron 51:561–573
10. Fowler CE, Aryal P, Suen KF, Slesinger PA (2007) Evidence for association of GABA(B) receptors with Kir3 channels and regulators of G protein signalling (RGS4) proteins. J Physiol 580:51–65

Chapter 17

The Use of Dansyl-Calmodulin to Study Interactions with Channels and Other Proteins

Alessandro Alaimo, Covadonga Malo, Pilar Areso, Kerman Aloria, Oscar Millet, and Alvaro Villarroel

Abstract

Steady-state fluorescence spectroscopy is a biophysical technique widely employed to characterize interactions between proteins in vitro. Only a few proteins naturally fluoresce in cells, but by covalently attaching fluorophores virtually all proteins can be monitored. One of the first extrinsic fluorescent probes to be developed, and that is still in use, is dansyl chloride. We have used this method to monitor the interaction of a variety of proteins, including ion channels, with the Ca^{2+}-dependent regulatory protein calmodulin. Here we describe the preparation and use of dansyl-calmodulin (D-CaM).

Key words Fluorescence spectroscopy, Protein purification, In vitro binding, Protein labelling

1 Introduction

Fluorescence spectroscopy is a widely used approach to analyze protein structure and function. Among the advantages of using fluorescence techniques are their high sensitivity, enabling minute amounts of sample material to be used, its noninvasive nature, the presence of natural intrinsic fluorophores in proteins, and the relatively simple equipment required to perform different experiments. These techniques are useful to study biochemical parameters, such as protein–protein interactions, conformational changes, metal binding, cellular localization, and much more. The intrinsic fluorescent probes in proteins are tyrosine and tryptophan residues, although the majority of studies using intrinsic protein fluorescence focus on tryptophan as it is almost always the dominant source of signal, and its fluorescence is much more sensitive to the environment than that of tyrosine. Unfortunately, there are usually only few tryptophan residues per protein, which means that the method senses only these few points in the proteins' structure.

Nikita Gamper (ed.), *Ion Channels: Methods and Protocols*, Methods in Molecular Biology, vol. 998,
DOI 10.1007/978-1-62703-351-0_17, © Springer Science+Business Media, LLC 2013

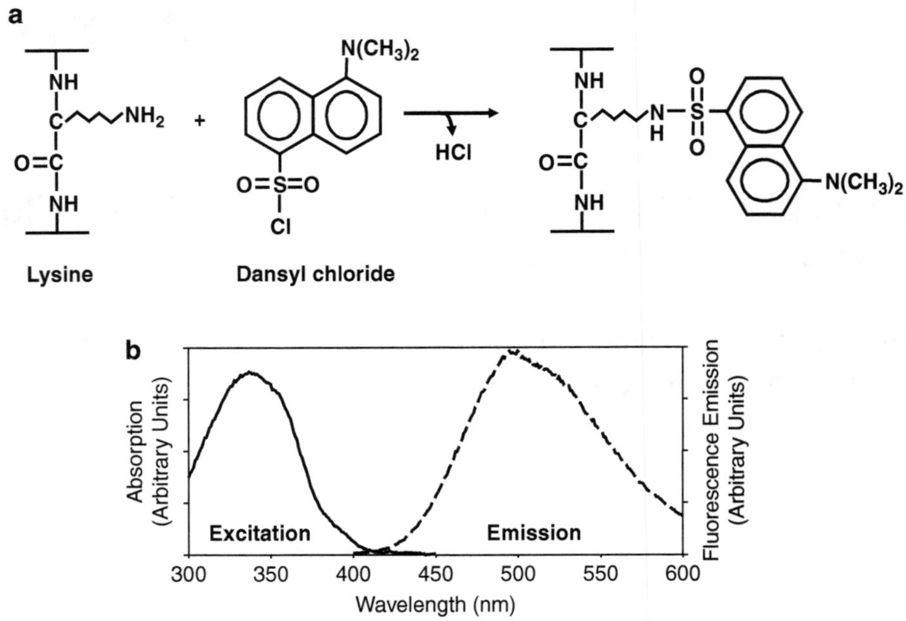

Fig. 1 Dansyl reactivity and its fluorescent properties. (**a**) Reaction scheme for the dansyl chloride labelling of a lysine residue. (**b**) Fluorescence emission spectrum of D-CaM (*solid line*, excitation at 340 nm) and excitation spectrum (*dotted line*, emission at 500 nm) in binding buffer

Furthermore, the signal is often weak, demanding the use of high protein concentrations. Thus, it is usually more convenient to use extrinsic fluorescence probes that can be covalently attached to a protein. There are many reagents available that generally react with amino or thiol groups, allowing the introduction of fluorophores that may detectably alter the spectral characteristics of the native protein. A commonly used fluorescent reagent is 1-dimethylaminonaphthalene-5-sulfonyl chloride (dansyl chloride or DNS-chloride), which was for a long time the most widely used fluorogenic derivatizing reagent for amino acid determination in proteins and peptides (1). Dansyl chloride is a probe that binds to functional amine groups to form a fluorescent sulphonamide (Fig. 1a), but it can also react with the phenol group of tyrosine (2). The resulting dansyl derivates have proven particularly useful to analyze protein interactions since the dansyl emission spectrum is greatly perturbed by the local environment.

In our laboratory, we have prepared dansyl-calmodulin (D-CaM) in order to characterize the interaction of calmodulin (CaM) with different components of the Kv7.2 channel CaM binding domain (Kv7.2 CBD). Since the intensity of the D-CaM emission spectrum is enhanced when the environment of the dansyl moiety becomes hydrophobic (3–5), this tool has proved useful to detect conformational changes as a consequence of interactions with Ca^{2+}, peptides or proteins. The changes in fluorescence can be

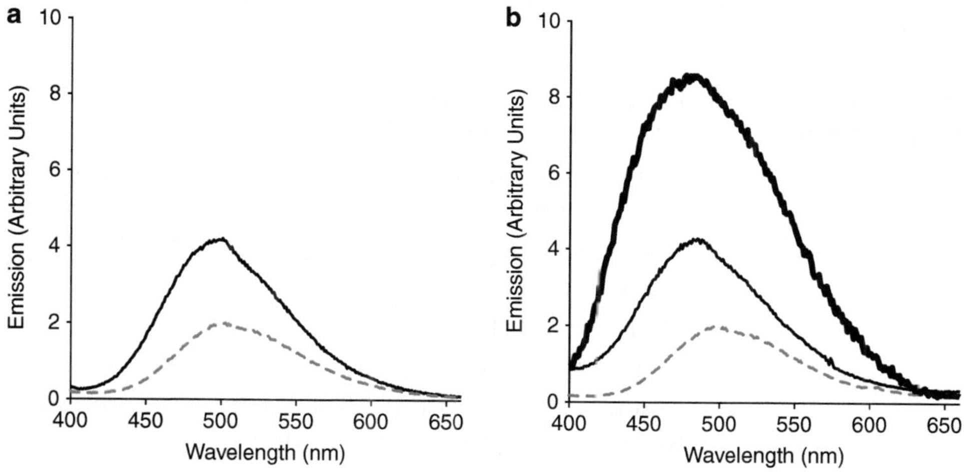

Fig. 2 Changes in the D-CaM fluorescence spectrum induced by peptide and/or calcium. (**a**) Emission spectrum of D-CaM alone (12.5 nM, *grey dotted* spectrum) and after addition of 200 nM GST-Kv7.2 CBD (*black solid* spectrum) in the absence of Ca^{2+} (10 mM EGTA). (**b**) Emission spectrum of D-CaM alone (12.5 nM, *grey dotted* spectrum), and after the addition of 2 μM free Ca^{2+} (*black solid* spectrum) and the successive addition of 200 nM GST-Kv7.2 CBD (*black bold* spectrum). Note the blue shift in the spectra in the presence of Ca^{2+}

detected in the low nanomolar range, with negligible interference from intrinsic aromatic residues to the emission window of the dansyl moiety (at ~500 nm; see Note 1 and Fig. 1b).

In addition to a large increase in fluorescence intensity, the peak emission shifts from ~500–510 nm to ~480–485 nm when D-CaM binds Ca^{2+}, towards a more prominent blue wavelength (Fig. 2). This can be explained by the alteration in CaM conformation known to occur when it binds Ca^{2+}, changing the environment of the dansyl moiety to a more hydrophobic location. This shift in emission spectra provides a very convenient way to detect Ca^{2+} contamination in protein and peptide samples (see Note 2). However, since D-CaM is very sensitive to its environment, care must be taken to avoid the contamination of some solvents that hinder fluorescent emission, such as glycerol.

Since D-CaM retains the physical and biological properties of native CaM (4, 6–8), it has been used to determine the binding constants and stoichiometry of proteins or peptides that interact with it (9–11). In general, yet not always, protein binding to D-CaM enhances its emission. However, some interactions do not affect emission (12) and there is even one report of reduced peak emission as a consequence of peptide binding (13).

Here we describe the preparation and purification of D-CaM, and the fluorescence experiments designed to study the interaction of CaM with its binding domain in the Kv7.2 channel.

2 Materials

2.1 Purification of GST Fusion Proteins

1. LB medium (in grams per liter of distilled water: Tryptone 10, yeast extract 5, and NaCl 5) sterilized by autoclaving.
2. 100 mg/mL ampicillin stock solution sterilized by filtration and stored at −20 °C.
3. 1 M isopropyl β-d-1-thiogalactopyranoside (IPTG) solution sterilized by filtration and stored at −20 °C.
4. Dithiothreitol (DTT).
5. GST buffer: 20 mM Tris–HCl (pH 7.4), 100 mM NaCl, 0.5% Triton X-100, 2 mM DTT, half a tablet of the protease inhibitor cocktail "1× Complete" (Roche).
6. Solubilization buffer: 20 mM Tris–HCl (pH 7.4), 100 mM NaCl, 2 mM DTT, and 6 M urea.
7. Refolding buffers: 20 mM Tris–HCl (pH 7.4), 100 mM NaCl, and 2 mM DTT, containing 4, 2, 1, 0.5, or 0 M urea. Filter all the buffers before use.
8. Glutathione-Sepharose 4B (GE Healthcare) and glutathione reduced (Sigma-Aldrich).
9. Elution buffer: 50 mM Tris–HCl (pH 8.5) and 15 mM glutathione reduced.
10. pGEX expression vectors (GE Healthcare).
11. BL21-DE3 (Novagen) and BL21-Codon plus (Agilent Technologies) competent cells.
12. General laboratory equipment: Spectrophotometer, orbital incubator, ultracentrifuge with fixed-angle rotor, microcentrifuge, ultrasonic probe sonicator, standard dialysis tubing, filters, protein concentrators, magnetic stirrer, small columns (1–5 mL bed volume), electrophoresis chamber and buffers, SDS-PAGE gels, and standard material for protein detection and staining.

2.2 Purification and Dansylation of Calmodulin

1. LB medium sterilized by autoclaving.
2. 100 mg/mL ampicillin stock solution sterilized by filtration and stored at −20 °C.
3. 1 M isopropyl β-d-1-thiogalactopyranoside (IPTG) solution sterilized by filtration and stored at −20 °C.
4. 100 mM phenylmethylsulfonylfluoride (PMSF, Sigma-Aldrich) stock solution in isopropanol stored at −20 °C. Add PMSF to buffers just before use.
5. Lysis buffer: 50 mM Tris–HCl (pH 7.5), 2 mM EDTA and 0.2 mM PMSF.
6. Phenyl-Sepharose CL-4B (Sigma-Aldrich).

7. CQ buffer: 50 mM Tris–HCl (pH 7.5), 100 mM NaCl and 5 mM $CaCl_2$.
8. CW buffer: 50 mM Tris–HCl (pH 7.5), 100 mM NaCl and 0.1 mM $CaCl_2$.
9. CHSW buffer: 50 mM Tris–HCl (pH 7.5), 500 mM NaCl and 0.1 mM $CaCl_2$.
10. CE buffer: 50 mM Tris–HCl (pH 7.5) and 1 mM EGTA.
11. Dans buffer: 100 mM Tris–HCl (pH 8.5) and 20 mM $CaCl_2$.
12. Dansyl chloride and Sephadex G-25 (Sigma-Aldrich).
13. pET-14b expression vector (Novagen).
14. BL21-DE3 competent cells (Novagen).
15. Ponceau Red (Sigma-Aldrich).
16. General laboratory equipment: see Subheading 2.1.

2.3 Fluorescence Experiments

1. Binding buffer: 25 mM Tris–HCl (pH 7.4), 120 mM KCl, 5 mM NaCl, 2 mM $MgCl_2$, and 10 mM EGTA.
2. Spectrofluorimeter, generally we use an Aminco Bowman series 2 (SLM Aminco).
3. Quartz cuvette with two transparent faces (3 mm light path, 100 μL volume).
4. Concentrated $CaCl_2$ solution (1–5 M).

3 Methods

3.1 Purification of GST Fusion Proteins

The method we use to produce recombinant proteins of the Kv7.2 CBD (12) is detailed here. The constructs are cloned into pGEX expression vectors and while the BL21-DE3 *Escherichia coli* is strain generally employed to transform the plasmids, we also recommend the use of BL21-Codon plus.

1. Grow the cells at 37 °C in 1 L of LB medium containing 100 μg/mL ampicillin until A_{600} = 0.6–0.8. Induce protein expression for 3 h at 30 °C with 0.3 mM IPTG.
2. Harvest the cells by centrifugation and resuspend the pellet in 20 mL of chilled GST buffer. After lysis by sonication, remove the cell debris by centrifugation for 30 min at 80,000 × *g* and 4 °C. Transfer the supernatant (soluble fraction) to a fresh tube while the insoluble cell lysate (inclusion bodies) is treated separately.
3. Resuspend the inclusion bodies in 10 mL GST buffer and then centrifuge again to remove any remaining soluble material. Repeat this operation three times. Dissolve the precipitate in 5–10 mL solubilization buffer for 30 min at 4 °C, mixing occasionally.

Centrifuge for 20 min at 20,000×*g* and 4 °C, and dilute the supernatant to a final concentration of ~1 mg/mL. Refold the proteins with urea concentration gradient dialysis. Dialyze the protein against 2 L of freshly made refolding buffer, gradually reducing the concentration of urea. At each urea concentration, dialyze the proteins for 8–10 h at 4 °C. Centrifuge at 30,000×*g* for 30 min at 4 °C to remove any aggregates and recover the supernatant (see Note 3).

4. Incubate the soluble fraction and refolded proteins separately with Glutathione Sepharose beads previously equilibrated with GST buffer. Wash the resin three times before transferring it to a disposable column, and elute the fusion proteins with 10–20 mL of elution buffer.
5. Analyze the fractions obtained on 10–15% SDS-PAGE gels and dialyze to remove the glutathione against a buffer of choice at 4 °C. Concentrate the proteins if needed and store at −20 °C. Determine the protein concentration using the Bradford method (14).

3.2 Purification and Dansylation of Calmodulin

The rat CaM gene cloned into the pET-14b expression vector is transformed in BL21-DE3 *E. coli* (see Note 4). The protocol employed to purify CaM has been adapted from the literature (15) and it yields large amounts of soluble protein (see below). Finally, to some extent we follow the instructions described in the literature to covalently attach a dansyl group to CaM (8).

3.2.1 Purification of Calmodulin

1. Grow the BL21-DE3 cells at 37 °C in 1 L of LB medium containing 100 μg/mL ampicillin to A_{600} = 0.8–1. Induce protein expression with 0.4 mM IPTG for 4–6 h at 37 °C.
2. Spin down the cells and wash the pellet twice with 50 mL of fresh Lysis buffer, resuspending it in the same buffer. Store the sample at −20 °C as 10 mL aliquots.
3. Thaw an aliquot on ice and sonicate it (3 cycles of 10 s on ice) before performing three freeze–thaw cycles using a dry ice–ethanol bath alternating with a 37 °C water bath. Centrifuge the sample in a microcentrifuge at 14,000×*g* for 15 min, and then heat the supernatant to 95 °C for 5 min before again centrifuging as above. This step takes advantage of CaM's elevated thermal stability.
4. Add $CaCl_2$ (5 mM final concentration) to the new supernatant and load at room temperature to a Phenyl–Sepharose column equilibrated with CQ buffer. Wash with 20 volumes of CW buffer then with 10 volumes of CHSW buffer, and elute with 20 volumes of CE buffer.
5. Analyze the fractions containing CaM on 15% SDS-PAGE (+5 mM EGTA) gels, dialyze, concentrate if needed and store

at −20 °C or lyophilize. The CaM concentration can be estimated from the absorbance at 276 nm, given that $\varepsilon_{276} = 3{,}030\ M^{-1}\ cm^{-1}$ (16) and $\varepsilon_{276,\ 1\%} = 1.8\ (g/100\ mL)^{-1}\ cm^{-1}$ (15). Alternatively use the Bradford method.

3.2.2 Dansylation of Calmodulin

1. Dilute CaM in Dans buffer to a final concentration of 1 mg/mL.
2. Dissolve dansyl chloride in acetone (2.17 mg/mL) and store at 4 or −20 °C in the dark. This sample is stable for many months.
3. Add 12.5 μL of dansyl chloride to 1 mL of the CaM solution to achieve a final dansyl chloride concentration of ~100 μM. Incubate at room temperature in the dark for 2 h, vortexing every 20 min.
4. To separate the dansylated CaM from unreacted dansyl, prepare ~1 mL of Sephadex G-25 in a disposable column and equilibrate ~250 mg of dry resin with distilled water. Load the D-CaM mixture and collect fractions of 50–100 μL. The excluded fraction, which corresponds to the first fractions eluted, is the conjugate D-CaM (see Note 5).
5. Quickly check the presence of the protein in the fractions by dot blotting on nitrocellulose and staining with Ponceau Red. Analyze the fractions in 15% SDS-PAGE gels and in particular, register the emission spectra of each sample (see below).
6. Collect the D-CaM fractions, concentrate if necessary and store as aliquots in the dark at −20 °C or lyophilized. We have found little change in the behavior of the conjugate when stored at −20 °C for several months or more.
7. Use the Bradford assay to determine the protein concentrations of D-CaM using unlabeled CaM as the standard. In addition, the concentration of D-CaM may be determined by UV absorption with a $\varepsilon_{320} = 3{,}400\ M^{-1}\ cm^{-1}$ (17).
8. Determine the concentration of dansyl moiety incorporated by spectroscopy (see Note 6) and, when possible, calculate the number of specific dansylated residues in a D-CaM molecule (see Note 7).

3.3 Fluorescence Experiments

In this section, we describe the preparation of the samples and the methods employed to perform fluorescence assays with D-CaM. We present three experiments: two are spectrophotometric titrations to measure fluorescence of a solution of D-CaM upon successive addition of ligand (Ca^{2+} and/or peptide), while the third is a competition assay. Generally, the emission spectra are recorded while adding increasing amounts of the target until no further change is observed. Finally, the data are analyzed to obtain information about the protein interactions, conformational changes in

CaM and parameters relating to affinity. For convenience, we will use the term "peptide" in this section to refer to the GST-Kv7.2 CBD employed in these assays.

1. Before dialysis, add EGTA (final concentration, 500 mM) to D-CaM sample to facilitate the elimination of Ca^{2+}. This is a very important step as free Ca^{2+} does not readily diffuse out of the dialysis bag. Dialyze D-CaM and the peptide against 2 L of binding buffer (without EGTA) for 24 h, and successively against 2 L of binding buffer complemented with 10 mM EGTA for 24 h (see Note 8).
2. Centrifuge all the proteins in a microcentrifuge at 14,000 × *g* for 10 min and check for the presence of aggregates by dynamic light scattering (DLS; see Note 3). Prepare and filter the binding buffer and carefully clean the cuvette. Care should be taken to avoid contamination with Ca^{2+}, cleaning the material with 10 mM EGTA buffered solutions. Leave the instrument on for several minutes prior to use to allow the light source and the electronics to stabilize, and then obtain a baseline fluorescence-emission scan of the binding buffer alone from 360 to 660 nm, keeping the excitation constant at 340 nm. Both the excitation and emission band-passes are 5 nm. Repeat the same operation with WT CaM as a control and then separately, with the peptide dissolved in binding buffer.
3. Put 100 μL of binding buffer in a cuvette, add sufficient stock D-CaM to achieve 12.5 nM and mix by drawing the solution into the pipette tip and expelling it several times (see Note 9). Avoid bubbles, since they will affect the fluorescence measurements. Place the cuvette in the fluorimeter and collect the emission spectra. As above, the excitation wavelength is 340 nm and the emission is registered from 400 to 660 nm. All the measurements are made at 25 °C and the spectra are corrected for the contribution of buffer. In these conditions the fluorescence spectrum of D-CaM should exhibit a maximum at ~500 nm (Fig. 2).
4. Calcium titration: Sequentially add aliquots of concentrated $CaCl_2$ to a cuvette containing 12.5 nM D-CaM (Table 1), and then record the emission spectra (see Note 10).

   After each addition of Ca^{2+}, mix the solution to assure homogeneity and obtain the fluorescence spectra 20–30 s after adding the sample (longer equilibration times do not improve the data). The Ca^{2+} is titrated to saturation, when no further changes in spectra are observable. This experiment shows that the fluorescence emission spectrum of D-CaM is markedly affected by Ca^{2+}, which causes a conformation change in CaM that can be observed as a "blue shift" in λ_{max} from ~500 to ~485 nm and an increase in intensity (of at least 100%; Fig. 2b).

Table 1
Calculated free Ca^{2+} concentrations as a function of total Ca^{2+} (EGTA 10 mM, pH 7.4)

(Ca^{2+}) total (mM)	(Ca^{2+}) free (μM)	(Ca^{2+}) total (mM)	(Ca^{2+}) free (μM)
0	0	8.385	0.408
2.498	0.024	8.634	0.499
4.246	0.055	8.784	0.571
4.995	0.075	9.033	0.741
5.244	0.083	9.182	0.894
5.743	0.102	9.431	1.325
6.242	0.127	9.581	1.827
6.741	0.159	9.628	2.000
7.239	0.203	9.829	4.539
7.488	0.231	9.979	19.77
7.987	0.310	10.22	231.4
8.136	0.341	10.47	4,735

Repeat the Ca^{2+} titration in the presence of a saturating concentration of peptide, for which we use 200 nM of the peptide to complex with D-CaM. This experiment is useful to determine how the Ca^{2+} binding affinity of CaM is affected in the CaM–peptide complex (7), which is achieved by plotting the fluorescence enhancement as a function of free (Ca^{2+}). A simple method is to plot F/F_0 (Fluorescence increase) as a function of free (Ca^{2+}), where F_0 is the initial fluorescence intensity in the absence of Ca^{2+} and F is the fluorescence intensity at each free (Ca^{2+}). To obtain F and F_0 from each spectrum, we average the fluorescence intensity in the 450–550 nm range. Alternatively, the maximum fluorescence can be normalized to 100% and the percentage fluorescence increase plotted as a function of the free (Ca^{2+}). The parameters of the Hill equation (Fluorescence increase $= A \times (\text{peptide})h/\text{EC}_{50}h + (\text{peptide})h$: where A is the maximal fluorescence increase and h is the Hill coefficient) are fitted to the data by curvilinear regression, enabling the apparent affinity (EC_{50} or concentration that gives half-maximal change in fluorescence emission intensity) and Hill coefficient to be accurately determined. Generally, 3–6 titrations are needed on average to fully analyze Ca^{2+} binding to D-CaM. Finally, this assay is useful to determine the free (Ca^{2+}) necessary to saturate D-CaM, as seen below.

5. Peptide titration in the presence or absence of Ca^{2+}: Add aliquots of peptide from a concentrated stock to a cuvette containing 12.5 nM D-CaM alone or in the presence of an excess of calcium (2 μM free Ca^{2+}) in binding buffer (Fig. 2 and see Note 11). After adding each aliquot, mix the solution, wait 20–30 s and record the emission spectra. In both conditions (+Ca^{2+} or −Ca^{2+}), adding the peptide to D-CaM increases the fluorescence intensity proportional to the amount of binding protein added.

   GST alone is used as a control to estimate the impact of sample dilution, both in the presence and absence of Ca^{2+}, which does not cause any change in the fluorescent emission (10, 18). Plot fluorescence enhancement against the peptide–D-CaM ratio or against (peptide) to obtain concentration-response curves, and the parameters of the Hill equation can be fitted to the data by curvilinear regression. To obtain the true affinities (K_d), repeat the titrations varying the initial concentration of D-CaM (6.25–200 nM) and generate the concentration-response curves at each D-CaM concentration. The apparent dissociation constants (EC_{50}) calculated from these curves can be plotted as a function of (D-CaM), and the true dissociation constants of the complexes can be derived by fitting the data with a linear regression and extrapolating the apparent dissociation constants to (D-CaM) equal to zero (see Note 12).

6. Competition assays: Not every ligand that interacts with CaM increases the fluorescent emission. In such cases, the interaction can be monitored using a competition assay in which the displacement of a ligand from its complex with D-CaM by a test peptide causes a reduction in dansyl fluorescence (12, 19). To perform these assays, add the peptide or other CaM binding protein to 12.5 nM D-CaM dissolved in binding buffer and record the dansyl emission spectra (see Note 13). The experiment may be performed in the presence (adding 2 μM free Ca^{2+} to D-CaM) or absence of Ca^{2+} (adding 10 mM EGTA), and the assay is most sensitive when using the concentration of target peptide that produces a 50% effect (e.g., at the EC_{50}). Titrate the complex with increasing concentrations of the second ligand and record the fluorescence spectra for each aliquot. Analyze the data by plotting the percentage reduction in fluorescence (taking the initial complex as 100%) against the (competing peptide) (Fig. 3).

4 Notes

1. Rat CaM does not contain tryptophan residues but there are five phenylalanines in the N-terminal domain and three in the C-terminal domain; there are also two tyrosine residues in

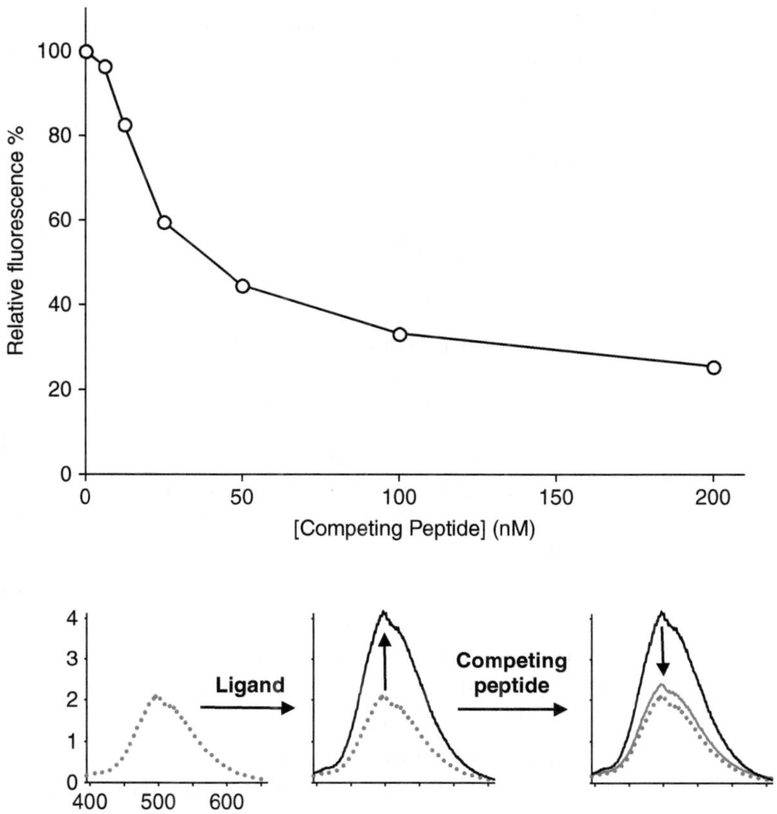

Fig. 3 Competition assay. Example of the competition of a peptide that does not cause an increase in D-CaM fluorescence emission. First, the baseline is obtained by mixing D-CaM with a ligand at a concentration corresponding to its calculated EC_{50}. The ligand is displaced from CaM by the increasing addition of a competing peptide. Below, schematic representation of the experiment

the C-terminal domain. Changes in the phenylalanine and tyrosine fluorescence intensity have been used to monitor Ca^{2+} binding in a single domain or in full-length CaM (20).

2. We have studied the interaction of D-CaM with recombinant proteins that contain the CaM-binding domains of SK2 channels, NR1a receptors, and neurogranin, none of which causes a blue shift in the emission spectra.

3. It is important to ensure that the sample protein or peptide does not aggregate. We routinely evaluate the dispersion of the samples by dynamic light scattering (DLS) using a Zetasizer Nano instrument (Malvern Instruments Ltd.). Only use samples in which the correlation function and the polydispersity index (<0.2) demonstrate that the proteins are monodispersed.

4. CaM from other sources can be used, but some modifications in the protein that will affect the results may be present.

Recombinant CaM lacks the first methionine residue and the numbering starts from alanine. CaM has nine methionine residues that can be naturally modified and that are sensitive to oxidation, resulting in methionine sulfoxide. The oxidatively modified form exhibits a dramatically lower affinity for Ca^{2+} and its interaction with ligands is affected (21).

5. To remove the excess of dansyl chloride it is advisable to dialyze the D-CaM samples in addition to performing the gel filtration described.
6. The dansyl incorporated per mol of CaM can be determined by absorption spectroscopy using the $\varepsilon_{335} = 3{,}980\ M^{-1}\ cm^{-1}$ (3) or $\varepsilon_{320} = 3{,}400\ M^{-1}\ cm^{-1}$ (17).
7. To determine the specific dansylated residues in CaM, tryptic digestion coupled to mass spectrometry or gas-phase protein sequencers has been employed (7, 8, 12). It has been reported that Lys75 (7, 8) or Lys115 (22) are dansylated. Through mass spectrometry we have detected the binding of up to four dansyl molecules per CaM. Furthermore, tandem mass spectrometry of the tryptic peptides indicates that Ala1 and Lys148 are certainly dansylated. We have not identified the other two dansylated residues yet but our data are compatible with them being Lys75 and Lys115. The $^{1}H,^{15}N$-Heteronuclear Single Quantum Coherence (HSQC) spectrum of calcified D-CaM (Fig. 4), that contains a peak for each unique proton attached to the heteronucleus being considered, clearly shows that the protein retains a similar folding as the wild type one. However, chemical shift perturbations are observed throughout the protein consistent with the idea that several dansylation events take place in both domains of CaM.
8. The initial Ca^{2+} concentration in the D-CaM sample is critical in these assays and it is important to remove as much Ca^{2+} as possible. Using the dialysis conditions described in this chapter, we determined the levels of contaminant Ca^{2+} to be less than 40 nM by inductively coupled plasma mass spectroscopy.
9. We have obtained the best results using 12.5 nM D-CaM, balancing the use of low sample concentration with the resolution of the spectra. The signal to noise ratio decreases at lower (D-CaM), reaching unacceptable levels below 6.25 nM. Nevertheless, extremely low (D-CaM), such as 3.3 nM, has been used to perform these assays (23).
10. The most important points for accurate Ca^{2+} titrations is the precise control of free (Ca^{2+}) using Ca^{2+} chelators like EGTA or EDTA. We prefer using EGTA because the experiments are carried out at pH 7.5 to mimic intracellular conditions, including physiological concentrations of magnesium. It is important to carefully control the pH because the chelating power is very

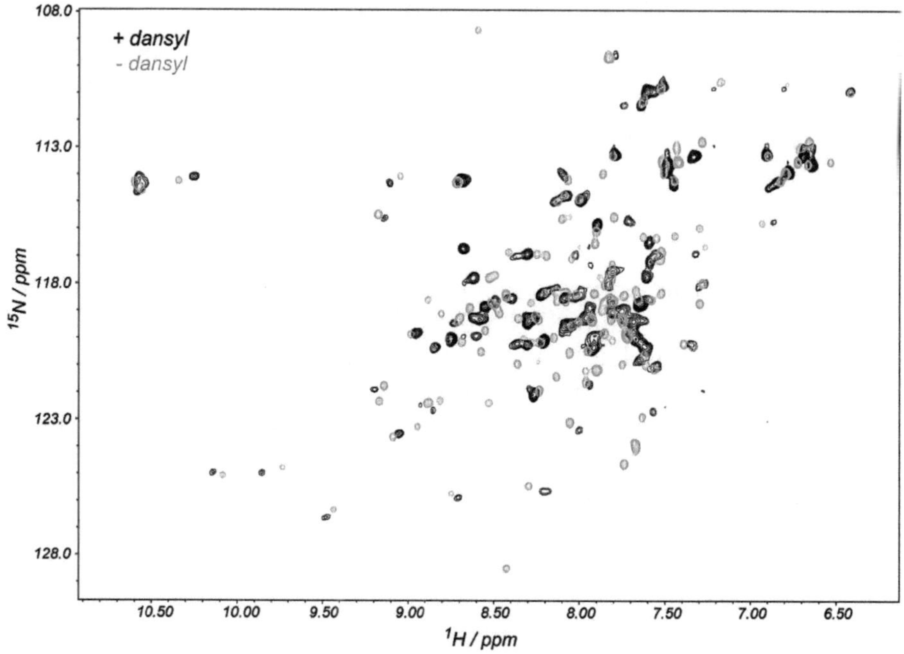

Fig. 4 ^{1}H,^{15}N-HSQC for dansylated (*black*) and wild type calmodulin (*grey*) in the presence of 5 mM calcium. The peak distribution is similar indicating that both proteins fold in a similar way. The changes in the chemical shift observed throughout the molecule are an indication of multiple dansylation sites in the two protein domains

sensitive to the concentration of protons, and sufficient Ca^{2+} buffer must be used (10 mM EGTA in our case) to ensure that the small amount of endogenous Ca^{2+} bound to the protein being titrated does not affect the free (Ca^{2+}). To calculate the free (Ca^{2+}) we used the Maxchelator program (http://maxchelator.stanford.es) and custom software. Several other programs are available to determine the free (Ca^{2+}) in function of the total Ca^{2+} added, for a given (EGTA) at different temperatures, ionic strengths and pH's.

11. Titrating D-CaM with Ca^{2+}, as described in this chapter, we found that 2 μM free (Ca^{2+}) is sufficient to saturate D-CaM at a concentration of 12.5 nM.

12. The apparent EC_{50} values of the ligand–CaM interaction depend on the concentration of D-CaM employed (10, 11, 24). This dependence can be attributed to the depletion of the free ligand, which is most pronounced at low ligand concentrations. It is possible to correct for depletion by determining the EC_{50} values over a range of D-CaM concentrations. For infinitely low (D-CaM), depletion should be nonexistent and the EC_{50} value should hence be a true affinity. If the stoichiometry is known, the K_d values can be estimated using a Scatchard plot analysis (9), although we recommend the previous method.

13. We have also performed the competition assay using GST-NMDA (NR1a, aa 818–922) and GST-Neurogranin (aa 1–78) as reporters of the interaction with CaM in the presence or absence of Ca^{2+}, respectively (12).

Acknowledgments

This work was supported by grants from the Spanish Ministry of Education (BFU2009-07581), the Spanish Ion Channel Initiative Consolider project (CSD2008-00005), and the Basque Government (SAIOTEK SA-2006/00023). A. Alaimo was partially funded by Fundación Biofísica Bizkaia. Proteomics Core Facility-SGIKER is a member of ProteoRed-ISCIII. Technical and human support provided by SGIKER (UPV/EHU, MICINN, GV/EJ, ERDF, and ESF) is gratefully acknowledged.

References

1. Walker JM (1994) The dansyl method for identifying N-terminal amino acids. Methods Mol Biol 32:321–328
2. Kincaid RL, Vaughan M (1986) Direct comparison of Ca^{2+} requirements for calmodulin interaction with and activation of protein phosphatase. Proc Natl Acad Sci U S A 83: 1193–1197
3. Johnson JD, Wittenauer LA (1983) A fluorescent calmodulin that reports the binding of hydrophobic inhibitory ligands. Biochem J 211:473–479
4. Kincaid RL, Vaughan M, Osborne JC Jr, Tkachuk VA (1982) Ca^{2+}-dependent interaction of 5-dimethylaminonaphthalene-1-sulfonyl-calmodulin with cyclic nucleotide phosphodiesterase, calcineurin, and troponin I. J Biol Chem 257:10638–10643
5. Olwin BB, Storm DR (1983) Preparation of fluorescent labeled calmodulins. Methods Enzymol 102:148–157
6. Malencik DA, Anderson SR (1983) Binding of hormones and neuropeptides by calmodulin. Biochemistry 22:1995–2001
7. Mori M, Konno T, Ozawa T, Murata M, Imoto K, Nagayama K (2000) Novel interaction of the voltage-dependent sodium channel (VDSC) with calmodulin: does VDSC acquire calmodulin-mediated Ca^{2+}-sensitivity? Biochemistry 39: 1316–1323
8. Torok K, Cowley DJ, Brandmeier BD, Howell S, Aitken A, Trentham DR (1998) Inhibition of calmodulin-activated smooth-muscle myosin light-chain kinase by calmodulin-binding peptides and fluorescent (phosphodiesterase-activating) calmodulin derivatives. Biochemistry 37:6188–6198
9. Ehlers MD, Zhang S, Bernhadt JP, Huganir RL (1996) Inactivation of NMDA receptors by direct interaction of calmodulin with the NR1 subunit. Cell 84:745–755
10. El Far O, Bofill-Cardona E, Airas JM, O'Connor V, Boehm S, Freissmuth M, Nanoff C, Betz H (2001) Mapping of calmodulin and Gbg binding domains within the C-terminal region of the metabotropic glutamate receptor 7A. J Biol Chem 276:30662–30669
11. Schleiff E, Schmitz A, McIlhinney RA, Manenti S, Vergères G (1996) Myristoylation does not modulate the properties of MARCKS-related protein (MRP) in solution. J Biol Chem 271: 26794–26802
12. Alaimo A, Gomez-Posada JC, Aivar P, Etxeberría A, Rodriguez-Alfaro JA, Areso P, Villarroel A (2009) Calmodulin activation limits the rate of KCNQ2 K+channel exit from the endoplasmic reticulum. J Biol Chem 284:20668–20675
13. Dick IE, Tadross MR, Liang H, Tay LH, Yang W, Yue DT (2008) A modular switch for spatial Ca^{2+} selectivity in the calmodulin regulation of CaV channels. Nature 451:830–834
14. Bradford MM (1976) A rapid and sensitive method for the quantitation of microgram quantities of protein utilizing the principle of protein-dye binding. Anal Biochem 72:248–254
15. Hayashi N, Matsubara M, Takasaki A, Titani K, Taniguchi H (1998) An expression system of rat calmodulin using T7 phage promoter in Escherichia coli. Protein Expr Purif 12:25–28

16. Wallace RW, Tallant EA, Cheung WY (1983) Assay of calmodulin by Ca^{2+}-dependent phosphodiesterase. Methods Enzymol 102: 39–47
17. Chen RF (1968) Dansyl labeled proteins: determination of extinction coefficient and number of bound residues with radioactive dansyl chloride. Anal Biochem 25:412–416
18. Yus-Nájera E, Santana-Castro I, Villarroel A (2002) The identification and characterization of a noncontinuous calmodulin-binding site in noninactivating voltage-dependent KCNQ potassium channels. J Biol Chem 277: 28545–28553
19. Pitt GS, Zuhlke RD, Hudmon A, Schulman H, Reuter H, Tsien RW (2001) Molecular basis of calmodulin tethering and Ca^{2+}-dependent inactivation of L-type Ca^{2+} channels. J Biol Chem 276:30794–30802
20. VanScyoc WS, Sorensen BR, Rusinova E, Laws WR, Ross JB, Shea MA (2002) Calcium binding to calmodulin mutants monitored by domain-specific intrinsic phenylalanine and tyrosine fluorescence. Biophys J 83:2767–2780
21. Gao J, Yao Y, Squier TC (2001) Oxidatively modified calmodulin binds to the plasma membrane Ca-ATPase in a nonproductive and conformationally disordered complex. Biophys J 80:1791–1801
22. Leclerc E, Corti C, Schmid H et al (1999) Serine/threonine phosphorylation of calmodulin modulates its interaction with the binding domains of target enzymes. Biochem J 344 (Pt 2):403–411
23. Liu M, Chen TY, Ahamed B, Li J, Yau KW (1994) Calcium-calmodulin modulation of the olfactory cyclic nucleotide-gated cation channel. Science 266:1348–1354
24. Turner JH, Gelasco AK, Raymond JR (2004) Calmodulin interacts with the third intracellular loop of the serotonin 5-hydroxytryptamine1A receptor at two distinct sites: putative role in receptor phosphorylation by protein kinase C. J Biol Chem 279:17027–17037

Chapter 18

Imaging and Quantification of Recycled K_{ATP} Channels

Christopher J. Cockcroft

Abstract

This chapter describes immunochemistry-based methods to investigate recycling of membrane proteins at the cell surface. Two methods are described, one qualitative and the other quantitative. Both methods consist of two rounds of extracellular antibody capture. Firstly, a primary antibody is captured by an extracellular epitope presented by the target membrane protein and is subsequently internalized. Secondly, the primary antibody-labelled protein is recycled back to the membrane where it is captured by a probe-conjugated secondary antibody. In the qualitative assay, the probe is a fluorophore, which can be imaged by fluorescence microscopy. In the quantitative assay, the probe is horse-radish peroxidase (HRP) and enzyme activity can be assayed by chemiluminescence.

Key words Endocytic recycling, K_{ATP}, Confocal microscopy, Chemiluminescence

1 Introduction

Recycling membrane proteins are internalized from the plasma membrane into early and sorting endosomes from where they can either return to the cell surface via a Rab4-dependent pathway (1) or enter the endosomal recycling compartment (ERC) from where their return to the plasma membrane occurs via Rab11-dependent pathways (2). A limited number of membrane proteins use a somewhat convoluted pathway, involving trafficking from the ERC to the cell surface via the trans-Golgi network (3). Trafficking mechanisms for recycling are extensively reviewed by Maxfield and McGraw (4) and Grant and Donaldson (5).

Recycling of internalized membrane proteins back to the cell surface has been reported for a range of membrane proteins including ion channels, receptors, and transporters. It represents an efficient mechanism by which prompt delivery of proteins to the plasma membrane occurs in response to changes in cell physiology. For example, recycling endosomes supply APMA receptor for long-term potentiation in neurons (6), β-adrenergic receptors for

Nikita Gamper (ed.), *Ion Channels: Methods and Protocols*, Methods in Molecular Biology, vol. 998,
DOI 10.1007/978-1-62703-351-0_18, © Springer Science+Business Media, LLC 2013

receptor resensitization (7), and GLUT-4 transporters for glucose uptake by adipose and muscle (8).

This chapter describes methods for imaging and quantification of recycled channels. Using the K_{ATP} channel as an example, we describe immunochemistry-based methods for investigating membrane protein recycling (9). These methods employ an antibody capture approach that requires an antibody that recognizes an extracellular epitope (native or engineered) of the membrane protein of interest. A similar approach has also been described for investigating endocytosis of K_{ATP} channels (10, 11). In the examples presented here, we also use an HA-epitope (YPYDVPDYA) engineered into the extracellular loop of the Kir6.2 subunit of the K_{ATP} channel (12). The HA-tagged Kir6.2 was co-expressed with the sulphonylurea receptor 1(SUR1) in HEK-293 cells to produce HA-tagged K_{ATP} channels (K_{ATP}-HA) at the cell surface (under normal conditions neither subunit is capable of independent surface expression). Internalized primary antibody-channel complexes can recycle back to the cell surface to capture the probe-conjugated secondary antibodies (see Note 1). For example, recycled channels can be detected by using either confocal fluorescent microscopy or image-captured fluorescent-conjugated secondary antibodies (Fig. 1); following these steps, cells can then be fixed and permeabilized to stain either non-recycled channels or intracellular organelle markers. Alternatively, recycling can be quantified using chemiluminescence to assay-captured HRP-conjugated secondary antibody by assaying enzyme activity (Fig. 2).

2 Materials

2.1 Preparation of Coverslips and Cells

1. No. 7 curved watchmaker forceps.
2. Borosilicate glass coverslips (13 mm).
3. HEK293 cell line (ECACC).
4. Dulbecco's phosphate buffered saline (DPBS).
5. Poly-L-lysine coating solution: 0.01% poly-L-lysine (70–150 kDa, Sigma), diluted to 0.001% in DPBS.
6. 70% ethanol.
7. Culture medium: DMEM + Glutamax-1, supplemented with 10% fetal bovine serum, penicillin (50 U), streptomycin (50 μg).
8. Trypsin solution: 0.05% trypsin–EDTA.
9. Suitable transfection reagent available from a range of suppliers (see Chapters 2, 4, and 5 of this book for more details on transfection techniques and reagents).

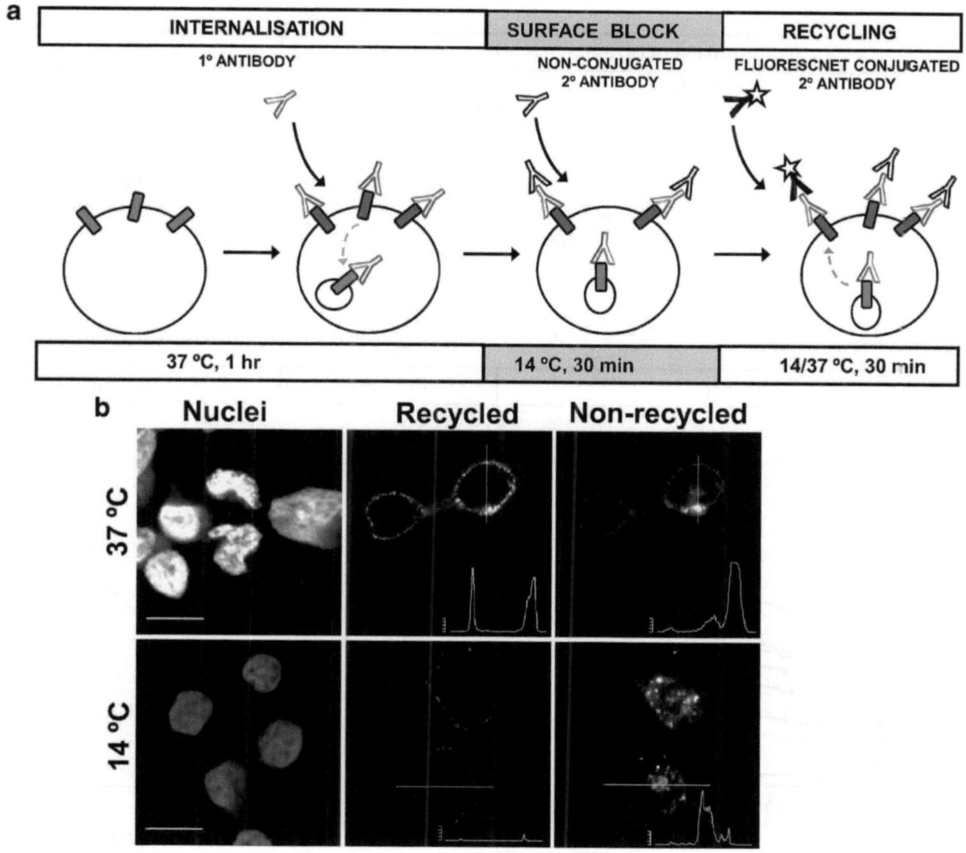

Fig. 1 Visualization of K_{ATP}-HA channel recycling by confocal fluorescent microscopy. (**a**) Schematic diagram showing the key stages of the antibody capture experiment. (**b**) Confocal images of cells expressing K_{ATP}-HA following fluorescent-conjugated antibody capture by recycled channels at the cell surface. K_{ATP}-HA channel recycling was detected after 30 min incubation at 37°C. Following fixation and permeabilization steps, non-recycled channels were stained using anti-rat Cy3 secondary antibody (1:500 dilution, 1.25 μg/mL). The cell nuclei were stained using DAPI (supplied in the mounting medium) (scale bar = 10 μm). Inset, grey value profiles of line scans through cross sections of cell images (scale bar = 50 grey value)

2.2 Imaging of Channel Recycling

1. PBS: 137 mM NaCl, 4.3 mM Na_2HPO_4, 1.47 mM KH_2PO_4, 2.7 mM KCl, 1 mM $CaCl_2$, 1 mM $MgCl_2$.
2. Staining support plate: 24-well plate lid, covered with Parafilm.
3. Blocking medium: DMEM, 10 mM HEPES (from 1 M stock solution, pH 7.4), 1% Ovalbumin (Sigma) (see Note 2).
4. Primary antibody: rat anti-HA high-affinity primary antibody 3F10 (Roche), diluted in blocking medium 1:500 (0.2 μg/mL) working concentration (see Note 3).
5. Nonconjugated secondary antibody: goat anti-rat IgG secondary antibody (Sigma), diluted in blocking medium 1:250 (4 μg/mL) working concentration.

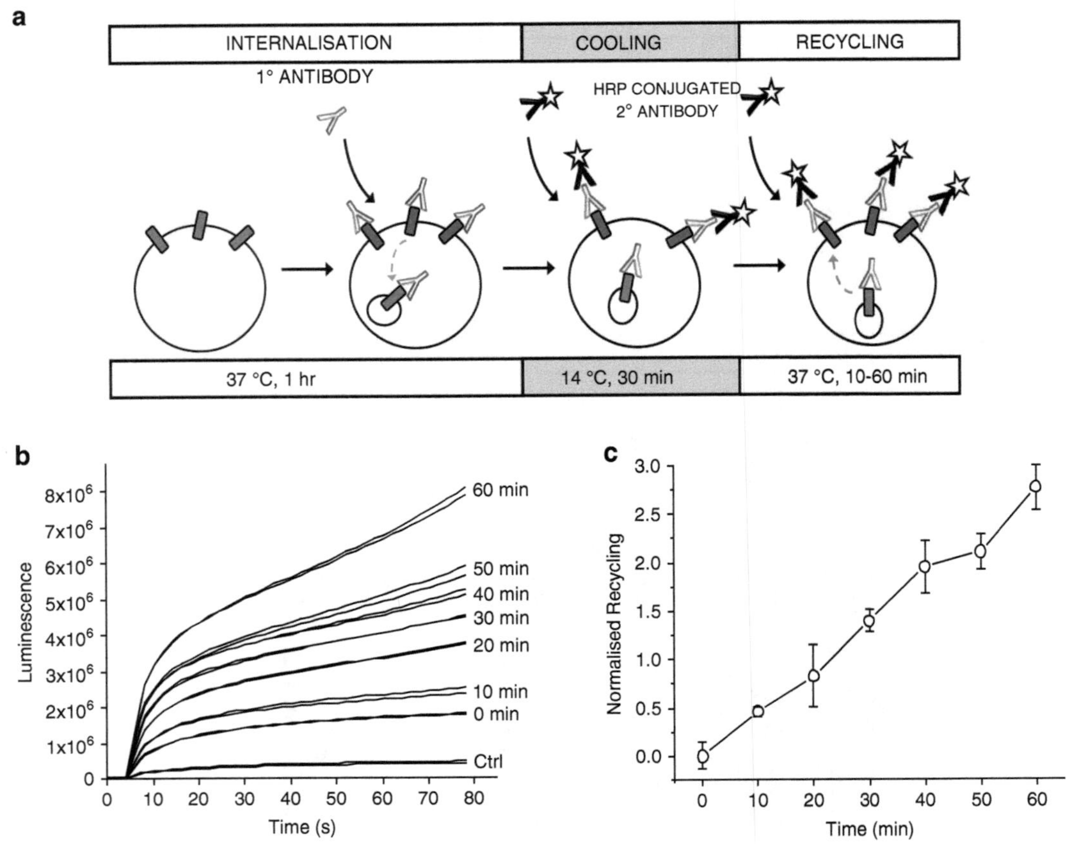

Fig. 2 Quantification of K_{ATP}-HA channel recycling channels by chemiluminescence. (**a**) Schematic diagram showing the key stages of the antibody labelling experiment. (**b**) Representative real-time chemiluminescence data from cell lysates following HRP-conjugated antibody capture by recycled channels at the cell surface after 0, 10, 20, 30, 40, 50, and 60 min at 37°C. Cells expressing SUR1alone were used as a control for nonspecific antibody binding (Ctrl). (**c**) The time-course plot showing that K_{ATP}-HA is continuously recycled in a time-dependent manner

6. Fluorophore-conjugated secondary antibody: goat anti-rat Alexa Fluor 488 IgG (molecular probes), diluted in blocking medium 1:500 (1.25 μg/mL) working concentration.
7. Fixing tray: an upturned lid of a 24-well plate.
8. Fixative: methanol/acetone 1:1 (Me:Ac), stored at −20°C.
9. Microscope glass slides (76 × 26 mm).
10. Mounting medium: DAPI Fluoromount G mounting medium (Southern Biotech) or equivalent.
11. Confocal microscope: fitted with relevant lasers and filters; we use Axiovert LSM510 (Zeiss) with 63×/1.4 oil DIC objective.
12. Images were analyzed using Image J (NIH).

2.3 Additional Materials for Quantification of Channel Recycling

1. HRP-conjugated secondary antibody: goat anti-rat HRP-conjugated IgG antibody (Sigma), diluted in blocking media to 1:500 (1–8 μg/mL) working concentration.
2. Lysis buffer: 1% Triton X100, 50 mM HEPES, 150 mM KCl, 1 mM $MgCl_2$, pH 7.4.
3. BCA reagent protein assay kit (Available from Thermo Scientific or Sigma, etc.).
4. Clear flat-bottomed 96-well microtiter plate.
5. White flat-bottomed 96-well microtiter plate.
6. HRP substrate: Lumigen PS atto luminol reagent solution A+B (1:1) (Lumigen) or Femto Super Signal (Thermo Scientific).
7. Plate reader: we use Varioskan Flash (Thermo Scientific) multimode plate reader (see Note 4).

3 Methods

3.1 Preparation of Cells

HEK-293 cells are grown on poly-lysine-coated coverslips placed in a 24-well dish. The cells are transfected with an expression vector construct containing the cDNA of interest engineered to express an extracellular HA-epitope. If necessary, additional cells should also be prepared for introducing controls for additional variables, such as transfection efficiency and antibody specificity.

1. Gently mix 200–300 coverslips with 20 mL poly-L-lysine coating solution for 2 h at room temperature. Rinse ten times with distilled water and store in 70% ethanol at 4°C until required.
2. In a tissue culture cabinet, using forceps, place a coverslip into each well of a 24-well plate, using as many wells as appropriate for the experiment. To speed up drying time, place each coverslip so that it rests against the wall of the well. Once dry, gently tap the plate so that the coverslips fall to the bottom of the well. Add 0.5 mL culture medium per well.
3. Remove the culture medium from HEK 293 cells grown in a T-25 flask to ~80% confluence (grown in a humid CO_2 incubator supplied with 5% CO_2/95% air). Rinse cells with 5–10 mL of DPBS, add 0.5 mL trypsin solution, and incubate until the cells are detached. Then add 6 mL of culture medium and resuspend gently.
4. Add 1–2 drops trypsinized cell suspension (~1×10^5 cells/mL) per well and incubate overnight (see Note 5).
5. Transfect the cells with plasmid DNA constructs using an appropriate transfection reagent, following the instructions of the manufacturer (see Note 5). Incubate for 48–72 h.

3.2 Imaging of Channel Recycling

This section describes a method for continuous labelling of recycled K_{ATP}-HA channels over a 30 min period using fluorophore-conjugated secondary antibody for fluorescent microscopy imaging of recycled channels (Fig. 1).

1. Pipette 30 μL droplets of primary antibody medium onto a staining support tray.
2. Remove each coverslip from the 24-well plate (step 5, Subheading 3.1) using forceps and rinse by gently dipping into a beaker of ice cold PBS (200 mL) for 5–10 s (see Note 6). Drain excess liquid from the coverslip by dabbing the edge carefully onto tissue paper, taking care not to completely dry out the cells on the coverslip.
3. Place each coverslip onto a 30 μL droplet of primary antibody, with the cells facing toward the droplet (see Note 7). Incubate at 37°C for 1 h to permit continuous labelling and internalization of surface channels (see Note 8).
4. Cool the primary antibody staining support plate to ≤14°C for 15 min to prevent further trafficking (see Note 9). On ice, remove each coverslip from the staining support tray and wash with chilled PBS (as in step 2, using fresh PBS).
5. Place each coverslip onto a droplet of prechilled nonconjugated secondary antibody (prepared before use, as step 1). Incubate at ≤14°C for 30 min to permit saturation of primary antibody-labelled surface channels.
6. Remove each coverslip from the nonconjugated secondary antibody and wash with chilled PBS (as in step 2, using fresh PBS).
7. Place each coverslip onto a droplet of Fluorophore-conjugated secondary antibody (prepared before use, as in step 1). Incubate for 30 min at either 37°C to permit recycling or return to ≤14°C conditions to prevent recycling.
8. Wash coverslips with chilled PBS to remove unbound secondary antibody (as in step 2, using fresh PBS).
9. Fix and permeabilize the cells in Me:Ac (dispense sufficient ice cold Me:Ac to fill a fixing tray). Submerge the coverslips in Me:Ac and incubate for 10 min on ice (with the cells facing upward). Gently perfuse the coverslips with PBS to remove MeAc (see Note 10).
10. Remove each coverslip from the fixing tray and wash with chilled PBS to remove any remaining fixative (as in step 2, using fresh PBS).

    Optional steps: the cells can now be stained for primary antibody-labelled non-recycled channels or specific cellular compartments, such as early endosomes (see Note 11).
11. Wash the coverslips by gently dipping into in a beaker of Milli-Q water to remove buffer salts.

12. Mount each coverslip onto a droplet (~6 μL) of mounting medium on a glass microscope slide (with the cells facing the droplet). Place the slides in a dark, dry place at room temperature overnight, and then store at 5°C until imaging.
13. Image the cells under a confocal florescent microscope.

3.3 Quantification of Channel Recycling

This section describes a method for quantifying K_{ATP}-HA channel recycling in a time-course experiment (Fig. 2). For this experiment, a full 24-well plate with coverslips is required. In the example given below, the plate layout consisted of 21 coverslips with cells expressing K_{ATP}, three for each time period (0, 10, 20, 30, 40, 50, and 60 min). Three coverslips with cells expressing only SUR1 (no Kir6.2-HA) were used as a negative control for nonspecific antibody binding.

Follow steps 1–3 *from Subheading* 3.2

4. Remove each coverslip from the primary antibody and wash with chilled PBS (as in step 2, Subheading 3.2, using fresh PBS).
5. Place each coverslip onto a 30 μL droplet of prechilled HRP-conjugated secondary antibody (prepared before use as in step 1 of Subheading 3.2). Incubate at ≤14°C for 30 min to label the remaining channels at the cell surface only.
6. Pipette 300 μL of lysis buffer into the wells of a 24-well plate and store on ice or 4°C throughout the experiment.
7. Remove a set of three coverslips for measurement of channels at the cell surface at the start of the time-course experiment (0 min recycling). Wash with chilled PBS (as in step 2 of Subheading 3.2, using fresh PBS), then place each coverslip into a well of lysis buffer (from step 6), and gently pipette up and down to dissociate the cells. Incubate overnight at 4°C with gentle shaking or rocking to complete lysis.
8. Transfer the remaining coverslips on the staining plate to 37°C. Remove sets of three coverslips at desired time intervals (e.g., 10 min intervals). Wash with chilled PBS (as in step 2, Subheading 3.2, using fresh PBS) and place each coverslip into a well of lysis buffer (from step 7).
9. Assay the amount of protein in the cell lysate with the BCA assay following the manufacturer's instructions (see Note 12).
10. Pipette 60 μL of each lysate sample (in duplicate) into a white microtiter 96-well plate. Assay chemiluminescence at 25°C using a luminometric equipped plate reader. After recording a baseline reading for 6 s, dispense 50 μL HRP substrate and continue to record for a 70 s.
11. Data analysis: subtract the mean luminescence endpoint value (final three measurements, after the addition of the HRP substrate)

from the mean baseline value (first three measurements, before addition of the HRP substrate) for each sample. Normalize this luminescence value to the protein content of the corresponding sample. Determine the mean and s.e.m. of the triplicates for each time point. Plot the mean ± s.e.m. against time to obtain a time-course graph.

4 Notes

1. Epitope tags other than HA can be used for your protein of interest (e.g. Myc tag).
2. The antibody concentrations used here is optimal for detection of Kir6.2-HA in our hands and can be used as a guideline for future experiments.
3. Alternatively, 1% BSA can also be used; ovalbumin is preferred because of its relatively low hydrophobic binding properties that can interfere with solubility of compounds used to disrupt recycling (e.g., Phorbol-12-myristate-13-acetate).
4. Alternative plate readers can be used that are capable of dispensing luminol into wells and measuring real-time or end-point luminescence.
5. For imaging experiments, the cells should not exceed ~50% confluency by the time of the experiment. Whereas for chemiluminescence experiments, cells should ideally reach 80% confluency. The generation of stable cell lines can be useful, especially for quantitative procedures where large numbers of coverslips maybe advantageous for multiple experiments.
6. Coverslips can also be washed by perfusion with PBS. This is useful for consistent high-throughput washes when using an entire 24-well plate. An upturned lid of a 24-well plate (with well rims) can be used as a tray for submersion of coverslips. PBS can be gently perfused and aspirated using a 20 mL syringe and a vacuum pump, respectively. Approximately 3 × 20 mL wash steps are sufficient to rinse the coverslips in the tray.
7. Cells should be maintained in a humid environment during incubation steps on the staining tray. Typically, a wet tissue in a plastic box with a loose lid is sufficient to prevent the coverslips from drying out during incubation steps.
8. The antibody incubation time can be extended to 2 h to increase the size of the labelled pool of internalized channels for slowly internalizing channels.
9. The acid strip approach is an alternative to the nonconjugated secondary antibody block step. This approach involves stripping the primary antibody from labelled channels at the cell surface. After the chilled PBS wash step, cells should be washed twice with chilled

acidic buffer (0.15 M NaCl, 0.5% acetic acid, pH 2.4), followed by neutralization with PBS and continuation of the protocol.

10. Removal of most Me:Ac and wash in PBS is important to prevent damage to the cells as a result of rapid solvent evaporation whilst handling the coverslips. Alternatively, fixation can be carried out in a 24-well plate, and removing Me:Ac by gradual dilution with PBS.
11. The non-recycled primary antibody-labelled pool of channels can be detected using an alternative anti-rat fluorophore-conjugated secondary antibody (e.g., Cy3). Cells can also be stained for a range of cellular markers, such as endosomal markers (EEA1/CD63) and Golgi (TGN46/Golgin).
12. The BCA absorbance values (OD_{570nm}) are sufficient for normalizing data to luminescence values. However, BSA standards (0–1 mg/mL) can also be used to estimate protein concentration in the lysate samples if required. Alternative protein concentration assay kits may also be suitable (e.g., Bradford assay).

Acknowledgement

The author wishes to thank Prof. A. Sivaprasadarao for advice on manuscript preparation and the MRC for funding support.

References

1. van der Sluijs P, Hull M, Webster P, Mâle P, Goud B, Mellman I (1992) The small GTP-binding protein rab4 controls an early sorting event on the endocytic pathway. Cell 70:729–740
2. Wilcke M, Johannes L, Galli T, Mayau V, Goud B, Salamero J (2000) Rab11 regulates the compartmentalization of early endosomes required for efficient transport from early endosomes to the trans-golgi network. J Cell Biol 151:1207–1220
3. Gokool S, Tattersall D, Seaman MN (2007) EHD1 interacts with retromer to stabilize SNX1 tubules and facilitate endosome-to-Golgi retrieval. Traffic 8:1873–1886
4. Maxfield FR, McGraw TE (2004) Endocytic recycling. Nat Rev Mol Cell Biol 5: 121–132
5. Grant BD, Donaldson JG (2009) Pathways and mechanisms of endocytic recycling. Nat Rev Mol Cell Biol 10:597–608
6. Park M, Penick EC, Edwards JG, Kauer JA, Ehlers MD (2004) Recycling endosomes supply AMPA receptors for LTP. Science 305: 1972–1975
7. Yu SS, Lefkowitz RJ, Hausdorff WP (1993) Beta-adrenergic receptor sequestration. A potential mechanism of receptor resensitization. J Biol Chem 268:337–341
8. Shewan AM, van Dam EM, Martin S, Luen TB, Hong W, Bryant NJ, James DE (2003) GLUT4 recycles via a trans-Golgi network (TGN) subdomain enriched in Syntaxins 6 and 16 but not TGN38: involvement of an acidic targeting motif. Mol Biol Cell 14:973–986
9. Manna PT, Smith AJ, Taneja TK, Howell GJ, Lippiat JD, Sivaprasadarao A (2009) Constitutive endocytic recycling and protein kinase C-mediated lysosomal degradation control K(ATP) channel surface density. J Biol Chem 285:5963–5973
10. Smith AJ, Sivaprasadarao A (2008) Investigation of K(ATP) channel endocytosis by immunofluorescence. In: Lippiat JD (ed) Potassium channels, methods in molecular biology, vol 491. Humana Press, New York, pp 69–77
11. Smith AJ, Sivaprasadarao A (2008) Chemiluminescence assays to investigate membrane expression and clathrin-mediated endocytosis of K(ATP) channels. In: Lippiat JD (ed) Potassium channels, methods in molecular biology, vol 491. Humana Press, New York, pp 63–68
12. Zerangue N, Schwappach B, Jan YN, Jan LY (1999) A new ER trafficking signal regulates the subunit stoichiometry of plasma membrane K(ATP) channels. Neuron 22:537–548

Part IV

Biochemical and Structure-Functional Approaches in Ion Channel Studies

Chapter 19

Generation of Antibodies That Are Externally Acting Isoform-Specific Inhibitors of Ion Channels

Jacqueline Naylor and David J. Beech

Abstract

There is demand for isoform-specific ion channel inhibitors as tools to investigate the biology of endogenous ion channels and validate them as targets in drug discovery programs. There is also hope that such inhibitors may be new therapeutic agents or provide the foundation for such agents. However, in practice, it is commonly experienced that inhibitors lack sufficient specificity, fail to distinguish between members of a class of ion channel, or have other (non-ion channel) off-target effects. Due to their extraordinary specificity, antibodies offer a potentially attractive strategy for overcoming these problems. Inhibitory antibodies acting at the extracellular face of ion channels are particularly attractive because there is enhanced possibility for specificity and intracellular delivery methods are not required. Here we describe experience with such an antibody approach and methodology for generating agents based on anti-peptide polyclonal antibodies.

Key words Ion channel, Transient receptor potential, Functional antibodies, E3 targeting

1 Introduction

Antibodies are proteins of the immune system that normally serve to recognize and bind foreign objects (antigens). Mutations in the antigen-binding site can produce a high number of variants with the ability to recognize a diverse range of targets. This feature can be exploited to create antibodies that display a high affinity for surface-exposed proteins such as ion channels. In addition to binding, the antibodies may modulate or down-regulate ion channel function by binding directly to the ion pore-forming subunit or to an associated regulatory subunit. An example of the success of this approach is antibody targeted to transient receptor potential polycystin 2 (TRPP2) which inhibits calcium entry evoked by shear stress (1). Another example is antibody to the N-terminus of stromal interaction molecule 1 (STIM1) which inhibits calcium entry associated with store-depletion (2). Both of these antibodies target

Nikita Gamper (ed.), *Ion Channels: Methods and Protocols*, Methods in Molecular Biology, vol. 998,
DOI 10.1007/978-1-62703-351-0_19, © Springer Science+Business Media, LLC 2013

extracellular peptides. Such extracellular targeting has advantages of ease of use in living cell/tissue studies and enhanced likelihood of specificity because only extracellular epitopes are available to the antibody when cells are intact. The antibody effects are relatively rapid (within 30 min) and thus they confer an additional advantage relative to other slower methods, such as gene modification or RNA interference, which carry a greater risk of compensatory effects and the associated more complex interpretation of the arising data. Channel-blocking antibodies can be targeted to intracellular regions, but this requires complex delivery methods in living cell situations. An antibody targeted to an intracellular epitope on the C-terminus of Kir2.1 was, for example, coupled to carrier peptides to allow passage into the cell; this approach was used to demonstrate a functional role of Kir2.1 in the rat retina (3).

In comparison with established ion channel modulators such as naturally occurring toxins or synthetic compounds, antibody blockers have the advantage of exceptionally high specificity and the ability to distinguish between ion channel subtypes. Such specificity was demonstrated for NESOpAb, a Na^+ channel ($Na_V1.5$) blocking antibody that selectively inhibits current carried by the neonatal splice variant of the channel, which differs from the adult form by six amino acids (4). NESOpAb may have therapeutic applications in breast cancer because the neonatal channel variant is up-regulated in cancer tissue (5). The potential of this type of antibody was also indicated by an antibody targeted to a region of the ion-conducting pore of the voltage-dependent Ca^{2+} channel (the α_{1A} subunit) that is only exposed upon depolarization (6).

The design of antibody channel inhibitors has been rationalized to specifically target the third extracellular loop (E3) of six membrane-spanning segment ion channels (also called the turret region of these ion channels) (7). This surface-exposed region of the proteins adjacent to the ion-conducting pore is often accessible and lacking in posttranslation modifications that could impede antibody binding. The E3 region was originally shown to be an important target in the voltage-gated K^+ channels $K_V1.2$ and $K_V3.1$ (8). Polyclonal antibodies (a population of antibodies each targeting a different epitope on the same antigen) directed to E3 in these channels inhibited whole cell current, providing evidence that $K_V1.2$ is functional in neuronal cells. Another example of E3-targeting is D-III, a polyclonal antibody which was found to be an inhibitor of P/Q-type voltage-gated Ca^{2+} channels in cerebellar granule cells and revealed a relationship between voltage-gated Ca^{2+} channels and the pathogenesis of para-neoplastic cerebellar ataxia (9).

To the best of our knowledge there are now thirteen E3-targeted ion channel inhibitor antibodies reported in the scientific literature (10). These antibodies are summarized in Fig. 1. An ion channel family that has been particularly targeted is the transient receptor

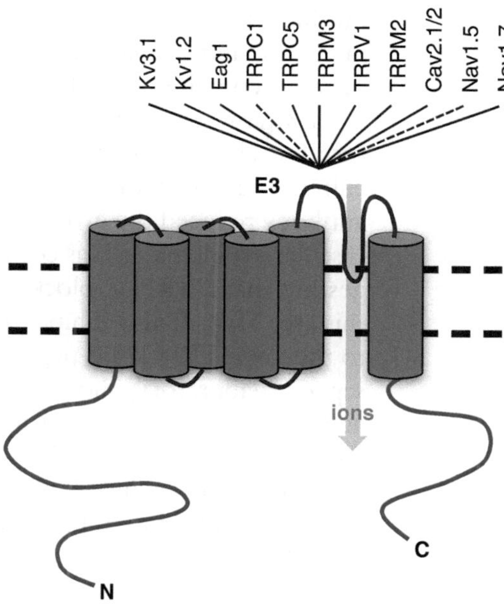

Fig. 1 Summary of E3-targeted extracellular blocking antibodies. Schematic represents the predicted structure of six transmembrane domain ion channels. Voltage-gated channels (Na_V, Ca_V) consist of four repeats. E3 refers to the third extracellular loop. *Dashed lines* indicate channels for which there are multiple antibodies targeted to different sequences within E3. References are Kv3.1, Kv1.2 (8), hEag (21), TRPC1 (11, 19), TRPC5 (7), TRPM3 (10), TRPV1 (20), TRPM2 (28), Ca_V2.1, Ca_V2.2 (9), Na_V1.5 (4), and Na_V1.7 (25)

potential (TRP) family of channels, perhaps because they often lack classical or specific pharmacology, their functions are quite often unclear, some knockout mice fail to show a clear phenotype, and there is hope that the channels can be important therapeutic targets.

TRP channels are a large and diverse family of nonselective cationic channels with structural resemblance to the *Shaker* K^+ channel. For many of these channels the lack of selective pharmacological tools has delayed channel characterization and determination of endogenous channel functions, increasing the demand for inhibitor antibodies. T1E3 antibody, directed against E3 of TRPC1, was used to provide evidence that TRPC1 is a contributor to Ca^{2+} entry in store-depleted cells (2, 11). Functional roles for TRPC5 channels in store-operated Ca^{2+} entry, cell motility, and secretion were elucidated with E3-targeted anti-TRPC5 antibody, T5E3 (7, 12, 13). The polyclonal antibody TM3E3 targeted to E3 of human TRPM3, a member of the melastatin subtype of TRP channel, partially suppressed TRPM3-dependent Ca^{2+} entry evoked by either sphingosine or pregnenolone sulfate (10). TM3E3 did not affect TRPC5, TRPV4, or TRPM2 but partially inhibited murine TRPM3 presumably because of the identical E3 sequences in mouse and human (14). The antibody provided evidence for negative coupling of TRPM3 to secretion (14, 15).

E3-targeted antibodies are applicable not only to cells in culture but also to studies of intact tissues. When applied to organ-cultured segments of human saphenous vein, anti-TRPC1 (T1E3) antibody reduced the growth of neointima, an event that underlies vascular problems such as coronary artery bypass graft failure, suggesting that a TRPC1 blocker could have therapeutic potential in this area (16). It also inhibited arterial contraction (17). Antibody to the same TRPC1 E3 peptide was generated independently by a separate laboratory, showing reproducibility of the E3 approach and revealing interrelationship of TRPC1 with Ca^{2+}-activated K^+ channels (18). There is also anti-TRPC1 antibody targeted to a different sequence in E3, which is commercially available and was suggested to be an inhibitor of store-operated Ca^{2+} entry in platelets (19).

So far, the majority of extracellular functional antibodies to ion channels have been polyclonal antibodies. There has been less success with monoclonal antibodies. Although polyclonal E3-targeted anti-TRPV1 antibodies were strong inhibitors of the channel, the monoclonal antibodies tested were ineffective (20). However, a monoclonal antibody targeted to the voltage-gated K^+ channel hEag1 was a successful inhibitor and was used to demonstrate a role of the channel in tumor cell growth (21). It can be expected that considerable effort will be required to identify the active antibody within a polyclonal pool.

Reagents that can potentiate or inhibit channel activity are important for the determination of physiological roles of ion channels, but once characterized and generated in a suitable format they may also have potential as therapeutic agents, perhaps paralleling successful therapeutic antibody developments towards other, non-ion channel, targets. Antibodies have been used to displace or prevent the binding of pathogenic autoantibodies to their aquaporin 4 targets (22), actions which could be useful in the development of treatments for the autoimmune disorder neuromyelitis optica. Humanized monoclonal antibodies are one option when pursuing such a direction, but an alternative may be the use of VHH fragments; heavy chain antibodies are found in camelids which contain only one variable domain (VH) and lack the Fc region that activates an immune response. The use of VHH fragments has been demonstrated for matrix-2 protein (M2), a single transmembrane domain proton channel in the influenza virus essential for viral function. The channel has a conserved extracellular region which makes it an attractive target as an alternative to other surface-exposed proteins such as hemagglutinin and neuraminidase. When anti-M2 VHH fragments were isolated from infected camels, the fragment M2-7A prevented influenza infection in mice (23). A further alternative is to develop intrabodies, which contain only the single-chain Fv fragment (scFv) of a full antibody yet still retain the ability to recognize and selectively bind a target antigen (24). They are designed to be retained within the cell, and are thus more suited to the disruption of intracellular signalling processes. Intrabodies have a reduced host response and their small size is advantageous, giving better access to partially hidden binding sites.

The number of therapeutic monoclonal antibodies entering clinical trials is rapidly increasing. Therefore, we hope to see potential therapeutic antibodies arising from the ion channel field. An early sign of success may be the suggested functional antibody to the $Na_V1.7$ channel (25).

Here we focus on methodology for generating ion channel inhibitor agents based on anti-peptide polyclonal antibodies (see Fig. 2).

2 Materials

2.1 ELISA Components

1. MAXI-SORP plates (NUNC).
2. Sodium carbonate buffer: 50 mM Na_2CO_3, pH 9.6.
3. Phosphate buffered saline (PBS): 2.5 mM Na_2HPO_4, 9 mM NaH_2PO_4, and 154 mM NaCl, pH 7.4.
4. PBS/Tween: 0.05% Tween-20 in PBS.
5. Blocking solution: 5% nonfat milk powder in PBS.
6. Antibody dilution buffer: 1% nonfat milk powder, 0.05% Tween in PBS.
7. Phosphate citrate buffer: 60 mM citric acid, 25 mM Na_2HPO_4.
8. ABTS reagent: 0.055% 2′2-azido-bis (3-ethylbenzthiazoline-6-sulfonic acid) and 0.01% H_2O_2 in phosphate citrate buffer.
9. HRP-conjugated secondary antibody.
10. Plate reader capable of reading absorbance at 405 nm.

2.2 Dialysis Components

1. Dialysis membranes (Scientific Laboratory Supplies Ltd, UK).
2. Sodium bicarbonate buffer: 0.1 M $NaHCO_3$.
3. PBS.

3 Methods

3.1 Epitope Selection and Peptide Design

Custom-made antisera can be supplied by numerous companies with routine expertise in this field, which can be beneficial over generating antibodies in-house. Regardless of the method chosen, there are several key steps to follow (see Fig. 2).

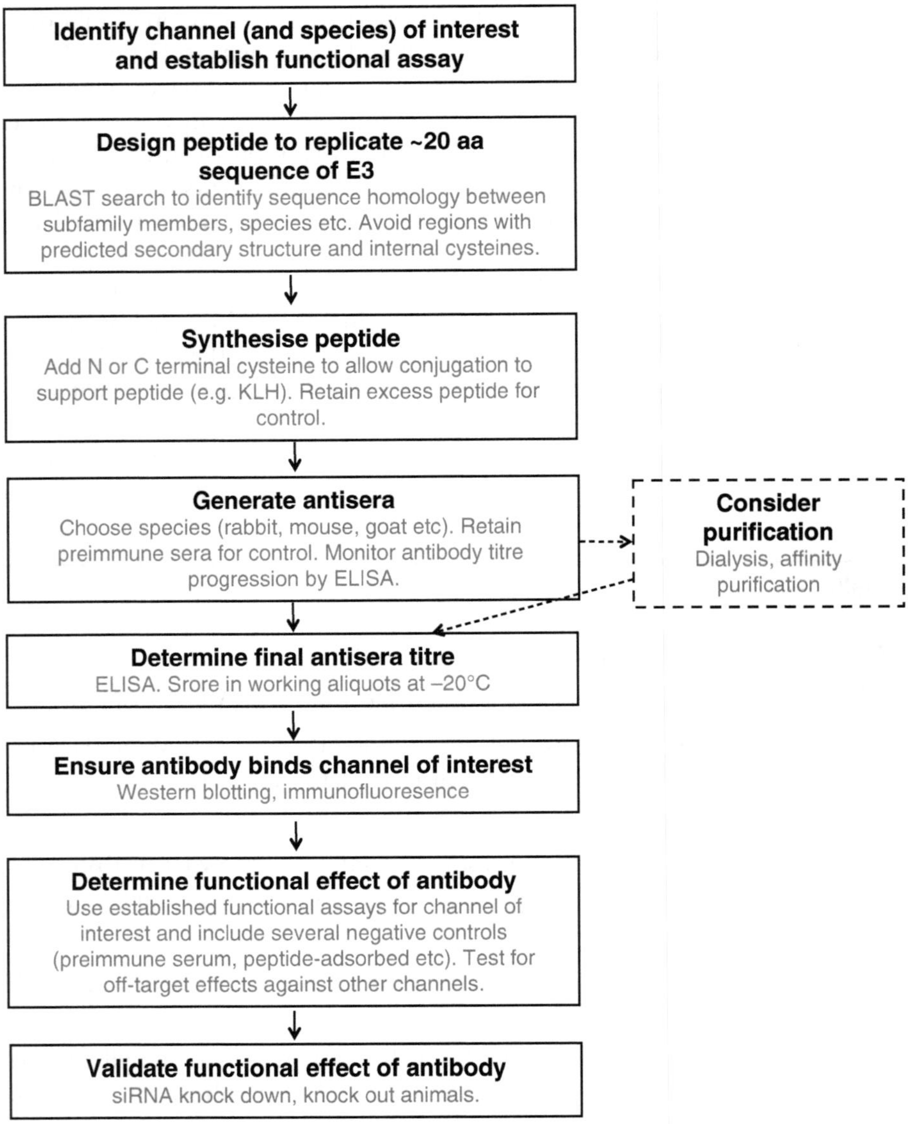

Fig. 2 Methods' schema. Summary of the methodology for generating anti-peptide polyclonal antibodies that function as ion channel inhibitor agents

1. Obtain the amino acid sequence of the ion channel target. Use Kyte-Doolittle hydrophobicity analysis (Protean Software, Lasergene) (26) to map surface-exposed and membrane-embedded regions of the protein in order to locate the third extracellular loop.
2. Design a ~20-mer peptide based on the E3 sequence. Suppliers of custom-made antisera will provide prediction tools to assist with this step (see Note 1).

3. Perform sequence alignment of the E3 peptide with related channels or subfamily members to determine sequence homology and whether your chosen peptide sequence is unique.
4. BLAST searches of public sequence databases (http://blast.ncbi.nlm.nih.gov/) will provide a broader outlook to identify other proteins that might contain the same sequence.
5. When synthesizing the peptide it is usual for manufacturers to include an N- or C-terminal cysteine residue to be conjugated to a support protein such as keyhole limpet hemacyanin (KLH). KLH increases the antigenicity of a peptide and is useful in affinity purification steps. You may wish to avoid E3 sequences that contain native cysteines in order to avoid complications from disulfide bridge formation with the terminal cysteine, although this is not essential.
6. Peptide synthesis should generate 20–30 mg of peptide, which will enable a standard immunization protocol (e.g., two rabbits in parallel) and provide excess peptide for use as a control in functional experiments and in affinity purification.

3.2 Generation of Antisera

1. Establish which species you require for antibody generation (see Note 2).
2. Obtain an initial bleed prior to immunization in order to collect preimmune sera, an important control for later experiments.
3. Following an initial immunization with antigenic peptide, standard immunization protocols are usually continued for at least 1 month with weekly injections, but can vary as antibody levels will vary animal to animal.
4. Intermittent test bleeds should be collected to monitor antibody titer using an enzyme-linked immunosorbent assay (ELISA).
5. Final-bleed antisera are usually stored in the presence of sodium azide as a preservative.

3.3 ELISA

ELISA is an important analytical test for determining the titer of antibodies present in a preparation of antiserum. If a titer is below 1:1,000 (i.e., dilutions of >1:1,000 fail to bind the antigenic peptide) further immunizations are advisable. Ideally the titer will be better than 1:10,000.

1. Coat wells of a 96-well MAXI-SORP plate with antigenic peptide diluted to 4 μg/mL in a buffer suitable for the peptide. Leave "no-peptide" control wells blank. Incubate overnight at 4 °C.
2. Wash wells three times with PBS/Tween to remove unbound peptide.

3. Add 300 μL blocking solution to each well. Incubate for 1 h at 37 °C.
4. Wash wells three times with PBS/Tween.
5. Prepare serial dilutions of antisera in antibody dilution buffer and add 50 μL to wells. Assay each concentration in duplicate. Incubate for 2 h at 37 °C to allow any antibody present to bind to the antigen.
6. Wash wells three times with PBS/Tween.
7. Add 50 μl of an HRP-conjugated secondary antibody to each well and incubate for 1 h at 37 °C.
8. Wash wells three times with PBS/Tween.
9. Apply 50 μL of ABTS reagent and incubate 30 min at RT in the dark.
10. Oxidation results in a color change which can be quantified at 405 nm in an absorption plate reader.
11. The intensity of the signal correlates positively with the amount of antibody present in the sample.

3.4 Antibody Dialysis and Purification

Antisera are usually preserved in the presence of sodium azide, which is included to protect antibody stocks against contaminating infections. The sodium azide may, however, have unwanted effects on live cells. Therefore dialysis of the antibody with buffer to remove the sodium azide is advisable in advance of live cell or tissue experiments.

1. Remove membrane from the roll and cut to 10 cm (wear gloves).
2. Boil for 10 min in 0.1 M sodium bicarbonate buffer.
3. Wash membrane in distilled water.
4. Boil again for 10 min in 0.1 M sodium bicarbonate buffer.
5. Wash membrane in distilled water. At this point membranes may be stored in 25% ethanol at 4 °C.
6. For dialysis, soak membranes in PBS until flexible.
7. Fill membrane with 1 mL of antiserum and clamp both ends.
8. Place into a beaker filled with PBS at a volume 500 times that of the antisera sample.
9. Leave for 48 h at 4 °C with gentle stirring to allow exchange.
10. During this time, change the PBS a total of three times.
11. Store dialyzed antibody in working aliquots −20 or −80 °C and minimize freeze/thaw cycles.

The final-bleed serum provided will contain a mixture of immunoglobulin isotypes, typically 5–10 mg/mL of IgG, of which 100 μg/mL will be specific antibody. If required, antisera can be subjected to further purification (see Note 3).

3.5 Functional Analysis and Important Controls

An important step is to investigate if your antibody binds to your ion channel target. This can be determined using western blotting and immunofluorescence staining. It is also advantageous to have access to the cDNA clone for your chosen protein because an established functional assay for your chosen channel will be important for determining whether your antibody has a blocking effect. Negative controls are important in these assays, and preimmune antiserum is one of these controls. It is presumed to contain the same components as the antiserum except for your peptide-specific antibody of interest. Another useful control is antigenic peptide pre-adsorbed antibody, i.e., antiserum pre-incubated with antigenic peptide, which saturates the peptide-specific binding site of your antibody, preventing it from binding to your target protein. Another is your antiserum after ~10 min at 100 °C, which denatures antibody components but leaves intact heat-resistant chemical factors. All of these controls are useful when checking for nonspecific effects (see Note 4).

As the antiserum is a finite supply, you should use your stocks carefully and it is important to use the optimum concentration of antibody or dilution of antiserum. You may not be able to generate a similar antiserum simply by injecting another set of animals with peptide as each individual animal will respond differently to the peptide antigen, producing a different combination of polyclonal antibodies (i.e., lot-to-lot variation). If you are assaying the activity of your ion channel of choice using live-cell assay (i.e., patch-clamp or Ca^{2+} imaging analysis), you can minimize the amount of antiserum required for an experiment by using a pre-incubation protocol in your experiments (e.g., pre-incubate cells with antibody but do not include the antibody in the perfusion medium during recordings). The pre-incubation can be important to enable enough time for antibody binding and functional effect, but the antibody remains bound to its target after washing (as in immunohistochemistry experiments) (see Note 5).

1. Prior to functional experiment, aspirate cell media.
2. Incubate live cells with fresh media containing antibody at the required concentration for 3.5 h at 37 °C.
3. For calcium imaging, include antibody during dye loading and washing steps, but not during recording.

If your antibody does have a blocking effect at your channel of interest, it is then advisable to test for off-target effects (e.g., on other channel family members) and to validate effects using independent methods such as RNA interference or gene-modified animals.

Methods that allow the rational design of specific blocking agents are necessary for the understanding of the critical roles ion channels play in health and disease. E3-targeting is a useful and versatile method for the production of such agents, which have the additional benefit of use in live cell and tissue experiments. However the use of antibodies is not without limitations.

4 Notes

1. Algorithms are available that can predict regions of a protein sequence that are likely to be highly antigenic and thus elicit the best immune response (27). It is also useful to be aware of secondary structure and posttranslational modification for the peptide region, as this may impede antibody access for the full protein.
2. You would usually choose a different species from the one you will use in your biological studies. Rabbits will generate a large volume of approximately 60 mL of antiserum and are commonly used. Mouse, goat, chicken, llama, etc. are alternatives.
3. Antisera can be purified by Protein A agarose columns. These consist of bacterial proteins that recognize the non-antigen-binding regions of antibodies, allowing the achievement of an immunoglobin-only sample. A purer sample can be achieved by affinity purification in which the antigenic peptide is coupled to a column. If antiserum is passed through the column, antibodies that bind to the antigenic peptide are preferentially retained. They can subsequently be eluted. It is important to note that problems may arise with such purification steps. For example, high-affinity components may be lost or preferentially selected depending on how you perform the assay. Alternatively, a high-affinity fraction may be selected that does not contain the active ion channel blocker. It is important to reassess the antibody titer by ELISA following dialysis or purification to determine whether there has been loss of activity.
4. Blocking antibodies, like other antibodies, can expose the investigator to the common technical difficulties experienced, for example, in western blotting and immunocytochemistry experiments. For this reason multiple experimental controls are necessary to minimize off-target effects and increase the likelihood that inhibition occurs because of an effect on the ion channel of interest.
5. For the majority of E3 antibodies, 50% channel inhibition is achieved in approximately 15 min. Such acute antibody effects allow for a shorter period of exposure, reducing nonspecific or off-target effects, but complete inhibition is rarely achieved with over-expressed channels. The mechanism of inhibition remains unknown, although at least for voltage-gated channels it would not appear to be due to channel internalization or down regulation of channels from the cell surface (6).

References

1. Nauli SM, Alenghat FJ, Luo Y, Williams E, Vassilev P, Li X, Elia AE, Lu W, Brown EM, Quinn SJ, Ingber DE, Zhou J (2003) Polycystins 1 and 2 mediate mechanosensation in the primary cilium of kidney cells. Nat Genet 33:129–137
2. Li J, Sukumar P, Milligan CJ, Kumar B, Ma ZY, Munsch CM, Jiang LH, Porter KE, Beech DJ (2008) Interactions, functions, and independence of plasma membrane STIM1 and TRPC1 in vascular smooth muscle cells. Circ Res 103:e97–e104
3. Raz-Prag D, Grimes WN, Fariss RN, Vijayasarathy C, Campos MM, Bush RA, Diamond JS, Sieving PA (2010) Probing potassium channel function in vivo by intracellular delivery of antibodies in a rat model of retinal neurodegeneration. Proc Natl Acad Sci U S A 107:12710–12715
4. Chioni AM, Fraser SP, Pani F, Foran P, Wilkin GP, Diss JK, Djamgoz MB (2005) A novel polyclonal antibody specific for the $Na_v1.5$ voltage-gated Na^+ channel 'neonatal' splice form. J Neurosci Methods 147:88–98
5. Fraser SP, Diss JK, Chioni AM, Mycielska ME, Pan H, Yamaci RF, Pani F, Siwy Z, Krasowska M, Grzywna Z, Brackenbury WJ, Theodorou D, Koyuturk M, Kaya H, Battaloglu E, De Bella MT, Slade MJ, Tolhurst R, Palmieri C, Jiang J, Latchman DS, Coombes RC, Djamgoz MB (2005) Voltage-gated sodium channel expression and potentiation of human breast cancer metastasis. Clin Cancer Res 11:5381–5389
6. Wyatt CN, Campbell V, Brodbeck J, Brice NL, Page KM, Berrow NS, Brickley K, Terracciano CM, Naqvi RU, MacLeod KT, Dolphin AC (1997) Voltage-dependent binding and calcium channel current inhibition by an anti-alpha 1D subunit antibody in rat dorsal root ganglion neurones and guinea-pig myocytes. J Physiol 502:307–319
7. Xu SZ, Zeng F, Lei M, Li J, Gao B, Xiong C, Sivaprasadarao A, Beech DJ (2005) Generation of functional ion-channel tools by E3 targeting. Nat Biotechnol 23:1289–1293
8. Zhou BY, Ma W, Huang XY (1998) Specific antibodies to the external vestibule of voltage-gated potassium channels block current. J Gen Physiol 111:555–563
9. Liao YJ, Safa P, Chen YR, Sobel RA, Boyden ES, Tsien RW (2008) Anti-Ca^{2+} channel antibody attenuates Ca^{2+} currents and mimics cerebellar ataxia in vivo. Proc Natl Acad Sci U S A 105:2705–2710
10. Naylor J, Milligan CJ, Zeng F, Jones C, Beech DJ (2008) Production of a specific extracellular inhibitor of TRPM3 channels. Br J Pharmacol 155:567–573
11. Xu SZ, Beech DJ (2001) TrpC1 is a membrane-spanning subunit of store-operated Ca^{2+} channels in native vascular smooth muscle cells. Circ Res 88:84–87
12. Xu SZ, Muraki K, Zeng F, Li J, Sukumar P, Shah S, Dedman AM, Flemming PK, McHugh D, Naylor J, Cheong A, Bateson AN, Munsch CM, Porter KE, Beech DJ (2006) A sphingosine-1-phosphate-activated calcium channel controlling vascular smooth muscle cell motility. Circ Res 98:1381–1389
13. Xu SZ, Sukumar P, Zeng F, Li J, Jairaman A, English A, Naylor J, Ciurtin C, Majeed Y, Milligan CJ, Bahnasi YM, Al-Shawaf E, Porter KE, Jiang LH, Emery P, Sivaprasadarao A, Beech DJ (2008) TRPC channel activation by extracellular thioredoxin. Nature 45(1):69–72
14. Naylor J, Li J, Milligan CJ, Zeng F, Sukumar P, Hou B, Sedo A, Yuldasheva N, Majeed Y, Beri D, Jiang S, Seymour VA, McKeown L, Kumar B, Harteneck C, O'Regan D, Wheatcroft SB, Kearney MT, Jones C, Porter KE, Beech DJ (2010) Pregnenolone sulphate- and cholesterol-regulated TRPM3 channels coupled to vascular smooth muscle secretion and contraction. Circ Res 106:1507–1515
15. Ciurtin C, Majeed Y, Naylor J, Sukumar P, English AA, Emery P, Beech DJ (2010) TRPM3 channel stimulated by pregnenolone sulphate in synovial fibroblasts and negatively coupled to hyaluronan. BMC Musculoskelet Disord 11:111
16. Kumar B, Dreja K, Shah SS, Cheong A, Xu SZ, Sukumar P, Naylor J, Forte A, Cipollaro M, McHugh D, Kingston PA, Heagerty AM, Munsch CM, Bergdahl A, Hultgardh-Nilsson A, Gomez MF, Porter KE, Hellstrand P, Beech DJ (2006) Upregulated TRPC1 channel in vascular injury in vivo and its role in human neointimal hyperplasia. Circ Res 98:557–563
17. Bergdahl A, Gomez MF, Dreja K, Xu SZ, Adner M, Beech DJ, Broman J, Hellstrand P, Sward K (2003) Cholesterol depletion impairs vascular reactivity to endothelin-1 by reducing store-operated Ca^{2+} entry dependent on TRPC1. Circ Res 93:839–847
18. Kwan HY, Shen B, Ma X, Kwok YC, Huang Y, Man YB, Yu S, Yao X (2009) TRPC1 associates with BK(Ca) channel to form a signal complex in vascular smooth muscle cells. Circ Res 104:670–678
19. Rosado JA, Brownlow SL, Sage SO (2002) Endogenously expressed Trp1 is involved in store-mediated Ca^{2+} entry by conformational coupling in human platelets. J Biol Chem 277:42157–42163
20. Klionsky L, Tamir R, Holzinger B, Bi X, Talvenheimo J, Kim H, Martin F, Louis JC,

Treanor JJ, Gavva NR (2006) A polyclonal antibody to the prepore loop of transient receptor potential vanilloid type 1 blocks channel activation. J Pharmacol Exp Ther 319:192–198
21. Gomez-Varela D, Zwick-Wallasch E, Knotgen H, Sanchez A, Hettmann T, Ossipov D, Weseloh R, Contreras-Jurado C, Rothe M, Stuhmer W, Pardo LA (2007) Monoclonal antibody blockade of the human Eag1 potassium channel function exerts antitumor activity. Cancer Res 67:7343–7349
22. Tradtrantip L, Zhang H, Saadoun S, Phuan PW, Lam C, Papadopoulos MC, Bennett JL, Verkman AS (2012) Anti-aquaporin-4 monoclonal antibody blocker therapy for neuromyelitis optica. Ann Neurol 71:314–322
23. Wei G, Meng W, Guo H, Pan W, Liu J, Peng T, Chen L, Chen CY (2011) Potent neutralization of influenza A virus by a single-domain antibody blocking M2 ion channel protein. PLoS One 6:e28309
24. Lo AS, Zhu Q, Marasco WA (2008) Intracellular antibodies (intrabodies) and their therapeutic potential. Handb Exp Pharmacol 181:343–373
25. Miller KM, De Ryck MR, Wolff CGJ, Lawson ADG, Finney HM, Baker TS (2011) Function modifying Nav1.7 antibodies. Patent number WO/2011/051350
26. Kyte J, Doolittle RF (1982) A simple method for displaying the hydropathic character if a protein. J Mol Biol 157:105–132
27. Kolaskar AS, Tongaonkar PS (1990) A semi-empirical method for prediction of antigenic determinants on protein antigens. FEBS Lett 276:172–174
28. Hecquet CM, Ahmmed GU, Vogel SM, Malik AB (2008) Role of TRPM2 channel in mediating H_2O_2-induced Ca^{2+} entry and endothelial hyperpermeability. Circ Res 102:347–355

Chapter 20

Site-Directed Mutagenesis to Study the Structure–Function Relationships of Ion Channels

Wei Yang and Lin-Hua Jiang

Abstract

Ion channels mediate a wide variety of physiological processes by forming small pores across the membranes that allow regulated flow of ions into or out of the cell. The primary linear sequences of ion channel proteins, like any proteins, are composed by 20 different amino acids, each of which is determined by specific triplet codon in their genes. Site-directed mutagenesis is a widely used molecular biology method to change the triplet in the coding sequence and thereby the amino acid residue in the protein sequence. Functional characterization of the ion channels carrying point mutations allows us to interrogate the structure–function relationships of the ion channels. Here, we will describe the site-directed mutagenesis procedures, in which the wide-type cDNA or plasmid is used as a template to synthesize the complementary mutation-containing DNAs from two mutagenic primers in the polymerase chain reaction.

Key words Ion channels, Structure–function relationships, Site-directed mutagenesis, PCR

1 Introduction

Ion channels are integral membrane proteins that form small pores to allow ions to cross the cell membranes and thereby transduce important signals for a variety of physiological functions. More than three hundreds of genes encoding ion channel proteins have been identified (1). Ion channels are known to undergo closed, open, inactivated, or desensitized states, each of which often has more than one step, but the gating mechanisms are still not well understood. X-ray crystallography in general provides a snapshot of the structure of proteins in one particular state or step, and it is notoriously challenging to determine the crystal structure of membrane proteins such as ion channels. Nonetheless, a handful of ion channels have been solved at the atomic level, including voltage-gated potassium channel (2–4), ionotropic glutamate receptor (5), acid-sensing ion channels (6, 7), and P2X receptors (8, 9). Such structural information is tremendously useful and become particularly powerful, when combined with the results from functional

Nikita Gamper (ed.), *Ion Channels: Methods and Protocols*, Methods in Molecular Biology, vol. 998,
DOI 10.1007/978-1-62703-351-0_20, © Springer Science+Business Media, LLC 2013

studies of ion channels carrying point mutations, to understand the fundamental principles or mechanisms by which the ion channels operate (e.g., (9, 10)). In addition, a number of naturally occurring mutations are causatively associated with diseases (e.g., (11)). To elucidate how the mutation alters the ion channel function is an important step towards to a good understanding of the mechanisms underlying the disease (e.g., (12, 13)).

Site-directed mutagenesis method is very useful to interrogate the structure–function relationships of an ion channel. Different mutational strategies have been described over the past decades (14–19). For example, residues in the region of interest can be individually replaced with cysteine, and modification of the introduced cysteine with cysteine-specific reagents may confer detectable changes in the ion channel function. Furthermore, two cysteines can be simultaneously introduced into two adjacent domains and their cross-linking via disulfide bonding may also alter the ion channel function. Such substituted cysteine accessibility and cross-linking methods have been proved to be useful in studying the ligand-binding, ion permeation, and channel-gating properties (see Chapter 21 of this book). Substitution of individual residues with tryptophan (or tryptophan scanning) can be used to determine whether a particular region faces to or interact with lipids within the membranes (e.g., (20)). Many ion channels are strongly modulated by metal cations (e.g., Zn^{2+} and H^{+}), post-translational modifications (e.g., phosphorylation and glycosylation), and interacting proteins. Functional studies in conjunction with site-directed mutagenesis can be used to pinpoint the binding or interacting sites (e.g., (21–24)). Reciprocal residue swapping can help to understand the contribution of species-specific residues to the differences in the functional and pharmacological properties of the ion channels (e.g., (25, 26)).

The site-directed mutagenesis method in principle allows to change, delete, or insert amino acid residues in the protein sequence. Figure 1 illustrates the widely used site-directed mutagenesis approach using polymerase chain reaction (PCR) to introduce point mutations. This method is to use double-stranded complementary DNA (cDNA) or usually a plasmid as a template and two synthetic mutagenic primers containing the desired mutation. The double strands of the cDNA template are firstly separated by denaturation at high temperature (Table 1). At reduced or annealing temperature (Table 1), the mutagenic primers bind to the target or complementary sequences in the cDNA template (step 1 in Fig. 1), and elongated into a new complementary strand with the desirable mutation, catalyzed by a high-fidelity DNA polymerase (step 2 in Fig. 1). The parental strands from the plasmid are methylated, but the newly synthesized strands are not methylated. DpnI endonuclease selectively only digests the methylated strands, and thus treatment with DpnI removes the parental

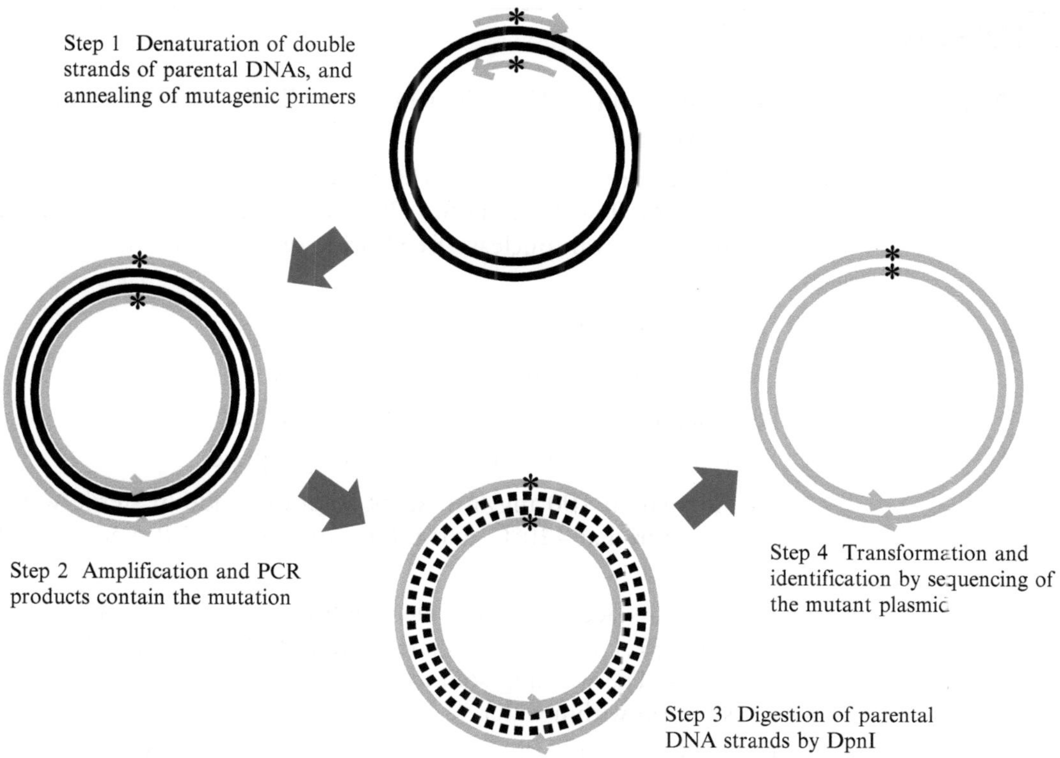

Fig. 1 Schematic diagram of the major steps in PCR-based site-directed mutagenesis

Table 1
PCR protocols used for site-directed mutagenesis

Segments	Cycle(s)	Temperature	Time
1	1	95°C	5 min
2	12–18[a]	95°C $T_m - 5$[b] 68°C	30 s 60 s 1 min/kb of plasmid length
3	1	72°C	10 min
4	1	4–8°C	Hold

[a]See Note 4
[b]See Note 5

strands, leaving the newly synthesized mutant strands intact (step 3 in Fig. 1). DpnI-treated DNA samples are then transformed into competent bacterial cells, and form colonies that should only harbor the plasmids with the desired mutation. The plasmids are subsequently purified and the mutation is further confirmed by DNA sequencing (step 4 in Fig. 1). Here, we will describe the experimental details of this method.

2 Materials

2.1 Reagents

1. High-fidelity DNA polymerase (e.g., Turbo or Ultra Pfu DNA polymerase).
2. 10× Pfu reaction buffer: 100 mM KCl, 100 mM $(NH_4)_2SO_4$, 200 mM Tris–HCl pH 8.8, 20 mM $MgSO_4$, 1% Triton X-100, and 1 mg/mL nuclease-free bovine serum albumin.
3. DpnI enzyme.
4. 2.5 mM dNTP mix.
5. Sense and antisense mutagenic primers.
6. Chemically competent *E. coli* cells (see Note 1).
7. Luria-Bertani (LB) media and LB agar plates.
8. Ampicillin: 100 mg/mL stock solution made in sterile water and stored at −20°C, and 60 μg/mL used in LB media or LB agar plates.
9. 50× Tris-acetate (TAE) electrophoresis solution: 242 g Tris base, 57.1 mL glacial acetic acid, and 100 mL 0.5 M EDTA pH 8. The working solution is prepared by diluting the stock solution within water.
10. Agarose gel for electrophoresis.
11. Ethidium bromide: 5 mg/mL stock solution made in water. Add 5 μL to 100 mL of agarose solution to visualize DNA in agarose gel using a transilluminator.

2.2 Equipment

1. PCR machine: e.g., Eppendorf Mastercycler or other commercial suppliers.
2. Agarose gel electrophoresis system: e.g., Bio-Rad or other commercial suppliers.
3. Transilluminator: e.g., Bio-Rad Gel Doc EZ system or other commercial suppliers.

2.3 Kits

1. Site-directed mutagenesis kit: e.g., Stratagene, Promega, Clonetech, or other commercial suppliers.
2. Plasmid mini-prep kit: e.g., Qiagen, Promega or other commercial suppliers.

3 Methods

3.1 Design of the Primers

1. The mutagenic primers should be 25–40 bases long, with a melting temperature (T_m) of ≥78°C. The following formula is commonly used to estimate the T_m value of the primers:

$$T_m = 81.5 + 0.41(\%GC) - 675 / N - \% \text{ mismatch}$$

where N is the primer length in bases, and values for % GC and % mismatch are whole numbers that are calculated as number of G/C bases or mismatch bases in the primer divided by primer length in bases and multiplied by 100. Ideally, the GC content is ~50% but this is difficult to achieve for certain mutations. The T_m for primers designed for insertion or deletion can be estimated by the modified formula:

$$T_m = 81.5 + 0.41(\%GC) - 675 / N$$

where N is the primer length in bases excluding the bases to be introduced or deleted.

2. The mutation site should locate in or close to the middle of the primer with 12–20 correct bases on each side. The primer should start and end with one or more G or C base to improve the efficiency of annealing and synthesis by DNA polymerases.
3. The primers need not to be 5′-phosphorylated but should be purified by fast polynucleotide liquid chromatography or polyacrylamide gel electrophoresis. Failure to purify the primers could significantly decrease the mutation efficiency. Synthesis and purification of mutagenic primers are standard services of many commercial suppliers.
4. It is recommended that the primers stock solution is prepared at 100 μM in DNase/RNase-free water or Tris–HCl (pH 8), aliquoted, and stored at −20°C until use. The working solution at 10 μM is prepared by further diluting the stock solution in DNase/RNase-free water.

3.2 PCR

1. Set up the following PCR sample in a sterile thin-wall Eppendorf tube for each mutation: 5 μL of 10× reaction buffer, 1–2 μL (50–100 ng) of cDNA template (see Note 2), 2 μL of 10 μM sense mutagenic primers (see Note 3), 2 μL of 10 μM antisense mutagenic primers, and 4 μL of 2.5 mM dNTP mix. Add sterile DNAase-free H_2O to a final volume of 49 μL. Finally, add 1 μL of 2.5 U/μL Pfu DNA polymerase.
2. Overlay the PCR sample in each tube with ~30 μL of mineral oil, if the thermal cycler does not have a hot-top assembly.
3. Use the PCR cycles as detailed in Table 1 (see Notes 4 and 5).
4. Add 1 μL of DpnI enzyme (10 U/μL) to each PCR sample, and mix gently and thoroughly by pipetting the samples several times before incubated at 37°C for 1 h to remove the cDNA templates (see Notes 6 and 7).

3.3 Transformation

1. Thaw competent cells, which are normally stored at −80°C, on ice just before use. Transfer 50–100 μL of competent cells into a 1.5 mL Eppendorf tube for each sample.
2. Add 1–3 μL of PCR sample to the competent cells, mix by gentle flicking (not pipetting), and incubate on ice for 30 min.

3. Heat-shock at 42°C for exactly 45 s by placing the Eppendorf tube in a preheated water bath, and immediately return on ice and incubate for another 2 min.
4. Add 800 μL of pre-warmed LB media and incubate in a shaking incubator at 37°C for 1 h at 225–250 rotations per min (rpm).
5. Collect bacterial cells by centrifugation for 2 min. Remove 700 μL of supernatant.
6. Resuspend cell pellet in the remaining solution by gentle pipetting, and spread on a pre-warmed LB agar plate containing appropriate antibiotics. Incubate the agar plate at 37°C for 16 h or overnight.
7. Check colonies on the agar plate (see Note 8), and streak 3–4 single colonies with sterile 10 μL tips, each colony into 5 mL of LB medium containing appropriate antibiotics in a 20 mL universal tube. Cap the tubes loosely and grow bacteria at 37°C in a shaking incubator at 225–250 rpm for 16–20 h.

3.4 Extraction of Plasmid and DNA Sequencing

1. Extract plasmids from bacterial culture using a plasmid purification kit, which can be obtained from many commercial suppliers, according to the manufacturer's instructions.
2. Check the plasmid quality by running 1 μL of purified plasmid sample on 1% agarose gel electrophoresis (see Note 6).
3. Send an aliquot of purified plasmid sample to be sequenced by a commercial service provider (see Note 9).
4. Analyze the DNA sequencing results against the original cDNA sequence to determine whether the desirable mutation is successfully introduced (see Note 10).

4 Notes

1. Competent cells are available from many commercial suppliers but can be prepared in the lab using calcium chloride protocol as described elsewhere (27).
2. If possible, the cDNA or protein-coding sequence should be prior subcloned into a mammalian expression vector (e.g., pcDNA3.1) in order to express the mutant protein in a heterologous expression system for subsequent functional studies.
3. It is important to put an excessive amount of the primers.
4. As a general guideline, use 14–16 cycles for change in single nucleotide mutation and 18 cycles for multiple nucleotide mutations.
5. The annealing temperature is a crucial factor in determining the PCR yield. As the starting point, use a temperature 5°C lower than the melting temperature (T_m). However, one should optimize experimentally, e.g., using a gradient PCR machine (Fig. 2).

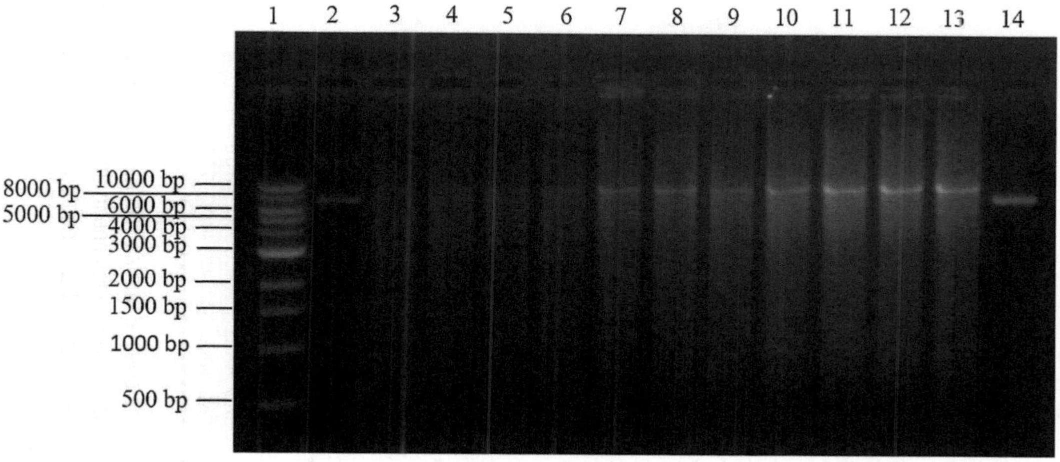

Fig. 2 An agarose gel shows the PCR products after DpnI treatment. *Lane 1*, the DNA ladder; *lanes 2* and *14*, the plasmids used as the template; *lanes 3–13*, PCR products using gradient PCRs with the annealing temperature from 56 to 67°C. The better PCR yield was observed with the annealing temperature higher than 60°C (*lanes 7–13*)

6. It is important to remove the parental DNAs as much as possible using DpnI. Any residual DNAs can result in unwanted colonies harboring the wild-type cDNA (the "false positive" colonies). If this problem becomes serious, treat the PCR sample with DpnI for 2 h and add more DpnI or both.
7. Run 5–10 μL of DpnI-treated PCR sample on an agarose gel to make sure sufficient amount of newly synthesized cDNAs for transformation. DNA samples in the range of 0.5–10 kb are well separated on a 1% (w/v) agarose gel. If there is no band on the agarose gel, increase the DNA templates when setting up the PCR sample, or alternatively reduce the annealing temperature, which however may increase nonspecific PCR products or random mutations. For smearing DNA, adjust the annealing temperature (see Note 5). If none of these changes help, design a new set of primers.
8. Lack of colony after transformation can arise from little or no DNA in the PCR sample, which can be easily ruled out by agarose gel electrophoresis (see Note 7). Another potential cause relates to the competency of bacterial cells used. Avoid repeated thawing and freezing of competent cells. Determine the competency by transforming a known amount of plasmids and repeat transformation using an increased amount of PCR samples using bacterial cells with reduced competency. Alternatively, switch to a different batch of competent cells.
9. It is necessary to determine the entire cDNA sequence, particularly the longer sequences which have a greater risk of unwanted mutations being introduced despite using high-fidelity or proof-reading polymerases.
10. There are several free online nucleotide sequence analysis programs. Figure 3 shows the nucleotide sequence alignment of the

```
TRPM2   TTCTTCTTCCTCTTCCTGCTGGCTGTGTGGGTGGTGTCCTTCGGGGTGGCCAAGCAGGCCATCCTCATCCACAACGAGCGCCGGGTGGAC 2908
E960Q   TTCTTCTTCCTCTTCCTGCTGGCTGTGTGGGTGGTGTCCTTCGGGGTGGCCAAGCAGGCCATCCTCATCCACAACCAGCGCCGGGTGGAC 168
D987E   TTCTTCTTCCTCTTCCTGCTGGCTGTGTGGGTGGTGTCCTTCGGGGTGGCCAAGCAGGCCATCCTCATCCACAACGAGCGCCGGGTGGAC 171
        *************************************************************************** **************

TRPM2   TGGCTGTTCCGAGGGGCCGTCTACCACTCCTACCTCACCATCTTCGGGCAGATCCCGGGCTACATCGACGGTGTGAACTTCAACCCGGAG 2998
E960Q   TGGCTGTTCCGAGGGGCCGTCTACCACTCCTACCTCACCATCTTCGGGCAGATCCCGGGCTACATCGACGGTGTGAACTTCAACCCGGAG 258
D987E   TGGCTGTTCCGAGGGGCCGTCTACCACTCCTACCTCACCATCTTCGGGCAGATCCCGGGCTACATCGAGGGTGTGAACTTCAACCCGGAG 261
        ******************************************************************** *********************

TRPM2   CACTGCAGCCCCAATGGCACCGACCCCTACAAGCCTAAGTGCCCCGAGAGCGACGCGACGCAGCAGAGGCCGGCCTTCCCTGAGTGGCTG 3088
E960Q   CACTGCAGCCCCAATGGCACCGACCCCTACAAGCCTAAGTGCCCCGAGAGCGACGCGACGCAGCAGAGGCCGGCCTTCCCTGAGTGGCTG 348
D987E   CACTGCAGCCCCAATGGCACCGACCCCTACAAGCCTAAGTGCCCCGAGAGCGACGCGACGCAGCAGAGGCCGGCCTTCCCTGAGTGGCTG 351
        ******************************************************************************************

TRPM2   ACGGTCCTCCTACTCTGCCTCTACCTGCTCTTCACCAACATCCTGCTGCTCAACCTCCTCATCGCCATGTTCAACTACACCTTCCAGCAG 3178
E960Q   ACGGTCCTCCTACTCTGCCTCTACCTGCTCTTCACCAACATCCTGCTGCTCAACCTCCTCATCGCCATGTTCAACTACACCTTCCAGCAG 438
D987E   ACGGTCCTCCTACTCTGCCTCTACCTGCTCTTCACCAACATCCTGCTGCTCAACCTCCTCATCGCCATGTTCAACTACACCTTCCAGCAG 441
        ******************************************************************************************

TRPM2   GTGCAGGAGCACACGGACCAGATTTGGAAGTTCCAGCGCCATGACCTGATCGAGGAGTACCACGGCCGCCCCGCCGCGCCGCCCCCCTTC 3268
E960Q   GTGCAGGAGCACACGGACCAGATTTGGAAGTTCCAGCGCCATGACCTGATCGAGGAGTACCACGGCCGCCCCGCCGCGCCGCCCCCCTTC 528
D987E   GTGCAGGAGCACACGGACCAGATTTGGAAGTTCCAGCGCCATGACCTGATCGAGGAGTACCACGGCCGCCCCGCCGCGCCGCCCCCCTTC 531
        ******************************************************************************************

TRPM2   ATCCTCCTCAGCCACCTGCAGCTCTTCATCAAGAGGGTGGTCCTGAAGACTCCGGCCAAGAGGCACAAGCAGCTCAAGAACAAGCTGGAG 3358
E960Q   ATCCTCCTCAGCCACCTGCAGCTCTTCATCAAGAGGGTGGTCCTGAAGACTCCGGCCAAGAGGCACAAGCAGCTCAAGAACAAGCTGGAG 618
D987E   ATCCTCCTCAGCCACCTGCAGCTCTTCATCAAGAGGGTGGTCCTGAAGACTCCGGCCAAGAGGCACAAGCAGCTCAAGAACAAGCTGGAG 621
        ******************************************************************************************

TRPM2   AAGAACGAGGAGGCGGCCCTGCTATCCTGGGAGATCTACCTGAAGGAGAACTACCTCCAGAACCGACAGTTCCAGCAAAAGCAGCGGCCC 3418
E960Q   AAGAACGAGGAGGCGGCCCTGCTATCCTGGGAGATCTACCTGAAGGAGAACTACCTCCAGAACCGACAGTTCCAGCAAAAGCAGCGGCCC 708
D987E   AAGAACGAGGAGGCGGCCCTGCTATCCTGGGAGATCTACCTGAAGGAGAACTACCTCCAGAACCGACAGTTCCAGCAAAAGCAGCGGCCC 711
        ******************************************************************************************
```

Fig. 3 Nucleotide sequence alignment of the DNA sequencing results for the E960Q and D987E two pore mutations against the original human TRPM2 cDNA sequence using the multiple sequence alignment ClustalW2 program. The numbers on the *right* indicate the position of the last nucleotide in each line in the human TRPM2 cDNA sequence, or the readable sequencing results for the E960Q or D987E mutation. All the nucleotide sequences are identical as denoted by the *asterisks* underneath, with the exception of changes in the G2894C for the E960Q mutation and the C2977G for the D987E mutation

sequencing results for the E960Q and D987E pore mutations in human TRPM2 (28) against the original cDNA sequence using the multiple sequence alignment ClustalW2 program available from the EMBL-EBI site (http://www.ebi.ac.uk/Tools/msa/clustalw2/).

References

1. Gabashvili IS, Sokolowski BH, Morton CC, Giersch AB (2007) Ion channel gene expression in the inner ear. J Assoc Res Otolaryngol 8:305–328
2. Jiang Y, Lee A, Chen J, Ruta V, Cadene M, Chait BT, MacKinnon R (2003) X-ray structure of a voltage-dependent K^+ channel. Nature 423:33–41
3. Jiang Y, Lee A, Chen J, Cadene M, Chait BT, MacKinnon R (2002) Crystal structure and mechanism of a calcium-gated potassium channel. Nature 417:515–522
4. Long SB, Campbell EB, Mackinnon R (2005) Crystal structure of a mammalian voltage-dependent Shaker family K^+ channel. Science 309:897–903
5. Sobolevsky AI, Rosconi MP, Gouaux E (2009) X-ray structure, symmetry and mechanism of an AMPA-subtype glutamate receptor. Nature 462:745–756
6. Jasti J, Furukawa H, Gonzales EB, Gouaux E (2007) Structure of acid-sensing ion channel 1 at 1.9 A resolution and low pH. Nature 449:316–323
7. Gonzales EB, Kawate T, Gouaux E (2009) Pore architecture and ion sites in acid-sensing ion channels and P2X receptors. Nature 460: 599–604
8. Kawate T, Michel JC, Birdsong WT, Gouaux E (2009) Crystal structure of the ATP-gated P2X (4) ion channel in the closed state. Nature 460:592–598
9. Hattori M, Gouaux E (2012) Molecular mechanism of ATP binding and ion channel activation in P2X receptors. Nature 485:207–212
10. Browne LE, Jiang LH, North RA (2010) New structure enlivens interest in P2X receptors. Trends Pharmacol Sci 31:229–237
11. Hermosura MC, Cui AM, Go RC, Davenportb B, Shetlera CM, Heizerb JW, Schmitzc C, Mocza G, Garrutod RM, Perraudb A-L (2008) Altered functional properties of a TRPM2 variant in Guamanian ALS and PD. Proc Natl Acad Sci U S A 105:18029–18034
12. Roger S, Mei ZZ, Baldwin JM, Dong L, Bradley H, Baldwin SA, Surprenant A, Jiang LH (2010) Single nucleotide polymorphisms that were identified in affective mood disorders affect ATP-activated P2X7 receptor functions. J Psychiatr Res 44:347–355
13. Sorge RE, Trang T, Dorfman R et al (2012) Genetically determined P2X7 receptor pore formation regulates variability in chronic pain sensitivity. Nat Med 18:595–599
14. Kunkel TA (1985) Rapid and efficient site-specific mutagenesis without phenotypic selection. Proc Natl Acad Sci 82:488–492
15. Taylor JW, Ott J, Eckstein F (1985) The rapid generation of oligonucleotide-directed mutations at high frequency using phosphorothioate modified DNA. Nucleic Acids Res 13:8765–8785
16. Weiner MP, Costa GL, Schoettlin W, Cline J, Mathur E, Bauer JC (1994) Site-directed mutagenesis of double-stranded DNA by the polymerase chain reaction. Gene 151:119–123
17. Vandeyar MA, Weiner MP, Hutton CJ, Batt CA (1988) A simple and rapid method for the selection of oligonucleotide-directed mutants. Gene 65:129–133
18. Ishii TM, Zerr P, Xia XM, Bond CT, Maylie J, Adelman JP (1998) Site-directed mutagenesis. Methods Enzymol 293:53–71
19. Nagy ZB, Felfoldi F, Tamas L, Puskás LG (2004) A one-tube, two-step polymerase chain reaction-based site-directed mutagenesis method with simple identification of the mutated product. Anal Biochem 324:301–303
20. Monks SA, Needleman DJ, Miller C (1999) Helical structure and packing orientation of the S2 segment in the Skaker K^+ channel. J Gen Physiol 113:415–423
21. Yang W, Manna PT, Zou J, Luo J, Beech DJ, Sivaprasadarao A, Jiang LH (2011) Zinc inactivates melastatin transient receptor potential 2 channels via the outer pore. J Biol Chem 286:23789–23798
22. Yang W, Zou J, Xia R, Vaal ML, Seymour VA, Luo J, Beech DJ, Jiang L-H (2010) State-dependent inhibition of TRPM2 channel by acidic pH. J Biol Chem 285:30411–30418
23. Liu X, Surprenant A, Mao HJ, Roger S, Xia R, Bradley H, Jiang L-H (2008) Identification of key residues coordinating functional inhibition of P2X7 receptors by zinc and copper. Mol Pharmacol 73:252–259

24. Liu X, Ma W, Surprenant A, Jiang L-H (2009) Identification of the amino acid residues in the extracellular domain of rat P2X(7) receptor involved in functional inhibition by acidic pH. Br J Pharmacol 156:135–142
25. Zou J, Yang W, Beech DJ, Jiang L-H (2011) A residue in the TRPM2 channel outer pore is crucial in determining species-dependent sensitivity to extracellular acidic pH. Pflugers Arch 462:293–302
26. Bradley HJ, Browne LE, Yang W, Jiang L-H (2011) Pharmacological properties of the rhesus macaque monkey P2X7 receptor. Br J Pharmacol 164:743–754
27. Hanahan D (1983) Studies on transformation of Escherichia coli with plasmids. J Mol Biol 166:557–580
28. Xia R, Mei ZZ, Mao HJ, Yang W, Dong L, Bradley H, Beech DJ, Jiang LH (2008) Identification of pore residues engaged in determining divalent cationic permeation in transient receptor potential melastatin subtype channel 2. J Biol Chem 283: 27426–27432

Chapter 21

Cysteine-Based Cross-Linking Approach to Study Inter-domain Interactions in Ion Channels

Lin-Hua Jiang

Abstract

Cysteine contains a highly reactive thiol group and therefore under oxidizing conditions a disulfide bond can form between a pair of cysteines that are juxtaposed in the close vicinity, which can be only reversed by reducing agents. These attributes have been elegantly exploited to study the functional role of an interaction or contact between two adjacent domains that are present in ion channels or virtually in any proteins, by introducing double cysteine substitutions at the domain interface and measuring changes in the ion channel functions arising from cross-linking the two substituted cysteines via formation of a disulfide bond. Here I describe this cysteine-based cross-linking approach.

Key words Disulfide bond, Inter-domain interaction, Double cysteine substitution, Cross-linking, Ion channel, Ligand binding, Gating

1 Introduction

Ion channels in the plasma membrane act as a ubiquitous and key mechanism of transporting ions across the membranes in both excitable and non-excitable cells, thereby altering cell membrane potential and/or ion homeostasis (1). Many ion channels are made of multiple membrane-spanning subunits that are entwined with each other surrounding a central aqueous ion-conducting pore. Examples include the cysteine-loop receptors (pentamers) such as nicotine acetylcholine and γ-aminobutyric acid receptors (nAChR and $GABA_AR$), ionotropic glutamate receptors (tetramers), ATP-gated P2X receptors (P2XR, trimers) (2, 3), voltage-gated K^+ (K_V) and inwardly rectifying K^+ (K_{ir}) channels (tetramers), transient receptor potential channels (tetramers) (4), and acid-sensing and epithelial Na^+ channels (trimers) (5). For the Ca_V and Na_V channels, the ion-conducting pore is constituted by four homologous domains of single subunit (4). An increasing number of ion channel

Nikita Gamper (ed.), *Ion Channels: Methods and Protocols*, Methods in Molecular Biology, vol. 998,
DOI 10.1007/978-1-62703-351-0_21, © Springer Science+Business Media, LLC 2013

structures have been solved at the atomic level and such structures reveal numerous interactions or contacts between domains located within the same or neighboring subunits. Some of these interactions have been demonstrated to be crucial in coordinating ligand binding, e.g., ACh binding in the nAChR (6), and ATP and Zn^{2+} binding in the P2XR (3), or mediating ion channel gating, e.g., in the $GABA_AR$ (7), K_V channels (8), and P2XR (9). Thus, a fuller understanding of inter-domain/subunit interactions in the closed, open, and desensitized states can help to provide mechanistic insights into the principles of how the ion channels function. Such insights will facilitate elucidation of the molecular mechanisms of diseases arising from mutational perturbance of ion channels, e.g., (10), and structure-guided design of therapeutic drugs.

Cysteine amino acid is one of the building blocks of virtually every protein including ion channels. Uniquely, cysteine contains a highly reactive thiol group. Thus, when a pair of cysteines is exposed to each other in the close vicinity, a disulfide bond can form between them, catalyzed by ambient oxygen or oxidizing agents. The thiol group in cysteine can also form disulfide bonds with many of the cysteine-modifying reagents (e.g., methanethiosulfonates). Such a unique chemical property of cysteine underpins the substituted cysteine accessibility method, which was elegantly developed by Karlin and his colleagues to study the ion permeation and gating properties in the nAChR (11, 12), and has since been used to study many other ion channels. In this method, single-cysteine substitution is introduced into the region of interest. One assumes that the residue replaced with cysteine occupies a functionally crucial position in the channel, formation of a disulfide bond between the introduced cysteine and cysteine-modifying reagents confers discernible changes in the ion channel functional properties.

This chapter describes cysteine-based cross-linking approach, which follows the same principle of disulfide bonding and extends the single substituted cysteine accessibility method. It is primarily used to probe the role of an interaction between two neighboring domains or regions, predicted on the basis of the atomic structures or educated guess, in determining the functional properties of ion channels. This approach thus involves introduction of double cysteine substitutions into two positions located on the opposite side of an interface between two domains, either from adjacent subunits (Fig. 1a) or within the same subunit (Fig. 1b), and takes the assumptions that if one or both of these positions or the surrounding regions are engaged in, e.g., ligand binding, permeation, or conformational changes accompanying the channel gating, cross-linking of the substituted cysteines as a result of forming a disulfide bond alters such ion channel functional properties. Such functional alterations can be reversed using dithiothreitol (DTT) or other reducing agents. It is easy to implement, and it is versatile,

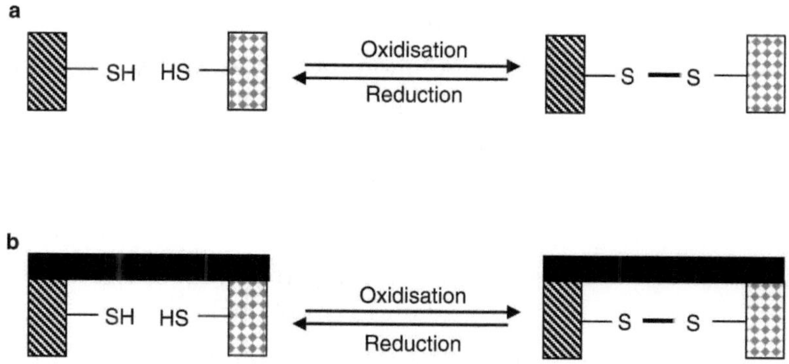

Fig. 1 Schematic diagrams illustrating cross-linking of two cysteines substituted into positions located in two different subunits (**a**) or two domains of the same protein (**b**)

i.e., applicable to virtually any ion channel types, and can yield important mechanistic understanding of the role of inter-domain interactions in coordinating ligand binding and mediating channel gating and ion permeation, as have been demonstrated in a number of ion channels (7–9, 13–18).

2 Materials

Materials required for site-directed mutagenesis are listed in Chapter 20 of this book.

2.1 Cell Culture and Transfection

1. Human embryonic kidney (HEK) 293 cells.
2. Dulbecco's modified Eagle medium containing 10% fetal bovine serum (DMEM culture medium).
3. Phosphate buffered saline (PBS).
4. 0.05% Trypsin/ethylenediaminetetraacetic acid (EDTA) solution.
5. 15-mL Falcon tubes.
6. 35-mm plastic Petri dishes.
7. 13-mm coverslips.
8. 1.5-mL Eppendorf tubes.
9. Transfection medium (e.g., OPTI-MEM-I).
10. Transfection reagents (e.g., Lipofectamine 2000).
11. Plasmids for green fluorescent protein (GFP) and ion channel of interest (see Note 1).

2.2 Electrophysiology

1. Patch-clamp recording rig with a fluorescent microscopy (see Chapter 7 of this book for further details).
2. Pipette puller.

3. Microforge.
4. Solution perfusion system (e.g., RSC-160 rapid solution change system from BioLogic).
5. Glass capillary.
6. 1-mL plastic syringe.
7. Syringe filter disk with 0.2-mm pores.
8. Internal/pipette solution: 145 mM NaF, 10 mM EGTA, and 10 mM HEPES, pH 7.3.
9. External/bath solution: 147 mM NaCl, 2 mM $CaCl_2$, 2 mM KCl, 1 mM $MgCl_2$, 13 mM glucose, and 10 mM HEPES, pH 7.3.
10. 10 mM DTT freshly prepared in external solution.

3. Methods

3.1 General Considerations

The starting point is obvious: which pair of residues is chosen to be replaced with cysteine? The decision can be straightforward if atomic structural information exists to indicate juxtaposition of two residues in two neighboring domains, and some functional evidence to suggest their importance in the ion channel functions. For example, structural models indicate the close vicinity of Asp^{57} and Asp^{149} residues in the extracellular domain and Lys^{279} residue in the extracellular transmembrane linker region in the $GABA_AR$ (7) or Glu^{63} in one subunit and Arg^{274} in the adjacent subunit in the extracellular domain of the P2X2R (17), and functional studies of the single-point mutant channels implied their engagement in the channel gating (7, 17). Cysteine-based cross-linking approach yielded unambiguous evidence to support a crucial role of the electrostatic interactions mediated by these residues in the channel gating (7, 17). On other occasions where structural information is limited or not available, the choice of double cysteine substitution is largely reduced to educated guess, even trial and error. For example, conserved charged residues including Lys^{68}, Lys^{70}, Lys^{190}, Arg^{292}, Arg^{305}, and Lys^{309} in the extracellular domain of P2X1R and equivalent residues in other P2XR are essential for P2XR activation by ATP (3). Double cysteine substitution introduced into two separate subunits and cross-linking of the co-expressed single-cysteine mutant subunits, in combination with Western blotting and functional studies, showed that K69C and K309C formed an inter-subunit disulfide (16). This provides evidence to indicate that Lys^{68} and Lys^{309} residues contribute to formation of a novel inter-subunit ATP binding site, which has been substantiated by the atomic structural determination (3). Another example is shown in Fig. 2; Val^{48} and Ile^{328} residues, located at the extracellular end of the first and the second transmembrane domains, respectively (Fig. 2),

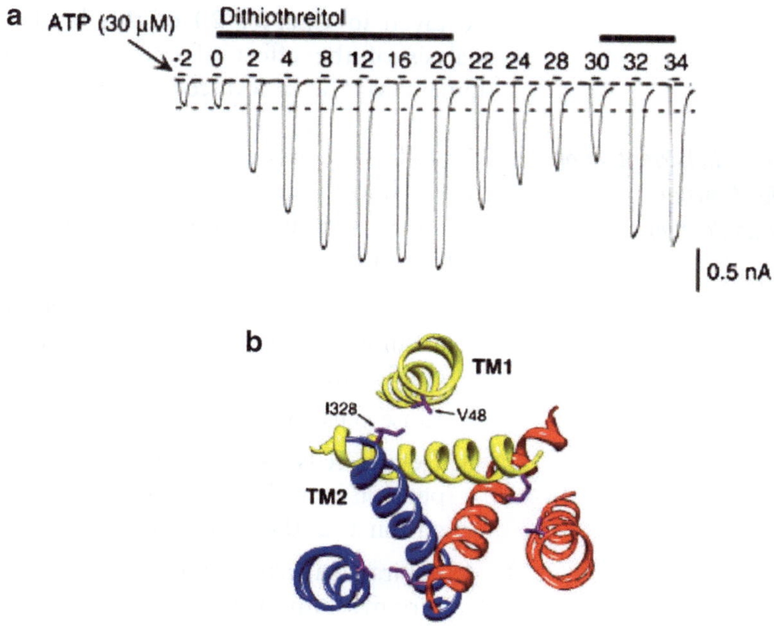

Fig. 2 Cross-linking of V48C in the first transmembrane domain (TM1) and I328C in the second transmembrane domain (TM2) impairs P2X2R activation. (**a**) A representative whole-cell recording of ATP-evoked inward currents in a HEK293 cell expressing [V48C, I328C] double-mutant rat P2X2R. The initial ATP-induced current was small but drastically increased by exposure to 10 mM DTT. The current gradually declined upon washing DTT and was reversed upon reapplication of DTT (modified from (13). This research was originally published in Journal of Biological Chemistry. Jiang LH et al. Amino acid residues involved in gating identified in the first membrane-spanning domain of the rat P2X2 receptor. 2001; 276: 14902–14908. © the American Society for Biochemistry and Molecular Biology). (**b**) A structural model of the transmembrane domains of the trimeric rat P2X2R in the closed state viewed from extracellular side, showing close juxtaposition of V48 residue in the TM1 on the periphery and I328 residue in the TM2 pore-forming domain (reproduced from (3) with permission)

were initially identified by single substituted cysteine accessibility method (13). Double cysteine substitution, in combination with trial and error in this case, led to discovery that formation of DTT-sensitive disulfide bonds prevented activation of the P2X2R carrying both V48C and I318C mutations (Fig. 2) (13). A further study, by cross-linking the cysteines substituted into the corresponding residues, revealed the subunit stoichiometry of the heteromeric P2X2/3R (14). In summary, the strategy of selecting the pair of residues for cysteine substitution strongly depends on the prior knowledge regarding the structure-function relationships of the ion channels and particularly the potential role of the domain interactions under investigation.

Detailed protocols for introduction of cysteines into the desired position of an ion channel protein using site-directed mutagenesis are given in Chapter 20 of this book. Below are protocols for analysis of the effect of chemical cross-linking of ion channel domains via the introduced cysteines on the ion channel function.

3.2 Cell Preparation for Transient Transfection

The following procedures should be conducted in a sterile tissue culture hood except indicated otherwise (refer to Chapters 2, 4 and 5 of this book for alternative transfection methods and optimization).

1. Grow cells in DMEM culture medium in a T25 flask media in a humid 37 °C, 5% CO_2 tissue culture incubator until 80% confluent.
2. Rinse cells with PBS.
3. Detach cells using trypsin–EDTA solution, transfer cell suspension in a 15-mL Falcon tube, and collect cells by centrifugation at 200 × g for 5 min.
4. Aspirate supernatant and resuspend cell pellet in 2 mL of culture medium by gentle pipetting.
5. Seed 0.2–0.3 mL of cell suspension in each 35-mm Petri dish, add 1.5 mL of culture medium, and leave in the incubator to grow until >80% confluent (see Note 2). Prepare one Petri dish of cells for each transfection.
6. For each transfection, dilute 0.1 μg of plasmid for GFP (see Note 3) and 1 μg of plasmid for the ion channel of interest into 100 μL of transfection medium in one 1.5-mL Eppendorf tube (see Note 4) and 3 μL Lipofectamine 2000 into 100 μL of transfection medium in another Eppendorf tube. Leave them in the tissue culture hood at room temperature for 5 min.
7. Combine plasmid and Lipofectamine 2000 containing transfection medium, and mix thoroughly by gentle pipetting. Leave it in the tissue culture hood at room temperature for 20 min.
8. Add 0.8 mL of culture medium into transfection medium.
9. Replace existing culture medium with transfection medium.
10. Return cells to the incubator and leave overnight (see Note 5).

3.3 Plating Cells for Patch-Clamp Recording

1. Aspirate culture medium.
2. Repeat steps 2–4 described above (see Subheading 3.2).
3. Resuspend cell pellet in 2 mL of fresh culture medium gently and thoroughly, seed 50 μL of cell suspension on each 13-mm circular coverslip placed in one 35-mm Petri dish (up to four coverslips per dish), and add 100–200 μl of fresh culture medium to dilute and disperse cells (see Note 6). Return cells to the incubator for 2–4 h to allow them to recover and adhere onto the coverslips.

4. Add 2 mL of culture medium in each Petri dish to immerse the coverslips. Use cells for patch-clamp recording straight away or maintain them in the incubator for use within the following 48 h.

3.4 Whole-Cell Patch-Clamp Recording

1. Make recording pipettes from glass capillary using a pipette puller, and if necessary, polish the tip using a microforge.
2. Backfill the recording pipette with the pipette solution (see Subheading 2.2), and place in the electrode holder.
3. Apply positive air pressure to the recording pipette via the side-port of the electrode holder (see Note 7).
4. Immerse the recording pipette into the external or bath solution, and offset the liquid junction potential (see Note 8).
5. Apply a 5 mV test pulse every 5 s from 0 mV holding potential to examine the recording pipette resistance, which should be in the range of 2–4 MΩ.
6. Select single GFP-positive cells under the fluorescent microscope.
7. Position the recording pipette closely to the cell and impale gently onto it using a micromanipulator.
8. Apply negative air pressure to the recording pipette to obtain a giga-ohm (GΩ) seal (see Note 9).
9. Set a negative holding potential (e.g., −60 mV), and cancel the pipette capacitance using amplifier fast capacitance (see Note 8).
10. Apply further negative air pressure to the recording pipette to break the membrane patch underneath the tip of the recording pipette to achieve the whole-cell figuration.
11. Compensate the access resistance (Rs) using whole-cell (slow) capacitance cancellation (see Note 8).
12. Once Rs is below an acceptable value (≤10 MΩ), and if required, carry out series resistance compensation (see Note 8; see also Chapter 7 of this book for more detailed protocol of the patch-clamp recording).
13. Begin recording. Figure 2a shows an example of whole-cell current patch-clamp recording made from a HEK293 cell expressing double cysteine mutant [V48C, I328C] of rat P2X2R at a holding potential of −60 mV. During this recording, 30 μM ATP was applied for 2 s every 2 or 4 min, using RSC-200 fast solution changer, to the cell to activate the mutant receptor channels. After two initial ATP applications, 10 mM dithiothreitol (DTT) was applied to the cell (denoted by 0 min), using RSC-200. The initial ATP-induced current was small but drastically increased and reached maximum after application of DTT for 20 min. The current gradually declined upon washing DTT, and was reversed upon reapplication of

DTT. Cysteines were introduced to replace valine residue at position 48 (Val^{48}) located at the extracellular end of the first transmembrane domain and isoleucine at position 328 (Ile^{328}) at the extracellular end of the second transmembrane domain, resulting in the [V48C, I328C] double cysteine mutant receptor. The results, from recording the double-mutant receptor-mediated currents like the one shown in Fig. 2a, suggest that Val^{48} and Ile^{328} residues appose side by side in the closed P2X2R receptor and cross-linking of V48C and I328C via a disulfide bond prevents the ion channel from fully opening (refer to (12) for further details and interpretation). The vicinity between Val^{48} and Ile^{328} residues (in green) is evident in a structural model of the rat P2X2R based on the zebrafish P2X4R atomic structure in the closed state (Fig. 2b) (3). The three subunits are indicated in blue, red, and yellow, with the TM2 forming the central ion-conducting pathway and TM1 be peripheral.

4 Notes

1. For some ion channel proteins, it is necessary to generate a cDNA construct expressing the "cysteine-less" version of the wild-type protein by replacing all endogenous cysteines with alanine (e.g., see ref. 8), if they can potentially interact or interfere with substituted cysteines.
2. A >80% cell confluency is ideally required for optimum transfection of HEK293 cells using Lipofectamine 2000. The cell confluency requirement may differ using transfection reagents from other suppliers.
3. GFP is often used as a selective marker for identification of positive transfectants. Instead of co-transfecting cells with two separate plasmids, the cDNAs for GFP and ion channel protein can be inserted into a bicistronic internal ribosome entry site (IRES) mammalian expression vector (19). Alternatively, GFP and ion channel protein can be expressed as a fusion protein where GFP is fused to the N- or C-terminus of the ion channel protein.
4. To express a heteromeric channel composed of two or more subunits, co-transfection with two or more plasmids encoding different subunits is required and the optimum ratio of plasmids needs to be determined experimentally (e.g., see ref. 14). Alternatively, if there are two cDNAs, they can be inserted into an IRES mammalian expression vector (see Note 3).
5. Prolonged incubation may be required for functional expression of some channel proteins. In such cases, it is better to replace transfection medium with culture medium on the following day.

6. To prepare cells for patch-clamp recording, cells need to be seeded as single cells. This can be achieved by gentle and through pipetting of the cell suspension, e.g., firstly using 1,000-μL pipette tips and then using 200-μL pipette tips. Alternatively, the cell suspension can be diluted in 10 mL instead of 2 mL, using 100–200 μL of cell suspension onto each coverslip.
7. This can be done usually via tubing that is connected to a 1-mL syringe. Positive pressure is applied by pushing the plunger; this helps prevent contamination of pipette solution from bath solution and attachment of debris to the pipette tip that makes it virtually impossible to obtain giga-ohm seal.
8. Compensations can be done using built-in programmes if the amplifier is computer driven (e.g., HEKA) or manually using appropriate compensation or cancellation knobs (e.g., as in Axo-patch 200B).
9. Suction can be applied using a 1-mL syringe (see Note 7). Sealing can be improved using negative holding potentials, but the optimum sealing procedures using different cell preparations, patch pipettes, and solutions can be only obtained experimentally.
10. Exposure to oxidizing conditions, such as Cu^{2+}: phenanthroline (Cu:Phe), is sometimes necessary to facilitate disulfide bond formation between a pair of cysteines over a longer distance (7), and the Cu:Phe-induced effects due to cysteine cross-linking are readily reversed with reducing agents such as DTT (7) or 2,3-dimercapto-1-propane-sulphonic acid (8). An alternative is to use Cd^{2+} ions to bridge the sulfur atoms to form a metal bridge (S-Cd^{2+}-S) (8).

References

1. Hille B (2001) Ion channels of excitable membranes, 3rd edn. Sinauer associates, Sunderland, MA
2. Khakh BS, North RA (2006) P2X receptors as cell-surface ATP sensors in health and disease. Nature 442:527–532
3. Browne LE, Jiang LH, North RA (2011) New structure enlivens interest in P2X receptors. Trends Pharmacol Sci 31:229–237
4. Yu FH, Yarov-Yarovoy V, Gutman GA, Catterall WA (2005) Overview of molecular relationships in the voltage-gated ion channel superfamily. Pharmacol Rev 57:387–395
5. Kashlan OB, Kleyman TR (2011) ENaC structure and function in the wake of a resolved structure of a family member. Am J Physiol Renal Physiol 301:F684–F696
6. Corringer PJ, Le Novère N, Changeux JP (2000) Nicotinic receptors at the amino acid level. Annu Rev Pharmacol Toxicol 40: 431–458
7. Kash TL, Jenkins A, Kelley JC, Trudell JR, Harrison NL (2003) Coupling of agonist binding to channel gating in the GABA(A) receptor. Nature 421:272–275
8. Elliott DJ, Neale EJ, Aziz Q, Dunham JP, Munsey TS, Hunter M, Sivaprasadarao A (2004) Molecular mechanism of voltage sensor movements in a potassium channel. EMBO J 23:4717–4726
9. Roberts JA, Allsopp RC, El Ajouz S, Vial C, Schmid R, Young MT, Evans RJ (2012) Agonist binding evokes extensive conformational changes in the extracellular domain of the

ATP-gated human P2X1 receptor ion channel. Proc Natl Acad Sci USA 109:4663–4667
10. Sorge RE, Trang T, Dorfman R, Smith SB, Beggs S, Ritchie J, Austin JS, Zaykin DV, Meulen HV, Costigan M, Herbert TA, Yarkoni-Abitbul M, Tichauer D, Livneh J, Gershon E, Zheng M, Tan K, John SL, Slade GD, Jordan J, Woolf CJ, Peltz G, Maixner W, Diatchenko L, Seltzer Z, Salter MW, Mogil JS (2012) Genetically determined P2X7 receptor pore formation regulates variability in chronic pain sensitivity. Nat Med 18:595–599
11. Akabas MH, Stauffer DA, Xu M, Karlin A (1992) Acetylcholine receptor channel structure probed in cysteine-substitution mutants. Science 258:307–310
12. Karlin A, Akabas MH (1998) Substituted-cysteine accessibility method. Methods Enzymol 293:123–145
13. Jiang LH, Rassendren F, Spelta V, Surprenant A, North RA (2001) Amino acid residues involved in gating identified in the first membrane-spanning domain of the rat P2X(2) receptor. J Biol Chem 276:14902–14908
14. Jiang LH, Kim M, Spelta V, Bo X, Surprenant A, North RA (2003) Subunit arrangement in P2X receptors. J Neurosci 23:8903–8910
15. Spelta V, Jiang LH, Bailey RJ, Surprenant A, North RA (2003) Interaction between cysteines introduced into each transmembrane domain of the rat P2X2 receptor. Br J Pharmacol 138:131–136
16. Marquez-Klaka B, Rettinger J, Bhargava Y, Eisele T, Nicke A (2007) Identification of an intersubunit cross-link between substituted cysteine residues located in the putative ATP binding site of the P2X1 receptor. J Neurosci 27:1456–1466
17. Jiang R, Martz A, Gonin S, Taly A, de Carvalho LP, Grutter T (2010) A putative extracellular salt bridge at the subunit interface contributes to the ion channel function of the ATP-gated P2X2 receptor. J Biol Chem 285: 15805–15815
18. Jiang R, Taly A, Lemoine D, Martz A, Cunrath O, Grutter T (2012) Tightening of the ATP-binding sites induces the opening of P2X receptor channels. EMBO J 31: 2134–2143
19. Trouet D, Nilius B, Voets T, Droogmans G, Eggermont J (1997) Use of a bicistronic GFP-expression vector to characterise ion channels after transfection in mammalian cells. Pflugers Arch 434:632–638

Chapter 22

Analysis of Ca^{2+}-Binding Sites in the MthK RCK Domain by X-Ray Crystallography

Frank J. Smith and Brad S. Rothberg

Abstract

Regulator of K^+ conductance (RCK) domains form a conserved class of ligand-binding domains that control the activity of a variety of prokaryotic and eukaryotic K^+ channels. Structural analysis of these domains by X-ray crystallography has provided insight toward mechanisms underlying ligand binding and channel gating, and thus the experimental strategies aimed at determining structures of liganded and unliganded forms of the domains may be useful in analysis of other ligand-binding domains. Here, we describe a basic strategy for crystallographic analysis of the RCK domain from the MthK channel, for determination of its Ca^{2+}-bound structure.

Key words Crystallization, Cytoplasmic domain, Calcium, Binding site, Potassium channel

1 Introduction

Large-conductance Ca^{2+}-activated K^+ channels (BK channels) are found in a wide range of tissues, and play a critical role in linking K^+ efflux to increases in cytoplasmic Ca^{2+} levels, thus tying Ca^{2+} signaling to electrical hyperpolarization of the cell membrane (1, 2). In electrically excitable nerve and muscle cells, this linkage provides an important feedback mechanism to allow for rapid repolarization of the membrane, promoting termination of action potentials and smooth muscle relaxation (3–7). While structural analysis of the mammalian BK channel can be technically complex and has so far yielded low-resolution structural information (8–10), the prokaryotic Ca^{2+}-activated K^+ channel, MthK, can be expressed and crystallized prodigiously, and has served as a model system for understanding gating mechanisms in these channels (11–16).

MthK and BK channels are regulated by Ca^{2+} binding to a conserved cytoplasmic domain, known as the regulator of K^+ conductance (RCK) domain (8, 9, 13, 14, 17–20). RCK domains are ubiquitous among prokaryotic and eukaryotic K^+ channels and

Nikita Gamper (ed.), *Ion Channels: Methods and Protocols*, Methods in Molecular Biology, vol. 998, DOI 10.1007/978-1-62703-351-0_22,

transporters, and control the activity of these proteins through the binding of cytoplasmic ligands such as nucleotides, Na^+, Ca^{2+}, Mg^{2+}, and H^+ (21–28). In MthK, channel activation occurs through Ca^{2+} binding to an octameric "gating ring" of RCK domains, which in turn is tethered to the pore-lining helices of the channel (13). Upon Ca^{2+} binding, the RCK domain undergoes a conformational change which is then translated to the pore facilitating K^+ conduction, though the structural interactions that underlie this conformational change are not entirely clear (11, 14). Our approach toward understanding the mechanisms of the conformational change has involved screening to determine crystallization conditions that may yield new conformations of the RCK domain, with the goal of solving high-resolution structures that will reveal the details of chemical bonds that underlie stabilization of a range of conformations, in the presence and absence of Ca^{2+}. Here, we describe a representative successful crystallization protocol, which enabled us to confirm the presence of newly determined Ca^{2+} binding sites in the RCK domain that underlie MthK channel activation.

2 Materials

All solutions are prepared using deionized, distilled water (ddH_2O).

2.1 Plasmids and *E. coli* Strains

1. *E. coli* strain BL21(DE3), purchased from Stratagene, La Jolla, CA.
2. MthK RCK plasmid: MthK RCK domain cDNA (residues M107-A336 in MthK protein sequence) in pET21a vector with D184N mutation generated using QuickChange (Stratagene, La Jolla, CA). The MthK RCK domain sequence was followed by a thrombin recognition/cleavage sequence (amino acids LVPRGS) and C-terminal hexahistidine tag, and the coding region was codon-optimized for *E. coli* expression.

2.2 Preparation and Transformation of BL21(DE3) *E. coli* Cells

1. Ampicillin (1,000× solution): 1 g per 10 mL H_2O, sterilize through 0.22-μm filter and store in 1 mL aliquots at −20°C.
2. LB broth: 25 g Luria-Bertani (LB) Broth powder (BD, Sparks, MD) per 1 L H_2O, autoclaved.
3. Agar plates with ampicillin: autoclave ten capsules of Agar B (Bio101, Carlsbad, CA) and 12.5 g LB broth powder per 500 mL H_2O. Allow LB/agar solution to cool to approximately 50°C, add 0.75 mL ampicillin per 500 mL LB/agar, and mix by swirling. Pour into plates, avoiding introduction of air bubbles. Allow agar to cool to room temperature and store at 4°C.

4. NZYM+Glu medium: add 100 μL of 20% glucose solution to 5 mL autoclaved, cooled NZYM medium (Bio 101 Carlsbad, CA).
5. Buffer B: 250 mM NaCl, 20 mM Tris–HCl, pH 8.0.
6. Sonication buffer: 250 mM NaCl, 50 mM Tris–HCl, pH 8.0.
7. Complete EGTA-free protease inhibitor (Roche, Indianapolis, IN).
8. Phenylmethanesulfonyl fluoride (PMSF) solution: 0.035 g in 1 mL acetone.
9. Thrombin solution: prepared by adding 800 μL Buffer B, 200 μL glycerol, to 20 U thrombin (Roche R&D, Indianapolis, IN); aliquot in 100 μL (2 U) fractions and store at −80°C.

2.3 Chromatography

1. $CoCl_2$ buffer.
2. Buffer B+400 mM imidazole (Sigma-Aldrich, St. Louis, MO).

2.4 Crystallization

1. 96-condition crystallization screening reagent kit (i.e., PEGs suite, Qiagen, Valencia, CA).
2. Polyethylene glycol (PEG) 3350 stock solution: PEG3350, 50% w/v.
3. 1 M 2-(*N*-morpholino)ethanesulfonic acid (MES) stock solutions, pH ranging from 5.9 to 6.5, in increments of 0.2 pH unit.
4. 1 M $CaCl_2$ stock solution.

2.5 Equipment

1. Large-probe sonicator (i.e., VirSonic, VirTis, Gardiner, NY).
2. Preparative liquid chromatography system (i.e., ÄKTA FPLC, GE Healthcare).
3. Superdex-200 10/300 column (GE Healthcare).
4. Metal-affinity chromatography column (i.e., HiTrap chelating HP (5 mL) column, GE Healthcare).
5. Peristaltic pump system (i.e., Model EP-1 Econo Pump, Bio-Rad).
6. UV Monitor (i.e., Model EM-1 Econo UV Monitor, Bio-Rad).
7. 50 kDa molecular-weight cutoff concentrator unit (e.g., Vivaspin 15, from Sartorius Biotech, Gottingen, Germany).
8. 96-well Greiner Sitting Drop Plate (Hampton Research, Aliso Viejo, CA).
9. 24-well, greased VDX crystallization plate (Hampton Research, Aliso Viejo, CA).

3 Methods

3.1 BL21(DE3) E. coli Transformation

1. Pre-warm one agar plate (with ampicillin) to room temperature.
2. Thaw BL21(DE3) cells on ice for 5 min.
3. Add 0.1 μg/μL MthK RCK plasmid solution to 25 μL BL21(DE3) cells and store on ice for 15 min.
4. Heat shock the cells for 45 s at 42°C.
5. Cool cells on ice immediately for 5 min.
6. Add 400 μL NZYM + Glu medium.
7. Incubate for 45 min at 37°C with vigorous shaking (on a shaker table at 225–250 rpm).
8. Plate the cell suspension on agar plate.
9. Incubate the plate overnight at 37°C.

3.2 MthK RCK BL21(DE3) Expression

1. Prepare 1 L LB broth in a 2 L baffled culture flask by adding 1 mL 1,000× ampicillin; wash colonies from the agar plate (from Subheading 3.1) using 5 mL LB broth and transfer into the flask containing LB medium.
2. Incubate flask at 37°C with vigorous shaking (180–200 rpm) and monitor cell growth by checking the optical density at 600 nm (OD_{600}).
3. Once OD_{600} reaches the range of 0.60–0.80, induce protein expression by adding 85 mg IPTG (pre-dissolved in 1 mL ddH_2O) to 1 L culture, and continue to incubate with shaking at 37°C for 4 h.
4. Following 4-h incubation, transfer suspension to centrifuge bottles and pellet bacterial cells by centrifugation at 7,200 × g for 10 min. Discard supernatant. The bacterial pellet can be stored overnight at 4°C, or up to 1 month at −80°C.

3.3 Lysis of BL21(DE3) Cells by Sonication

1. Add 25 mL sonication buffer to the cell pellet (from Subheading 3.2) and resuspend completely. Transfer cell suspension to stainless steel beaker. Add one tablet of complete EGTA-free protease inhibitor to the cell suspension; incubate on ice.
2. Prepare PMSF solution.
3. Sonicate using large-probe sonicator starting at 50% power (setting of 5.0) for 1 min, on ice. Add 100 μL PMSF solution, and allow to cool for 1 min.
4. Repeat sonication for 1 min at 5.0 two times with 1 min of cooling between sonication rounds; increase power by 10% (1 step) in each subsequent sonication round, alternating between 1-min sonication and 1-min cooling, until setting of 9.0 is reached.

5. Transfer lysate into centrifuge tubes.
6. Centrifuge at 30,500 × *g* (16,000 rpm using a Sorvall SS-34 rotor) for 45 min.
7. Transfer cleared lysate to beaker and add imidazole to yield final imidazole concentration of 20 mM.

3.4 Affinity Chromatography

1. Prepare HiTrap chelating HP (5 mL) column (see Note 1):
   (a) Perfuse with 20–30 mL ddH_2O at 2.5 mL/min.
   (b) Perfuse with 20–30 mL 0.1 M $CoCl_2$ buffer at 2.5 mL/min.
   (c) Perfuse column with 20–30 mL ddH_2O at 2.5 mL/min.
   (d) Perfuse column with 20–30 mL Buffer B + 400 mM imidazole at 2.5 mL/min.
   (e) Perfuse column with 20–30 mL Buffer B + 20 mM imidazole at 2.5 mL/min.
2. Load cleared lysate (from Subheading 3.3) onto column at 4.5 mL/min. Monitor protein elution by UV detection (see Note 1).
3. Once lysate is loaded onto column, perfuse column with Buffer B + 20 mM imidazole buffer at 4.5 mL/min until unbound protein is eluted, as indicated by reduction in UV signal to baseline level (approximately 20–30 mL).
4. Once a baseline is reached, perfuse column with Buffer B + 400 mM imidazole buffer to elute bound protein. Monitor UV signal, and collect protein upon increase of UV signal in a sterile 50 mL conical tube. Stop collection once UV signal decreases to baseline level (around 5 mL).
5. Immediately add 4 U thrombin (two aliquots of thrombin solution) to eluate and incubate 2 h at room temperature with gentle shaking.
6. Filter protein using 0.22 μm SPIN-X tubes for 1 min in refrigerated desktop centrifuge at 18,000 × *g*, at 4°C.

3.5 Gel Filtration Chromatography, Concentration, and Protein Storage

1. Equilibrate Superdex-200 10/300 column with two column volumes of degassed Buffer B, at a flow rate of 1 mL/min (see Note 2).
2. Load protein onto column via 1 mL sample loop and elute with Buffer B, at a flow rate of 0.5 mL/min. Collect fractions beginning at approximately 11 mL elution volume, and continue collection until the UV signal falls to baseline, typically at approximately 13 mL elution volume. If automated fraction collection is used, pool fractions collected during the major elution peak, which is centered at approximately 12 mL elution volume.
3. Measure protein concentration using Bradford or similar assay.

4. Prepare a 50 kDa molecular-weight cutoff concentrator unit by adding 5 mL Buffer B and centrifuging at 2,000 × *g* for 10 min.
5. Add protein solution (pooled fractions from step 2) to concentrator and centrifuge at 3,100 × *g* until volume yields a final protein concentration of 6 mg/mL.
6. Aliquot 100 μL portions of concentrated protein solution into 1.5 mL tubes; store protein in liquid nitrogen.

3.6 Setting Initial Crystallization Screen in 96-Well Sitting-Drop Format

1. Prepare crystallization screen in 96-well format by pipetting 100 μL of each crystallization solution from the PEGs suite into each reservoir of 96-well Greiner sitting-drop plate. Use Greiner plates that enable setting of three drops per well. This and subsequent steps are facilitated through use of multichannel pipettors (see Note 3).
2. Pipette 1 μL of thawed protein solution into each of the three drop compartments in each well. A total of approximately 300 μL of protein solution will be required for a 96-well screen, with three drops per well (at 6 mg/mL, this is equal to around 1.8 mg purified protein).
3. To each protein drop in a given well, transfer 0.8 μL of the corresponding reservoir solution.
4. To the drop at the lower left of each well, add 0.2 μL of ddH_2O; to the drop at the upper left of each well, add 0.2 μL of 200 mM $CaCl_2$; then to the drop at the upper right of each well, add 0.2 μL of 500 mM $CaCl_2$.
5. Store tray in a location not prone to frequent temperature perturbations or excessive movement/vibrations.
6. Periodically monitor crystallization drops using a microscope, noting crystal drop phenomena (see Note 4).

3.7 Optimization of Crystallization in Polyethylene Glycol (PEG) 3350 in 24-Well Hanging-Drop Format

1. Add components to the appropriate reservoir chambers of a 24-well, greased VDX crystallization plate, as described in Table 1.
2. Following preparation of reservoir solutions, pipette 1.0 μL of protein to the center of a clean, 22 mm plastic coverslip; then add 1.0 μL of the appropriate reservoir solution to the protein drop.
3. Gently flip the coverslip and firmly seal the coverslip to the greased top of the well, with the protein drop hanging above the reservoir solution.
4. Repeat the process with each well. This process may be facilitated by pipetting protein solution onto each of six coverslips, following by transfer of the appropriate well solutions and sealing of each coverslip sequentially.

5. Store tray in a location not prone to frequent temperature perturbations or excessive movement/vibrations.
6. Periodically monitor crystallization drops using a microscope, noting crystal drop phenomena; an example of successful crystal formation is illustrated in Fig. 1 (see Note 5).
7. Finally, crystals may be mounted on nylon loops and rapidly frozen and then stored in liquid nitrogen for subsequent X-ray diffraction experiments. An example of a mounted crystal is shown in Fig. 2 (see Note 6).

4 Notes

1. Perfusion of the HiTrap metal-affinity chromatography column and fraction collection is facilitated by the use of a peristaltic pump coupled to a UV detector (i.e., Model EP-1 Econo-Pump and Model EM-1 UV Monitor, Bio-Rad), with the UV detector signal output connected to a chart recorder.
2. Buffers used in gel filtration chromatography steps should be degassed by dispensing buffer into a sidearm flask, sealing the top with a rubber stopper, and connecting the sidearm to a vacuum (via a vacuum trap) for a minimum of 1 h. Alternatively, this could be performed immediately subsequent to filtration using a vacuum-filtration unit, by leaving the buffer in the collection bottle with the sidearm connected to vacuum for at least 1 h after filtration. Gel filtration chromatography is performed using a preparative fast protein liquid chromatography apparatus (i.e., ÄKTA FPLC system, GE Healthcare), with elution monitored by UV detection. This chromatography procedure is facilitated through use of automated fraction collection.
3. Crystallization screens can alternatively be set in hanging-drop format.
4. For this crystallization experiment, we were particularly interested in discovering conditions in which crystal growth was more successful in drops containing 50 mM $CaCl_2$, whereas no crystal growth was noted in the drop with no added $CaCl_2$ in the same well. It should be further noted that calcium (and other divalent cations) is particularly prone to formation of insoluble complexes and salt crystals with anions that are commonly included as additives in commercially available screens. These anions include phosphate, sulfate, and citrate. Distinguishing between calcium salt crystals and protein crystals can be facilitated by setting a "control" tray, in which the screen is set by mixing the well solution with the calcium-containing protein buffer (i.e., with no protein). In this control screen, protein crystals will not form—only salt crystals.

Table 1
Example of a pos sible grid screen for optimization of crystal growth of the MthK RCK domain with D184N mutation in 50 mM $CaCl_2$, using PEG3350 as precipitant, at pH ranging from 5.9 to 6.3

	PEG3350, 16%	18%	20%	22%	24%	26%
pH 5.9	320 μl PEG3350 (50% stock) 100 μl 1 M MES pH 5.9 50 μl 1 M $CaCl_2$ 530 μl ddH_2O	360 μl PEG3350 (50% stock) 100 μl 1 M MES pH 5.9 50 μl 1 M $CaCl_2$ 490 μl ddH_2O	400 μl PEG3350 (50% stock) 100 μl 1 M MES pH 5.9 50 μl 1 M $CaCl_2$ 450 μl ddH_2O	440 μl PEG3350 (50% stock) 100 μl 1 M MES pH 5.9 50 μl 1 M $CaCl_2$ 410 μl ddH_2O	480 μl PEG3350 (50% stock) 100 μl 1 M MES pH 5.9 50 μl 1 M $CaCl_2$ 370 μl ddH_2O	520 μl PEG3350 (50% stock) 100 μl 1 M MES pH 5.9 50 μl 1 M $CaCl_2$ 330 μl ddH_2O
pH 6.1	320 μl PEG3350 (50% stock) 100 μl 1 M MES pH 6.1 50 μl 1 M $CaCl_2$ 560 μl ddH_2O	360 μl PEG3350 (50% stock) 100 μl 1 M MES pH 6.1 50 μl 1 M $CaCl_2$ 490 μl ddH_2O	400 μl PEG3350 (50% stock) 100 μl 1 M MES pH 6.1 50 μl 1 M $CaCl_2$ 450 μl ddH_2O	440 μl PEG3350 (50% stock) 100 μl 1 M MES pH 6.1 50 μl 1 M $CaCl_2$ 410 μl ddH_2O	480 μl PEG3350 (50% stock) 100 μl 1 M MES pH 6.1 50 μl 1 M $CaCl_2$ 370 μl ddH_2O	520 μl PEG3350 (50% stock) 100 μl 1 M MES pH 6.1 50 μl 1 M $CaCl_2$ 330 μl ddH_2O
pH 6.3	320 μl PEG3350 (50% stock) 100 μl 1 M MES pH 6.3 50 μl 1 M $CaCl_2$ 530 μl ddH_2O	360 μl PEG3350 (50% stock) 100 μl 1 M MES pH 6.3 50 μl 1 M $CaCl_2$ 490 μl ddH_2O	400 μl PEG3350 (50% stock) 100 μl 1 M MES pH 6.3 50 μl 1 M $CaCl_2$ 450 μl ddH_2O	440 μl PEG3350 (50% stock) 100 μl 1 M MES pH 6.3 50 μl 1 M $CaCl_2$ 410 μl ddH_2O	480 μl PEG3350 (50% stock) 100 μl 1 M MES pH 6.3 50 μl 1 M $CaCl_2$ 370 μl ddH_2O	520 μl PEG3350 (50% stock) 100 μl 1 M MES pH 6.3 50 μl 1 M $CaCl_2$ 330 μl ddH_2O
pH 6.5	320 μl PEG3350 (50% stock) 100 μl 1 M MES pH 6.5 50 μl 1 M $CaCl_2$ 530 μl ddH_2O	360 μl PEG3350 (50% stock) 100 μl 1 M MES pH 6.5 50 μl 1 M $CaCl_2$ 490 μl ddH_2O	400 μl PEG3350 (50% stock) 100 μl 1 M MES pH 6.5 50 μl 1 M $CaCl_2$ 450 μl ddH_2O	440 μl PEG3350 (50% stock) 100 μl 1 M MES pH 6.5 50 μl 1 M $CaCl_2$ 410 μl ddH_2O	480 μl PEG3350 (50% stock) 100 μl 1 M MES pH 6.5 50 μl 1 M $CaCl_2$ 370 μl ddH_2O	520 μl PEG3350 (50% stock) 100 μl 1 M MES pH 6.5 50 μl 1 M $CaCl_2$ 330 μl ddH_2O

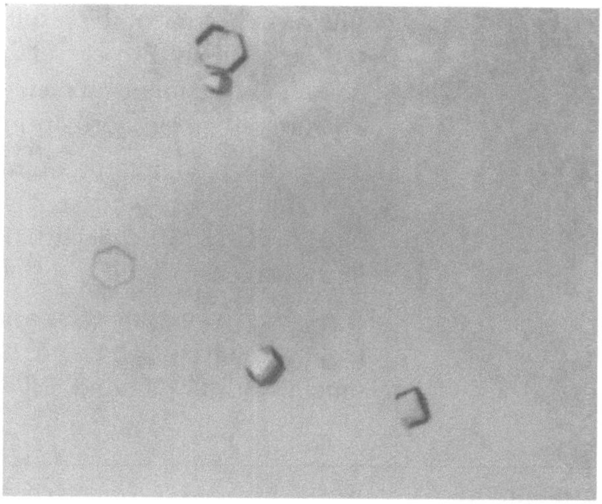

Fig. 1 Examples of MthK D184N RCK domain crystals grown in the presence of calcium. Crystals were grown in a hanging drop, over a reservoir solution containing 20% PEG3350, 20 mM $CaCl_2$, and 100 mM MES pH 5.9. These crystals display a hexagonal morphology, and formed in 2–3 weeks under these conditions

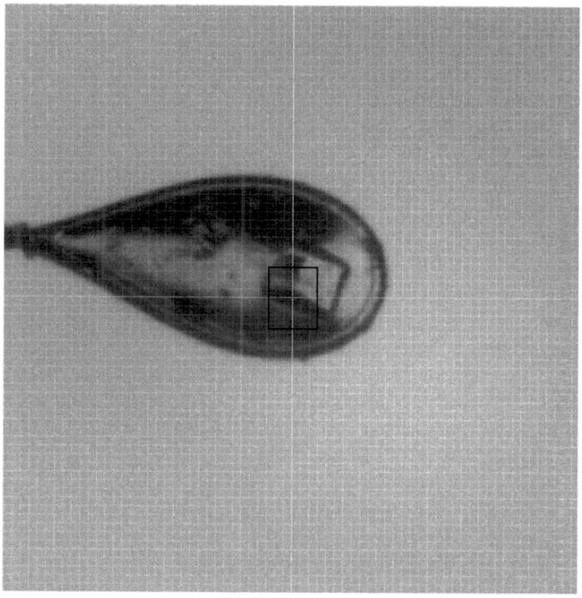

Fig. 2 MthK D184N RCK domain crystal, mounted in a nylon loop and frozen in a nitrogen stream at 100 K. This crystal was grown in a hanging drop, over a reservoir solution containing 20% PEG3350, 100 mM $CaCl_2$, and 100 mM MES pH 5.9 (The photograph was captured at the National Synchrotron Light Source, beamline X25. The dimension (see planaons) of the black box is 100 × 150 μm)

The morphology of the resulting salt crystals in each condition can be compared to crystals which form in the experimental (protein-containing) screen to determine the likelihood of calcium salt crystal formation.

5. The grid screen presented can be further modified to optimize crystal growth conditions, by systematically varying the range of pH, PEG3350 concentration, or $CaCl_2$ concentration used in the screen.
6. A crystal grown using this protocol was used in diffraction experiments to solve the structure of the MthK RCK domain D184N mutant with Ca^{2+} bound, PDB accession number 3RBX (14).

Acknowledgements

We wish to thank Karin Abarca-Heidemann, Andrew S. Thomson, Matthew M. Callaghan, and Elsie Samakai for technical expertise and helpful discussions. This work was supported by NIH grant R01 GM68523 to B.S.R.

References

1. Vergara C, Latorre R, Marrion NV, Adelman JP (1998) Calcium-activated potassium channels. Curr Opin Neurobiol 8:321–329
2. Hille B (2001) Ion channels of excitable membranes, 3rd edn. Sinauer Associates, Inc., Sunderland, MA
3. Jaffe DB, Wang B, Brenner R (2011) Shaping of action potentials by type I and type II large-conductance Ca^{2+}-activated K^+ channels. Neuroscience 192:205–218
4. Brayden JE, Nelson MT (1992) Regulation of arterial tone by activation of calcium-dependent potassium channels. Science 256:532–535
5. Herrera GM, Heppner TJ, Nelson MT (2000) Regulation of urinary bladder smooth muscle contractions by ryanodine receptors and BK and SK channels. Am J Physiol Regul Integr Comp Physiol 279:R60–R68
6. Storm JF (1987) Action potential repolarization and a fast after-hyperpolarization in rat hippocampal pyramidal cells. J Physiol 385:733–759
7. Lancaster B, Nicoll RA (1987) Properties of two calcium-activated hyperpolarizations in rat hippocampal neurones. J Physiol 389:187–203
8. Yuan P, Leonetti MD, Pico AR, Hsiung Y, MacKinnon R (2010) Structure of the human BK channel Ca^{2+}-activation apparatus at 3.0 A resolution. Science 329:182–186
9. Yuan P, Leonetti MD, Hsiung Y, MacKinnon R (2011) Open structure of the Ca2+ gating ring in the high-conductance Ca^{2+}-activated K^+ channel. Nature 481:94–97
10. Wu Y, Yang Y, Ye S, Jiang Y (2010) Structure of the gating ring from the human large-conductance Ca^{2+}-gated K^+ channel. Nature 466:393–397
11. Ye S, Li Y, Chen L, Jiang Y (2006) Crystal structures of a ligand-free MthK gating ring: insights into the ligand gating mechanism of K^+ channels. Cell 126:1161–1173
12. Dong J, Shi N, Berke I, Chen L, Jiang Y (2005) Structures of the MthK RCK domain and the effect of Ca^{2+} on gating ring stability. J Biol Chem 280:41716–41724
13. Jiang Y, Lee A, Chen J, Cadene M, Chait BT, MacKinnon R (2002) Crystal structure and mechanism of a calcium-gated potassium channel. Nature 417:515–522
14. Pau VP, Smith FJ, Taylor AB, Parfenova LV, Samakai E, Callaghan MM, Abarca-Heidemann K, Hart PJ, Rothberg BS (2011) Structure and function of multiple Ca^{2+}-binding sites in a K^+ channel regulator of K^+ conductance (RCK) domain. Proc Natl Acad Sci U S A 108:17684–17689
15. Niu X, Qian X, Magleby KL (2004) Linker-gating ring complex as passive spring and Ca(2+)-dependent machine for a voltage- and Ca^{2+}-activated potassium channel. Neuron 42:745–756
16. Hou S, Xu R, Heinemann SH, Hoshi T (2008) Reciprocal regulation of the Ca^{2+} and H+sensitivity in the SLO1 BK channel conferred by

the RCK1 domain. Nat Struct Mol Biol 15:403–410
17. Schreiber M, Salkoff L (1997) A novel calcium-sensing domain in the BK channel. Biophys J 73:1355–1363
18. Wei A, Solaro C, Lingle C, Salkoff L (1994) Calcium sensitivity of BK-type KCa channels determined by a separable domain. Neuron 13:671–681
19. Bao L, Kaldany C, Holmstrand EC, Cox DH (2004) Mapping the BKCa channel's "Ca^{2+} bowl": side-chains essential for Ca^{2+} sensing. J Gen Physiol 123:475–489
20. Bao L, Rapin AM, Holmstrand EC, Cox DH (2002) Elimination of the BK(Ca) channel's high-affinity Ca^{2+} sensitivity. J Gen Physiol 120:173–189
21. Kuo MMC, Haynes WJ, Loukin SH, Kung C, Saimi Y (2005) Prokaryotic K^+ channels: From crystal structures to diversity. FEMS Microbiol Rev 29:961–985
22. Kroning N, Willenborg M, Tholema N, Hanelt I, Schmid R, Bakker EP (2007) ATP binding to the KTN/RCK subunit KtrA from the K^+-uptake system KtrAB of Vibrio alginolyticus: Its role in the formation of the KtrAB complex and its requirement in vivo. J Biol Chem 282:14018–14027
23. Roosild TP, Castronovo S, Miller S, Li C, Rasmussen T, Bartlett W, Gunasekera B, Choe S, Booth IR (2009) KTN (RCK) domains regulate K^+ channels and transporters by controlling the dimer-hinge conformation. Structure 17:893–903
24. Zhang X, Solaro CR, Lingle CJ (2001) Allosteric regulation of BK channel gating by Ca^{2+} and Mg^{2+} through a nonselective, low affinity divalent cation site. J Gen Physiol 118:607–636
25. Xia XM, Zeng X, Lingle CJ (2002) Multiple regulatory sites in large-conductance calcium-activated potassium channels. Nature 418: 880–884
26. Zeng XH, Xia XM, Lingle CJ (2005) Divalent cation sensitivity of BK channel activation supports the existence of three distinct binding sites. J Gen Physiol 125:273–286
27. Zhang X, Zeng X, Lingle CJ (2006) Slo3 K^+ channels: voltage and pH dependence of macroscopic currents. J Gen Physiol 128: 317–336
28. Yuan A, Santi CM, Wei A, Wang ZW, Pollak K, Nonet M, Kaczmarek L, Crowder CM, Salkoff L (2003) The sodium-activated potassium channel is encoded by a member of the Slo gene family. Neuron 37:765–773

Chapter 23

Isotope Labeling Strategies for Analysis of an Ion Channel Cytoplasmic Domain by NMR Spectroscopy

Karin Abarca-Heidemann, Elke Duchardt-Ferner, Jens Woehnert, and Brad S. Rothberg

Abstract

As large, multimeric, integral membrane proteins, ion channels pose technical challenges to analysis by NMR spectroscopy. Here we present a strategy to overcome some of these technical hurdles, using a representative ion channel modulatory domain, the regulator of K^+ conductance (RCK) domain from a K^+ channel cloned from *Thermoplasma volcanium*. By introducing a mutation to limit the stoichiometry of the octameric RCK domain "gating ring" complex to its dimeric building block, NMR spectral resolution can be greatly improved. Here we present protocols for efficient production of highly deuterated, uniformly ^{15}N-labeled protein, as well as protein containing ^{15}N-labeling to specific amino acid types. These labeling strategies can be applied to improve spectral resolution and facilitate sequential resonance assignments.

Key words Cytoplasmic domain, Resonance, Ligand, Potassium channel

1 Introduction

Analysis of ion channel structure has, over the last 10 years, become an indispensible component of molecular physiology (1–3). Yet while structural analysis of ion channels seems to be dominated by the static (but remarkably detailed) pictures provided by X-ray crystallography, a quantitative understanding of channel conformational dynamics through nuclear magnetic resonance (NMR) spectroscopy has remained relatively limited. This inequity may arise in part from technical hurdles posed by ion channels and their modulatory domains, which are frequently beyond the size limit inherent to solution NMR, except (notably) in the cases of KcsA and the glutamate receptor ligand-binding domain (4–8). Large proteins yield broad spectral lines and consequently have an intrinsically low signal-to-noise ratio. In addition, the large number of amino acids that makes up a large protein increases spectral overlap. Both of these phenomena contribute to difficulties in making

Nikita Gamper (ed.), *Ion Channels: Methods and Protocols*, Methods in Molecular Biology, vol. 998,
DOI 10.1007/978-1-62703-351-0_23, © Springer Science+Business Media, LLC 2013

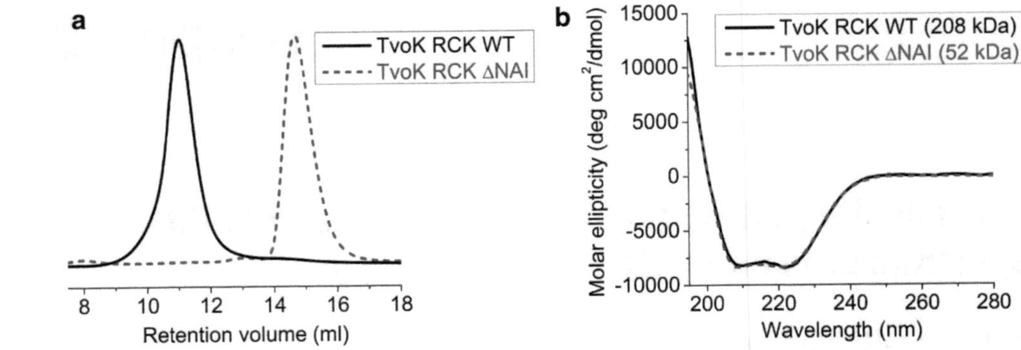

Fig. 1 Comparison of structural properties of the TvoK WT RCK domain and RCK(ΔNAI) mutant. (**a**) Gel-filtration elution profiles (measured by absorbance at 280 nm) of the wild-type RCK domain and RCK(ΔNAI) mutant. The wild-type domain complex elutes from the Superdex-200 column at ~11 mL, consistent with a ~200 kDa octamer, whereas the RCK(ΔNAI) mutant elutes at ~15 mL, consistent with a ~50 kDa dimer. (**b**) Circular dichroism spectra for the TvoK RCK WT and RCK(ΔNAI) are nearly identical to one another. Together these results are consistent with the (ΔNAI) mutation disrupting assembly of dimers to form an octameric complex, but leaving secondary and tertiary structure intact

unambiguous sequence-specific resonance assignments, which are critical to gleaning useful structural information from NMR experiments (9, 10).

While these technical issues are a fact of life in the NMR community, numerous strategies have been developed to extract useful information on the dynamics of large, multisubunit proteins; these include improvements in instrumentation, data acquisition protocols, and novel isotopic labeling strategies (10–13). Here we present an approach toward overcoming some of these hurdles in the case of a conserved class of K^+ channel ligand-binding domains, the regulator of K^+ conductance (RCK) domain. RCK domains are observed to form dimers or tandem "pseudo-dimers" (14–18); in turn, these RCK dimers assemble to form a radially fourfold symmetrical "gating ring" complex, which is typically tethered or bound to a K^+ channel or K^+ transporter (18–24). As the mass of an individual RCK domain can range from 25 to 30 kDa, it becomes clear that the mass of an overall octameric complex can be well over 200 kDa, beyond the reach of detailed NMR analysis.

Our approach toward this problem is similar, in principle, to the approach taken toward NMR analysis of the GroEL-GroES chaperone complex (25). In our case, we exploit a mutation (deletion of N196-I198; "ΔNAI") that inhibits formation of the octameric complex, and results in the formation of a stable, soluble 52 kDa RCK dimer in which the secondary and tertiary structures of the domain appear to be unaltered (Fig. 1). By reducing the number of subunits in the RCK domain assembly, spectral quality is greatly enhanced (Fig. 2). Resonance assignment can be further facilitated by isotopic labeling of specific amino acid types, as described below

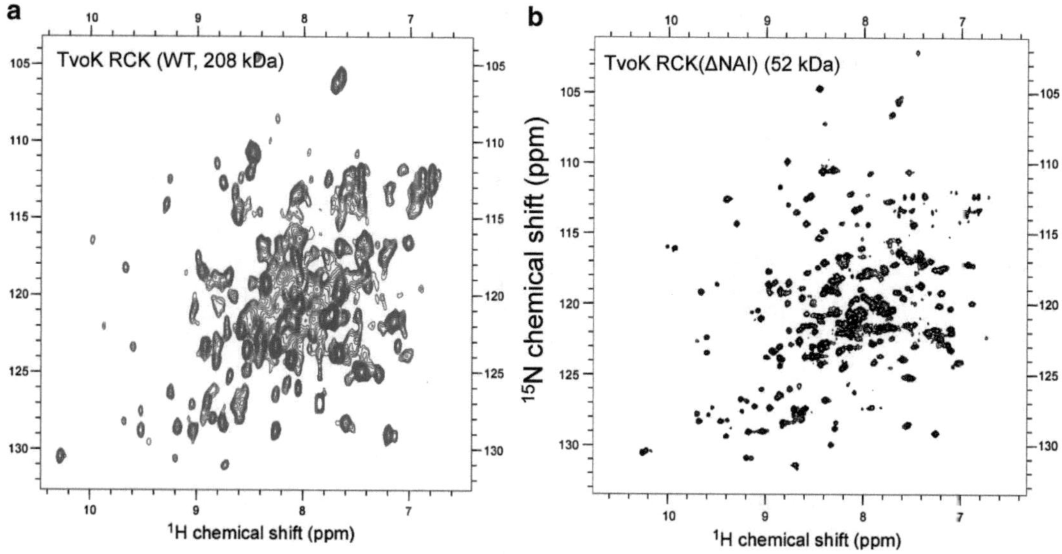

Fig. 2 $^{15}N,^{1}H$-TROSY-HSQC spectra obtained from deuterated, uniformly ^{15}N-labeled protein. (**a**) TvoK RCK WT: 150 mM KCl, 50 mM HEPES, 2 mM β-mercaptoethanol, pH 7.5, 10% D_2O, 47°C; (**b**) TvoK RCK(ΔNAI): 50 mM KCl, 50 mM MES, 2 mM β-mercaptoethanol, pH 6.5, 10% D_2O, 37°C. Both spectra were acquired using a Bruker Avance 900 MHz spectrometer equipped with a cryoprobe. The ΔNAI (dimer) spectrum exhibits sharper, more clearly distinguished crosspeaks than the spectrum from the octameric complex

(Fig. 3), combined with additional multidimensional experiments (see Note 1). These strategies provide the initial steps toward practical NMR analysis of a large ion channel domain.

2 Materials

All solutions are prepared using deionized, distilled water (ddH_2O).

2.1 Protein Expression

1. Competent *E. coli* strain BL21(DE3) (Stratagene, La Jolla, CA).
2. *E. coli* strain CT19: derivative of BL21(DE3) carrying the mutations aspC, ilvE, trpB, and tyrB, kindly provided by Dr. David Waugh, National Cancer Institute, Frederick, MD.
3. TvoK RCK domain cDNA (N-terminal methionine followed by residues R121-G348 in TvoK protein sequence), followed by thrombin recognition/cleavage sequence (LVPRGS) and C-terminal hexahistidine tag in pET21a vector, codon optimized for *E. coli* expression.
4. TvoK RCK mutant cDNA (RCK(ΔNAI)), generated using QuickChange (Stratagene, La Jolla, CA).
5. SOC medium: 0.5% yeast extract; 2% tryptone, 10 mM NaCl, 2.5 mM KCl, 10 mM $MgCl_2$, 10 mM $MgSO_4$, 20 mM glucose.
6. Luria-Bertani (LB) medium (BD, Sparks, MD): 1.0% tryptone, 0.5% yeast extract, 1.0% NaCl, pH 7.0.

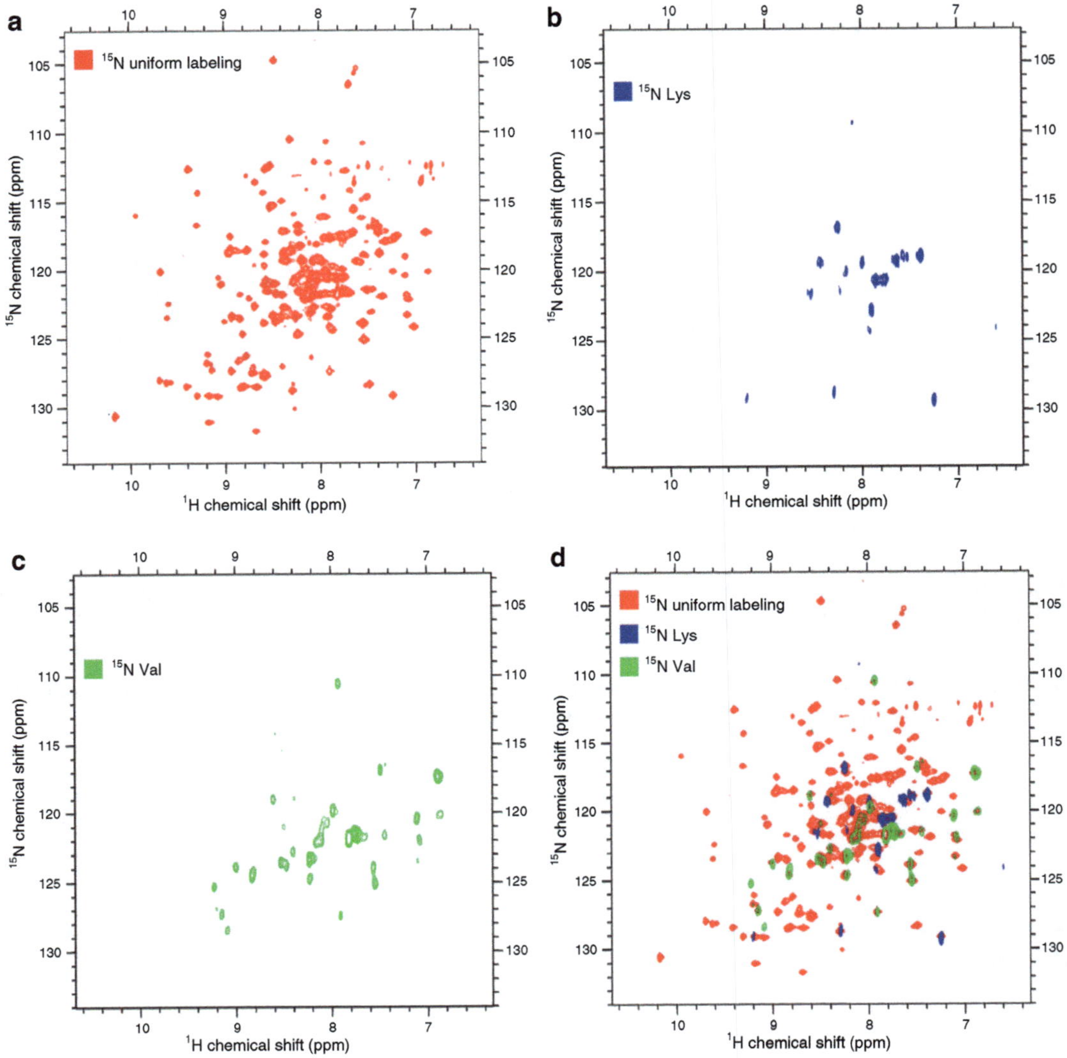

Fig. 3 ^{15}N,^{1}H-TROSY-HSQC spectra obtained from deuterated, uniformly ^{15}N-labeled and uniformly deuterated, amino acid-specific ^{15}N-labeled RCK(ΔNAI). (**a**) Spectrum from deuterated, uniformly ^{15}N-labeled protein; (**b**) deuterated, ^{15}N-Lys labeled protein; and (**c**) deuterated, ^{15}N-Val labeled protein. (**d**) Overlay of spectra in (**a**–**c**), illustrating superimposition of crosspeaks arising from Lys and Val N-H groups with those observed with uniform ^{15}N-labeling. Experimental conditions are the same as those described for Fig. 2b

7. Transformation buffer 1 (TfB1): 30 mM potassium acetate, 100 mM RbCl, 10 mM $CaCl_2$, 50 mM $MnCl_2$, 15% glycerol (v/v), pH 5.8.
8. Transformation buffer 2 (TfB2): 10 mM MOPS, 75 mM $CaCl_2$, 10 mM RbCl, 15% glycerol (v/v), pH 6.5.
9. Ampicillin 1,000× solution: 1 g ampicillin (Sigma, St. Louis, MO) per 10 mL H_2O, filtered through 0.22 μm filter, stored at −20°C.

10. Kanamycin (Sigma, St. Louis, MO): 200× solution, 100 mg per 10 mL H_2O, filtered through 0.22 μm filter, stored at −20°C.
11. Isopropyl-β-D-1-thiogalactoside (IPTG; Sigma, St. Louis, MO).
12. Vitamin mix, 500×: per 100 mL, 250 mg thiamin, 500 mg biotin, 500 mg choline chloride, 500 mg folic acid, 500 mg niacinamide, 500 mg D-panthothenic acid, 50 mg riboflavin. Store aliquots at −20°C.
13. Trace elements, 2,000×: per 100 mL, 2 g $FeCl_2$, 25.6 mg H_3BO_3, 7.2 mg $CoCl_2$, 1.6 mg $CuCl_2$, 136 mg $ZnCl_2$, 242 mg $NaMoO_4$, 16 mg $MnCl_2$. Store aliquots at −20°C.
14. Low glucose M9 medium: 47.7 mM Na_2HPO_4, 22 mM KH_2PO_4, 8.5 mM NaCl, 18.6 mM NH_4Cl. After autoclaving, add 2 mM $MgSO_4$, 5.5 mM glucose, 100 μM $CaCl_2$, 0.1 mL vitamin mix (500×), 2 mL trace elements (2,000×) 0.5 mL, pH 7.4.
15. ^{15}N,^{2}H-Celtone: Celtone Base Powder (D, 97%+; ^{15}N 98%+; Cambridge Isotope Laboratories, Andover, MA).
16. ^{2}H-only-Celtone: Celtone Base Powder (D, 97%+; Cambridge Isotope Laboratories, Andover, MA).
17. Spectra 9 medium (D, 97%; Cambridge Isotope Laboratories, Andover, MA).
18. α-^{15}N-Lys: L-Lysine:2HCl (ALPHA-^{15}N, 95–99%; Cambridge Isotope Laboratories, Andover, MA).
19. ^{15}N-Val: L-Valine (^{15}N 98%; Cambridge Isotope Laboratories, Andover, MA).
20. ^{15}N-Leu: L-Leucine (^{15}N 98%; Cambridge Isotope Laboratories, Andover, MA).
21. ^{15}N,^{2}H-Lys: L-Lysine:HCl (D9, 98%; $^{15}N2$ 98%; Cambridge Isotope Laboratories, Andover, MA).

2.2 Protein Purification

1. Sonication buffer: 250 mM KCl, 10% glycerol, 50 mM Tris–HCl pH 8.0 (prepared on the day before use, pH measured at RT, buffer stored at 4°C).
2. Protease inhibitor cocktail (e.g., Complete™ EDTA-free Protease Inhibitor Cocktail Tablets, Roche, Indianapolis, IN).
3. Phenylmethylsulfonyl fluoride (PMSF; Sigma-Aldrich, St. Louis, MO).
4. 5 mL HiTrap Chelating column (GE Healthcare, Piscataway, NJ).
5. Protein buffer B: 250 mM KCl, 20 mM Tris–HCl pH 8.0 (prepared on the day before use, pH measured at RT, buffer stored at 4°C).
6. HiTrap elution buffer: 250 mM KCl, 500 mM imidazole, 20 mM Tris–HCl pH 8.0, (imidazole is added immediately

before elution, with adjustment of pH to 8.0 after addition of imidazole).

7. HiTrap chelating column activation buffer: 100 mM $CoCl_2$.
8. HiTrap chelating column stripping buffer: 0.02 M sodium phosphate, 0.5 M NaCl, 50 mM EDTA, pH 7.2.
9. Centrifuge tube filters, 0.22 μm (i.e., Spin-X centrifuge tube filters, Corning, Tewksbury, MA).
10. Superdex-200 10/300 size exclusion column (GE Healthcare, Piscataway, NJ).
11. Thrombin solution: Prepared by adding 800 μL protein purification buffer and 200 μL glycerol to 20 U thrombin (Roche R&D, Indianapolis, IN); aliquot in 100 μL (2 U) fractions and store at −80°C.
12. Amicon Ultra-15 centrifugal filter units, 10 kDa molecular weight cutoff (EMD Millipore, Billerica, MA).
13. Slide-A-Lyzer mini dialysis cassettes, 10 kDa molecular weight cutoff (Pierce, Rockford, IL).

2.3 NMR Experiments

1. NMR buffer: 50 mM KCl, 2 mM β-mercaptoethanol, 50 mM Mes pH 6.5.

2.4 Equipment

1. Large-probe sonicator (i.e., VirSonic, VirTis, Gardiner, NY).
2. Preparative liquid chromatography system (i.e., ÄKTA FPLC, GE Healthcare).
3. Superdex-200 10/300 column (GE Healthcare).
4. Metal-affinity chromatography column (i.e., HiTrap chelating HP (5 mL) column, GE Healthcare).
5. Peristaltic pump system (i.e., Model EP-1 Econo Pump, Bio-Rad).
6. UV Monitor (i.e., Model EM-1 Econo UV Monitor, Bio-Rad).

3 Methods

3.1 Preparation of Competent CT19 E. coli (RbCl Method)

1. Inoculate 6 mL LB medium with CT19 *E. coli* cells and incubate the culture overnight at 37°C.
2. Add 0.8 mL of the overnight culture to 80 mL pre-warmed LB containing 20 mM $MgCl_2$ and grow culture at 30–37°C until OD600 ~ 0.4–0.5.
3. Chill the culture on ice at least for 10 min.
4. Spin at 2,500 × *g* for 15 min at 4°C.
5. Resuspend cells by gentle swirling in 32 mL ice-cold TfB1.
6. Incubate the cell suspension on ice for 10 min.

7. Spin cells at 2,500 × *g* for 15 min at 4°C.
8. Resuspend the pellet gently in 3.2 mL ice-cold TfB2 and keep the suspension on ice for 15 min.
9. Aliquot the cell suspension at 50 μL per sterile prechilled microcentrifuge tube.
10. Rapidly freeze the cell suspension in liquid nitrogen, and store the tubes at −80°C.

3.2 Transformation of Competent Bacteria and Generation of Starter Culture (See Note 2)

1. Thaw one aliquot of 50 μl competent cells (either BL21(DE3) or CT19) on ice for 5 min.
2. Add 10 μL of ice-cold 0.1 μg/μL plasmid DNA solution to the competent cell suspension; mix gently.
3. Incubate the tube on ice for 10 min.
4. Heat-shock the cells for 45 s at 42°C in a water bath.
5. Cool the tube immediately on ice for 10 min.
6. Add 250 μL SOC medium to the tube.
7. Incubate with vigorous shaking for 45 min at 37°C.
8. Plate the cell suspension on an LB-agar plate containing ampicillin; for CT19 cells, use an LB-agar plate containing kanamycin in addition to ampicillin.
9. Incubate the plate overnight at 37°C.
10. Use several colonies from the plate to inoculate 50 mL LB medium containing 50 μL 1,000× ampicillin; for CT19 cells, also include 0.25 mL of 200× kanamycin in addition to ampicillin.
11. Incubate the 50 mL starter culture overnight at 37°C with vigorous shaking.
12. Pellet the cells at 1,500 × *g* for 10 min at room temperature.
13. Wash the pellet two times using M9 medium or LB/ampicillin, depending on labeling method.

3.3 Protein Expression Using Isotope-Labeled Amino Acids

3.3.1 Expression of Deuterated, Uniformly ^{15}N-Labeled RCK(ΔNAI)

1. Resuspend pelleted BL21(DE3) transformants from overnight starter culture (Subheading 3.2) in 5 mL of low glucose M9 medium.
2. Use 1 mL of this suspension to inoculate 250 mL low glucose M9 medium containing 0.25 mL of 1,000× ampicillin, 0.5 mL vitamin mix, and 0.125 mL trace elements.
3. Incubate cells with vigorous shaking at 37°C until OD600 ~ 0.6.
4. Add 550 mg of ^{15}N, ^{2}H-Celtone.
5. Continue to incubate cells with vigorous shaking at 37°C until OD600 ~ 0.8 (~25 min).
6. Induce protein expression by adding IPTG (to 0.4 mM final concentration in the culture).

7. Continue to incubate cells with vigorous shaking at 37°C for 4 h.
8. Harvest cells after 4 h of expression by centrifuging at 5,000 × *g* for 15 min at RT.
9. Store cell pellet at −80°C.

3.3.2 Expression of Deuterated, Single Amino Acid ^{15}N-Labeled RCK(ΔNAI)

1. Resuspend pelleted BL21(DE3) transformants from overnight starter culture (Subheading 3.2) in 5 mL of low glucose M9 medium.
2. Use 1 mL of this suspension to inoculate 250 mL low glucose M9 medium containing 0.25 mL of 1,000× ampicillin, 0.5 mL vitamin mix, and 0.125 mL trace elements.
3. Incubated culture with vigorous shaking at 37°C until OD600 ~ 0.6.
4. Add 550 mg of ^{2}H-only-Celtone.
5. Continue incubation of culture with vigorous shaking at 37°C until OD600 ~ 0.7 (~20 min).
6. Add a fivefold excess (with respect to the content of the ^{2}H-only-Celtone) of the ^{15}N-labeled amino acid of interest. For example, to incorporate ^{15}N-labeled lysine (Fig. 3b), we add 250 mg α-^{15}N-Lys; to incorporate ^{15}N-labeled valine (Fig. 3c), we add 100 mg ^{15}N-Val.
7. Continue incubation of culture with vigorous shaking at 37°C until OD600 ~ 0.8.
8. Induce protein expression by adding IPTG (0.4 mM final concentration in culture).
9. Continue incubation of culture with vigorous shaking at 37°C for 3 h. Pellet cells by centrifugation at 5,000 × *g* for 15 min at RT.
10. Store cell pellet at −80°C.

3.3.3 Expression of Deuterated, ^{15}N-Lys Labeled RCK WT

1. Resuspend pelleted BL21(DE3) transformants from overnight starter culture (Subheading 3.2) in 6 mL of LB containing 6 μL of 1,000× ampicillin.
2. Use 3 mL of this suspension to inoculate 250 mL LB medium containing 250 μL of 1,000× ampicillin.
3. Incubate cells with vigorous shaking at 37°C until OD600 ~ 1.
4. Pellet cells at 1,500 × *g* for 15 min at RT.
5. Wash pellet twice with 30 mL of Spectra 9 medium.
6. Resuspend the pellet in 250 mL Spectra 9 medium containing 250 μL of 1,000× ampicillin.
7. Incubate cells with vigorous shaking at 37°C until OD600 ~ 0.7.
8. Add 135 mg of $^{15}N,^{2}H$-Lys.
9. Incubate cells with vigorous shaking at 37°C for 30 min.

10. Induce protein expression by adding IPTG (0.8 mM final concentration in culture).
11. Incubate cells with vigorous shaking at 37°C for 8 h.
12. Harvest cells after 8 h of expression by centrifuging at 5,000 ×*g* for 15 min at RT.
13. Store cell pellet at −80°C.

3.3.4 Expression of Deuterated, ^{15}N-Leu Labeled RCK(ΔNAI) in CT19 Cells (See Note 3)

1. Resuspend pelleted CT19 transformants from overnight starter culture (Subheading 3.2) in 6 mL of LB medium containing 6 µL 1,000× ampicillin and 30 µL 200× kanamycin.
2. Use this suspension to inoculate 1 L LB medium containing 1 mL 1,000× ampicillin and 5 mL 200× kanamycin.
3. Incubate cells with vigorous shaking at 37°C until OD600 ~ 0.7.
4. Pellet cells at 1,500 ×*g* for 10 min at RT.
5. Resuspend the pellet into 250 mL low glucose M9 medium containing 250 µL 1,000× ampicillin, 1.25 mL 200× kanamycin, 0.5 mL vitamin mix, 0.125 mL trace elements, 1 g ^{2}H-only-Celtone, and 100 mg ^{15}N-Leu.
6. Incubate cells with vigorous shaking at 37°C until OD600 reaches ~0.8.
7. Induce protein expression by adding IPTG (to 0.8 mM final concentration in the culture).
8. Incubate cells with vigorous shaking at 37°C overnight; pellet cells by centrifuging at 5,000 ×*g* for 15 min at RT.
9. Store cell pellet at −80°C.

3.4 Protein Purification (See Note 4)

1. Resuspend bacterial pellet on ice in 50 mL sonication buffer. Transfer cell suspension to stainless steel beaker. Add protease inhibitor cocktail to the cell suspension; incubate on ice.
2. Prepare PMSF solution (0.035 g in 1 mL acetone).
3. Sonicate using large-probe sonicator, starting at 50% power (setting of 5.0) for 1 min, on ice. Add 100 µL PMSF solution, and allow to cool for 1 min.
4. Repeat sonication for 1 min at 5.0 two times with 1 min of cooling between sonication rounds; increase power by 10% (1 step) in each subsequent sonication round, alternating between 1 min sonication and 1 min cooling, until setting of 9.0 is reached.
5. Pellet insoluble matter from lysate for 45 min at 30,500 ×*g* at 4°C.
6. Transfer the cleared lysate (supernatant) to a clean beaker and add imidazole to 20 mM (final concentration). Adjust the pH to 8.0, if necessary.

7. Load the cleared lysate to a $CoCl_2$-activated and pre-equilibrated HiTrap chelating column (GE Healthcare) according to the manufacturer instructions.
8. Wash the column with protein buffer B containing 20 mM imidazole until the absorbance returns to baseline.
9. Elute the bound protein with 5 column volumes of elution buffer.
10. Directly after elution, add 2 U of thrombin per 1 mg of protein, for enzymatic cleavage of the hexahistidine tag. Incubate with gentle shaking at RT for 2 h.
11. During thrombin digestion, equilibrate Superdex-200 10/300 size exclusion column (GE Healthcare) with 2 column volumes of degassed protein buffer B, at a flow rate of 1 mL/min (see Note 5).
12. Following thrombin digest, filter protein solution using 0.22 μm centrifuge tube filter for 1 min in refrigerated desktop centrifuge at 18,000 × *g* at 4°C.
13. Load protein onto column via 1 mL sample loop and elute with protein buffer B, at a flow rate of 0.5 mL/min. Collect appropriate fractions according to elution peak (see Fig. 1).
14. Measure protein concentration using Bradford or similar assay.
15. Concentrate samples to 10 mg/mL using centrifugal concentrator unit, 10 kDa molecular weight cutoff.
16. Aliquot protein in microcentrifuge tubes and freeze in liquid nitrogen.

3.5 Protein Preparation for NMR Experiments

1. Thaw protein on the day before NMR experiment.
2. Dialyze protein overnight into NMR buffer using mini-dialysis cassette.
3. Measure protein concentration using Bradford or other assay.
4. Concentrate protein to 15 mg/mL using centrifugal concentrator unit, 10 kDa molecular weight cutoff.
5. To 250 μL protein sample, add 25 μL D_2O. The sample is now prepared for transfer to an NMR sample tube and subsequent acquisition of ^{15}N-HMQC or other spectra.

4 Notes

1. We do not address NMR data collection or analysis in this chapter; however, there are many excellent articles that describe pulse protocols and data acquisition strategies for high molecular weight proteins (26–28).

2. For optimal high-yield expression in minimal media, always use freshly transformed competent cells. The expression constructs used in these experiments have been codon optimized for expression in *E. coli*; in our case, codon-optimization resulted in a large improvement in protein recovery using a small culture volume, which reduces the expense in consumption of isotope-labeled media, and also enables high protein yield with short induction times, which minimizes scrambling of the ^{15}N isotope label introduced by a specific amino acid through metabolic transamination ("scrambling"). We have been able to typically recover 50 mg of purified protein per liter of cell culture. The protein expression methods were modified after (29).
3. CT19 is an *E. coli* strain that contains mutations that reduce the function of aminotransferase enzymes as follows: aspC, aspartate aminotransferase; ilvE, branched chain amino acid aminotransferase; trpB, tryptophan synthase; and tyrB, aromatic amino acid aminotransferase. Use of this strain in expression of specific ^{15}N-labeled amino acids thus greatly reduces "scrambling," or transfer of the ^{15}N label among amino acids through bacterial aminotransferase-catalyzed reactions. In our hands, this cell strain did not express high amounts of labeled protein, unless grown as described below.
4. It is advisable to subsequently analyze each purification step by Coomassie-stained SDS-PAGE.
5. Gel-filtration chromatography is performed using a fast protein liquid chromatography apparatus, with elution monitored by UV detection. This chromatography procedure is facilitated through use of automated fraction collection.

Acknowledgements

We wish to thank Andrew Hinck for technical advice and helpful discussions in the early stages of this work. This research was supported by NIH grant R01 GM68523 to B.S.R.

References

1. McCoy JG, Nimigean CM (2012) Structural correlates of selectivity and inactivation in potassium channels. Biochim Biophys Acta 1818:272–285
2. Chung S-H, Andersen OS, Krishnamurthy V (2007) Biological membrane ion channels: dynamics, structure, and applications. Springer, New York
3. Kew JNC, Davies CH (2009) Ion channels: from structure to function, 2nd edn. Oxford University Press, Oxford
4. Imai S, Osawa M, Takeuchi K, Shimada I (2010) Structural basis underlying the dual gate properties of KcsA. Proc Natl Acad Sci U S A 107:6216–6221
5. Maltsev AS, Oswald RE (2010) Hydrophobic side chain dynamics of a glutamate receptor ligand binding domain. J Biol Chem 285: 10154–10162
6. Fenwick MK, Oswald RE (2008) NMR spectroscopy of the ligand-binding core of ionotropic glutamate receptor 2 bound to

5-substituted willardiine partial agonists. J Mol Biol 378:673–685
7. Ma D, Tillman TS, Tang P, Meirovitch E, Eckenhoff R, Carnini A, Xu Y (2008) NMR studies of a channel protein without membranes: structure and dynamics of water-solubilized KcsA. Proc Natl Acad Sci U S A 105:16537–16542
8. Maltsev AS, Ahmed AH, Fenwick MK, Jane DE, Oswald RE (2008) Mechanism of partial agonism at the GluR2 AMPA receptor: Measurements of lobe orientation in solution. Biochemistry 47:10600–10610
9. Kay LE, Gardner KH (1997) Solution NMR spectroscopy beyond 25 kDa. Curr Opin Struct Biol 7:722–731
10. Tugarinov V, Kanelis V, Kay LE (2006) Isotope labeling strategies for the study of high-molecular-weight proteins by solution NMR spectroscopy. Nat Protoc 1:749–754
11. Tugarinov V, Kay LE (2003) Ile, Leu, and Val methyl assignments of the 723-residue malate synthase G using a new labeling strategy and novel NMR methods. J Am Chem Soc 125: 13868–13878
12. Tugarinov V, Kay LE (2004) An isotope labeling strategy for methyl TROSY spectroscopy. J Biomol NMR 28:165–172
13. Tugarinov V, Ollerenshaw JE, Kay LE (2005) Probing side-chain dynamics in high molecular weight proteins by deuterium NMR spin relaxation: an application to an 82-kDa enzyme. J Am Chem Soc 127:8214–8225
14. Jiang Y, Pico A, Cadene M, Chait BT, MacKinnon R (2001) Structure of the RCK domain from the E. coli K^+ channel and demonstration of its presence in the human BK channel. Neuron 29:593–601
15. Dong J, Shi N, Berke I, Chen L, Jiang Y (2005) Structures of the MthK RCK domain and the effect of Ca^{2+} on gating ring stability. J Biol Chem 280:41716–41724
16. Roosild TP, Castronovo S, Miller S, Li C, Rasmussen T, Bartlett W, Gunasekera B, Choe S, Booth IR (2009) KTN (RCK) domains regulate K^+ channels and transporters by controlling the dimer-hinge conformation. Structure 17:893–903
17. Dvir H, Valera E, Choe S (2010) Structure of the MthK RCK in complex with cadmium. J Struct Biol 171:231–237
18. Yuan P, Leonetti MD, Pico AR, Hsiung Y, MacKinnon R (2010) Structure of the human BK channel Ca^{2+}-activation apparatus at 3.0 A resolution. Science 329:182–186
19. Jiang Y, Lee A, Chen J, Cadene M, Chait BT, MacKinnon R (2002) Crystal structure and mechanism of a calcium-gated potassium channel. Nature 417:515–522
20. Albright RA, Ibar JL, Kim CU, Gruner SM, Morais-Cabral JH (2006) The RCK domain of the KtrAB K^+ transporter: multiple conformations of an octameric ring. Cell 126:1147–1159
21. Ye S, Li Y, Chen L, Jiang Y (2006) Crystal structures of a ligand-free MthK gating ring: insights into the ligand gating mechanism of K^+ channels. Cell 126:1161–1173
22. Parfenova LV, Abarca-Heidemann K, Crane BM, Rothberg BS (2007) Molecular architecture and divalent cation activation of TvoK, a prokaryotic potassium channel. J Biol Chem 282:24302–24309
23. Wu Y, Yang Y, Ye S, Jiang Y (2010) Structure of the gating ring from the human large-conductance Ca^{2+}-gated K^+ channel. Nature 466: 393–397
24. Pau VP, Smith FJ, Taylor AB, Parfenova LV, Samakai E, Callaghan MM, Abarca-Heidemann K, Hart PJ, Rothberg BS (2011) Structure and function of multiple Ca^{2+}-binding sites in a K^+ channel regulator of K^+ conductance (RCK) domain. Proc Natl Acad Sci U S A 108: 17684–17689
25. Fiaux J, Bertelsen EB, Horwich AL, Wuthrich K (2002) NMR analysis of a 900 K GroEL GroES complex. Nature 418:207–211
26. Mantylahti S, Koskela O, Jiang P, Permi P (2010) MQ-HNCO-TROSY for the measurement of scalar and residual dipolar couplings in larger proteins: application to a 557-residue IgFLNa16-21. J Biomol NMR 47:183–194
27. Guo C, Zhang D, Tugarinov V (2008) An NMR experiment for simultaneous TROSY-based detection of amide and methyl groups in large proteins. J Am Chem Soc 130: 10872–10873
28. Korzhnev DM, Kloiber K, Kanelis V, Tugarinov V, Kay LE (2004) Probing slow dynamics in high molecular weight proteins by methyl-TROSY NMR spectroscopy: application to a 723-residue enzyme. J Am Chem Soc 126: 3964–3973
29. Fiaux J, Bertelsen EB, Horwich AL, Wuthrich K (2004) Uniform and residue-specific ^{15}N-labeling of proteins on a highly deuterated background. J Biomol NMR 29:289–297

Part V

Studying Ion Channels in Native Tissues

Chapter 24

Recording Dendritic Ion Channel Properties and Function from Cortical Neurons

Mala M. Shah

Abstract

Dendrites emerging from the cell bodies of neurons receive the majority of synaptic inputs. They possess a plethora of ion channels that are essential for the processing of these synaptic signals. To fully understand how dendritic ion channels influence neuronal information processing, various patch-clamp techniques that allow electrophysiological recordings to be made directly from dendrites have been developed. In this chapter, I describe one such method that is suitable for making electrophysiological recordings from the apical dendrites of hippocampal and cortical pyramidal neurons.

Key words Dendrites, Electrophysiology, Ion channels, Cortex

1 Introduction

Dendrites are extensive processes emerging from the cell body of neurons. They cover a vast surface area and receive most of the synaptic inputs impinging onto neurons. Dendrites also possess a large variety of ligand-gated and voltage-gated ion channels (1, 2). Often, the density and characteristics of these dendritic ion channels differ from those present at the soma. These ion channels play an important role in determining the shape and integration of synaptic inputs (1–3). To fully understand synaptic signal processing, it is critical and essential to evaluate their properties and function.

There are a number of methods including immunohistochemistry and imaging that would provide valuable information on the location and potential function of dendritic ion channels. Perhaps the most direct method, though, for assessing dendritic ion channel biophysical properties and their effects on dendritic activity is electrophysiological recording (4). The advent of the patch-clamp technique together with advances in microscopy has made it possible to make electrophysiological recordings under visual guidance from the dendrites of many neurons such as hippocampal and

Nikita Gamper (ed.), *Ion Channels: Methods and Protocols*, Methods in Molecular Biology, vol. 998,
DOI 10.1007/978-1-62703-351-0_24, © Springer Science+Business Media, LLC 2013

cortical pyramidal cells in the acute brain slice preparation. The most common approach, which I will describe in this chapter, involves the use of infrared differential interference contrast (IR-DIC) optics. This generally allows dendrites that are greater than 1 μm in diameter to be patched. Newer techniques involving the use of confocal and two-photon laser microscopy to patch smaller diameter dendrites such as basal dendrites are also being developed (5). The ability to patch dendrites has resulted in considerable new and interesting information about the contribution of dendritic ion channels to neuronal cell excitability as well as neural network activity.

2 Materials

Prepare all solutions using deionized 18 MΩ water at room temperature and store at 4°C (unless otherwise indicated). All waste materials should be disposed of according to local institutional waste disposal regulations.

2.1 Solutions

1. Cutting solution: 110 mM choline chloride, 2.5 mM KCl, 1.25 mM NaH_2PO_4, 25 mM $NaHCO_3$, 0.5 mM $CaCl_2$, 7 mM $MgCl_2$, and 10 mM dextrose; bubbled with 95%O_2/5% CO_2 to obtain pH of 7.2; ~290 mOsm (see Note 1).
2. External solution: 125 mM NaCl, 2.5 mM KCl, 1.25 mM NaH_2PO_4, 25 mM $NaHCO_3$, 2 mM $CaCl_2$, 2 mM $MgCl_2$, 10 mM dextrose; bubbled with 95% O_2/5% CO_2 to obtain pH 7.2; ~300 mOsm (see Note 2).
3. Standard internal recording solution for whole-cell recordings: 120 mM $KMeSO_4$, 20 mM KCl, 10 mM HEPES, 2 mM $MgCl_2$, 0.2 mM EGTA, 4 mM Na_2-ATP, 0.3 mM Tris-GTP, 14 Tris-phosphocreatine; pH was adjusted to 7.3 with KOH; ~310 mOsm (see Note 3).
4. *Cell-attached recording solution (mM)*: 140 KCl, 10 HEPES, 1 $MgCl_2$, 2 $CaCl_2$; pH adjusted to 7.3 (see Note 4).

2.2 Slice Preparation

1. Surgical instruments.
2. Beakers and other glassware.
3. Vibratome slicer.
4. Water bath.
5. Carbogen.

2.3 Patch-Clamp Recording

1. Pipette puller.
2. Borosilicate glass capillaries.

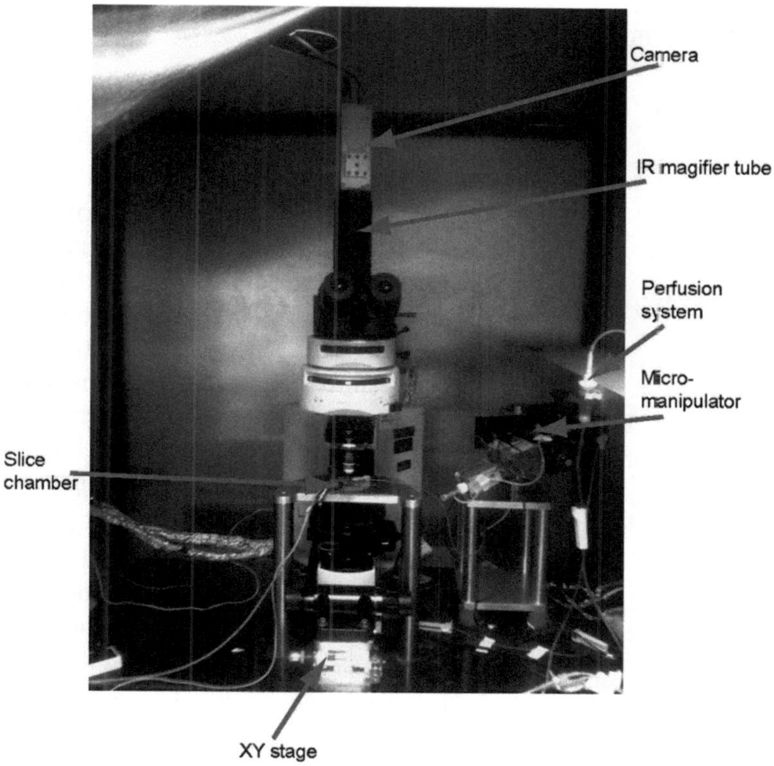

Fig. 1 An image of a typical electrophysiology setup that is used for making patch-clamp recordings from dendrites

3. Patch-clamp rig. To view slices, it is preferable to use an upright microscope equipped with differential interference contrast (DIC) and high-magnification objective lens such as 60× objective. The upright microscope is typically placed on a vibration-isolation table and surrounded by a faraday cage. The slices are placed in a slice chamber, which is positioned between the objective and the condenser of the microscope. The bottom of the slice chamber is often a glass coverslip. Because dendrites are very thin, it is often necessary to use further magnifying lens such as a 4× magnification infrared tube. The microscope is coupled to a camera suitable for detecting infrared images, which is connected to a video monitor (either a black and white analog or a digital high-resolution screen). The microscope sits on an XY stage that allows it to be moved independently of the patch-pipettes or the slice chamber (see Fig. 1 for an image of a typical setup; refer also to the Chapter 7 of this book).

3 Methods

3.1 Slicing Procedure

1. Make and freeze ~400 mL of cutting solution until ~45% of it is ice. Depending on the freezer, this can take between 45 min and 1 h.
2. In the meantime, lay out the tools for surgery, make the external solution (usually 1 L), set up the slicer [including cleaning the blade (e.g., Gillette blade) used for cutting slices with acetone and/or ethanol], and prepare an incubation chamber for storing slices. This usually consists of a beaker containing either an interface chamber or a submerged chamber with cotton mesh on it in which the slices rest. Add external solution to the chamber and bubble it with 95% O_2/5% CO_2. The solution in this chamber is pre-warmed to 37°C in a water bath prior to slices being added.
3. When the cutting solution is adequately frozen, take it out of the freezer and use a handheld blender or equivalent to make a slushy icy solution. This solution is bubbled with 95% O_2/5% CO_2 and kept on ice for the remainder of the procedure.
4. Terminally anesthetize the animal in accordance with local ethical guidelines (e.g., with an injectable anesthetic such as ketamine/xylazine solution).
5. When the animal is fully anesthetized (pinching the paw produces no reflexes), cut open the chest cavity and intracardially perfuse the slushy cutting solution until the liver turns pale. Decapitate the head and speedily remove the brain into the ice-cold cutting solution. Appropriate regulations for animal experimentation must be followed.
6. Place the brain on a suitable cold surface (block of cold metal or petri dish with a sylgard bottom) and submerge it in ice-cold cutting solution. For hippocampal slices from which recordings from CA1 pyramidal dendrites can be made, first hemi-sect the brain. Each half is then placed on the cut (medial surface) and a cut at approximately 30°C to the base of the half-brain is made on the dorsal surface. If the experiment involves patching entorhinal cortex pyramidal cell dendrites, then a few millimeters parallel to the ventral surface is cut from the dorsal side (see Note 5).
7. This cut surface is then glued onto the slicing stage of a vibratome using a thing film of cyanoacrylate glue or equivalent. The slicing stage can be cooled prior, though this is not necessary.
8. Transfer the slicing stage into the slicing chamber of the vibratome. Submerge the blocks of tissue with ice-cold cutting solution. The solution in the chamber is also bubbled with 95% O_2/5% CO_2.
9. Lower the blade for cutting slices to the appropriate position. Cut thin tissue slices (200–400 μm) by advancing the blade at

an appropriate speed, vibration amplitude, and frequency such that the tissue is not compressed or displaced (see Note 6).

10. Place the slices in the incubation chamber containing pre-warmed external solution and bubble continuously with 95% O_2/5% CO_2. After a period of 30 min to 1 h the slices can be used for electrophysiological recordings.

3.2 Dendritic Electrophysiological Recordings

1. Pull patch pipettes using thick-walled borosilicate glass with a suitable patch pipette puller.
2. Add the slice to the slice chamber being perfused at a constant rate (1–2 mL/min) with external solution bubbled with 95%O_2/5% CO_2. The slice should be orientated such that the dendrites to be patched are perpendicular to the patch pipette (Fig. 2).
3. The slices are held down in the slice chamber using a "harp." This is a horseshoe-shaped flat metal (either silver or platinum) onto which thin threads are glued.

Fig. 2 A differential interference contrast (DIC) image of hippocampal CA1 pyramidal cell somata and dendrites. Note that the patch pipettes are placed perpendicular to the orientation of the neuron. The vertical scale bar represents 20 μm

4. Once the slice is in the chamber and the external solution is flowing at a constant rate, lower the objective into the fluid and focus on the surface of the slice. It might be necessary to optimize the optics at this stage to obtain a crisp image (for details on adjusting DIC optics, see ref. 4).
5. Identify somata and then attempt to follow the dendrites (Fig. 2; see Note 7).
6. Fill a patch-pipette with internal solution, which has been filtered to remove any debris. Insert the pipette into a holder securely attached to a micromanipulator.
7. Apply positive pressure to the pipette via tubing connected to the pipette holder and use a three-way tap or a switchable valve to hold the pressure in the pipette.
8. Lower the pipette to the identified dendrite. The pressure added to the pipette will help clear the slice debris. A healthy dendrite will not sway if the pipette is brought next to it.
9. Zero any offset caused between the pipette and the ground wire (earth) using the patch-clamp amplifier.
10. Gently lower the pipette onto the top of the dendrite. If the dendrite is healthy, a dimple will form. Release the pressure and obtain a giga-ohm seal (see Note 8).
11. For cell-attached recordings, no further steps are required. If whole-cell electrophysiological recordings are to be made, then hyperpolarizing voltage or current should be applied, before applying further gentle suction to break through the membrane. A healthy CA1 hippocampal or entorhinal cortical pyramidal cell dendrite typically has resting membrane potentials between ~–65 and –70 mV.

4 Notes

1. Additional types of solutions such as those containing sucrose instead of choline chloride can also be utilized (4).
2. If appropriate for the experiment, this solution can be further supplemented by other inhibitors of ligand-gated and voltage-gated ion channels.
3. Ten times concentrated stocks of Na_2-ATP, Tris-GTP, and Tris-phosphocreatine can be preprepared and stored at –20°C. If this option is chosen, then the internal solution without these should be made in 90% of the volume.
4. Depending on the ion channel current that is to be isolated, this basic solution is then supplemented with other inhibitors. To maintain the osmolarity of the internal solution to ~310 mOsm, the concentration of KCl can be reduced.

5. The angle at which the brain is cut varies depending on the region of interest. For more details on preparing slices in which dendrites of other regions of the brain are preserved, see ref. 4.
6. It might also be necessary to remove tough tissue such as white matter to ensure that slices are smoothly cut. In between slices, raise the blade by 50–100 μm above the surface of the tissue before moving back to the start position.
7. Avoid sharp contrast dendrites, as these are often sick. Healthy dendrites are very flat and sometimes difficult to visualize. To aid with visualization of the dendrites, it is sometimes helpful to first patch the soma using internal solution containing a fluorescent dye (e.g., Alexa Fluor 594) and fill the cell with the dye. Dendrites can then be easily viewed using a fluorescent microscope.
8. To obtain a giga-Ohm seal, it might be necessary to apply some gentle suction pressure. Too much pressure can damage the dendrite.

References

1. Nusser Z (2009) Variability in the subcellular distribution of ion channels increases neuronal diversity. Trends Neurosci 32:267–274
2. Shah MM, Hammond RS, Hoffman DA (2010) Dendritic ion channel trafficking and plasticity. Trends Neurosci 33:307–316
3. Sjostrom PJ, Rancz EA, Roth A, Hausser M (2008) Dendritic excitability and synaptic plasticity. Physiol Rev 88:769–840
4. Davie JT, Kole MH, Letzkus JJ, Rancz EA, Spruston N, Stuart GJ, Häusser M (2006) Dendritic patch-clamp recording. Nat Protoc 1:1235–1247
5. Nevian T, Larkum ME, Polsky A, Schiller J (2007) Properties of basal dendrites of layer 5 pyramidal neurons: a direct patch-clamp recording study. Nat Neurosci 10:206–214

Chapter 25

M-Current Recording from Acute DRG Slices

Kirstin E. Rose, Sylvain Gigout, and Nikita Gamper

Abstract

Electrophysiological recordings from an acutely sliced preparation provide information on ionic currents and excitability of native neurons under near physiological conditions. Although this technique is commonly used on central nervous system structures such as spinal cord and brain, structures within the peripheral nervous system (including sensory ganglia and fibers) have proven to be much more difficult to study in acute preparations. Here we describe a method for patch-clamp recordings from rat dorsal root ganglion (DRG) slices.

Key words Dorsal root ganglion, Patch-clamp recording, Slice preparation, Ion channel, M-type K^+ channel

1 Introduction

Peripheral somatosensory neurons sense and transmit all types of tactile, temperature, and chemical information from the environment and viscera to the CNS. A subset of these neurons (known as "nociceptive" neurons) specifically respond to tissue damage and can mediate sensation of pain. Therefore, understanding the mechanisms controlling the excitability of somatosensory neurons holds the key to the understanding of somatic sensation and pain. Extensive research worldwide uses electrophysiological recordings from these sensory neurons in order to study their excitability; however, a large share of such recordings is routinely carried out using dissociated and cultured sensory neuron preparations. This is in contrast to the research in the CNS where the "gold standard" is a recording from the acute slices of a particular brain or spinal cord region. Several anatomical features of the sensory ganglia make slice recording particularly difficult including the small size of the ganglia and "wrapping" of individual neuronal cell bodies by a satellite glia cell sheath. However, cultured neurons are subject to various short- and long-term changes (such as axotomy, enzymatic treatment,

Nikita Gamper (ed.), *Ion Channels: Methods and Protocols*, Methods in Molecular Biology, vol. 998,
DOI 10.1007/978-1-62703-351-0_25, © Springer Science+Business Media, LLC 2013

mechanical stress, artificial environment) which can trigger differentiation of neurons from their native state and thus compromise the results of the investigation. In this chapter we describe the DRG slice preparation for electrophysiological recordings with an aim to allow recordings from an acute, enzyme-free preparation.

Here we have adopted the technique developed in previous studies (1) to record M-type K^+ current from the acutely sliced DRG preparation using the perforated patch-clamp approach. M current is a slow voltage-gated K^+ current with a negative activation threshold (negative to-60 mV) which is increasingly recognized as one of the major regulators of nociceptive neuron resting membrane potential and excitability (2–9). M current is conducted by the members of Kv7 family of K^+ channel proteins with five known subunits (Kv7.1–7.5). The Kv7.2 M channel subunit has recently been shown to be present at cell bodies of nociceptive DRG neurons (8, 9), nodes of Ranvier of myelinated sensory fibers (10, 11), as well as in unnmyelinated fibers and free nociceptive nerve endings in the skin (7); there is also evidence for the expression of other M channel subunits (Kv7.3 and Kv7.5) throughout the peripheral nociceptive pathways (8–10). Additionally, there is accumulating evidence for the presence of functional M channels in nociceptive fibers and nerve endings (5, 7, 12). However most patch-clamp recordings of M channels in sensory neurons were performed using DRG cell cultures (2–4, 8). Therefore, in order to investigate the presence of functional M channels in acute sensory neurons in their native environment, we have utilized the perforated patch-clamp technique to record from an acutely sliced DRG preparation. The perforated patch-clamp technique is a further step to preserve native tissue properties as it prevents the dialysis of the intracellular milieu (see Chapter 11 of this book); this is particularly relevant for the M-current recording as M channel activity can run down very rapidly in conventional whole-cell patch-clamp (13).

2 Materials

1. Extracellular solution (ES) 124 mM NaCl, 26 mM $NaHCO_3$, 10 mM Glucose, 3 mM KCl, 2 mM $MgSO_4$, 2.5 mM NaH_2PO_4, 2 mM $CaCl_2$. ES is continuously bubbled with carbogen gas (95% O_2, 5% CO_2) during experiments.
2. Agar: 2% w/v in ES prepared fresh every day, usually 0.5 g in 25 mL.
3. Hanks' balanced salt solution (HBSS).
4. Intracellular solution (IS): 140 mM KCl, 1 mM $MgCl_2$, 10 mM HEPES, 10 mM EGTA, 1 mM $CaCl_2$, 1 mM ATP, 0.1 mM GTP, pH 7.3.

5. Amphotericin B dissolved in dimethyl sulfoxide (DMSO) to obtain a stock solution 40–60 mg/mL that could be stored for ~7 days at 4°C. Every day the working solution is prepared fresh by diluting 50–100 μL of stock solution in 900–950 μL of IS, to obtain a maximum final amphotericin B concentration of 400 μg/mL (for detailed advice on the preparation of the amphotericin B solutions refer to Chapter 11 of this book).
6. (Optional) Lucifer yellow (Sigma): 0.5 mg Lucifer yellow can be added to 1 mL of amphotericin B containing IS just prior to pipette filling.
7. Pipettes manufactured from filamented borosilicate glass capillaries (GC15OF-10, Harvard apparatus, Kent) by using a conventional microelectrode puller (e.g., DMZ-Universal Puller, Zeitz Instrumente). The optimal resistance of the pipette (when filled with the IS) is 3–5 MΩ.
8. Binocular stereo dissecting microscope.
9. Dissection tools: scissors, fine scissors, watchmakers forceps with very fine tips, scalpel, and Pasteur pipettes.
10. Vibrating blade microtome (i.e., Leica VT100S or the Microm HM 650 V).
11. Cyanoacrylate glue.
12. Slice recording setup: a vibration isolation table, an upright microscope equipped with IR-DIC, a microscope stage, water-immersion objectives, a video camera, a video monitor, a patch-clamp amplifier, a recording equipment, micromanipulators, pipette holders, and a perfusion system (see Chapter 7 of this book for further advice on the patch-clamp rig assembly). An exemplary rig is depicted on Fig. 1.

3 Methods

An exemplary set of tools and major steps of the procedures outlined in Subheadings 3.1 and 3.2 are depicted in the Fig. 2.

3.1 Removal and Embedding of DRG

1. Wash all tools (Fig. 2a) and sterilize with 70% v/v ethanol/deionized water prior to tissue removal.
2. Following humane sacrifice of rat to local ethical standards, place the body on a dissection board with the dorsal surface facing uppermost. Cut away the skin overlying the lumbar region and proceed to carry out a laminectomy of the exposed lumbar region. Specifically, remove the entire posterior backbone along with overlying ligaments and muscles, which will normally include DRGs L3–L6.

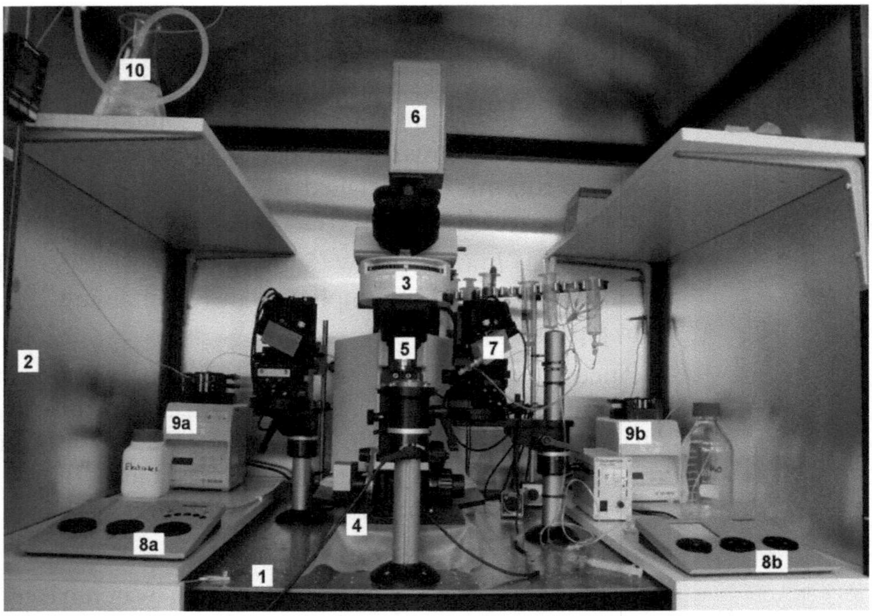

Fig. 1 Patch-clamp setup for slice recording. (**1**) Vibration isolation table; (**2**) Faraday cage; (**3**) upright microscope equipped with IR-DIC; (**4**) microscope stage; (**5**) water-immersion objectives; (**6**) video camera (in our case, connected to a monitor located in a rack beside the patch-clamp setup); (**7**) headstage and pipette holder attached to a micromanipulator; (**8**) micromanipulators remote control panels (one controlling the microscope stage (**8a**) and another one controlling the pipette (**8b**)), mounted on a bench separated from the vibration isolation table; (**9**) 2 peristaltic pumps [one controlling inflow (**9a**) and the second the outflow (**9b**)]; (**10**) erlenmeyer containing carbogen-bubbled ES

3. Pin the excised section onto a dissection board with the anterior surface facing uppermost. Make cuts through the lamina body of the spinal vertebrae, starting at the anterior end, until the left and right sides of the column are separated. Using fine forceps remove the spinal cord and meninges to expose the DRGs under the binocular stereo dissecting microscope. They should look like little pearls at regular intervals along the column (Fig. 2b).
4. Use fine forceps to carefully remove the L4 and L5 DRGs (see Note 1). Place the DRG into HBSS buffer on ice (see Note 2). While in buffer on ice, remove any attached nerves and roots using fine scissors (see Note 3).
5. Prepare a 2% w/v agar solution by diluting for example 0.5 g agar in 25 mL ES. Heat the solution using a microwave owen and stop the heating just before the solution boils. Pour liquid 2% w/v agar into a small weighing boat and carefully place DRG into agar using fine forceps (see Note 4). Place weighing boat containing embedded DRGs on ice for 5–10 min to allow agar to set.

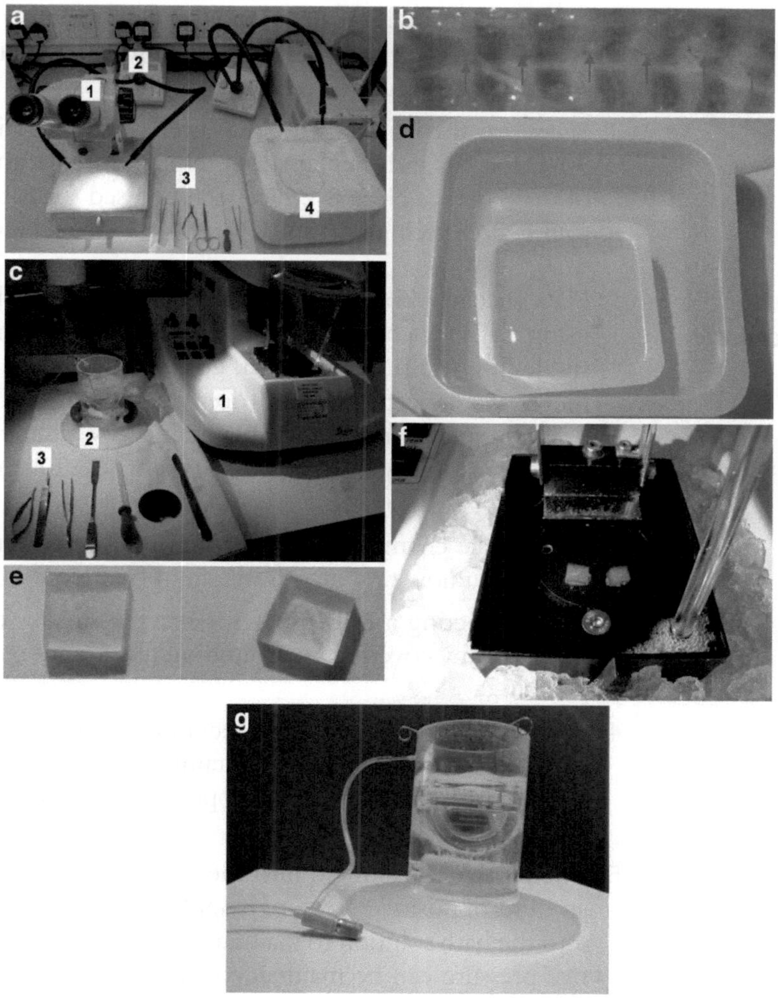

Fig. 2 Major steps of the procedure to prepare DRG slices. (**a**) General view of the setup for removal of DRGs. (**1**) Binocular stereo dissecting microscope; (**2**) light source; (**3**) dissecting tools (from left to right: 2 watch-maker forceps with very fine tips, fine scissors, scissors, Pasteur pipette, flat tip tweezers); (**4**) 2 ES-containing Petri dishes (one large to keep the spine and one small to keep the DRGs) placed on ice. (**b**) View of DRGs (indicated by the *red arrows*) still within the cut open vertebral column (with spinal cord removed) observed with binocular stereo dissecting microscope. (**c**) General view of the slicing setup. (**1**) Vibrating blade micro-tome; (**2**) brain slice keeper; (**3**) tools (from *left* to *right*: fine scissors, watchmaker forceps with very fine tips, flat tip tweezers, spatula to prepare the agar solution, Pasteur pipette to transfer the slices to the slice keeper, specimen disc, scalpel). (**d**) DRGs are embedded in cooled agar in a weighing boat placed on cold ES. (**e**) Blocks of agar (each containing one DRG) dried by placing them on filter paper. (**f**) Each block of agar is glued on the specimen disc, placed in the buffer tray (containing cold carbogen-bubbled ES) of the vibratome and cut into slices. (**g**) Lateral view of a slice keeper containing ES continuously "bubbled" with carbogen. The flow rate of the gas is adjusted with the needle valve

3.2 Slicing and Slice Incubation

1. Using a scalpel cut cubes of agar containing one DRG and glue onto the specimen disc (see Note 5).
2. Slice cube containing DRG on a vibroslicer (i.e., vibrating blade microtome Leica VT100S) in ES solution on ice. Set the

vibratome to a high level of vibration and low speed level (see Note 6). We slice newborn rat DRG at 190 μm and adult rat DRG at 250 μm (see Note 7).

3. Slices are then placed into ES after slicing and incubated at 30°C for 1 h and then at room temperature for the remainder of the day (see Note 8). Slices can be used for patching for up to 8 h after slicing (that is, GΩ seals are still achievable).

3.3 Perforated Patch-Clamp of Acute DRG Slice

1. Place slice in recording chamber (see Note 9) and hold in place using a slice anchor usually made from a high-density material stable in water (i.e., platinum, gold, or tungsten). Our experiments were performed at room temperature. Optionally, a desired temperature can be maintained by a temperature controller heating solution prior its entry into the recording chamber.
2. Perfuse the chamber with ES continuously saturated with carbogen at a flow rate of 3–5 mL/min (see Note 10).
3. When selecting a cell to patch generally smaller cells that did not have any covering/surrounding glia are more amenable to patching.
4. For perforated patch-clamp technique pipettes are backfilled with amphotericin solution containing Lucifer yellow (optional; see Note 11) and then tip is dipped in amphotericin-free IS for 30 s to 1 min.
5. The pipette is mounted into the headstage amplifier and small amount of positive pressure is applied before the pipette enters into the bath solution to maintain a clean electrode tip; the positive pressure can be maintained with bulldog clip up until the contact with the cell surface is made (see below and Note 12). The pipette is maneuvered down through bath until it is approximately level with the cell of interest.
6. Once in close approximation to the cell, the pipette is slowly advanced onto cell membrane until pipette resistance reaches ~10 MΩ (see Note 13).
7. Positive pipette pressure is then released and negative pressure is progressively applied to pipette by mouth until GΩ seal is achieved. Negative pressure is then slowly released.
8. The pipette capacitance (C_{fast}) is then compensated using an amplifier/software. The cell membrane capacitance (C_{slow}) will start increasing with time as the amphotericin perforation developed. This can be monitored on the oscilloscope online; the perforation usually takes up to 10 min (see Chapter 11 of this book for more details).
9. Once membrane capacitance has reached a plateau it is cancelled with the use of the amplifier/software. The method normally

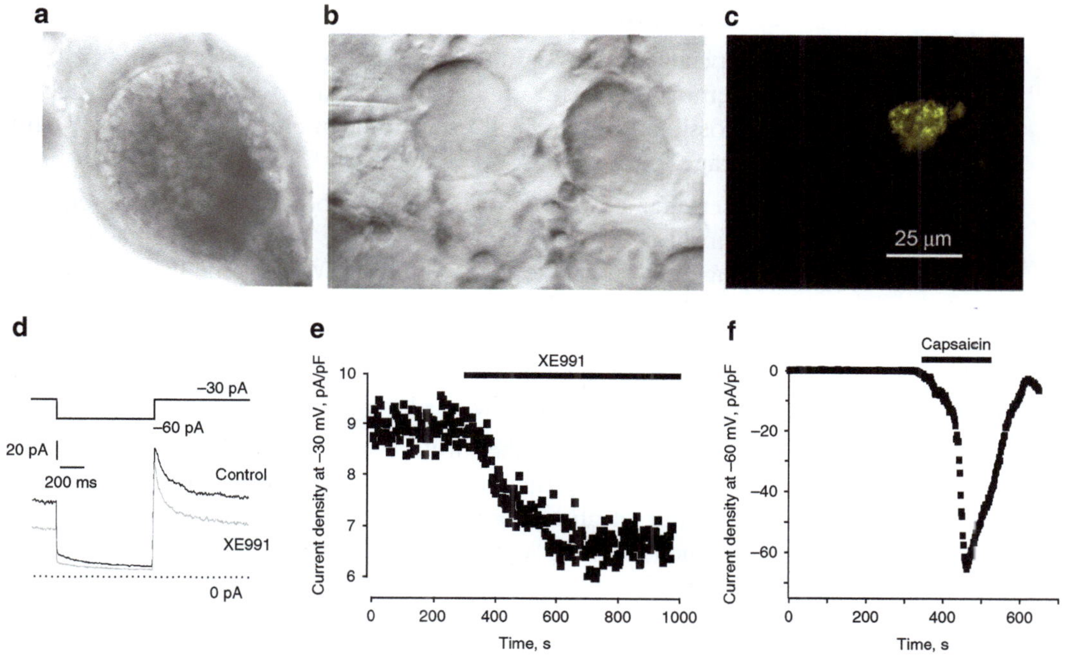

Fig. 3 Patch-clamp recording from DRG slice. (**a**) Low-magnification (×10) micrograph of the DRG. (**b**) Micrograph depicting DRG slice with a patch-clamp pipette making a GΩ seal with a DRG neuron. (**c**) Fluorescence micrograph of the DRG slice taken after the perforated patch recording was completed and the recorded cell was filled with Lucifer yellow through the patch pipette (see Note 11). (**d**) Perforated patch recording of M-like current from a small-diameter neuron in an acute DRG slice in the absence (control) or presence (XE991) of specific M channel blocker XE991 (3 μM); voltage protocol is depicted above. (**e**) Time course of M-current inhibition by 3 μM XE991 (*black bar*). (**f**) Activation of TRPV1 channel with capsaicin (1 μM); perforated patch recording, currents were elicited with the same voltage pulse shown in (**d**), plotted is a steady-state current at −60 mV. Panels (**d**) and (**e**) are modified from (9) with permission

gives an access resistance in the range of 10–35 MΩ; we aim to achieve the access resistance below 15 MΩ to avoid large voltage error (see Note 14).

10. Once perforation is complete, the recording can be started. Example recordings of M-like K^+ currents and capsaicin-induced TRPV1 currents from the DRG neuron of 7-day-old rat (see Note 15) are given in Fig. 3.

4 Notes

1. Although DRGs from all spinal levels can be used, it is often beneficial to use the same identified ganglia throughout the study for consistency. In our studies we used L4 and L5 DRGs that give rise to the sciatic nerve. To distinguish L4 and L5 DRG relative to surrounding DRGs note that the L4 and L5 DRGs are considerably larger in comparison to neighboring DRGs.

2. Alternatively ganglia can be placed in ES.
3. Specifically, for adult rat DRGs, care needs to be taken to remove DRG sheath and attached axons, as close to ganglia as possible, as these will prevent clean cutting of slices at a later stage.
4. Let agar cool down slightly (to about 39°C) before embedding DRGs; the temperature of the agar can be tested by placing agar on back of hand or wrist. When embedded in hot agar the DRGs can be damaged and the resulting slices can be of poor quality. On the other hand, when pushed into colder, more solid agar, the DRGs can be damaged as they become compressed within the harder agar.
5. For best results make sure the specimen disc and agar cube are dry (by placing them on filter paper) before applying glue. The amount of glue should be moderate; when too much glue is applied, it can cover the agar block which renders the cutting more difficult.
6. DRGs are tough. Therefore we suggest adjusting the blade speed (forward movement) to an extremely low value (1–4 mm/min) to avoid pushing the tissue. The highest available frequency should be selected for the vibration of the cutting blade.
7. The first and last DRG slices should be discarded since they are generally the most damaged during cutting and it is difficult to recognize which side is covered with agar.
8. The temperature and time of incubation vary between different labs. We have found that placing slices at 37°C for 10 min also produced good quality slices. An incubation chamber (submerged or interface type) providing sufficient oxygenation of the tissue should be chosen (Fig. 2g).
9. We use a 1,000 μL perfusion chamber. A small volume of the recording chamber allows a rapid solution exchange.
10. Two peristaltic pumps could be used (Fig. 1): independent inflow and outflow facilitate the adjustment of the liquid height in the recording chamber. It is advisable to use oxygen-impermeable tubing (e.g., Teflon) for perfusion. Alternatively a gravity-fed perfusion system can also be used.
11. Lucifer yellow (Sigma, final concentration 0.5 mg/mL) can be included into the IS to check the integrity of perforated patch-clamp recordings. If a cell was labelled with Lucifer yellow dye during any time point of the experiment, it is an indication that the perforated patch-clamp configuration was lost and conventional whole cell resulted, where amphotericin gains access to intracellular contents of the cell and perforates plasma membrane throughout. In addition, Lucifer yellow can be used for

post-recording identification of the neurons using complimentary immunostaining. In such a case, a breakthrough into the whole-cell mode is required at the end of recording (Fig. 25.3d).

12. Pressure to pipette can be carried out by mouth or syringe and a bulldog clip is used to seal off tubing to maintain pressure. Optionally a manometer can be used to monitor the pressure.
13. Only cells from the top surface of the slice could be patched using the DRG slice preparation. Generally, it is not possible to patch from larger neurons on the surface as they are more tightly wrapped in glia as previously reported (14) preventing a tight contact between the tip of the electrode and the cell membrane.
14. According to Ohm's law, the larger is access resistance and the larger is the recorded current, the higher is the error in voltage applied by the amplifier. Therefore if the recorded currents are large (i.e., in nA range) and the access resistance is also large (i.e., above 10 MΩ), a series resistance compensation is necessary (this issue is further discussed in Chapters 7 and 11 of this book).
15. In older animals the neuronal surface is firmly ensheathed by satellite glial cells which preclude the experimenter from gaining access to neuronal membrane; the sheath also impairs the visualization of the neuronal membrane. We found that a reasonable success rate (at least one recording a day) can be achieved from DRG slices from rat pups up to 7 days old. We have obtained very few recordings from the adult rat DRGs (9) but the success rate of these recordings was negligible.

Acknowledgments

This work was supported by the MRC, BBSRC, and Wellcome Trust.

References

1. Safronov BV, Bischoff U, Vogel W (1996) Single voltage-gated K^+ channels and their functions in small dorsal root ganglion neurones of rat. J Physiol 493:393–408
2. Crozier RA, Ajit SK, Kaftan EJ, Pausch MH (2007) MrgD activation inhibits KCNQ/M-currents and contributes to enhanced neuronal excitability. J Neurosci 27:4492–4496
3. Linley JE, Ooi L, Pettinger L, Kirton H, Boyle JP, Peers C, Gamper N (2012) Reactive oxygen species are second messengers of neurokinin signaling in peripheral sensory neurons. Proc Natl Acad Sci USA 109:E1578–E576
4. Linley JE, Pettinger L, Huang D, Gamper N (2012) M channel enhancers and physiological M channel block. J Physiol 590:793–807
5. Linley JE, Rose K, Patil M, Robertson B, Akopian AN, Gamper N (2008) Inhibition of M current in sensory neurons by exogenous proteases: a signaling pathway mediating inflammatory nociception. J Neurosci 28:11240–11249

6. Mucha M, Ooi L, Linley JE, Mordaka P, Dalle C, Robertson B, Gamper N, Wood IC (2010) Transcriptional control of KCNQ channel genes and the regulation of neuronal excitability. J Neurosci 30:13235–13245
7. Passmore GM, Reilly JM, Thakur M, Keasberry VN, Marsh SJ, Dickenson AH, Brown DA (2012) Functional significance of M-type potassium channels in nociceptive cutaneous sensory endings. Front Mol Neurosci 5:63
8. Passmore GM, Selyanko AA, Mistry M, Al-Qatari M, Marsh SJ, Matthews EA, Dickenson AH, Brown TA, Burbidge SA, Main M, Brown DA (2003) KCNQ/M currents in sensory neurons: significance for pain therapy. J Neurosci 23:7227–7236
9. Rose K, Ooi L, Dalle C, Robertson B, Wood IC, Gamper N (2011) Transcriptional repression of the M channel subunit Kv7.2 in chronic nerve injury. Pain 152:742–754
10. King CH, Scherer SS (2012) Kv7.5 is the primary Kv7 subunit expressed in C-fibers. J Comp Neurol 520:1940–1950
11. Devaux JJ, Kleopa KA, Cooper EC, Scherer SS (2004) KCNQ2 is a nodal K^+ channel. J Neurosci 24:1236–1244
12. Roza C, Lopez-Garcia JA (2008) Retigabine, the specific KCNQ channel opener, blocks ectopic discharges in axotomized sensory fibres. Pain 138:537–545
13. Shen W, Hamilton SE, Nathanson NM, Surmeier DJ (2005) Cholinergic suppression of KCNQ channel currents enhances excitability of striatal medium spiny neurons. J Neurosci 25:7449–7458
14. Scholz A, Vogel W (2000) Tetrodotoxin-resistant action potentials in dorsal root ganglion neurons are blocked by local anesthetics. Pain 89:47–52

Chapter 26

Studying Ion Channels in Human Erythrocytes by Direct and Indirect Means

Stephan M. Huber

Abstract

Mature human or mouse erythrocytes functionally express an unexpected diversity of ion channels that endows these small enucleated cells with a toolkit for electrosignaling. Being largely dormant under resting conditions, these ion channels enable erythrocytes to quickly respond to internal or external stimuli. They are integral modules of complex programs such as oxygen-regulated ATP release or stress-induced programmed erythrocyte death. This article summarizes electrophysiologial and non-electrophysiological methods to analyze erythrocyte ion channels and provides protocols for channel activation.

Key words Gardos channels (IK, SK4, $K_{Ca}3.1$), ClC-2, Organic osmolyte and anion channels, Protein kinase A-activated anion channels, Ca^{2+}-permeable nonselective cation channels, Patch-clamp, Isosmotic hemolysis, Flow-cytometry, Tracer flux, ATP release

1 Introduction

The obvious task of erythrocytes is the transport of blood oxygen and carbon dioxide. Hemoglobin and band 3 anion exchanger are the most abundant proteins in the erythrocyte cytosol and membrane, respectively. Because of this high abundance and the substantial absence of intracellular organelles, mature human erythrocytes are commonly simplified to hemoglobin-containing sacks. In sharp contrast to this view, emerging numbers of signaling molecules such as functional surface receptors, kinases, and released factors are identified in mature erythrocyte. These signaling molecules build up functional signaling cascade participating in the cross talk between erythrocytes and other blood cells or the endothelium. As a matter of fact, erythrocytes fulfill many further functions beyond gas transport such as regulation of vascular tone (1).

Therefore, it is no wonder that mature mammalian erythrocytes express a diversity of ion channels similar to nucleated cells. In particular, functional expression of CFTR (2–4), ClC-2 (5, 6)

Nikita Gamper (ed.), *Ion Channels: Methods and Protocols*, Methods in Molecular Biology, vol. 998,
DOI 10.1007/978-1-62703-351-0_26,

TMEM16A (7), 18 pS anion channels (8), 80 pS outwardly rectifying organic osmolyte and anion channels (9, 10), peripheral-type benzodiazepine receptor/voltage-dependent anion channels (VDAC)-like maxi anion channels (11, 12), acid-sensitive anion channels (13), Gardos $K_{Ca}3.1$ K^+ channels (14), TRPC6 (15) and possibly TRPC3 nonselective cation channels (16), acetylcholine-stimulated cation channels (17), and pannexin-1 ATP-release channels (18) has been demonstrated in human (mouse) erythrocytes. In addition, expression of tetrodotoxin-sensitive Na^+ channels $Na_v1.4$ and $Na_v1.7$ and ionotropic purinergic receptors $P2X_1$, $P2X_4$, and $P2X_7$ is evident from erythrocyte ghost immunoblotting or reticulocyte mRNA analysis (19).

These ion channels are largely silent under resting conditions. The whole-cell conductance of unstimulated human erythrocytes recorded by patch-clamp fast whole-cell mode is in the range of few pS at the very most and no ion channel gating can be observed (20). The resting Cl^- conductance of the erythrocyte membrane, which is approximately 100-fold greater than the resting cation conductance, is most probably generated by a conductive operation mode of the AE1 (band 3) anion Cl^-/HCO_3^- exchanger (21).

Although being electrically very tight under resting conditions, the erythrocyte membrane can generate conductances in the nS range upon various signals. Strong activators of erythrocyte ion channel activity are bacterial toxins (7, 14, 20) and the intraerythrocytic amplification of the protozoa *Plasmodium* (1) suggesting functional significance of erythrocyte channels in the pathophysiology of septicemia and malaria. Experimentally, *Plasmodium*-induced ion channel activation can be mimicked in uninfected erythrocytes by applying oxidative stress (5, 9, 22). In addition, erythrocyte channel activation is stimulated by hyperosmotic erythrocyte shrinkage (23, 24), Cl^- depletion (22, 23, 25), ligation of receptors (26, 27), and kinases (8). This chapter provides protocols for both the activation and the study of erythrocyte ion channels for those who are interested in the physiology and pathophysiology of erythrocyte ion channels.

2 Materials

2.1 Cells

1. Human erythrocytes freshly drawn from healthy donors by vein or finger tip puncture, collected in a heparinized vial, and stored at a hematocrit of about 5% after washing trice in NaCl solution (see below) at 8°C.
2. Mouse erythrocytes drawn by retro-orbital puncture, collected, washed, and stored as described above.

2.2 External Solutions

1. NaCl solution: 125 mM NaCl, 5 mM KCl, 5 mM D-glucose, 1 mM $MgCl_2$, 1 mM $CaCl_2$, 32 mM HEPES; pH 7.4 with NaOH.
2. KCl solution: 130 mM KCl, 5 mM D-glucose, 1 mM $MgCl_2$, 1 mM $CaCl_2$, 32 mM HEPES; pH 7.4 with KOH.
3. Na-D-gluconate solution: 125 mM Na-D-gluconate, 5 mM K-D-gluconate, 5 mM D-glucose, 2 mM Ca-D-$gluconate_2$, 32 mM HEPES; pH 7.4 with NaOH.
4. NMDG-D-gluconate solution: titrate 150 mM *N*-methyl-D-glucamine (NMDG), 2 mM Ca-D-$gluconate_2$ with D-gluconic acid to pH 7.4.
5. NMDG-Cl bath solution: titrate 180 mM NMDG, 2 mM $CaCl_2$ with HCl to pH 7.4.
6. Glucose-free NaCl solution: 125 mM NaCl, 5 mM KCl, 2.5 mM $CaCl_2$, 1 mM $MgCl_2$, 32 mM HEPES; pH 7.4 with NaOH.
7. Sorbitol-containing solution: 100 mM sorbitol, 90 mM NaCl, 1 mM $CaCl_2$, 1 mM $MgCl_2$, 10 mM HEPES; pH 7.4 with NaOH.
8. Ca^{2+}-free NaCl solution: 125 mM NaCl, 5 mM KCl, 5 mM D-glucose, 1 mM $MgCl_2$, 0.5 mM EGTA, 32 mM HEPES; pH 7.4 with NaOH.
9. Elevated Ca^{2+}-containing NaCl solution: 125 mM NaCl, 5 mM KCl, 5 mM D-glucose, 2 mM $CaCl_2$, 1 mM $MgCl_2$, 32 mM HEPES; pH 7.4 with NaOH.
10. Hypotonic lysis buffer: 30 mM KCl, 1 mM $MgCl_2$, 3 mM Mg-ATP, 0.5 mM EGTA, 10 mM HEPES; pH 7.4 with KOH.
11. Isosmotic sorbitol solution: 290 mM sorbitol, 5 mM HEPES; pH 7.4 with NaOH.
12. Glucose-free KCl/NaCl solution: 80 mM KCl, 50 mM NaCl, 32 mM HEPES; pH 7.4 with KOH.
13. Hypotonic sorbitol solution: 200 mM sorbitol, 5 mM HEPES; pH 7.4 with NaOH.
14. KCl/NaCl solution: 80 mM KCl, 50 mM NaCl, 5 mM D-glucose, 0.2 mM $MgCl_2$, 32 mM HEPES; pH 7.4 with KOH.

2.3 Pipette Solutions

1. K-gluconate/KCl pipette solution: 60 mM K-D-gluconate, 80 mM KCl, 1 mM EGTA, 5 mM $MgCl_2$, 1 mM Mg-ATP, and 5 mM HEPES; pH 7.4 with KOH.
2. K-gluconate pipette solution: 140 mM K-D-gluconate, 1 mM EGTA, 5 mM $MgCl_2$, 1 mM Mg-ATP, and 5 mM HEPES; pH 7.4 with KOH.

3. KCl pipette solution: 140 mM KCl, 1 mM K-EGTA, 5 mM $MgCl_2$, 1 mM Mg-ATP, and 5 mM HEPES; pH 7.4 with KOH.

2.4 Other Solutions and Reagents

1. Poly-L-lysine (1:100 in phosphate buffered saline for 10 min).
2. Triarylmethane-34 (TRAM-34).
3. *Tert*-Butylhydroperoxide (tBHP).
4. 5-Nitro-2-(3-phenylpropylamino)-benzoic acid, NPPB.
5. Ionomycin.
6. Sodium (meta)arsenite ($NaAsO_2$).
7. Ca^{2+}-pump inhibitor Na_3VO_4.
8. Fluo-3/AM or Fluo-4/AM.
9. Amiloride.
10. Ethylisopropylamiloride, EIPA.
11. 1-Ethyl-2-benzimidazolinone, 1-EBIO.
12. Gramicidin D.
13. Valinomycin.
14. Luciferin–luciferase assay kit (Roche Diagnostics, Mannheim, Germany).
15. ELISA kit for the detection of trace amounts of free hemoglobin (i.e., E-90HM; Immunology Consultants Laboratory, Newberg, OR, USA).
16. Inosine.
17. Radioactive tracer $^{45}Ca^{2+}$ (handling requires radioisotope laboratory).
18. Trichloroacetic acid (6%).

2.5 Parasite Culture

1. Laboratory strains of *Plasmodium falciparum* such as BINH (28) kept in continuous culture (requires biosafety level 2 practice and facilities).
2. RPMI medium supplemented with 0.5% Albumax (Life Technologies, Invitrogen).
3. Gas: 90% N_2, 5% CO_2, 5% O_2.

2.6 Equipment

1. Patch-clamp rig (see Chapter 7 by Jonnathan Lipipat) with bath superfusion and 37°C heating.
2. Flow cytometer (e.g., Becton Dickinson FACS Calibur).
3. Luminometer.
4. ELISA plate reader.
5. β-scintillation counter.

3 Methods

3.1 Patch-Clamp Recording in Human and Mouse Erythrocytes

1. Mount Petri dish in your patch-clamp setup, superfuse with NaCl solution and preheat the system to 37°C.
2. Earth the bath solution via salt bridge filled with NaCl solution.
3. Stop superfusion and add few microliter of erythrocyte suspension.
4. Let the cells sediment and attach to the dish bottom. Remove nonattached cells by fast superfusion (see Note 1).
5. Apply a constant superfusion of about 200 μL/min at 37°C.
6. Use borosilicate pipettes with resistances of 8–14 MΩ.
7. If necessary, improve giga-ohm seal formation (to 10–100 GΩ) by applying negative pipette potential (−30 mV) at constant (low) negative pipette pressure for up to 5 min and/or transiently increase the bath $CaCl_2$ concentration to 5 mM.
8. If your system is mechanically vulnerable, lift the giga-ohm sealed cell from the bottom (not possible with poly-L-lysineated dish).
9. In cell-attached mode, continuously control for spontaneous entry into whole-cell mode by microscopy (see below and Fig. 1).
10. For entry into the whole-cell mode, rupture the membrane by additional suction and/or brief electrical pulses (700 mV during 100–200 μs). The whole-cell recording configuration is indicated by a minute increase in capacitance and a simultaneous bleaching of the erythrocyte due to dialysis of hemoglobin by the pipette solution (see Fig. 1 and Note 2).
11. Since whole-cell currents are low, control for formation of an excised outside-out patch by monitoring the presence of a “ghost” at the pipette tip (see Fig. 1c).
12. For patch excision, increase erythrocyte attachment to the dish bottom by pre-coating the dish with poly-L-lysine.
13. Elicit currents by applying standard voltage pulse protocols from −100 mV to +100 mV. Adapt protocol if high positive voltages (≥ +80 mV) result in loss of seal resistance. Sample and low-pass filter currents at 10 and 3 kHz, respectively.
14. Estimate the leak current fraction by changing the electrochemical equilibrium of the charge-carrying ion species and/or by pharmacologically inhibition of the involved ion channels (see Note 3).
15. Compute liquid junction potentials between pipette and NaCl bath solution and between NaCl salt bridge and further bath solution as reported (29) and correct voltages.

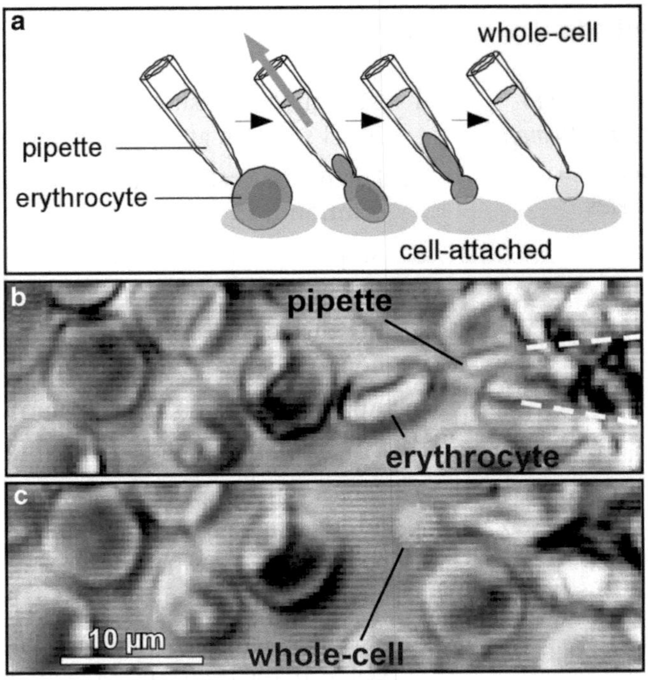

Fig. 1 Whole-cell patch-clamp recording in human erythrocytes. (**a**) The aspiration of the erythrocyte membrane during formation of the whole-cell recording mode is redrawn schematically. (**b**, **c**) Light phase-contrast micrographs taken before (**b**) and after applying negative pressure to the pipette lumen (**c**). Note that the erythrocyte bleached after the rupture of the membrane (**c**)

3.2 Activation of Dormant Erythrocyte Channels

3.2.1 Activation of Gardos $K_{Ca}3.1$ K^+ Channels (See Note 4)

1. Record human erythrocytes at 37°C in cell-attached mode with K-gluconate/KCl pipette solution and NaCl bath solution at 37°C under continuous superfusion and −30 mV pipette potential.
2. Superfuse the Ca^{2+} ionophore ionomycin (1 µM) in NaCl bath solution.
3. Instantaneous echinocyte formation of the recorded erythrocyte and bystander cells indicates the ionomycin activity.
4. Test for unitary current transitions at positive pipette potential.
5. If single channels are apparent, superfuse KCl solution (see Note 5) and record single channels at various voltages. Go back in NaCl solution and break into whole-cell mode.
6. If single channels are not apparent, stay in NaCl solution and directly break into whole-cell mode. Moderately negative current reversal potentials indicate Gardos channel activation (see Fig. 2).
7. Switch now to KCl solution. Gardos whole-cell currents exhibit typical inward rectification in symmetrical KCl solution (not shown in Fig. 2).

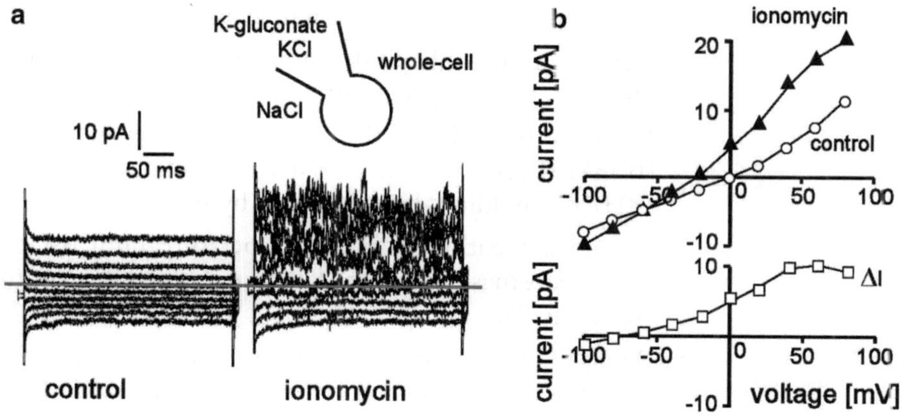

Fig. 2 Ca^{2+}-permeabilization of the erythrocyte membrane induces activation of Gardos K^+ channels. (**a**) Whole-cell current traces and (**b**) *I/V* relationships (**b**; *upper plot*) of a human erythrocyte recorded with K-gluconate/KCl pipette and standard NaCl bath solution under control conditions (**a**, *left traces*; and **b**, *open circles*) and upon bath application of the Ca^{2+} ionophore ionomycin (1 μM; **a**, *right traces*; and **b**, *closed triangles*). (**b**) *Lower plot* shows the ionomycin-induced current fraction that reverses at K electrochemical equilibrium

8. Confirm Gardos channel activity by applying inhibitors (e.g., triarylmethane-34, TRAM-34, 2 μM).

3.2.2 Activation of Nonselective Cation Channels (See Note 6)

1. Record human erythrocytes in fast whole-cell mode with K-gluconate pipette solution and NaCl bath solution at 37°C under continuous superfusion and −30 mV holding potential.
2. Replace NaCl bath by Na-D-gluconate bath solution.
3. Superfuse Na-D-gluconate bath solution for several min until increase in whole-cell currents occurs.
4. Replace Na-D-gluconate bath by NMDG-D-gluconate or NMDG-Cl bath solution (see Note 7).
5. Replace $NMDG^+$ in the bath by Na^+ and D-gluconate$^-$ by Cl^- to re-inactivate the current.

3.2.3 Activation of Organic Osmolyte and Anion Channels (See Note 8)

1. Wash twice and suspend freshly drawn human erythrocytes in glucose-free NaCl solution to a hematocrit of 5% and store them at 8°C for 1–7 days in order to energy deplete the cells and to lower the oxidative defense.
2. Wash aliquots of 500 μL cell suspension with glucose-free NaCl solution, pellet the cells (500 × *g* for 5 min), and discard the supernatants.
3. Resuspend every 20 s a cell pellet in 1 mL of glucose-free NaCl solution containing *tert*-butylhydroperoxide (tBHP, 0 or 1 mM). Prepare additional oxidized samples for isosmotic hemolysis control (see Subheading 3.3.4).

4. Incubate for 5 min at room temperature.
5. Centrifuge the cells (500 × *g* for 5 min at room temperature). Do not remove supernatant.
6. Compare the color of the erythrocyte pellet between control (0 mM tBHP) and the first started oxidizing sample (1 mM tBHP) by holding the pellets side by side.
7. Extend incubation at room temperature until the oxidizing pellet becomes only just darker red than the control cells (see Note 9).
8. Remove quickly and quantitatively (water-jet pump) the supernatant of the samples and instantaneously resuspend the pellets in 400 μL NaCl solution containing 100 μM Na-ATP (see Note 10). These steps must be performed within <20 s and the samples processed in the order of oxidation begin so that each sample receives identical oxidation time.
9. Resuspend the remaining oxidized pellets (isosmotic hemolysis control) in 400 μL of isosmotic sorbitol solution (see Subheading 3.3.4) further containing the organic osmolyte and anion channel inhibitor 5-nitro-2-(3-phenylpropylamino)-benzoic acid, NPPB (0 or 50 μM).
10. Incubate all samples at 37°C under slight agitation for 1.5–3 h.
11. Stop incubation and harvest the cells in NaCl solution at that time when a significant fraction of the sorbitol-incubated cells have lysed isosmotically in a NPPB-sensitive manner (see Subheading 3.3.4).
12. Wash harvested cells twice in NaCl solution and analyze the organic osmolyte and anion channels in single channels and whole-cell recording by the use of K-gluconate or KCl pipette solution (see Note 11).

3.2.4 Activation of ClC-2 Cl^- Channels (See Note 12)

1. Infect human erythrocytes (blood group 0) with laboratory strains of *Plasmodium falciparum* and culture the infected cells asynchronously at a hematocrit of 5% and a parasitemia of 2–10% in RPMI 1640 medium supplemented with Albumax II (0.5%) in an atmosphere with reduced oxygen tension.
2. Mount a Petri dish in your patch-clamp setup and start superfusion with NaCl solution at 37°C. Transfer a 10 μL aliquot of the infected culture directly in the NaCl solution. Prior to that, stop superfusion.
3. Restart superfusion when erythrocytes have been sedimented and attached to the dish bottom.
4. Identify erythrocytes parasitized with the trophozoite stage of *Plasmodium falciparum* by light microscopy (see Note 13).

5. Record trophozoite-parasitized erythrocyte under continuous superfusion in fast whole-cell mode at 37°C with NaCl pipette solution.
6. Replace the NaCl bath solution by sorbitol-containing bath solution.
7. Induce cell swelling and shrinkage iso-ionically by decreasing and increasing the sorbitol concentration of the bath to 0 and 250 mM; swelling activates and shrinkage inactivates ClC-2, respectively (see Note 14).

3.2.5 Activation of 18 pS Cl⁻ Channels (See Note 15)

1. Resuspend washed human erythrocytes in NaCl solution additionally containing sodium (meta)arsenite ($NaAsO_2$, 1 mM) to a hematocrit of 1–5%.
2. Incubate for 1–3 h at 37°C.
3. Transfer 10 μL aliquots in your patch-clamp Petri dish superfused at 37°C with NaCl solution.
4. Analyze 18 pS channels by single-channel and whole-cell recording at 37°C by the use of K-gluconate or KCl pipette solution (see Fig. 3).

3.3 Indirect Measures of Erythrocyte Ion Channels

3.3.1 Ca^{2+}-Permeable Channels Measured by Ca^{2+}-Sensitive Fluorescence Dyes (See Note 16): Protocol for Intact Erythrocytes

1. Wash human erythrocytes twice in Ca^{2+}-free NaCl solution.
2. Resuspend the erythrocytes in Ca^{2+}-free NaCl solution (100 μL in a FACS tube, adjust to a hematocrit of about 5%) containing the membrane-permeable AM ester of Fluo-3 (or Fluo-4; 2 μM, Calbiochem).
3. Load the erythrocytes with Fluo-3/AM (Fluo-4/AM) for 30 min at 37°C.
4. Wash two times and resuspend erythrocytes in 100 μL of Ca^{2+}-free NaCl solution (washing steps may be omitted).
5. Add nine volumes (900 μL) of pre-warmed (37°C) elevated Ca^{2+}-containing NaCl solution further containing 0 or 1 mM of the Ca^{2+}-pump inhibitor Na_3VO_4 at time zero.
6. Measure at 37°C time course of changes in Fluo-3 (Fluo-4) fluorescence intensities (e.g., by a FACS Calibur flow cytometer, Becton Dickinson, at FL-1, excitation at 488 nm, emission at 530 nm).
7. Plot fluorescence intensity against time and calculate slope of fluorescence intensity increase during the initial linear course.
8. Control in every experimental condition for fluorescence dye loading by Ca^{2+}-permeabilizing the erythrocyte membrane (e.g., by 1 μM of ionomycin added to the elevated Ca^{2+}-containing NaCl solution).
9. Control for fluorescence quenching or autofluorescence of drugs/vehicle if applicable.

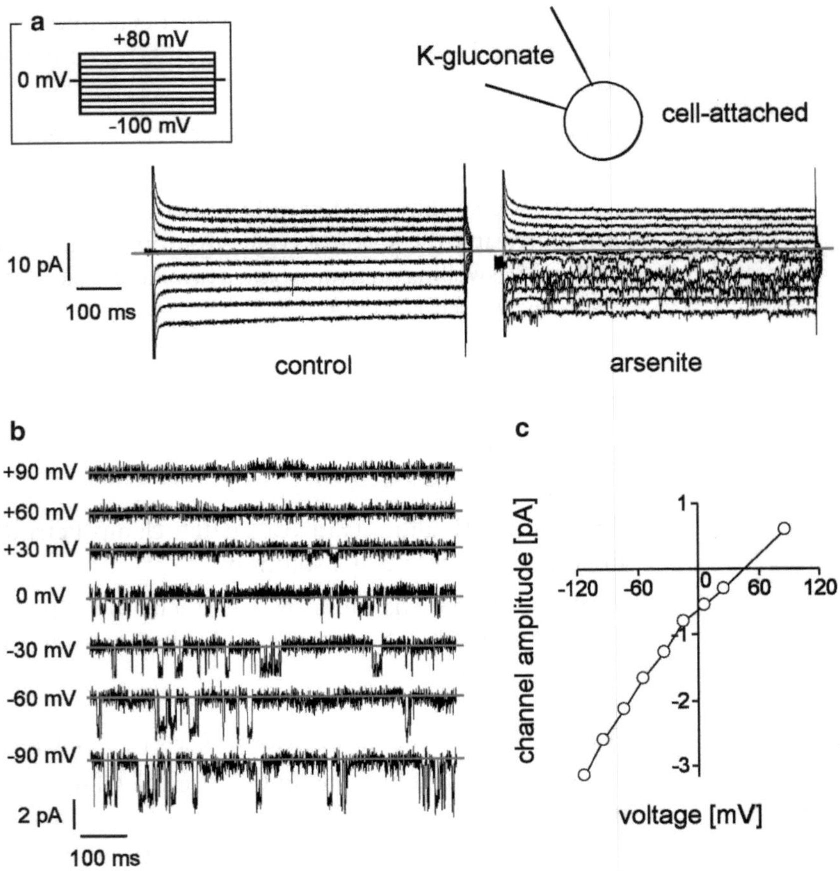

Fig. 3 Activation of the 18 pS Cl^- channels by arsenite. (**a**) Macroscopic cell-attached currents of a control erythrocyte (*left*) and an erythrocyte pretreated for 3 h with arsenite (1 mM, *right*). Currents were elicited by the indicated pulse protocol (*inset*) and recorded with K-gluconate pipette and NaCl bath solution. (**b**) Single-channel tracings recorded at different voltages as in (**a**). (**c**) Relationship between channel amplitude and voltage

3.3.2 Ca^{2+}-Permeable Channels Measured by Ca^{2+}-Sensitive Fluorescence Dyes: Protocol for Erythrocyte Ghosts

1. Wash erythrocyte three times in Ca^{2+}-free NaCl solution.
2. Lyse and pellet human erythrocytes (70 μL pellet) on ice in 1 mL of hypotonic lysis buffer additionally containing 2 μM Fluo-3/AM or Fluo-4/AM.
3. Entrap antibodies, signaling proteins or other membrane-impermeable molecules of interest during hypotonic hemolysis by adding them to the lysis buffer.
4. Spin down non-lysed erythrocytes (2,000 × *g* for 5 min at 4°C).
5. Harvest erythrocyte ghosts (25,000 × *g* for 20 min at 4°C).
6. Resuspend ghosts in 300 μL of 37°C pre-warmed elevated $CaCl_2$-containing NaCl solution at time zero and measure time-dependent increase in fluorescence intensity as described

above in the absence and presence of a Ca^{2+}-pump inhibitor (1 mM Na_3VO_4) or cation channel blockers (e.g., 1 mM amiloride, or 50 μM ethylisopropylamiloride, EIPA).

7. For negative and positive control resuspend the pelleted ghosts in Ca^{2+}-free NaCl solution and in Ca^{2+}-containing NaCl solution additionally containing the Ca^{2+} ionophore ionomycin (1 μM), respectively.

3.3.3 Gardos K^+ Channels Measured by Cell Volume Changes (See Note 17)

1. Resuspend washed erythrocytes in NaCl solution to a hematocrit of 1% (500 μL aliquots in FACS tubes) and place them on 37°C.
2. Add the same volume (500 μL) of 37°C pre-warmed NaCl solution additionally containing Gardos channel activator (e.g., ionomycin, 2 μM, or 1-ethyl-2-benzimidazolinone, 1-EBIO, 200 μM) or inhibitor (e.g., TRAM-34, 2 μM) at time zero.
3. Determine every 60 s forward scatter by flow cytometry as a measure of erythrocyte volume and calculate slope of forward scatter decline.

3.3.4 Organic Osmolyte and Anion Channels Measured by Isosmotic Hemolysis (See Note 18)

1. Resuspend pelleted erythrocytes (25 μL aliquots) in 400 μL of isosmotic sorbitol solution further containing inhibitors of the organic osmolyte and anion channels (e.g., NPPB, 0 and 50 μM).
2. For determination of background hemolysis, immediately spin down (500 × *g* for 5 min) and harvest supernatant (350 μL).
3. Heat the remaining aliquots to 37°C and incubate under slight shaking.
4. Harvest the supernatants (350 μL) of the remaining aliquots at different time points.
5. Determine the free hemoglobin concentration in the supernatants photometrically (e.g., absorption at 546 nm).
6. For standard curve, hemolyze 5, 10, 15, 20, and 25 μL pelleted erythrocytes in 400 μL of distilled water, spin down, harvest 350 μL supernatant, and determine the absorption.
7. Run further hemoglobin standard curves in the presence of the applied inhibitors to control for impairment of the photometry by the drugs.

3.3.5 ClC-2 Chloride Channels Measured by Cell Volume Changes (See Note 19)

1. Wash (three times) and equilibrate erythrocytes (5% hematocrit) for 15 min with glucose-free KCl/NaCl solution (100 μL) containing 4 μM gramicidin D, 2 μM valinomycin, and 50 μM NPPB in order to Na^+-permeabilize and to K^+-permeabilize the erythrocyte membrane and to inhibit organic osmolyte and anion channels, respectively.

2. Add at time zero 9 volumes (900 μL) of a hypotonic sorbitol solution additionally containing 4 μM gramicidin D, 2 μM valinomycin, and 50 μM NPPB.
3. Determine every 60 s forward scatter by flow cytometry as a measure of erythrocyte volume and calculate slope of forward scatter decline.

3.3.6 Pannexin-1 and VDAC ATP Channels Measured by Extracellular ATP Accumulation (See Note 20)

1. Wash (three times) and resuspend erythrocytes (5% hematocrit) in NaCl solution.
2. Place 200 μL aliquots in a 96-well plate on ice.
3. Let the erythrocytes sediment.
4. To avoid ATP release induced by mechanical perturbation, replace carefully 180 μL supernatant by the same volume of ice-cold NaCl solution containing inducer (e.g., 0 or 200 μM of *tert*-butylhydroperoxide) or/and inhibitors of ATP release (e.g., 0 or 50 μM of NPPB).
5. Heat the cells to 37°C.
6. Harvest immediately after re-sedimentation (background values) and every 60 min thereafter the supernatants (100 μL).
7. Spin down supernatants (500 × *g* for 5 min) to remove residual erythrocytes immediately after harvesting; again harvest supernatant (90 μL) and freeze it at −25°C.
8. Quantify the extracellular ATP concentration of the aliquots (50 μL) by a luciferin–luciferase assay kit and a luminometer according to the supplied experimental protocol.
9. Subtract mean background values and calculate ATP release per number of cells and time.
10. Run further ATP standard curves in the presence of the applied inducers/inhibitors to control for impairment of the luciferin–luciferase reaction by the drugs.
11. Determine the hemoglobin concentration of the remaining supernatant volume by ELISA according to the instructions of the supplier to control for hemolysis-mediated contamination by intracellular ATP.

3.3.7 Tracer Flux (See Note 21) Protocol for $^{45}Ca^{2+}$ Uptake

1. Wash erythrocytes in KCl/NaCl solution (see Note 22).
2. Pre-incubate the cells (10% hematocrit) for 10 min at 37°C in 400 μL of KCl/NaCl solution supplemented with 10 mM inosine (to replenish the erythrocytic nucleoside pool which diminishes by ATP release) and 1 mM Na_3VO_4 (to inhibit the Ca^{2+}-ATPase).
3. Add 400 μL of the same solution (37°C) additionally containing the radioactive tracer (~2 μCi/mL $^{45}Ca^{2+}$) and 0.4 mM $CaCl_2$ and incubate at 37°C under slight agitation.

4. Remove 100 μL of cell suspension at time point zero and every 10 min after careful resuspension of the erythrocytes and deliver them into ice-cold KCl/NaCl solution (900 μL).
5. Centrifuge at 4,000 × *g* for 10 s and discard supernatant.
6. Wash the pellet in 1 mL of the same solution and centrifuge and discard the supernatant.
7. Lyse and deproteinize the cell pellet by addition of 250 μL of trichloroacetic acid (6%) and finally centrifuge (17,000 × *g* for 5 min).
8. Count the radioactivity of the supernatant by a β-scintillation counter.

4 Notes

1. For superfusion, we use 3 cm cell culture Petri dishes and a ring-shaped dish insert which besides reducing the bath volume to <200 μL carries the superfusion and heating unit. Erythrocytes remain sticking to the plastic surface of the dish at moderate superfusion flows. If superfusion chambers with glass surfaces are used, erythrocyte adherence can be achieved by coating the surface (e.g., with poly-L-lysine).
2. Unstimulated human or mouse erythrocytes usually show no spontaneous ion channel activity in whole-cell mode. Macroscopic whole-cell currents are therefore often not larger than macroscopic (leakage) currents in the cell-attached mode before. Changes in capacitance upon entry into whole-cell mode are in the range of 1 pF and usually too low to safely define whole-cell recording mode.
3. Current leakage of erythrocytes shows many features of biological currents, i.e., voltage dependence, time-dependent activation/inactivation rectification, and somehow nonselective cation specificity (e.g., replacement of bath Na^+ by $NMDG^+$ decreases inward current leakage without large change of current reversal potential). Moreover, current leakage may change with modification of the divalent cation concentration in the bath and with applied voltage (negative pipette potentials may decrease and positive potentials increase current leakage). Since in whole-cell mode, transmembrane currents of unstimulated erythrocytes are usually smaller than the current leakage, the biological current fraction must be extracted experimentally, e.g., by an agonist-activated or inhibitor-sensitive current fraction. The biological nature of this current fraction is also confirmed (and additionally its ion selectivity) if (1) the activated/inhibited current fraction exhibits a reversal potential close to electrochemical equilibrium E of the charge-carrying

ion and (2) the electrochemical equilibrium E has been set to voltages clearly different from zero (by the use of asymmetrical solutions, see Fig. 2b, here the ionomycin-stimulated current fraction reverses at E_K).

4. Gardos (IK, SK4, $K_{Ca}3.1$) channels are Ca^{2+}-activated inwardly rectifying and intermediate conductance K^+ channels. Reported Gardos whole-cell currents in human (20) and mouse erythrocytes (14) suggest only low number of functional channel copies (see Fig. 2). In the provided protocol, Gardos activation is stimulated by Ca^{2+}-permeabilization of the plasma membrane. Alternatively, channel activator 1-ethyl-2-benzimidazolinone (1-EBIO, 100 μM (14)) can be used.
5. Activation of Gardos channels hyperpolarizes the erythrocyte membrane potential from about −7 mV to high negative voltages. Membrane potential is zeroed by a high KCl concentration in the bath solution which is required to define the applied voltage in cell-attached mode.
6. Ca^{2+}-permeable nonselective cation channels with a permselectivity for monovalent cations of $Cs^+ > K^+ > Na^+ = Li^+ >> NMDG^+$ are reversibly activated in human erythrocytes by oxidative stress (1 mM *tert*-butylhydroperoxide for 15 min) or decrease of the extracellular Cl^- concentration (half maximal effect at an extracellular Cl^- concentration of about 25 mM Cl^-) (22). These channels are probably identical to those activated by prostaglandin E_2 (100 nM) or hyperosmotic cell shrinkage (23, 24, 26). These channels are inhibited by amiloride (1 mM (22)) and ethylisopropylamiloride (EIPA, 10 μM). Hereinafter is the Cl^- depletion protocol which produces very reproducibly channel activation.
7. Decrease of inward current and shift of reversal potential to high negative voltages upon replacement of bath Na^+ by the less permeable $NMDG^+$ indicates (1) the biological nature of the current and (2) the cation selectivity of the Cl^- removal-stimulated current fraction. It further excludes the possibility that the activated current is simply due to an increase in current leakage.
8. Activation of the organic osmolyte and anion channels in *Plasmodium*-infected erythrocytes (9, 10, 30, 31) develops over several hours suggesting complex signaling and biochemical channel modification. Similarly, oxidation-stimulated channel activation takes several hours (9, 32). Here, a protocol for human erythrocytes is provided (for mouse erythrocytes, see (27))
9. The oxidation of the erythrocytes is the most critical step. Brown or even black pellets indicate over-oxidation and erythrocyte death.

10. Autocrine purinergic signaling via ATP release and P2Y receptors facilitates the induction of active organic osmolyte and anion channels (27).
11. Unusual features of organic osmolyte and anion channel-generated whole-cell currents are that a negative holding potential (−30 mV) induces time-dependent inactivation of the inward currents and that micromolar concentrations of serum albumin stimulate outward rectification (33, 34).
12. Human and mouse erythrocytes express volume-sensitive ClC-2 Cl^- channels which are not activated by cell volume changes in unstimulated erythrocytes. Infection by the malaria parasite *Plasmodium* or oxidative stress "resurrects" the dormant channels which are then reversibly activated by cell swelling and inactivated by cell shrinkage. ClC-2 channels have very low single-channel conductance (3–4 pS) and are best recorded by fast whole-cell mode where they exhibit typical inward rectification and time-dependent activation at negative voltages (5, 6). Here a protocol for *Plasmodium*-infected human erythrocytes is given. This protocol can also be used to activate the organic osmolyte and anion channels (see Subheading 3.2.3).
13. Trophozoites detoxify hem as hemozoin which is clearly visible as black dot in the food vacuole of the intraerythrocytic parasite. These "dots" typically moves by Brownian motion.
14. The ClC-2 generated whole-cell current fraction is sensitive to Zn^{2+} (1 mM).
15. The 18 pS anion channel of human erythrocytes is activated in inside-out patches by protein kinase A (8). Sodium (meta) arsenite confers oxidative stress to cells and perturbs intracellular phosphorylation/dephosphorylation pathways by targeting several kinases (35). Here a simple protocol is provided to activate 18 pS anion channels by pretreating the cells with arsenite. Figure 3 gives an example of such activation.
16. Erythrocytes lack intracellular Ca^{2+} stores. Hence, steady-state cytosolic free Ca^{2+} concentration [$(Ca^{2+})_i$] depends solely on Ca^{2+} entry and extrusion across the plasma membrane as well as on cytosolic Ca^{2+} buffering by Ca^{2+}-binding proteins. The latter have high affinity but low capacity and low resting Ca^{2+} concentrations are mainly maintained by the activity of the powerful Ca^{2+} pump in the erythrocyte membrane. Thus, perturbations of the pump leak equilibrium results in instantaneous changes in $[Ca^{2+}]_i$ which can be measured by Ca^{2+}-specific dyes such as Fluo-3 or Fluo-4. Increase in fluorescence (expressed here as a slope of the fluorescence increase with time) in Ca^{2+}-depleted erythrocytes after re-addition of extra-

cellular Ca^{2+} reflects channel-mediated Ca^{2+} entry when determined in the presence of a Ca^{2+}-pump inhibitor. This protocol can easily be extended for measuring Ca^{2+} entry into erythrocyte ghosts which allows inclusion of membrane-impermeable molecules in the ghosts during ghost preparation (15).

17. The resting background permeability of human erythrocytes for Cl^- exceeds that for K^+ (see Subheading 1). Under physiological conditions, Ca^{2+}-mediated activation of Gardos channels hyperpolarizes the membrane potential from about −7 mV towards K^+ equilibrium potential (−85 mV) which imposes a high outwardly directed driving force for Cl^-. This results in rapid shrinkage by the loss of KCl and osmotically obliged water.
18. Organic osmolyte and anion channels of human erythrocytes are permeable for neutral organic osmolytes such as sorbitol (30). Cells with active channels hemolyze when suspended in isosmotic solutions of sorbitol (or other channel-permeating neutral organic osmolytes). The time required for hemolysis of 50% of the cells ($t_{1/2}$) in a given solute species is a measure of channel activity. On the other hand, $t_{1/2}$ in different solutes at a given channel activity reflects the relative permeability of the solutes. Finally, anion channel inhibitors (such as micromolar concentrations of NPPB) delay or even block hemolysis which allows pharmacological characterization of the organic osmolyte and anion channels (9).
19. Efflux of Cl^-, counter cations, and osmotically obliged water results in erythrocyte shrinkage provided that outwardly directed chemical gradients for these ions are applying. Shrinkage is limited and therefore is reflecting Cl^- channel activity if the cation permeability exceeds that for Cl^- by far. Experimentally, this is accomplished by permeabilizing the erythrocyte membrane with cation-specific ionophores (5). Inducing efflux in hyposmotic solution transiently swells the erythrocytes which increases the activity of swelling-induced Cl^- channels such as ClC-2. Different anion channel types can be distinguished by applying anion channel inhibitors. For instance, the organic osmolyte and anion channels are inhibited by NPPB (50 μM) while the ClC-2 Cl^- channels are inhibited by $ZnCl_2$ (1 mM).
20. Erythrocytes release ATP upon a variety of stimuli (e.g., hypoxia, hypercapnia, mechanical deformation) via a pannexin-1-dependent (18, 36) and a pannexin-1-independent (probably VDAC-dependent) (12, 37) conductive pathways. Activity of these channels can be monitored by determining the accumulation of ATP in the medium (27, 31).

21. Uptake of radioactive tracers can be used to assess erythrocyte ion channel activity. To delineate channels from other transporter-mediated tracer uptake, fractions are determined either that are sensitive to specific ion channel blockers or that are insensitive to an inhibitor cocktail which blocks all putative non-channel-generated transports. Nuclides such as $^{86}Ru^{+}$, $^{45}Ca^{2+}$, $^{22}Na^{+}$, and $^{36}Cl^{-}$ are used as tracers for K^{+}- Ca^{2+}-, Na^{+}-, and Cl^{-}-permeable channels, respectively. Uptake rates are calculated from the slope of the linear phase of time-dependent uptake. Uptake rates [μmol/h] are usually given for 1 mM extracellular tracer and 10^{13} erythrocytes which correspond to 1 L of erythrocyte pellet. The tracer flux techniques have been described elsewhere in detail (38). Hereinafter is an example protocol for the uptake of $^{45}Ca^{2+}$ (39).
22. High K^{+} concentrations are used to avoid alteration of erythrocyte membrane potential or volume.

Acknowledgments

This work was supported by a Wilhelm-Sander-Stiftung grant (2011.083.1) and by the DFG International Graduate School 1302.

References

1. Huber SM, Duranton C, Lang F (2005) Patch-clamp analysis of the "new permeability pathways" in malaria-infected erythrocytes. Int Rev Cytol 246:59–134
2. Schillers H (2008) Imaging CFTR in its native environment. Pflugers Arch 456:163–177
3. Verloo P, Kocken CH, Van der Wel A, Tilly BC, Hogema BM, Sinaasappel M, Thomas AW, De Jonge HR (2004) Plasmodium falciparum-activated chloride channels are defective in erythrocytes from cystic fibrosis patients. J Biol Chem 279:10316–10322
4. Lange T, Jungmann P, Haberle J, Falk S, Duebbers A, Bruns R, Ebner A, Hinterdorfer P, Oberleithner H, Schillers H (2006) Reduced number of CFTR molecules in erythrocyte plasma membrane of cystic fibrosis patients. Mol Membr Biol 23:317–323
5. Huber SM, Duranton C, Henke G, Van De Sand C, Heussler V, Shumilina E, Sandu CD, Tanneur V, Brand V, Kasinathan RS, Lang KS, Kremsner PG, Hubner CA, Rust MB, Dedek K, Jentsch TJ, Lang F (2004) Plasmodium induces swelling-activated ClC-2 anion channels in the host erythrocyte. J Biol Chem 279:41444–41452
6. Shumilina E, Huber SM (2011) ClC-2 channels in erythrocytes. Open Biol J 4:18–26
7. Skals M, Jensen UB, Ousingsawat J, Kunzelmann K, Leipziger J, Praetorius HA (2010) Escherichia coli alpha-hemolysin triggers shrinkage of erythrocytes via K(Ca)3.1 and TMEM16A channels with subsequent phosphatidylserine exposure. J Biol Chem 285:15557–15565
8. Egee S, Lapaix F, Decherf G, Staines HM, Ellory JC, Doerig C, Thomas SL (2002) A stretch-activated anion channel is up-regulated by the malaria parasite Plasmodium falciparum. J Physiol 542:795–801
9. Huber SM, Uhlemann AC, Gamper NL, Duranton C, Kremsner PG, Lang F (2002) Plasmodium falciparum activates endogenous Cl⁻ channels of human erythrocytes by membrane oxidation. EMBO J 21:22–30
10. Huber SM, Lang C, Lang F, Duranton C (2008) Organic osmolyte channels in malaria-infected erythrocytes. Biochem Biophys Res Commun 376:514–518
11. Bouyer G, Cueff A, Egee S, Kmiecik J, Maksimova Y, Glogowska E, Gallagher PG, Thomas SL (2011) Erythrocyte peripheral type benzodiazepine receptor/voltage-dependent anion channels are upregulated by Plasmodium falciparum. Blood 118:2305–2312

12. Sridharan M, Bowles EA, Richards JP, Krantic M, Davis KL, Dietrich KA, Stephenson AH, Ellsworth ML, Sprague RS (2011) Prostacyclin Receptor-Mediated ATP Release from Erythrocytes Requires the Voltage-Dependent Anion Channel (VDAC). Am J Physiol Heart Circ Physiol 302:H553–H559
13. Kucherenko YV, Morsdorf D, Lang F (2009) Acid-sensitive outwardly rectifying anion channels in human erythrocytes. J Membr Biol 230:1–10
14. Foller M, Bobbala D, Koka S, Boini KM, Mahmud H, Kasinathan RS, Shumilina E, Amann K, Beranek G, Sausbier U, Ruth P, Sausbier M, Lang F, Huber SM (2010) Functional significance of the intermediate conductance Ca^{2+}-activated K^{+} channel for the short-term survival of injured erythrocytes. Pflugers Arch 460:1029–1044
15. Foller M, Kasinathan RS, Koka S, Lang C, Shumilina E, Birnbaumer L, Lang F, Huber SM (2008) TRPC6 contributes to the Ca^{2+} leak of human erythrocytes. Cell Physiol Biochem 21:183–192
16. Hirschler-Laszkiewicz I, Tong Q, Conrad K, Zhang W, Flint WW, Barber AJ, Barber DL, Cheung JY, Miller BA (2009) TRPC3 activation by erythropoietin is modulated by TRPC6. J Biol Chem 284:4567–4581
17. Bennekou P (1993) The voltage-gated non-selective cation channel from human red cells is sensitive to acetylcholine. Biochim Biophys Acta 1147:165–167
18. Locovei S, Bao L, Dahl G (2006) Pannexin 1 in erythrocytes: function without a gap. Proc Natl Acad Sci U S A 103:7655–7659
19. Hoffman JF, Dodson A, Wickrema A, Dib-Hajj SD (2004) Tetrodotoxin-sensitive Na+ channels and muscarinic and purinergic receptors identified in human erythroid progenitor cells and red blood cell ghosts. Proc Natl Acad Sci U S A 101:12370–12374
20. Lang PA, Kaiser S, Myssina S, Birka C, Weinstock C, Northoff H, Wieder T, Lang F, Huber SM (2004) Effect of Vibrio parahaemolyticus haemolysin on human erythrocytes. Cell Microbiol 6:391–400
21. Alper SL, Vandorpe DH, Peters LL, Brugnara C (2008) Reduced DIDS-sensitive chloride conductance in Ae1-/- mouse erythrocytes. Blood Cells Mol Dis 41:22–34
22. Duranton C, Huber SM, Lang F (2002) Oxidation induces a Cl^{-} dependent cation conductance in human red blood cells. J Physiol 539:847–855
23. Huber SM, Gamper N, Lang F (2001) Chloride conductance and volume-regulatory nonselective cation conductance in human red blood cell ghosts. Pflugers Arch 441:551–558
24. Lang KS, Duranton C, Poehlmann H, Myssina S, Bauer C, Lang F, Wieder T, Huber SM (2003) Cation channels trigger apoptotic death of erythrocytes. Cell Death Differ 10:249–256
25. Barksmann TL, Kristensen BI, Christophersen P, Bennekou P (2004) Pharmacology of the human red cell voltage-dependent cation channel; Part I. Activation by clotrimazole and analogues. Blood Cells Mol Dis 32:384–388
26. Lang PA, Kempe DS, Myssina S, Tanneur V, Birka C, Laufer S, Lang F, Wieder T, Huber SM (2005) PGE(2) in the regulation of programmed erythrocyte death. Cell Death Differ 12:415–428
27. Tanneur V, Duranton C, Brand VB, Sandu CD, Akkaya C, Kasinathan RS, Gachet C, Sluyter R, Barden JA, Wiley JS, Lang F, Huber SM (2006) Purinoceptors are involved in the induction of an osmolyte permeability in malaria-infected and oxidized human erythrocytes. FASEB J 20:133–135
28. Binh VQ, Luty AJ, Kremsner PG (1997) Differential effects of human serum and cells on the growth of Plasmodium falciparum adapted to serum-free in vitro culture conditions. Am J Trop Med Hyg 57:594–600
29. Barry PH, Lynch JW (1991) Liquid junction potentials and small cell effects in patch-clamp analysis. J Membr Biol 121:101–117
30. Duranton C, Huber SM, Tanneur V, Brand VB, Akkaya C, Shumilina EV, Sandu CD, Lang F (2004) Organic osmolyte permeabilities of the malaria-induced anion conductances in human erythrocytes. J Gen Physiol 123:417–426
31. Akkaya C, Shumilina E, Bobballa D, Brand VB, Mahmud H, Lang F, Huber SM (2009) The Plasmodium falciparum-induced anion channel of human erythrocytes is an ATP-release pathway. Pflugers Arch 457:1035–1047
32. Staines HM, Alkhalil A, Allen RJ, De Jonge HR, Derbyshire E, Egee S, Ginsburg H, Hill DA, Huber SM, Kirk K, Lang F, Lisk G, Oteng E, Pillai AD, Rayavara K, Rouhani S, Saliba KJ, Shen C, Solomon T, Thomas SL, Verloo P, Desai SA (2007) Electrophysiological studies of malaria parasite-infected erythrocytes: current status. Int J Parasitol 37:475–482
33. Staines HM, Powell T, Ellory JC, Egee S, Lapaix F, Decherf G, Thomas SL, Duranton C, Lang F, Huber SM (2003) Modulation of whole-cell currents in Plasmodium falciparum-infected human red blood cells by holding potential and serum. J Physiol 552:177–183
34. Duranton C, Tanneur V, Lang C, Brand VB, Koka S, Kasinathan RS, Dorsch M, Hedrich HJ, Baumeister S, Lingelbach K, Lang F, Huber SM (2008) A high specificity and affinity

interaction with serum albumin stimulates an anion conductance in malaria-infected erythrocytes. Cell Physiol Biochem 22:395–404

35. Ventura-Lima J, Bogo MR, Monserrat JM (2011) Arsenic toxicity in mammals and aquatic animals: a comparative biochemical approach. Ecotoxicol Environ Saf 74:211–218
36. Sridharan M, Adderley SP, Bowles EA, Egan TM, Stephenson AH, Ellsworth ML, Sprague RS (2010) Pannexin 1 is the conduit for low oxygen tension-induced ATP release from human erythrocytes. Am J Physiol Heart Circ Physiol 299:H1146–H1152
37. Qiu F, Wang J, Spray DC, Scemes E, Dahl G (2011) Two non-vesicular ATP release pathways in the mouse erythrocyte membrane. FEBS Lett 585:3430–3435
38. Kirk K (2001) Membrane transport in the malaria-infected erythrocyte. Physiol Rev 81:495–537
39. Lang PA, Kasinathan RS, Brand VB, Duranton C, Lang C, Koka S, Shumilina E, Kempe DS, Tanneur V, Akel A, Lang KS, Foller M, Kun JF, Kremsner PG, Wesselborg S, Laufer S, Clemen CS, Herr C, Noegel AA, Wieder T, Gulbins E, Lang F, Huber SM (2009) Accelerated clearance of Plasmodium-infected erythrocytes in sickle cell trait and annexin-A7 deficiency. Cell Physiol Biochem 24:415–428

Chapter 27

Recording Ion Channels in Isolated, Split-Opened Tubules

Elena Mironova, Vladislav Bugay, Oleh Pochynyuk, Alexander Staruschenko, and James D. Stockand

Abstract

Ion channels play key roles in physiology. They function as protein transducers able to transform stimuli and chemical gradients into electrical signals. They also are critical for cell signaling and play a particularly important role in epithelial transport acting as gateways for the movement of electrolytes across epithelial cell membranes. Experimental limitations, though, have hampered the recording of ion channel activity in many types of tissue. This has slowed progress in understanding the cellular and physiological function of these channels with most function inferred from in vitro systems and cell culture models. In many cases, such inferences have clouded rather than clarified the picture. Here, we describe a contemporary method for isolating and patch-clamping renal tubules for ex vivo analysis of ion channel function in native tissue. Focus is placed on quantifying the activity of the epithelial Na^+ channel (ENaC) in the aldosterone-sensitive distal nephron (ASDN). This isolated, split-open tubule preparation enables recording of renal ion channels in the close to native environment under the control of native cell signaling pathways and receptors. When combined with complementary measurements of organ and system function, and contemporary molecular genetics and pharmacology used to manipulate function and regulation, patch-clamping renal channels in the isolated, split-open tubule enables understanding to emerge about the physiological function of these key proteins from the molecule to the whole animal.

Key words Patch clamp, Isolated renal tubules, Ion channel recording, Collecting duct, ENaC

1 Introduction

Ion channels are key intrinsic membrane proteins. They are responsible for the diversity of electrical signaling found in every living cell. In addition, electrolytes and water, the latter of which follows the movement of electrolytes via osmosis, are transported across epithelial barriers of the integumentary system, gastrointestinal tract, the kidney, secretory glands, and many other organs, in part, via ion channels. Vectorial ion transport across epithelia is a consequence of the selective expression of ion channels and transporters in apical and basolateral membranes of epithelial cells. Understanding the basics of ion channel biophysics and function as well as regulation is

Nikita Gamper (ed.), *Ion Channels: Methods and Protocols*, Methods in Molecular Biology, vol. 998, DOI 10.1007/978-1-62703-351-0_27, © Springer Science+Business Media, LLC 2013

critical to understanding the physiological role of these important proteins and how, when dysfunctional, they contribute to pathology. The recent elucidation of many human diseases caused by ion channel dysfunction provides fascinating insights into the diverse roles ion channels serve (1–9). However, our limited ability to record ion channel activity in native tissue has limited discovery of channel function in physiological and pathological conditions. This is particularly true for ion channels in epithelia, such as those expressed in renal tubule cells critical to the fine-tuning of the electrolyte and water concentrations of plasma and urine.

Most electrophysiological studies of ion channels have been performed in either cell culture or recombinant systems (10–13). Because in vivo function must be extrapolated from findings in these artificial systems, such studies often produce results that differ from the channels' real function in native cells. Importantly, the patch-clamp method, which is often used to quantify channel activity in cultured cells, also provides an opportunity, when applied to native tissue, to explore the properties of ion channels in their more native environment. We describe here an isolated, split-open tubule preparation that is suitable for ex vivo analysis of renal ion channels. This preparation allows access to both the apical and basolateral membranes of tubule epithelial cells. Here we describe key aspects of this methodology, including mechanical isolation of the aldosterone-sensitive distal nephron, preparation of these tubules for patch-clamp analysis, and the recording of ion channel activity in the cell-attached configuration. Focus is placed on recording the activity of the epithelial Na^+ channel, ENaC, in the apical membrane of principal cells in the aldosterone-sensitive distal nephron.

The power of in vivo and ex vivo analysis of ion channel function is that it preserves the native setting and control of these critical proteins within the cell and retains many of the emergent properties inherent to real tissue. From these types of studies, structural and functional details, as well as understanding of regulation, can be obtained under normal and pathological conditions. Moreover, combining this readout of channel activity with genetically altered mice enables definition and clarification of the in vivo function of proteins that have been studied in vitro (14–18). In addition to being a tool for understanding the physiological function of particular proteins, mutant mice can also be used to model human diseases. Applying ex vivo analysis of ion channel function to preparations prepared from these animals then also enables the exploration of ion channels as causative agents and/or targets for disease.

Here we focus on a murine kidney preparation. However, this approach can be, and has been, used to study renal channels in other mammals. For instance, we have successfully used this approach with minor changes to investigate ENaC in rat and canine kidneys (*unpublished data*). In combination with other contemporary cell biology methods, measurements of organ and system function, and incorporation of genetically modified animals, this electrophysiological approach represents a powerful tool to study electrolyte transport in specific nephron segments, and enables precise understanding of the physiological role of specific channel proteins along the nephron (19–23). Knowledge gained from such studies in recent years has been instrumental in increasing our understanding of basic and clinical aspects of renal disease.

2 Materials

2.1 Mouse Sacrifice and Kidney Isolation

1. Six-week-old C57BL/6J male mice (this mouse strain is broadly commercially available, for instance from Jackson Laboratory, USA; strain code 000664) (see Note 1).
2. CO_2 gas tank and hermetic chamber for sacrifice.
3. Cold Hanks balanced salt solution (HBSS) (Sigma Aldrich, USA).
4. 6-cm-diameter plastic Petri dishes (BD Falcon, USA).
5. Standard straight forceps (Fine Science Tools, USA).
6. Straight surgical scissors (Fine Science Tools, USA).
7. Nitrile gloves (Kimberly-Clark, USA).

2.2 Isolation of Renal Tubules

1. Cold HBSS (Sigma Aldrich, USA).
2. One freshly harvested kidney kept on ice in HBSS.
3. 10-cm-diameter plastic Petri dishes (TPP, Fisher Scientific, USA).
4. Single-edge steel blade (American line, USA).
5. 6-cm-diameter plastic Petri dishes (BD Falcon, USA).
6. Two Dumont #4 forceps (Fine Science Tools, USA).
7. Stereo microscope (Nikon SMZ 645, Melville, NY, USA) (see Note 2).
8. 18 × 18—Cover glass #2 cover glass (Fisher Scientific, USA) cut into 5 × 5 mm chips.
9. Diamond scriber with diamond tips (Techni-Tool, USA).
10. 0.01% solution of poly-D-lysine (Sigma Aldrich, USA).

2.3 Single-Channel Analysis of ENaC Activity in Isolated Tubules Using the Patch-Clamp Method

1. Patch clamp amplifier (we use Axopatch 200B Molecular Devices., Downingtown, PA, USA; see Note 3).
2. Digitizer, i.e., Digidata 1322A or 1400 A/D board (Molecular Devices) interfaced with a PC running appropriate data acquisition and analysis software (i.e., pClamp 9.2 or newer software suite from Molecular Devices).

3. Precision micromanipulator (we use MP-285 from Sutter Instr. Co., Novato, CA, USA) and mechanical micromanipulator (i.e., NMN-21 Narishige, East Meadow, NY, USA).
4. Vibration isolation table with Faraday cage (i.e., Tech. Manufacturing Co., Peabody, MA, USA).
5. Inverted microscope (i.e., Nikon TE2000-U, Melville, NY, USA).
6. Micropipette puller (i.e., Model P-97 Flaming/Brown puller, Sutter Instrument Co., USA).
7. Micro-forge (MF-830 Narishige, East Meadow, NY, USA).
8. Borosilicate glass capillaries (World Precision Instruments, Sarasota, FL, USA) pulled and forged to 7–10 mΩ for cell-attached patch-clamp recording.
9. Fast exchange recording/perfusion chamber (we use model RC-22, Warner Instruments, USA).
10. Multichannel valve perfusion system (i.e., Valve Bank II, AutoMake Scientific, USA).
11. Pipette solution: 140 mM LiCl, 2 mM $MgCl_2$, and 10 mM HEPES (pH 7.4) (see Note 4).
12. Extracellular bathing solution: 150 mM NaCl, 5 mM KCl, 1 mM $CaCl_2$, 2 mM $MgCl_2$, 5 mM glucose, and 10 mM HEPES (pH 7.4) (see Note 4).
13. Adjustable volume pipette (10–100 μL) with appropriate tips (Eppendorf Research plus 100 μL, Eppendorf, USA) for drug application.
14. Eight-pole low-pass Bessel filter (LPF-8, Warner Instr. Corp. Hamden, CT, USA).

3 Methods

Isolating tubules for patch-clamp analysis has much in common with isolating them for perfusion and microelectrode work (19, 24–27). Similarly, patch-clamp analysis of channels in native cells of isolated tubules is similar to that of immortalized and freshly isolated cells held in culture (28–33).

To be able to study ENaC in ASDN isolated from normal and genetically altered mice, the technique used to isolate tubules was modified to allow patch clamping of the apical plasma membrane of tubule cells. This modification combines splitting open the tubule followed by patch clamping of individual cells. In addition to the electrophysiological measurements described below, the modified approach for isolation may be useful for several other applications, e.g., immunohistochemical, biochemical, and molecular analysis of this tissue. Figure 1 illustrates the steps required to mechanically

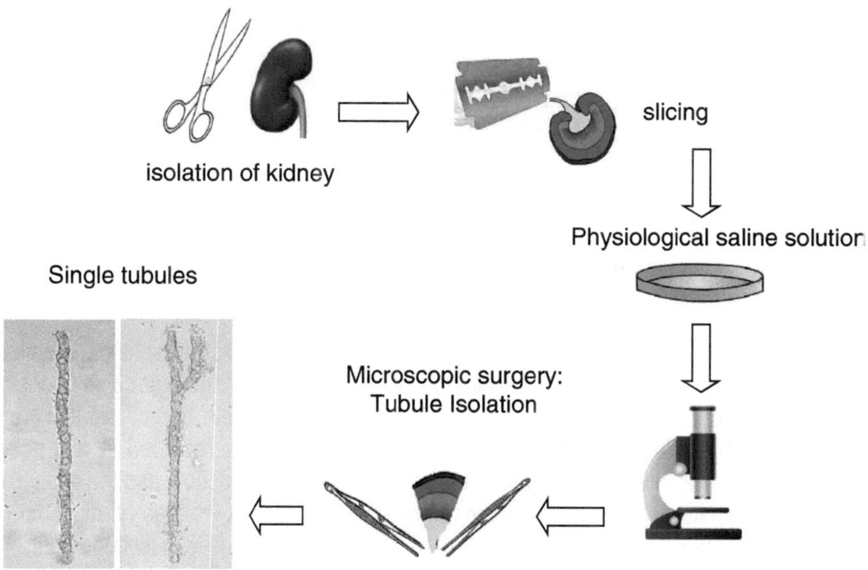

Fig. 1 Schematic illustration of the method used to isolate cortical collecting ducts for patch-clamp analysis. The kidney is isolated from the mouse and then cut into thin slices (<1 mm). Segments of interest are then mechanically isolated from these slices with microdissection typically using watchmaker forceps and a stereomicroscope

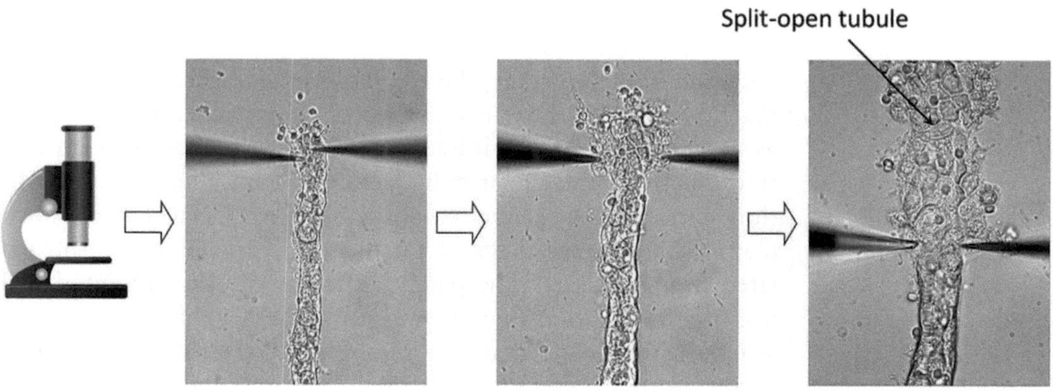

Fig. 2 Shown here is a segment of the cortical collecting duct. The top of this segment has been split open to allow patch-clamp access to the apical membranes of lining epithelial cells

isolate tubules from the kidney without enzymatic treatment. Opening of the tubule with two micropipettes is shown in Fig. 2. Enzymatic techniques have been commonly used for the preparation of isolated tubules as a model for many biochemical and physiologic investigations. However, they are not recommended for electrophysiology study since changes during digestion may lead to drastic disturbances in protein function. Destroying the normal cellular environment and connections with other cells may also result in conditions different than normally seen in physiology.

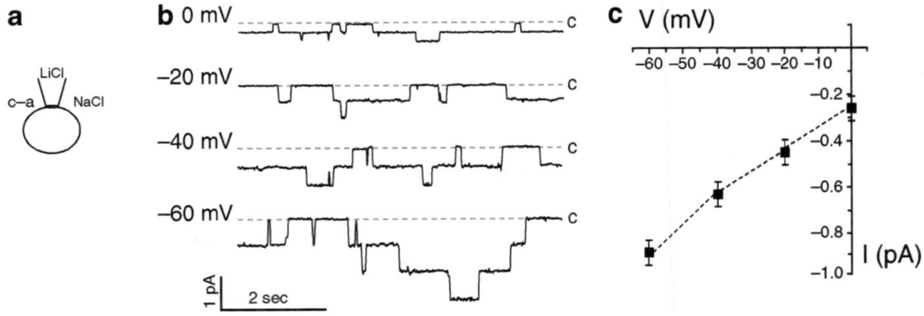

Fig. 3 Recording ENaC at the apical membrane of a principal cell in an isolated, split-open collecting duct. (**a**) Patch-clamp recording in cell-attached configuration. Major ions in bath and pipette solutions are shown. (**b**) Representative single-channel current traces for ENaC from a principal cell. All recordings were performed in the cell-attached configuration in the voltage-clamp mode. Current was recorded at test potentials that ranged from 0 to −60 mV. Inward Li^+ currents are depicted as downward deflection, and the dashed lines show the 0 current level (closed state) at each voltage. (**c**) Single-channel current–voltage relationship for ENaC in cell-attached patches that were made on the apical membrane of principal cells in isolated split-open mouse collecting ducts

Thus, mechanical dissociation of tubules, which preserves tissue and cellular integrity protecting native structure and surroundings, is recommended.

There are four main patch-clamp configurations: cell attach, inside-out, outside-out, and whole cell (34) (See also Chapter 7 of this book). We mostly use in our experiments the cell-attached configuration. However, these other configurations are also compatible with this preparation. With the cell-attached configuration, as shown in Fig. 3a, the intracellular face of the channel is exposed to the native intracellular milieu and the extracellular face to solution in the recording pipette. Figure 3b shows ENaC current traces from a cell-attached patch of a native principal cell in a freshly isolated mouse ASDN. The patched membrane was presented with test potentials that ranged from 0 to −60 mV. This seal contained at least three ENaC channels. Figure 3c shows the single-channel current–voltage relation for ENaC in mouse principal cells in freshly isolated ASDN. Similar current–voltage relationships are observed in rat tubules (32, 35–37). From this relationship, the conductance of ENaC can be determinate. In cell-attached patches, ENaC has a conductance of 4–5 pS with 140 mM NaCl in the pipette (38). To increase channel conductance without affecting any other channel property, the permeant ion in the patch pipette, Na^+, was replaced by Li^+. The latter permeates through ENaC better, increasing channel conductance.

As shown in Fig. 4 the effect of acute application of regulatory factors on ENaC activity can be studied in freshly isolated renal tubules. Figure 4 shows representative current traces for ENaC in native principal cells before and after application of 1 μM of arginine vasopressin (AVP) (Fig. 4a) and 10 μM ATP (Fig. 4b).

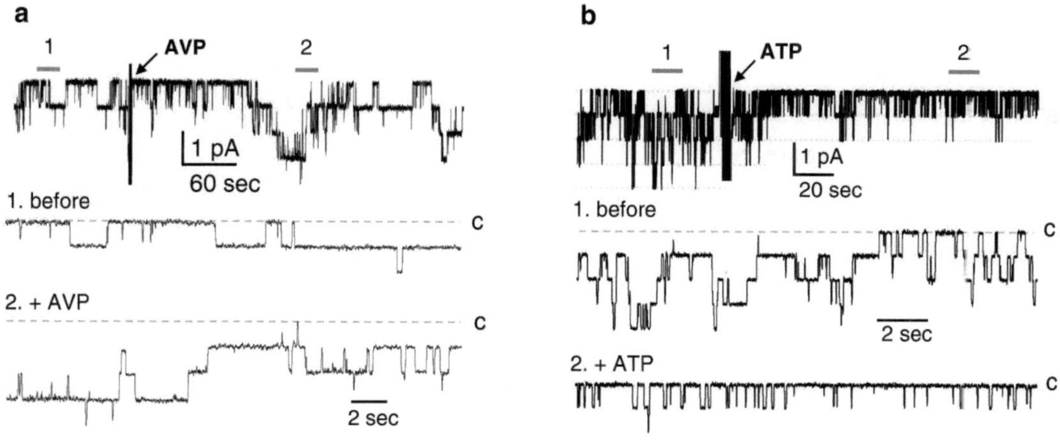

Fig. 4 Activation of ENaC in principal cells of isolated cortical collecting ducts by arginine vasopressin (AVP) (**a**) and inhibition by ATP (**b**). Shown here are representative single-channel current traces for ENaC in a cell-attached patch from a principal cell before and after application of 1 μM AVP (**a**) and 10 μM ATP (**b**). *Arrows* indicate addition of AVP and ATP to the external bath solution. These patches were clamped to a holding potential of $-V_p = -60$ mV. Closed state noted with *c* and areas below 1 (before) and 2 (after addition of AVP or ATP) shown at an expanded time scale below. Data originally presented in (39, 46) and reproduced here with permission

As is clear in these representative current traces and as we previously demonstrated, AVP increases ENaC activity (39–44). In contrast, ATP decreases ENaC activity (31, 45, 46).

The mechanism of ENaC regulation by dietary salt intake was studied in isolated tubules from animals maintained with different salt diets for at least 1 week (14–16). Experiments were performed on mice (Fig. 5) and rats (Fig. 6) kept on low (<0.01%) Na^+, regular (0.32%) Na^+, and high (2%) Na^+ diets. As clear from the current traces in Figs. 5a and 6a, ENaC activity is inversely related to dietary salt intake. In addition to revealing biophysical characteristics, single-channel resolution also provides information about the number of active channels within a patched membrane (*N*) and the average probability that a channel will be open, commonly referred to as open probability (P_o). Channel activity is routinely reported as NP_o. In addition to *N*, information can be gleamed about channel density by quantifying the frequency (*f*) of observing a channel where *f*= patches with at least one active channel of that type/total number of viable seals for that condition. As demonstrated in Figs. 5b and 6b salt diet affected both P_o and *N*.

Successful isolation and splitting of renal tubules and electrophysiological analyses is detailed as described below:

3.1 Mouse Kidney Isolation

1. Make sure that your handling of animals adheres to the appropriate ethical guidance and legislation (see Note 5).
2. Animal work is performed using protective gloves under appropriate safety conditions.

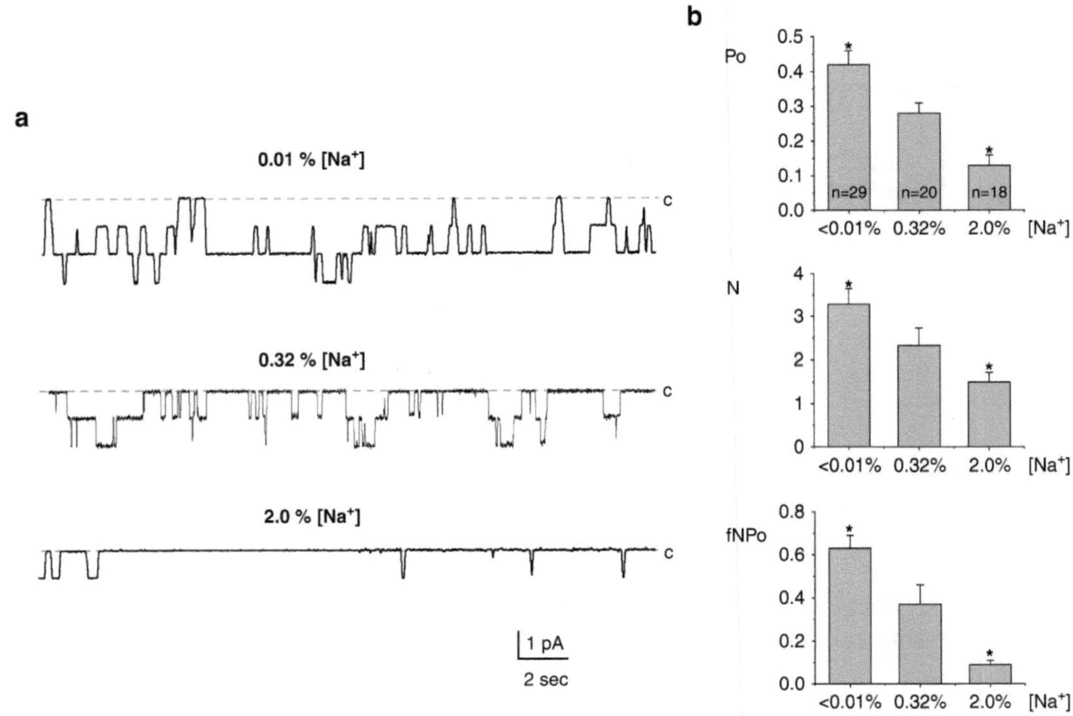

Fig. 5 ENaC activity and density are controlled by dietary NaCl intake in mice. (**a**) Representative current traces from cell-attached patches containing ENaC from mice kept on low (<0.01%)-Na^+ (*top*), regular (0.32%)-Na^+ (*middle*), and high (2%)-Na^+ (*bottom*) diets. C denotes the closed state. (**b**) Summary of ENaC P_0, *fNPo*, and *N* for mice kept on low, regular, and high-salt diets. *$P<0.05$ vs. regular (0.32%)-Na^+ diet; data are expressed as means ± SE. Numbers inside bars indicate number of experiments. Figure originally presented in (45) and reproduced here with permission

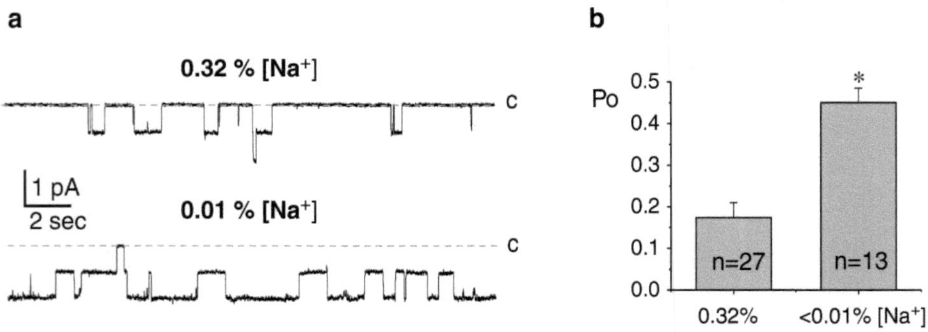

Fig. 6 ENaC activity and density are controlled by dietary NaCl intake in rats. (**a**) Representative current traces from cell-attached patches containing ENaC from Sprague–Dawley rats kept on low (<0.01%)-Na^+ (*bottom*) and regular (0.32%)-Na^+ (*top*) diets. C denotes the closed state. (**b**) Summary of ENaC P_0 for rats kept on low (<0.01%)-Na^+ and regular (0.32%)-Na^+ diets. *$P<0.05$ vs. regular (0.32%)-Na^+ diet; data are expressed as means ± SE. Numbers inside bars indicate number of experiments. Figure originally presented in (32) and reproduced here with permission

3. 6–8-week-old male mice are sacrificed by CO_2 administration followed by cervical dislocation (see Note 6).
4. The animal is placed on a surgical table. Using forceps and scissors make an incision exposing the chest cavity. Retractors can be used to keep the incision open (see Note 7).
5. The kidney is then removed and immediately put on ice in a 6-cm-diameter plastic Petri dish with approximately 4 mL of the HBSS solution.

3.2 Isolation of Mouse Renal Tubules

1. The kidney is taken from the HBSS solution and decapsulated using forceps. Figure 1 illustrates the steps of this procedure for reference.
2. The decapsulated kidney is cut into thin slices (<1 mm) with a razor blade in a 10-cm-diameter plastic Petri dish.
3. One of the slices is placed into a new 6-cm-diameter plastic Petri dish with approximately 3 mL of fresh ice-cold HBSS solution. This Petri dish is put under the stereomicroscope.
4. Using two watchmaker forceps the slice is divided into sections using blunt dissection. Each section should contain approximately 10–20 tubules. From the cortical part of each section, the collecting tubules of individual nephrons are carefully isolated (see Note 8).
5. Isolated tubules are transferred to poly-D-lysine pre-coated 5 × 5 mm glass chips (see Note 9).

3.3 Splitting Open Tubules

1. A chip containing tubules is placed into the recording chamber affixed to the stage of an inverted microscope. The chip is perfused with bath (extracellular) solution for 2–3 min to remove cell debris (see Note 10).
2. The tubule is then split open with two sharp micropipettes controlled with separate micromanipulators to gain access to the apical plasma membrane of epithelial cells within the tubule.
3. The tubule is oriented in the chamber so that its midline is perpendicular to the front of the microscope.
4. The tubule is gently pinned with the left micropipette, taking care not to press through to the bottom cells. With the left pipette stationary, the top surface of the tubule is scraped away with the right pipette to open it (see Fig. 2).
5. When the tubule is opened the left pipette is moved a small distance closer to the front, and step number 4 is repeated. Continue to split open the tubule as needed for your experiment.
6. The apical membranes of cells within the tubule are cleaned with a suction pipette under microscopic observation. With this cleaning procedure, the success rate in obtaining gigaohm seals increases.

3.4 Single-Channel Analysis of ENaC Activity Using the Patch-Clamp Method

1. Prepare bath (extracellular) and pipette solutions, and patch pipettes (see Note 11).
2. The fire-polished glass patch pipette is slowly lowered to the surface of the selected cell (see Note 12) and a high-resistance (>1 gΩ) seal is formed between the pipette and plasma membrane by applying negative pressure to the back of the recording pipette usually through gentle suction. Seal formation is assessed by monitoring pipette resistance with resistance going from ~7–10 mΩ to more than 8–10 gΩ upon successfully forming the cell-attached seal.
3. The patch membrane can then be voltage clamped (keeping the voltage constant) to observe changes in current or current clamped (keeping current constant) to observe changes in membrane voltage. For our cell-attached voltage-clamp studies, currents were low-pass filtered at 100 Hz by an eight-pole Bessel filter and digitized and stored on a PC hard drive using the Digidata interface (see Note 3).

4 Notes

1. It is also possible to use other mouse strains and gender. This approach also can be used to study renal channels in rat, rabbit, canine, and other mammals.
2. The stereomicroscope Nikon SMZ 645 is not produced anymore. However Nikon AMZ 745 or any other comparable model should be sufficient for tubule isolation.
3. There are many different patch-clamp amplifiers, A/D acquisition boards and programs, micromanipulators, isolation tables, microscopes, pipette pullers, perfusion chambers, etc. available for patch-clamp analysis. The user should use those best suited to their experiments and personal preference. For a detailed description of the patch-clamp method, the reader is directed towards the excellent book by B. Sakmann and E. Neher (34); see also Chapter 7 of this book.
4. The pipette and extracellular bathing solutions should be appropriate for the channel of interest. The choice of solutions depends on experiments conditions.
5. Non-survival surgeries are performed in septic but clean conditions. Instruments used should be clean, but not necessary sterile. All solutions used in this preparation should be sterile and kept on ice.
6. Carbon dioxide (CO_2) inhalation is a common method of euthanasia used for rats and mice. Without pre-charging the chamber, place the animals in the chamber and introduce 100% CO_2 at a slow rate so as to minimize distress. Animals should

be exposed to CO_2 for at least 5 min. It is also possible to use other euthanasia techniques. Upon completion of the procedure, death must be insured by cervical dislocation or other methods approved by the Institutional Ethics (or equivalent) Committee.

7. A retractor helps in keeping the incision open and to keep the skin and tissue away from organs. The user should use those best suited to their experiments and personal preference.
8. Isolated tubules can be identified by morphology under the stereomicroscope. The ASDN can be identified as merging of connecting tubule into cortical collecting duct. Proximal tubules in the cortex can be distinguished by their cloudy pale color and very broad structure. In contrast, distal elements (cortical ascending limb, distal convoluted tubule, cortical collecting duct, and loop of Henle) are more transparent and narrower. Cortical collecting duct can be readily distinguished by the ability to see individual cells within the tubule (26, 27).
9. Construct glass chips as follows: coverslips are placed onto a hard glass surface and held in place with plastic overlays. One coverslip is scored with a scriber pen with a diamond tip in a 4×4 grid. Scorings are tapped to separate chips. The resulting glass pieces are approximately 5×5 mm and are held in a 35 mm tissue culture dish until use. Cover glass chips are coated with poly-D-lysine a day before experiments. Poly-D-lysine is used to facilitate adherence of tubules to cover glass. This, in some instances, can be particularly important for patch-clamp experiments performed with constant bath perfusion. We do not recommend using coverslips with lysine coating that are more than 5–7 days old. Cell-Tak (BD Bioscience) or other adhesive materials can be used as alternative to poly-D-lysine.
10. Other chips with tubules should be kept on ice in a Petri dish in HBSS solution until needed. The number of chips and tubules depends on experimental needs, but the researcher should take into account that the experimental window for using freshly isolated tubules is less than 2–3 h.
11. The various configurations of the patch clamp technique may be used to record ion channel activity in renal tubules. The interior of the pipette should be filled with an appropriate solution. We use a standard physiological saline solution for the bath while performing cell-attached recordings. We recommend checking the osmolarity of pipette solutions which should be 295±5 mOsm. After preparation, the pipette solution should be aliquoted and stored in a freezer before experiments to prevent degradation of labile ingredients.
12. The collecting ducts contain two main cell types, principal and intercalated cells. For our study we used principal cells, which are the more abundant type.

Acknowledgments

Research from the author's laboratories is supported by the NIH grants R01 DK59594, R01 DK087460, R01 DK070571 (to J.D.S.), and R01HL108880 (to A.S.).

References

1. Biervert C, Schroeder BC, Kubisch C, Berkovic SF, Propping P, Jentsch TJ, Steinlein OK (1998) A potassium channel mutation in neonatal human epilepsy. Science 279:403–406
2. Bonny O, Hummler E (2000) Dysfunction of epithelial sodium transport: from human to mouse. Kidney Int 57:1313–1318
3. Hummler E, Horisberger JD (1999) Genetic disorders of membrane transport. V. The epithelial sodium channel and its implication in human diseases. Am J Physiol 276:G567–G571
4. Rossier BC, Pradervand S, Schild L, Hummler E (2002) Epithelial sodium channel and the control of sodium balance: interaction between genetic and environmental factors. Annu Rev Physiol 64:877–897
5. Schild L (1996) The ENaC channel as the primary determinant of two human diseases: Liddle syndrome and pseudohypoaldosteronism. Nephrologie 17:395–400
6. Schild L (2004) The epithelial sodium channel: from molecule to disease. Rev Physiol Biochem Pharmacol 151:93–107
7. Simon DB, Bindra RS, Mansfield TA, Nelson-Williams C, Mendonca E, Stone R, Schurman S, Nayir A, Alpay H, Bakkaloglu A, Rodriguez-Soriano J, Morales JM, Sanjad SA, Taylor CM, Pilz D, Brem A, Trachtman H, Griswold W, Richard GA, John E, Lifton RP (1997) Mutations in the chloride channel gene, CLCNKB, cause Bartter's syndrome type III. Nat Genet 17:171–178
8. Snyder PM, Price MP, McDonald FJ, Adams CM, Volk KA, Zeiher BG, Stokes JB, Welsh MJ (1995) Mechanism by which Liddle's syndrome mutations increase activity of a human epithelial Na^+ channel. Cell 83:969–978
9. Wollnik B, Schroeder BC, Kubisch C, Esperer HD, Wieacker P, Jentsch TJ (1997) Pathophysiological mechanisms of dominant and recessive KVLQT1 K^+ channel mutations found in inherited cardiac arrhythmias. Hum Mol Genet 6:1943–1949
10. Berjukow S, Doring F, Froschmayr M, Grabner M, Glossmann H, Hering S (1996) Endogenous calcium channels in human embryonic kidney (HEK293) cells. Br J Pharmacol 118:748–754
11. Gamper N, Stockand JD, Shapiro MS (2005) The use of Chinese hamster ovary (CHO) cells in the study of ion channels. J Pharmacol Toxicol Methods 51:177–185
12. Robbins J, Trouslard J, Marsh SJ, Brown DA (1992) Kinetic and pharmacological properties of the M-current in rodent neuroblastoma × glioma hybrid cells. J Physiol 451:159–185
13. Staruschenko A, Booth RE, Pochynyuk O, Stockand JD, Tong Q (2006) Functional reconstitution of the human epithelial Na^+ channel in a mammalian expression system. Methods Mol Biol 337:3–13
14. Bugaj V, Mironova E, Kohan DE, Stockand JD (2012) Collecting duct-specific endothelin B receptor knockout increases ENaC activity. Am J Physiol Cell Physiol 302:C188–C194
15. Mironova E, Peti-Peterdi J, Bugaj V, Stockand JD (2011) Diminished paracrine regulation of the epithelial Na^+ channel by purinergic signaling in mice lacking connexin 30. J Biol Chem 286:1054–1060
16. Pochynyuk O, Bugaj V, Rieg T, Insel PA, Mironova E, Vallon V, Stockand JD (2008) Paracrine regulation of the epithelial Na^+ channel in the mammalian collecting duct by purinergic P2Y2 receptor tone. J Biol Chem 283:36599–36607
17. Roos KP, Strait KA, Raphael KL, Blount MA, Kohan DE (2012) Collecting duct-specific knockout of adenylyl cyclase type VI causes a urinary concentration defect in mice. Am J Physiol Renal Physiol 302:F78–F84
18. Rubera I, Hummler E, Beermann F (2009) Transgenic mice and their impact on kidney research. Pflugers Arch 458:211–222
19. Bailey MA, Giebisch G, Abbiati T, Aronson PS, Gawenis LR, Shull GE, Wang T (2004) NHE2-mediated bicarbonate reabsorption in the distal tubule of NHE3 null mice. J Physiol 561: 765–775
20. Huang DY, Osswald H, Vallon V (2000) Sodium reabsorption in thick ascending limb of Henle's loop: effect of potassium channel blockade in vivo. Br J Pharmacol 130:1255–1262
21. Thomson SC, Rieg T, Miracle C, Mansoury H, Whaley J, Vallon V, Singh P (2012) Acute and chronic effects of SGLT2 blockade on glomerular and tubular function in the early diabetic rat. Am J Physiol Regul Integr Comp Physiol 302:R75–R83

22. Vallon V (2009) Micropuncturing the nephron. Pflugers Arch 458:189–201
23. Wang T, Hropot M, Aronson PS, Giebisch G (2001) Role of NHE isoforms in mediating bicarbonate reabsorption along the nephron. Am J Physiol Renal Physiol 281:F1117–F1122
24. Kirchner KA (1990) Greater loop chloride uptake contributes to blunted pressure natriuresis in Dahl salt sensitive rats. J Am Soc Nephrol 1:180–186
25. Roman RJ, Kaldunski ML (1991) Enhanced chloride reabsorption in the loop of Henle in Dahl salt-sensitive rats. Hypertension 17: 1018–1024
26. Wagner CA, Lukewille U, Valles P, Breton S, Brown D, Giebisch GH, Geibel JP (2003) A rapid enzymatic method for the isolation of defined kidney tubule fragments from mouse. Pflugers Arch 446:623–632
27. Schafer JA, Watkins ML, Li L, Herter P, Haxelmans S, Schlatter E (1997) A simplified method for isolation of large numbers of defined nephron segments. Am J Physiol Renal Physiol 273:F650–F657
28. Pavlov TS, Chahdi A, Ilatovskaya DV, Levchenko V, Vandewalle A, Pochynyuk O, Sorokin A, Staruschenko A (2010) Endothelin-1 inhibits the epithelial Na^+ channel through βPix/14-3-3/Nedd4-2. J Am Soc Nephrol 21:833–843
29. Pavlov TS, Ilatovskaya DV, Levchenko V, Mattson DL, Roman RJ, Staruschenko A (2011) Effects of cytochrome P450 metabolites of arachidonic acid on the epithelial sodium channel (ENaC). Am J Physiol Renal Physiol 301:F672–F681
30. Pochynyuk O, Tong Q, Medina J, Vandewalle A, Staruschenko A, Bugaj V, Stockand JD (2007) Molecular determinants of PI(4,5)P2 and PI(3,4,5)P3 regulation of the epithelial Na^+ channel. J Gen Physiol 130:399–413
31. Pochynyuk O, Bugaj V, Vandewalle A, Stockand JD (2008) Purinergic control of apical plasma membrane PI(4,5)P2 levels sets ENaC activity in principal cells. Am J Physiol Renal Physiol 294:F38–F46
32. Staruschenko A, Pochynyuk O, Vandewalle A, Bugaj V, Stockand JD (2007) Acute regulation of the epithelial Na+channel by phosphatidylinositide 3-OH kinase signaling in native collecting duct principal cells. J Am Soc Nephrol 18:1652–1661
33. Yu L, Helms MN, Yue Q, Eaton DC (2008) Single-channel analysis of functional epithelial sodium channel (ENaC) stability at the apical membrane of A6 distal kidney cells. Am J Physiol Renal Physiol 295:F1519–F1527
34. Sakmann, B. and Neher, E. (1995) Single-channel recording. Springer Second Ed, 1–722.
35. Bugaj V, Pochynyuk O, Mironova E, Vandewalle A, Medina JL, Stockand JD (2008) Regulation of the epithelial Na^+ channel by endothelin-1 in rat collecting duct. Am J Physiol Renal Physiol 295:F1063–F1070
36. Karpushev AV, Levchenko V, Ilatovskaya DV, Pavlov TS, Staruschenko A (2011) Novel role of Rac1/WAVE signaling mechanism in regulation of the epithelial Na^+ channel. Hypertension 57:996–1002
37. Sun P, Lin DH, Yue P, Jiang H, Gotlinger KH, Schwartzman ML, Falck JR, Goli M, Wang WH (2010) High potassium intake enhances the inhibitory effect of 11,12-EET on ENaC. J Am Soc Nephrol 21:1667–1677
38. Palmer LG, Frindt G (1986) Amiloride-sensitive Na channels from the apical membrane of the rat cortical collecting tubule. Proc Natl Acad Sci USA 83:2767–2770
39. Bugaj V, Pochynyuk O, Stockand JD (2009) Activation of the epithelial Na^+ channel in the collecting duct by vasopressin contributes to water reabsorption. Am J Physiol Renal Physiol 297:F1411–F1418
40. Chen L, Williams SK, Schafer JA (1990) Differences in synergistic actions of vasopressin and deoxycorticosterone in rat and rabbit CCD. Am J Physiol Renal Physiol 259: F147–F156
41. Frindt G, Burg MB (1972) Effect of vasopressin on sodium transport in renal cortical collecting tubules. Kidney Int 1:224–231
42. Frindt G, Palmer LG (1996) Regulation of Na channels in the rat cortical collecting tubule: effects of cAMP and methyl donors. Am J Physiol Renal Physiol 271:F1086–F1092
43. Hawk CT, Li L, Schafer JA (1996) AVP and aldosterone at physiological concentrations have synergistic effects on Na+transport in rat CCD. Kidney Int 57:S35–S41
44. Helman SI, Grantham JJ, Burg MB (1971) Effect of vasopressin on electrical resistance of renal cortical collecting tubules. Am J Physiol 220:1825–1832
45. Pochynyuk O, Rieg T, Bugaj V, Schroth J, Fridman A, Boss GR, Insel PA, Stockand JD, Vallon V (2010) Dietary Na^+ inhibits the open probability of the epithelial sodium channel in the kidney by enhancing apical P2Y2-receptor tone. FASEB J 24:2056–2065
46. Stockand JD, Mironova E, Bugaj V, Rieg T, Insel PA, Vallon V, Peti-Peterdi J, Pochynyuk O (2010) Purinergic inhibition of ENaC produces aldosterone escape. J Am Soc Nephrol 21:1903–1911

Chapter 28

Single-Channel Analysis of TRPC Channels in the Podocytes of Freshly Isolated Glomeruli

Daria V. Ilatovskaya and Alexander Staruschenko

Abstract

One of the most important functions of the kidney is the filtration of the blood that takes place in the glomeruli. Glomerular epithelial cells (podocytes) have several functions, including regulation of the filtration process and glomerular basement membrane turnover. Dysfunction of podocytes is a major cause of glomerular kidney diseases. Gain-of-function mutations in the TRPC6 channel underlie a subset of familial forms of focal segmental glomerulosclerosis (FSGS). While growing evidence supports an important role of TRPC channels in podocytes, the regulation of these channels has yet to be investigated in freshly isolated glomeruli. Native settings in glomeruli provide, by all means, the most appropriate as well as one of the most challenging environments to study ion channel regulation. Thus, it is important to develop new methods that would better reflect the native settings of the podocytes. To address this question, we have established an experimental approach that allows studying podocytes in the freshly isolated decapsulated intact glomeruli. Here we describe the preparation of the rat glomeruli for patch-clamping, focusing on special conditions required for single-channel analysis of TRPC channels. Several tricks useful for cell-attached patch-clamping of the glomerular podocytes and solutions appropriate for registration of the TRPC channels are also provided.

Key words TRPC channels, TPC6, Glomeruli, Podocytes, Patch clamp, FSGS

1 Introduction

One of the main functions of the kidney is blood filtration that is tightly controlled by renal glomeruli. Glomerular podocytes are the central components of the renal filtration barrier. Podocytes are very dynamic cells which, at least in part, play a role in the pathogenesis of most proteinuric glomerulopathies (1, 2).

Transient receptor potential canonical (TRPC) channels belong to the larger superfamily of the TRP channels. TRPC channels play an important role in the pathogenesis of renal and cardiovascular diseases (3–6). Several members of the TRPC family, including TRPC6, are essential components of the podocyte slit diaphragm, where they are integrated into a signaling complex that interacts

Nikita Gamper (ed.), *Ion Channels: Methods and Protocols*, Methods in Molecular Biology, vol. 998,
DOI 10.1007/978-1-62703-351-0_28, © Springer Science+Business Media, LLC 2013

with a number of podocyte structural proteins, including nephrin, podocin, alpha-actinin-4, and calcineurin (7–10). It was demonstrated that TRPC6 protein is the genetic impetus for an autosomal dominant form of focal segmental glomerulosclerosis (FSGS) (11–13). The known TRPC6 gain-of-function mutations cause an increase in calcium current in podocytes.

A number of studies provided new insights into the role of TRPC channels, and particularly TRPC6, in the functionality of the glomeruli and highlighted their role in mediating concomitant diseases (11, 12, 14). Most of these studies were conducted on the cultured podocytes (15–17) or in the recombinant systems (18, 19) which, although being quite sustainable, cannot provide the comprehensive setting for understanding the regulation of the channels under physiological conditions. The lack of such studies has not been a result of their unimportance, but rather resulted from the fact that electrophysiological recording of endogenous ion channels in their native surrounding is a complicated procedure requiring a combination of pure isolation of glomeruli, separation of endogenous currents, electrophysiological skills, etc. Gloy et al., who have developed a technique allowing to measure membrane voltages and ion conductances of podocytes in isolated glomeruli (20), stated later describing the approach they developed that "it is important to develop new methods that better reflect the in vivo situation of podocytes, but studying podocytes function in situ is methodologically very difficult" (2).

Taking into account the emerging need to comprehend the functionality of ion channels of the podocytes, we have attempted to perform single-channel analysis of endogenous TRPC-like ion channels in the podocytes of isolated decapsulated glomeruli on the basis of the technique suggested by Gloy and colleagues (2, 20). The method described here consists of several steps including removal and perfusion of the rat kidney, isolation of the glomeruli fraction by sequentially pushing the kidney cortex through the steel sieves of different mesh, preparation of the glomeruli for patch-clamping, and electrophysiological cell-attached recordings themselves.

In contrast to Gloy et al., who removed the Bowman's capsule mechanically with a small broken glass pipette, we used sieving of the glomeruli as it was described previously (21–23). This approach allowed to avoid a time-consuming and potentially podocytes-damaging step of manual removal of the glomerular basement membrane. Fluorescence and electron microscopy techniques confirmed the integrity of podocytes in the isolated glomeruli after this isolation. Thus, the technical approach developed by us and described here could be used to search for physiologically relevant mechanisms of various ion channels regulation in both physiological and pathophysiological conditions and opens new directions of research into glomerular diseases.

2 Materials

2.1 Rat Kidney Isolation Surgery and Perfusion

1. Eight-week-old Sprague–Dawley male rat (we used internal MCW colony; however this rat strain is broadly commercially available (for instance, Charles River, USA, strain code 400)).
2. 25 mL of the cold Hanks Balanced Salt Solution (HBSS) (Sigma-Aldrich, USA).
3. 100 μL of pentobarbital (Sigma-Aldrich, USA).
4. 10 mL of the physiological salt solution (saline) (Sigma-Aldrich, USA).
5. Polyethylene tubing (PE50, Sigma-Aldrich, USA).
6. Syringe pump-based perfusion system (Harvard Apparatus, USA).
7. 1×2 teeth tip 12 cm Adson forceps (Fine Science Tools, USA).
8. Straight sharp 13 cm surgical scissors (Fine Science Tools, USA).
9. Straight 10 cm Graefe forceps (Fine Science Tools, USA).
10. Straight 12.5 cm Halsted-Mosquito hemostats (Fine Science Tools, USA).
11. Schwartz 26 mm sharp-bend micro serrefines (Fine Science Tools, USA).
12. Straight 11 cm Dumont #4 forceps (Fine Science Tools, USA).
13. Surgical suture, braided silk (Surgical Specialties Corporation).
14. Temperature controlled surgical table for rodents (Harvard Apparatus, USA).
15. Nitrile gloves (Kimberly-Clark, USA).
16. Binocular microscope (Nikon Eclipse TS100, Nikon, USA).

2.2 Isolation of the Rat Glomeruli

1. One freshly removed perfused kidney of an 8-week-old Sprague–Dawley male rat, kept on ice in HBSS.
2. Single-edge steel razor blade (Fisher Sci, USA).
3. Steel sieves (#100 (150 μm), 140 (106 μm) (Fisher Sci, USA)) and mesh 200 (screen for CD-1, Sigma-Aldrich, USA).
4. 5 mL of 30% albumin solution from bovine serum (BSA) (Sigma-Aldrich, USA).
5. 25 mL of culture media solution RPMI 1640 without antibiotics or FBS (Sigma-Aldrich, USA).
6. Steel "spoonulet" lab spoon (Fisher Sci, USA).
7. 6.75 cm straight Mayo dissecting scissors (Fisher Sci, USA).

8. Two 10-cm diameter plastic Petri dishes (TPP, Switzerland).
9. Five Fisherbrand Urisystem disposable plastic transfer pipettes (Fisher Sci, USA).
10. One 15 mL plastic conical Falcon tube (Santa Cruz, USA).

2.3 Single-Channel Analysis of TRPC Channel Activity Using Patch-Clamp Method

1. 25 × 25 Cover glass #2 (Corning, USA) cut into 5 × 5 mm pieces.
2. 0.01% solution of MW 70,000–150,000 poly-D-lysine (Sigma-Aldrich, USA).
3. Multiclamp 700B patch-clamp amplifier (Molecular Devices, USA).
4. Digidata 1440A analog-to-digital board (Molecular Devices, USA) interfaced with a personal computer running the *pClamp 10.2* software suite (Molecular Devices, USA).
5. MP-225 motorized micromanipulator (Sutter Instrument Co., Novato, USA).
6. Microvibration isolation table equipped with Faraday cage (TMC, USA).
7. Inverted microscope (Nikon Eclipse T*i*, USA).
8. Model P-87 Flaming/Brown micropipette puller (Sutter Instrument Co., USA).
9. MF-830 microforge (Narishige, Japan).
10. Borosilicate glass capillaries (World Precision Instruments, USA) pulled and forged to 7–10 MΩ for cell-attached patch-clamp recordings.
11. Recording/perfusion chamber RC-26 (Warner Instruments, USA).
12. Multichannel valve perfusion system (Valve Bank II, AutoMake Scientific, USA).
13. Intracellular pipette solution: 126 mM NaCl, 1.5 mM $CaCl_2$, 10 mM HEPES, 10 glucose; pH 7.4 (all salts and chemicals purchased from Sigma-Aldrich, USA, unless noted otherwise).
14. Extracellular bathing solution: 135 mM NaAsp, 1 mM $CaCl_2$, 10 mM HEPES, 2 mM $MgCl_2$, 10 mM glucose; pH 7.4.
15. Chloride and potassium channels inhibitors in their final concentrations: 100 μM niflumic acid (Sigma-Aldrich, USA) or DIDS (Sigma-Aldrich, USA), 10 mM TEA chloride (Tocris, USA), 10 nM iberiotoxin (Sigma-Aldrich, USA), 10 μM nicardipine (Sigma-Aldrich, USA).
16. Adjustable volume pipette (10–100 μL) with appropriate tips (Eppendorf Research plus 100 μL, Eppendorf, USA).

3 Methods

For the successful isolation of rat kidneys, male rats of approximately 8 weeks old are used. The kidneys are flushed with PBS to clean out blood and urine. In addition to the electrophysiological measurements described below, this procedure may be useful for several other applications, e.g., immunohistochemical, biochemical, or molecular analysis of the tissue, etc. Figures 1 and 2 demonstrate steps required to isolate the glomeruli. Figure 3 shows the electron microscopy of the freshly isolated glomerulus fixed with 2% glutaraldehyde buffered in 0.1 M cacodylate (pH 7.4). As seen on these images acquired by the transmission electron microscopy, a normal configuration of podocytes with coordinated pattern of foot processes was observed (Fig. 3). With the help of the patch clamp technique, we have examined the biophysical properties of native TRPC channels in the podocytes of the freshly isolated decapsulated glomeruli. Our results demonstrate that it is possible to study functional channels in these intact glomeruli. The use of specific inhibitors within the patch pipette together with the composition of the pipette and bath solutions and the parameters of the current–voltage dependencies allowed distinguishing two channel types with distinct TRPC properties. However the development of this approach and identification of the precise composition of TRPC channels in podocytes require additional studies.

3.1 Rat Kidney Isolation Surgery and Perfusion

1. Appropriate ethical approvals need to be obtained before the work can be carried out; i.e., in our experiments, animal use and welfare adhered to the NIH *Guide for the Care and Use of Laboratory Animals* following a protocol reviewed and approved by the IACUC at the Medical College of Wisconsin.
2. 8-week-old male rats are anesthetized with pentobarbital *i.p.* injections (50 mg/kg) (see Notes 1 and 2). The rat is restrained manually and a 25 gauge or smaller needle attached to a syringe is inserted into the lower right quadrant of the abdomen. Before injecting, the syringe plunger is withdrawn to ensure that the needle has not entered a blood vessel or possibly the bowel. While anesthetized, the monitoring of anesthetic depth via assessments of toe-pinch withdrawal, respiratory rate, and related observational methods are utilized.
3. The anesthetized animal is brought to the animal preparation room where it is weighed and inspected.
4. After anesthesia the animal is placed on a temperature-controlled surgical table before a midline incision of the abdomen is made (see Fig. 1 for the schematic illustration of the following procedure).
5. The abdominal aorta is then catheterized with a polyethylene tubing (PE50), the vessel clamp is removed, and chilled

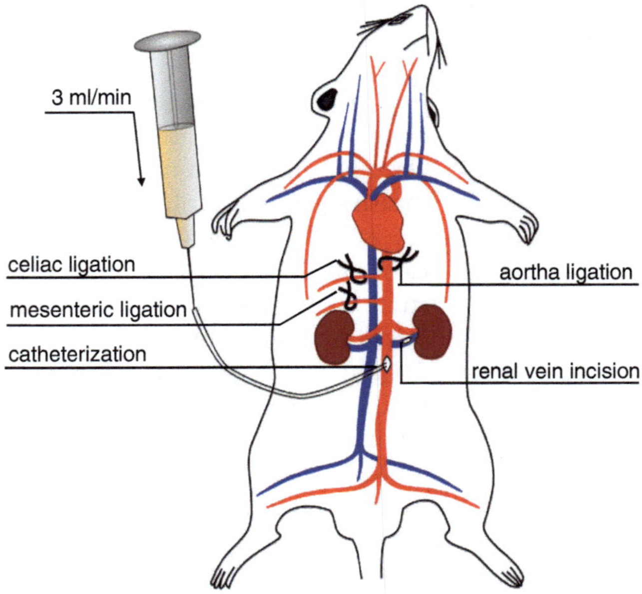

Fig. 1 Schematic illustration of the rat kidney perfusion protocol

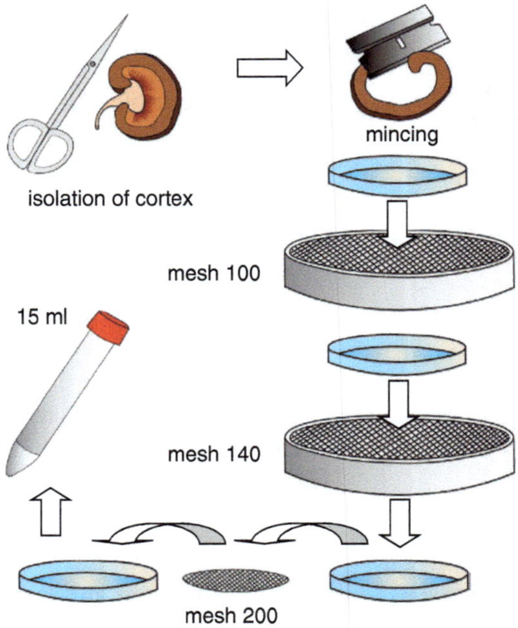

Fig. 2 Scheme of the glomeruli isolation protocol. Kidney cortex was isolated from the Sprague–Dawley rat kidney and then homogenized with a blade. The homogeneous cortex fraction was pushed through the consecutive steel sieves of the different mesh size and then collected into a Petri dish

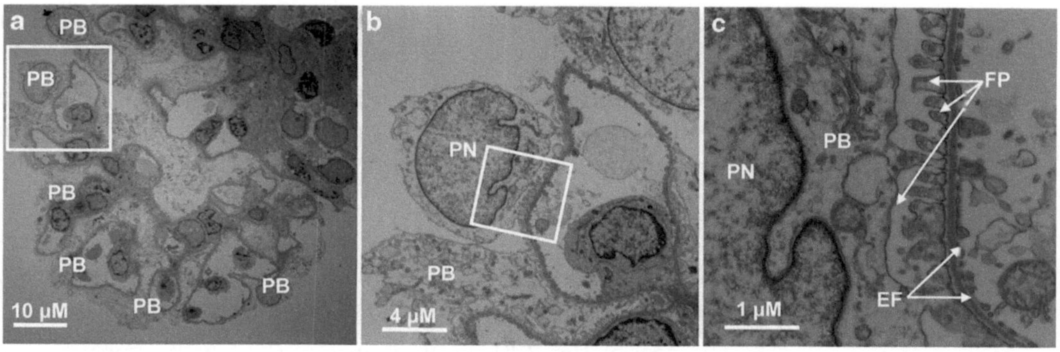

Fig. 3 Electron microscopy of freshly isolated decapsulated glomerulus. Fragments of the decapsulated glomerulus were visualized with electron microscopy at 1,500×, 5,000×, and 20,000× (*a*, *b*, and *c*, respectively; scale bars are shown). *PB* podocyte body, *PN* podocyte nucleus, *EF* endothelial fenestrations, *FP* foot processes (reproduced from ref. 24 with permission from Elsevier)

physiological salt solution (PBS) infused for 2–3 min at a rate of 3 mL/min.

6. A clamp is placed just below the renal arteries around the abdominal aorta. The abdominal aorta is then ligated below the clamp and the celiac and superior mesenteric arteries also ligated (see Note 3 for a reference to other methods of the kidney perfusion).
7. The vena cava near the renal veins is incised to prevent pressure buildup from fluid.
8. After 2–3 min of infusion, the kidneys are excised and the diaphragm is cut to euthanize the animal.
9. The kidneys are put on ice into a 50 mL tube with 25 mL of the HBSS solution.

3.2 Isolation of the Rat Glomeruli

1. Before the experiment, fresh solution of the RPMI 1640 with BSA is prepared adding 5 mL of 30% BSA to 25 mL of the media (see Note 4) (from here on referred to as *solution A*).
2. The kidney is taken from the HBSS solution and decapsulated with the help of the surgical scissors. See Fig. 2 that illustrates all the steps of the procedure for reference.
3. The decapsulated kidney is coronary cut in two halves with a razor blade (see Note 5) and then cortex is isolated and placed into approximately 4 mL of the ice-cold solution A.
4. The kidney is minced with a razor blade and mixed with a fine transfer pipette to obtain a homogenous substance (see Note 6).
5. The homogenate from step 4 is transferred onto the top of the 100 mesh sieve (see Note 7 that comments on the mesh size)

and is pushed through the sieve with a spoon-shaped end of the spatula and is collected into a Petri dish (see Note 8).

6. The solution from a Petri dish from step 5 is transferred onto the top of the presoaked 140 mesh steel sieve, is let to flow through the sieve by gravity force, and is collected into a new Petri dish.
7. The solution collected at step 6 is filtered through the presoaked 200 mesh sieve and the filtrate is discarded (see Note 9). The glomeruli fraction sediments on the top of the sieve.
8. The top of the 200 mesh sieve is washed with approximately 15 mL of the solution A, and then glomeruli fractions are collected into a 15 mL tube and stored on ice (see Notes 10 and 11).

3.3 Single-Channel Analysis of the TRPC Channel Activity Using Patch-Clamp Method

1. Immediately after isolation, the 15 mL tube with the fraction of the glomeruli (see Note 12) is left on ice for up to 20 min to let the glomeruli settle down and concentrate at the bottom of the tube (see Note 13).
2. After sedimentation the supernatant is removed and the rest of the solution containing the concentrated fraction of the glomeruli is kept on ice.
3. Before the patch-clamp experiment, the solution containing glomeruli is mixed and 50 μL of obtained solution are removed by a pipettor and placed on the poly-D-lysine pre-coated 5 × 5 mm cover glasses (see Note 14); the chip is left at room temperature for 5–10 min to let the glomeruli settle down and attach to the surface. As demonstrated in Fig. 3, the podocytes of isolated glomeruli are intact.
4. The glass chip placed into the recording chamber of the patch-clamp setup is filled with the extracellular bathing solution and the chamber is perfused with the solution for 2–3 min to remove the unattached glomeruli and ensure the replacement of the BSA-containing solution with the bath solution.
5. Directly before the patch-clamp experiments the pipette solution used for the single-channel recordings is complimented with various inhibitors to prevent the registration of the unrelated chloride, calcium, and potassium currents: 100 μM niflumic acid or DIDS, 10 mM TEA, 10 nM iberiotoxin, 10 nM nicardipine (see Note 15).
6. The patch pipette is slowly lowered to the surface of the glomerulus (see Note 16) and a high-resistance seal is formed with application of gentle suction. The formation of a gigaohm seal (>1 GΩ) is monitored by the pipette resistance that is increasing from 7 to 10 MΩ to more than 1 GΩ after the suction is applied. Figure 4 illustrates the decapsulated rat glomerulus in the patch-clamp setup with a patch pipette attached to a podocyte on the edge of the glomerulus.

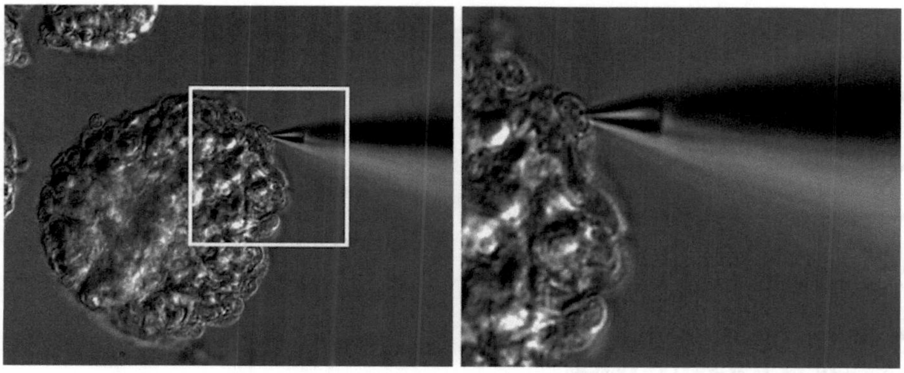

Fig. 4 A representative image of a decapsulated rat glomerulus in the patch-clamp setup with a patch pipette attached to a podocyte on the edge of the glomerulus (40× and a close-up image) (reproduced from ref. 24 with permission from Elsevier)

7. Upon the formation of a high-resistance seal and compensation of the offsets, the cell membrane patch isolated by the pipette is voltage clamped with the potential set in the software. The currents evoked by the potentials are recorded and stored in the PC hard drive. For the cell-attached measurements, the currents are low-passed at 300 Hz by an eight-pole Bessel filter and digitized at 1 kHz.
8. The membrane is voltage clamped at −60 mV test potential using Clampex in the gap-free mode. If the channels activity is present then a current–voltage dependency is recorded. Figure 5a, b illustrates the activity of the two distinct types of ion channels registered in the cell-attached conditions at different potentials (from −120 to +70 mV) in the membrane of the podocytes of the freshly isolated glomeruli. The data revealed that the channels recorded were active throughout the range of holding potentials tested (see Notes 17 and 18). The current–voltage dependency shown at Fig. 5d allowed distinguishing two channel types. The identified TRPC-like channels were characterized by different conductances and displayed similar gating properties. Data analysis revealed two major channel populations with the conductance values of 10.3 ± 0.8 pS ($n = 11$) and 20.1 ± 0.7 pS ($n = 10$) (Fig. 5d). Both channels had a reversal potential (E_{rev}) of 0 mV. We termed these channels as "big" and "small" TRPC-like channels, respectively. Different conductances were sometimes found to coexist in the same patch as illustrated in Fig. 5c (24).
9. The cell-attached configuration allows applying the drugs and observing the changes in the channels activity in the native preparation of the intact glomerulus. Figure 6 shows the modulation of the TRPC-like channels in the podocytes as represented by inhibition by 500 μM of the nonsteroid anti-inflammatory drug diclofenac (see Notes 19–21).

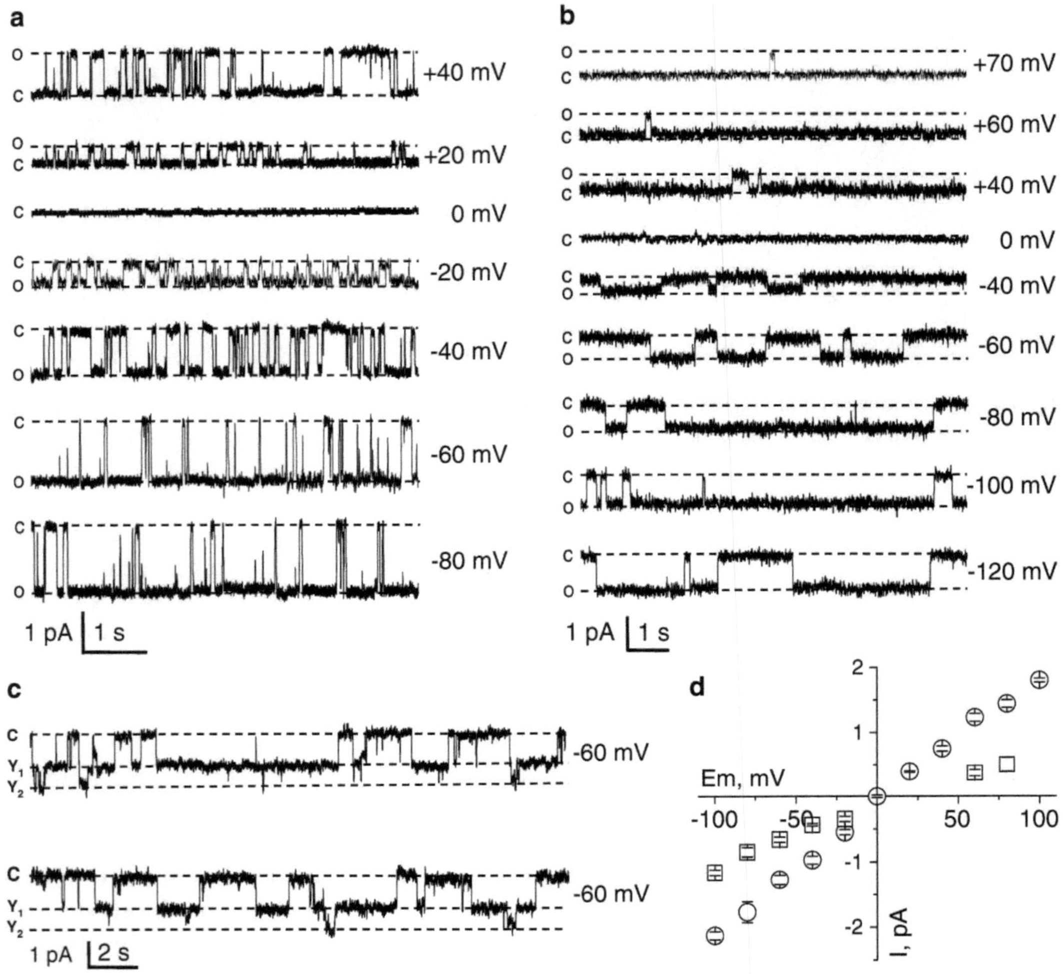

Fig. 5 Single-channel biophysical properties of TRPC-like channels identified in podocytes of freshly isolated rat glomeruli. All recordings were performed in cell-attached configuration in voltage-clamp mode. Representative current traces of "big" (**a**) and "small" (**b**) channels identified in the podocytes. The holding membrane potentials are indicated near the traces. *c* and *o* denote closed and open current levels, respectively. (**c**) Typical current traces held at −60 mV demonstrating coexistence of the two types of the channels in the same patch. Y_i denotes the open current levels of channels with different conductances. (**d**) Single-channel current–voltage relationship for two types of channels identified. Values are means ± SEM of at least four experiments (reproduced from ref. 24 with permission from Elsevier)

4 Notes

1. Non-survival surgeries are performed in septic but clean conditions. Instruments used should be clean, but not necessary sterile. All solutions used in this preparation should be sterile and kept on ice.
2. Anesthesia can also be performed using isoflurane gas. In the case of inhalant anesthesia, the rat is placed in a transparent

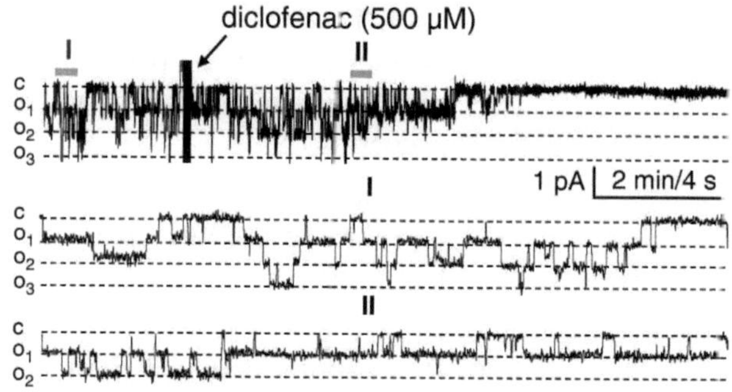

Fig. 6 Diclofenac modulate the activity of TRPC channels in podocytes of isolated glomeruli. Representative experiment from a cell-attached patch containing TRPC channels. Continuous current trace as well as the fragments of the traces at the expanded time scale are shown. *Arrows* demonstrate addition of diclofenac to the external bath solution. The patch was held at a −60 mV test potential during the course of the experiment. The *c* and o_i denote closed and open current levels, respectively (figure reproduced from ref. 24 with permission from Elsevier)

induction chamber. Isoflurane is delivered to the chamber via a precision vaporizer and compressed O_2. For induction, the percentage of isoflurane may be as high as 5%. Once the animal is unconscious, it is removed from the chamber and laced on a heated surgical table and a nose cone applied to continue delivery of anesthetic. The nose cone is attached to the vaporizer and oxygen source. At this point the concentration is reduced to that level which maintains the proper plane of anesthesia; typically this is between 1 and 3%.

3. There are other slightly different methods of kidney perfusion than we described in this procedure. Here we provide a brief description of two additional methods.

   (a) Ligate the distal abdominal aorta and distal inferior cava vein and clip the abdominal aorta and inferior cava vein by vessel clamp below the renal arteries and veins. Then catheterize the abdominal aorta and ligate the mesenteric and celiac arteries and abdominal aorta above the renal artery. After that remove the clamp and cut the inferior cava vein to ensure the flow of the perfusion solution and start perfusing the kidney.

   (b) Another method of kidney perfusion suggests that the distal abdominal aorta and distal inferior cava vein, as well as the superior mesenteric and celiac arteries, should be ligated. Then the thoracic aorta is to be catheterized with venous retention needles (24 G) and a hole should be cut in the inferior cava vein to ensure venous drainage.

4. We have tested different solutions for isolation of the glomeruli to lower the cost of this procedure (i.e., RPMI media with

FBS, DMEM media with FBS) and got a lower yield of the glomeruli, and the preparation was very difficult to patch; the glomeruli seemed to be too soft compared to original preparation.

5. We suppose that it might be more convenient to cut the kidney coronary in the ratio 1:3; this will allow removing the medulla from one piece, only. The second part will consist of the cortex only.
6. It is supposed that 2–3 mL of the solution A is added to the fraction isolated after every step and then it is mixed with a fine transfer pipette until homogeneity is reached.
7. The mesh size of the sieves listed in this chapter is adjusted for the 8-week-old male Sprague–Dawley rats; if you are going to isolate the glomeruli from mice or rats of other age, we would recommend adjusting the mesh size accordingly. The probable starting point for adjusting the mesh size for mice can be found in the works of Akis et al. and Rops et al. (25, 26).
8. Steel sieves need to be soaked in the solution A. For large sieves, a drop of the solution is to be spread on the top of the sieve in the area you are going to use (up to 6–7 cm in diameter), and the small sieve is to be fully soaked in the solution A. The flow of the homogenate can be facilitated with the spoon-shaped end of the spatula. The homogenate that stayed on the reverse side of the sieve and did not get into the Petri dish can be collected from the reverse side of the sieve with a razor blade. Homogenate pushed through a new sieve always requires being collected into a new Petri dish.
9. We would recommend being very careful with the last sieve as the pure fraction of the glomeruli rests of the top of it and it is likely to lose them at this step, so it would be reasonable to seep the solution onto it slowly, drop by drop.
10. After the glomeruli fraction settles down to the bottom of the tube, we recommend removing most of the supernatant and leaving about 800 μL of the solution in the tube.
11. Takemoto et al. (27) suggested that the glomeruli of the mice can be efficiently isolated with the Dynabeads. In brief, the kidney is perfused with the solution containing magnetic Dynabeads, then minced, digested with collagenase, pushed through the sieve, and then the Dynabeads-containing glomeruli are separated with a magnetic particle concentrator. However, this approach seems to be more time-consuming and disturbs the native setting of the glomerulus with the Dynabeads.
12. The fraction that is isolated with this preparation is highly pure and consists mostly of the decapsulated glomeruli; however, up to 10% of the glomeruli in the solution might still have a capsule.

The final mixture contains very low quantity of other nephron segments. We (24) and others (21) successfully used this preparation for other applications such as Western blotting.

13. The glomeruli fraction can be spun (5 min at 500 × *g*) to facilitate the sedimentation if the preparation is used for biochemical assays. However, we do not recommend this step if preparation is to be used for patch-clamp experiments as spinning makes the glomeruli more difficult to patch.
14. Glass coverslips cut into pieces of approximately 5 × 5 mm are covered with poly-D-lysine before the experiment. We do not recommend using coverslips with lysine coating that are more than 3 days old.
15. 100 μM niflumic acid or DIDS are added to block Ca^{2+}-activated Cl^- channels; 10 mM TEA is used to inhibit the activity of the large-conductance Ca^{2+}-dependent K^+ channel; 10 nM iberiotoxin blocks the Ca^{2+}-activated K^+ channels; and 10 μM nicardipine inhibits the N-type Ca^{2+} channels (24, 28). We recommend keeping the pipette solution on ice to maintain the inhibitors active.
16. The podocytes are found on the surface of the glomerulus; the most amenable for patching podocytes are usually found on the edges of the glomerulus and have an oval prominent form on its surface (see Fig. 4 for reference).
17. In our experiments the next day after isolation of the glomerular fraction, we succeeded to obtain stable patches with channels with approximately the same activity as on the same day of isolation. We have not tried to keep the glomeruli any longer, but it is possible that they might be vital for up to 1 week if they are kept at 4 °C. If this works, the same approach can be successfully used for over- or down-regulation of proteins of interest.
18. Active channels were observed in up to 30% of the cell-attached patches obtained, whereas in approximately 50% of the patches that initially showed no channel activity, channels were activated by respective drugs (data not shown). Thus, there is a possibility that majority of channels are silent in this preparation.
19. As reported by Gloy and colleagues, whole-cell current measurements are also possible on the freshly isolated preparation of the glomeruli (20).
20. The patches once formed remained stable and the activity of the single channels was monitored for up to 20 min; without application of the drugs no spontaneous activation or rundown of the channel activity and changes of the channel gating was observed.

21. Biophysical properties, inhibition by SKF-096365 (blocker of TRP cation channels (29), insensitivity to the inhibitors within the patch pipette, and the parameters of the current–voltage dependence (Fig. 5) strongly suggest that the channels recorded in the membrane of the freshly isolated rat glomeruli are the members of the TRPC superfamily. However, several types of TRPC channels are identified in the podocytes, and formation of homo- and heteromeric complexes of these subunits are proposed (10, 30–33). Thus, precise identification of TRPC composition remains to be identified.

Acknowledgments

We wish to thank Dr. Allen W. Cowley, Jr., and Dr. David R. Harder (Medical College of Wisconsin) for helpful discussion and help with development of this approach. Vladislav Levchenko, Debebe Gebremedhin, Robert P. Ryan, Glenn Slocum, Christine Duris, and Clive Wells (all Medical College of Wisconsin) are recognized for excellent technical assistance and initial help with described experiments. This research was supported by the American Diabetes Association Grant 1-10-BS-168 and the National Institutes of Health grant HL108880.

References

1. Patrakka J, Tryggvason K (2009) New insights into the role of podocytes in proteinuria. Nat Rev Nephrol 5:463–468
2. Pavenstadt H, Bek M (2002) Podocyte electrophysiology, in vivo and in vitro. Microsc Res Tech 57:224–227
3. Abramowitz J, Birnbaumer L (2009) Physiology and pathophysiology of canonical transient receptor potential channels. FASEB J 23:297–328
4. Kiselyov K, Patterson RL (2009) The integrative function of TRPC channels. Front Biosci 14:45–58
5. Watanabe H, Murakami M, Ohba T, Ono K, Ito H (2009) The pathological role of transient receptor potential channels in heart disease. Circ J 73:419–427
6. Woudenberg-Vrenken TE, Bindels RJ, Hoenderop JG (2009) The role of transient receptor potential channels in kidney disease. Nat Rev Nephrol 5:441–449
7. Kim JY, Saffen D (2005) Activation of M1 muscarinic acetylcholine receptors stimulates the formation of a multiprotein complex centered on TRPC6 channels. J Biol Chem 280:32035–32047
8. Moller CC, Flesche J, Reiser J (2009) Sensitizing the Slit Diaphragm with TRPC6 ion channels. J Am Soc Nephrol 20:950–953
9. Moller CC, Wei C, Altintas MM, Li J, Greka A, Ohse T, Pippin JW, Rastaldi MP, Wawersik S, Schiavi S, Henger A, Kretzler M, Shankland SJ, Reiser J (2007) Induction of TRPC6 channel in acquired forms of proteinuric kidney disease. J Am Soc Nephrol 18:29–36
10. Dietrich A, Chubanov V, Gudermann T (2010) Renal TRPathies. J Am Soc Nephrol 21:736–744
11. Reiser J, Polu KR, Moller CC, Kenlan P, Altintas MM, Wei C, Faul C, Herbert S, Villegas I, vila-Casado C, McGee M, Sugimoto H, Brown D, Kalluri R, Mundel P, Smith PL, Clapham DE, Pollak MR (2005) TRPC6 is a glomerular slit diaphragm-associated channel required for normal renal function. Nat Genet 37:739–744
12. Winn MP, Conlon PJ, Lynn KL, Farrington MK, Creazzo T, Hawkins AF, Daskalakis N, Kwan SY, Ebersviller S, Burchette JL, Pericak-Vance MA, Howell DN, Vance JM, Rosenberg PB (2005) A mutation in the TRPC6 cation channel causes familial focal segmental glomerulosclerosis. Science 308:1801–1804
13. Heeringa SF, Moller CC, Du J, Yue L, Hinkes B, Chernin G, Vlangos CN, Hoyer PF, Reiser J, Hildebrandt F (2009) A novel TRPC6 mutation that causes childhood FSGS. PLoS One 4:e7771

14. Winn MP, Daskalakis N, Spurney RF, Middleton JP (2006) Unexpected role of TRPC6 channel in familial nephrotic syndrome: does it have clinical implications? J Am Soc Nephrol 17:378–387
15. Eckel J, Lavin PJ, Finch EA, Mukerji N, Burch J, Gbadegesin R, Wu G, Bowling B, Byrd A, Hall G, Sparks M, Zhang ZS, Homstad A, Barisoni L, Birbaumer L, Rosenberg P, Winn MP (2011) TRPC6 enhances angiotensin II-induced albuminuria. J Am Soc Nephrol 22:526–535
16. Chen S, He FF, Wang H, Fang Z, Shao N, Tian XJ, Liu JS, Zhu ZH, Wang YM, Wang S, Huang K, Zhang C (2011) Calcium entry via TRPC6 mediates albumin overload-induced endoplasmic reticulum stress and apoptosis in podocytes. Cell Calcium 50:523–529
17. Krall P, Canales CP, Kairath P, Carmona-Mora P, Molina J, Carpio JD, Ruiz P, Mezzano SA, Li J, Wei C, Reiser J, Young JI, Walz K (2010) Podocyte-specific overexpression of wild type or mutant trpc6 in mice is sufficient to cause glomerular disease. PLoS One 5:e12859
18. Schlondorff J, del Camino D, Carrasquillo R, Lacey V, Pollak MR (2009) TRPC6 mutations associated with focal segmental glomerulosclerosis cause constitutive activation of NFAT-dependent transcription. Am J Physiol Cell Physiol 296:C558–C569
19. Kanda S, Harita Y, Shibagaki Y, Sekine T, Igarashi T, Inoue T, Hattori S (2011) Tyrosine phosphorylation-dependent activation of TRPC6 regulated by PLC-gamma1 and nephrin: effect of mutations associated with focal segmental glomerulosclerosis. Mol Biol Cell 22:1824–1835
20. Gloy J, Henger A, Fischer KG, Nitschke R, Mundel P, Bleich M, Schollmeyer P, Greger R, Pavenstadt H (1997) Angiotensin II depolarizes podocytes in the intact glomerulus of the Rat. J Clin Invest 99:2772–2781
21. Dahly-Vernon AJ, Sharma M, McCarthy ET, Savin VJ, Ledbetter SR, Roman RJ (2005) Transforming growth factor-beta, 20-HETE interaction, and glomerular injury in Dahl salt-sensitive rats. Hypertension 45:643–648
22. Sharma R, Khanna A, Sharma M, Savin VJ (2000) Transforming growth factor-beta1 increases albumin permeability of isolated rat glomeruli via hydroxyl radicals. Kidney Int 58:131–136
23. Savin VJ, Sharma R, Lovell HB, Welling DJ (1992) Measurement of albumin reflection coefficient with isolated rat glomeruli. J Am Soc Nephrol 3:1260–1269
24. Ilatovskaya DV, Levchenko V, Ryan RP, Cowley AW Jr, Staruschenko A (2011) NSAIDs acutely inhibit TRPC channels in freshly isolated rat glomeruli. Biochem Biophys Res Commun 408:242–247
25. Akis N, Madaio MP (2004) Isolation, culture, and characterization of endothelial cells from mouse glomeruli. Kidney Int 65:2223–2227
26. Rops AL, van der Vlag J, Jacobs CW, Dijkman HB, Lensen JF, Wijnhoven TJ, van den Heuvel LP, van Kuppevelt TH, Berden JH (2004) Isolation and characterization of conditionally immortalized mouse glomerular endothelial cell lines. Kidney Int 66:2193–2201
27. Takemoto M, Asker N, Gerhardt H, Lundkvist A, Johansson BR, Saito Y, Betsholtz C (2002) A new method for large scale isolation of kidney glomeruli from mice. Am J Pathol 161:799–805
28. Sours S, Du J, Chu S, Ding M, Zhou XJ, Ma R (2006) Expression of canonical transient receptor potential (TRPC) proteins in human glomerular mesangial cells. Am J Physiol Renal Physiol 290:F1507–F1515
29. Merritt JE, Armstrong WP, Benham CD, Hallam TJ, Jacob R, Jaxa-Chamiec A, Leigh BK, McCarthy SA, Moores KE, Rink TJ (1990) SK&F 96365, a novel inhibitor of receptor-mediated calcium entry. Biochem J 271:515–522
30. Dryer SE, Reiser J (2010) TRPC6 channels and their binding partners in podocytes: role in glomerular filtration and pathophysiology. Am J Physiol Renal Physiol 299:F689–F701
31. Goel M, Sinkins WG, Zuo CD, Estacion M, Schilling WP (2006) Identification and localization of TRPC channels in the rat kidney. Am J Physiol Renal Physiol 290:F1241–F1252
32. Greka A, Mundel P (2011) Balancing Calcium Signals through TRPC5 and TRPC6 in Podocytes. J Am Soc Nephrol 22:1969–80
33. Tian D, Jacobo SM, Billing D, Rozkalne A, Gage SD, Anagnostou T, Pavenstadt H, Hsu HH, Schlondorff J, Ramos A, Greka A (2010) Antagonistic regulation of actin dynamics and cell motility by TRPC5 and TRPC6 channels. Sci Signal 3:ra77

Chapter 29

Ca^{2+} Imaging as a Tool to Assess TRP Channel Function in Murine Distal Nephrons

Mykola Mamenko, Oleg Zaika, Roger G. O'Neil, and Oleh Pochynyuk

Abstract

Transient receptor potential (TRP) channels are expressed in almost every segment of renal nephron from the glomerulus to the inner medullary collecting duct. Serving as a route for Ca^{2+} entry from the intratubular space into cells in response to external cues, TRP channels modulate water–electrolyte transport, thus determining functional properties of the renal tubule. In this chapter, we discuss technical aspects of using Ca^{2+} imaging to monitor activity of TRP channels in situ, namely, in the freshly isolated distal nephrons, with a special emphasis on the mechanosensitive TRPV4 channel and its role in tubular flow sensing.

Key words Collecting duct, Connecting tubule, Fura-2, Mechanosensitivity, Shear stress

1 Introduction

Water and electrolytes transport in the distal part of the renal nephron determines urinary volume and composition. This tubular segment is comprised of the connecting tubule and the collecting duct and has two morphologically and functionally distinct cell types: principal cells (PC, 70% of total cell number) and intercalated cells (IC, ~30% of total cell number). PC cells are involved in regulation of Na^+, K^+, and water handling whereas IC cells control acid–base balance. The distal nephron is heavily influenced by circulating levels of hormones, such as aldosterone, vasopressin, and atrial natriuretic peptide (1–4). In addition to these systemic endocrine inputs, dynamic changes in tubular fluid flow and composition also affect functional properties of the distal nephron cells. This is thought to be a physiologically relevant local mechanism for regulation of water and electrolyte transport in this segment that complements systemic hormonal control (2, 5–7). Numerous observations are consistent with the idea that elevations in $[Ca^{2+}]_i$ in distal nephron cells serve to mediate adaptations of the transport rates in response to mechanical stimuli (6, 8–17). The primary cilia of epithelial cells were proposed to play a critical role in flow-mediated

Nikita Gamper (ed.), *Ion Channels: Methods and Protocols*, Methods in Molecular Biology, vol. 998,
DOI 10.1007/978-1-62703-351-0_29, © Springer Science+Business Media, LLC 2013

cellular responses (14, 18). However, ICs, which lack the primary cilia, respond to flow changes with comparable increases in $[Ca^{2+}]_i$ as PCs which possess primary cilia (11). Furthermore, mechanosensitive Ca^{2+} responses are abolished upon Ca^{2+} removal from extracellular media (14, 19). This suggests that mechanotransduction in the distal nephron involves direct Ca^{2+} entry via Ca^{2+}-permeable channels/conduits at the apical/luminal plasma membrane.

Transient receptor potential (TRP) channels belong to one of the most versatile superfamily of ion channels. TRP channels are known to participate in cellular adaptations to a variety of environmental stimuli, including thermo-sensation, chemo-sensation, and mechanical forces (reviewed in ref. 20). TRPs are Ca^{2+}-permeable channels, but can also conduct other physiologically relevant cations (Na^+, K^+, Mg^{2+}) with various degrees of selectivity (20). Activation of these channels drives Ca^{2+} entry from the extracellular media, thus, likely contributing to the elevation of $[Ca^{2+}]_i$ in response to environmental changes (20). Several TRP channels, including TRPC3, TRPC6, TRPV4-6, and TRPP2, can be detected with immunohistochemistry in the native distal nephron cells and distal nephron-derived cultures (14, 21–23). Among these channels, TRPV4 has routinely been shown to be activated by mechanical stimuli in many cell types (23–27). Indeed, recent experimental evidence points to a pivotal role of TRPV4 and likely TRPV4–TRPP2 heteromers in responses to elevations in tubular flow in distal nephron cells (19, 28).

Conventional patch clamp is the most direct experimental methodology to assess functional status of ion channels at the plasma membrane (29). This approach, though, has serious technical difficulties when applied to study mechanosensitive TRP channels in native epithelia, including that of the distal nephron. Electrical coupling between cells within epithelial monolayer and/or high constitutive background currents preclude obtaining high-resistant (~GΩ) seals in both "whole-cell" and "perforated-patch" configurations and thus, accurate voltage clamping. Furthermore, basal activity of mechanosensitive TRP channels is generally low. Direct mechanical stimulations to enhance channel activity are also difficult to perform. This makes single-channel recordings of mechanosensitive channels in isolated nephrons in "cell-attached" configuration an inefficient and daunting task. In contrast, Fura-2-based ratiometric imaging provides a reliable way to probe TRP channels activity by monitoring changes of $[Ca^{2+}]_i$ in individual cells as a reflection of Ca^{2+} entry through the channels.

Since developed by M. Burg and colleagues (30, 31), the technique to perfuse isolated segments of renal tubule has been instrumental to study regulation of transport of water and solutes as well as $[Ca^{2+}]_i$ dynamics by pharmacological and mechanical stimuli. This technique allows perfusing the tubular lumen at physiologically relevant rates presented in the intact nephron and distinguishing

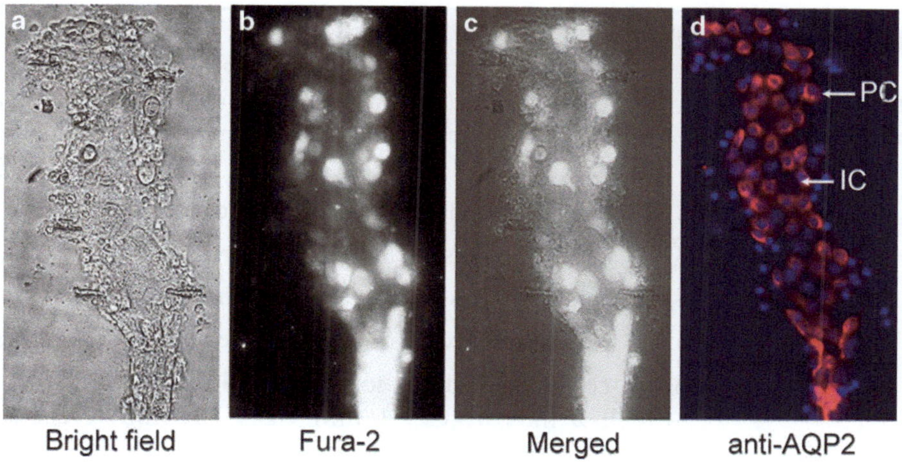

Fig. 1 Ca^{2+} imaging in freshly isolated split-opened distal nephron. Representative micrographs of the same split-opened distal nephron taken with bright-field illumination (**a**), excitation at 380 nm after loading with Fura-2 (**b**), the combined illumination with visible light and 380 nm (**c**), and after fixation and staining of the distal nephron with anti-AQP2 antibodies (**d**). Nuclear DAPI staining is also shown (*pseudocolor blue*). Examples of AQP2-positive principal cells (PC) and AQP2-negative intercalated cells (IC) are pointed with *arrows*

contributions of channels/transporters and membrane receptors from apical versus basolateral sides. Precise separation of fluorescent $[Ca^{2+}]_i$ signals from individual cells within a perfused tubule is, though, not readily feasible and often requires expensive instrumentations, such as a fast-scanning confocal microscope. Furthermore, perfusion with high flow rates causes a prominent increase (~20%) in tubular diameter and, as a consequence, significant changes in volume (32). This phenomenon is greatly alleviated in the intact renal tissue due to rigid support from the surrounding nephrons. Tubular stretching by high perfusion rates leads to elevation of $[Ca^{2+}]_i$ per se distorting cellular responses to luminal shear stress (32). Split-opening of an isolated renal tubule allows to gain access to the apical membrane of epithelial cells and to circumvent many technical pitfalls present in perfused tubules. First, Ca^{2+} signal from individual cells can be routinely separated using conventional wide-field fluorescent microscopy setup (Fig. 1). Second, split-opened tubules do not undergo lateral stretch even during elevated flow perfusion rates due to support provided by a coverslip. Third, the luminal shear stress can be precisely estimated and quantified using a parallel plate perfusion chamber (see Subheading 3.3). In addition, split-opened tubular segment can be further subjected to immunohistochemical staining with specific markers (i.e., aquaporin 2, AQP2) to separate fluorescent Ca^{2+} signals from different cell types (i.e., PC versus IC; see Subheading 3.4).

In summary, monitoring $[Ca^{2+}]_i$ dynamics in split-opened distal nephrons represents a convenient and physiologically relevant way to assess mechanosensitivity of the distal nephron and to probe involvement of Ca^{2+}-permeable TRP channels in this process.

2 Materials

2.1 Rodent Kidney Extraction

1. A rat or a mouse (6–8 weeks of age) of wild-type (such as S/D and C57BL/6, respectively) or a genetically modified strain of interest.
2. CO_2 gas tank and a hermetic chamber for sacrifice.
3. Dissecting scissors (Walter Stern, Port Washington, NY, USA).
4. Specimen forceps (Walter Stern, Port Washington, NY, USA).
5. Gloves (Microflex Tranquility powder-free nitrile gloves).
6. Ice-cold balanced salt solution (BSS): 150 mM NaCl, 5 mM KCl, 2 mM $MgCl_2$, 1 mM $CaCl_2$, 10 mM Hepes, 5 mM Glucose; pH = 7.4.
7. 60 × 15 mm plastic tissue culture dishes (Sarstedt, Newton, NC, USA).

2.2 Isolation and Split-Opening of Distal Nephrons

1. Nikon SMZ645 stereomicroscope (or equivalent from other vendors).
2. Dumont SS fine forceps (Fine Science Tools, Foster City, CA, USA).
3. 60 × 15 mm plastic tissue culture dishes (Sarstedt, Newton, NC, USA).
4. Balanced salt solution (BSS, see Subheading 2.1, item 6).
5. 5 × 5 mm coverslips, cut from square 22 × 22 mm, No. 1.5 cover glasses (Thermo Fisher Scientific, Pittsburgh, PA, USA).
6. 0.01% solution of poly-L-lysine (Sigma-Aldrich Corp., St. Louis, MO, USA).
7. Single-edge razor blades (American Line 66–0089, available from numerous vendors).
8. Freshly isolated murine kidney.
9. Inverted microscope (Nikon Eclipse T*i*-U microscope or analogue).
10. Two micromanipulators (Sutter Instrument Company, Novato, CA, USA; MP-285 motorized micromanipulators or analogue).
11. Glass borosilicate pipettes (World Precision Instruments, Sarasota, FL, USA).

2.3 Monitoring $[Ca^{2+}]_i$ in Individual Cells of Freshly Isolated Distal Nephrons

1. 35 × 10 mm plastic tissue culture dishes (Sarstedt, Newton, NC, USA).
2. Balanced salt solution (see Subheading 2.1, item 6).
3. Stock solution of 100 μg Fura-2/AM (EMD Biosciences/Calbiochem, La Jolla, CA, USA) dissolved in 50 μL dimethyl sulfoxide (DMSO, Sigma-Aldrich Corp., St. Louis, MO, USA).

4. Inverted microscope (Nikon Eclipse T*i*-U or analogue), with a 40× CFI Super Fluor (0.9 NA) objective (Nikon Instruments, Melville, NY, USA).
5. Fast exchange 0.5 mL recording chamber (RC-24 N, Warner Instruments, Hamden, CT, USA) or a proflow shear flow chamber (PFC-1, Warner Instruments, Hamden, CT, USA) or analogues.
6. Multichannel valve perfusion system (VC-8P eight channel perfusion valve control system by Warner Instruments, Hamden, CT, USA, or analogue).
7. Fluorescence light source, we use Lambda LS stand-alone xenon lighting system (Sutter Instrument Company, Novato, CA, USA) with 150 W Xenon Lamp (Hamanatsu Corp., Kista, Sweden).
8. ET FURA2 filter set (Chroma Technologies Corp., Bellows Falls, VT, USA) with 340/380 nm excitation filters, 400 nm long pass dichroic mirror and 510/80 nm emission filter with Lambda 10-C optical filter changing system (Sutter Instrument Company, Novato, CA, USA).
9. CCD camera (we use QIClick digital monochrome 12-bit non-cooled CCD camera by Q-Imaging, Surrey, BC, Canada; Model: QIClick-F-M-12).
10. Nikon NIS elements 4.00 or alternative software with capacity to run camera and switch filters (InCa, Intracellular Imaging, Inc., Cincinnati, OH).

2.4 Immunostaining of Distal Nephron Cells and Visualization of the Tubules with Immunofluorescence

1. 12-well microtiter plate (BD Falcon, Franklin Lakes, NJ, USA) or 35 × 10 mm diameter plastic tissue culture dishes (Sarstedt, Newton, NC, USA).
2. Phosphate buffer saline (PBS), pH = 7.4 (made of 10× concentrate, Sigma-Aldrich Corp., St. Louis, MO, USA).
3. 4% paraformaldehyde (Sigma-Aldrich Corp., St. Louis, MO, USA) solution in PBS (pH = 7.4).
4. 0.1% Triton X-100 (Sigma-Aldrich Corp., St. Louis, MO, USA) solution in PBS (pH = 7.4).
5. 0.05% Tween 20 in PBS (PBS-Tween buffer, also available as a dry powder from Sigma-Aldrich, USA).
6. Normal goat serum (NGS, Jackson Immunoresearch, West Grove, PA, USA).
7. Anti-AQP2 antibodies labeled with ATTO-550 fluorescent dye (Alomone labs, Jerusalem, Israel).
8. 300 nM solution of 4′,6-diamidino-2-phenylindole (DAPI, Sigma-Aldrich Corp., St. Louis, MO, USA) in PBS.
9. Mounting medium (Thermo Scientific, Pittsburgh, PA, USA).

10. Microscope glass slides (VWR, Radnor, PA, USA).
11. Nikon Eclipse T*i*-U inverted microscope with a 40× CFI Super Fluor (0.9 NA) objective (Nikon Instruments, Melville, NY, USA), a metal halide lamp (Nikon Instruments) and UV-2E/C (Nikon Instruments) and G-2E/C filter sets (Nikon Instruments).
12. CCD camera (QIClick digital monochrome 12-bit non-cooled CCD camera by Q-Imaging, Surrey, BC, Canada; Model: QIClick-F-M-12).
13. Nikon NIS elements 4.00 or alternative software allowing recording the images from a CCD camera.

3 Methods

Using freshly isolated renal tubular fragments, while technically more challenging, provides an invaluable opportunity to investigate functional properties of ion transport and signaling mechanisms in physiologically relevant tissue and native environment. Moreover, it allows to establish a link between dietary (food and water intake), pharmacological (systemic injections/treatment with a drug and/or hormone), and genetic (knockout or knock-in different genes) manipulations at the level of whole organism with changes in functional status of a particular molecule, such as a TRP channel. Culturing of epithelial cells, while proved to be a reasonable and simpler approach, often leads to phenotypic changes in expression patterns of ion channels and transporters. Therefore, the experimental evidence obtained in these systems might be distorted and require careful reevaluation in in vivo and ex vivo studies.

3.1 Preparations of Split-Opened Distal Nephrons

1. Preparation of native murine tubular segments starts with extraction of kidneys from sacrificed animals. This procedure is simple and requires basic knowledge about rat and mouse anatomy. For experiments, rats or mice 6–8 weeks old are commonly used (see Note 1). We recommend contacting veterinarians to comply with specific local/national ethic requirements. Typically, rats and mice are sacrificed by decapitation or administration of CO_2. Unless necessary for experimental design, sterile conditions for kidneys extraction are welcomed, but not absolutely required. It is sufficient to work in latex gloves and to treat surgical table and instruments with alcohol. After incision of the body cavity, kidneys are removed and placed onto 60 × 15 mm tissue culture dish with ice-cold BSS.
2. Mechanical isolation of individual distal nephrons from murine kidney is much more difficult to perform. It requires sufficient practicing and coordination. The detailed description of the

procedure is not covered in this chapter. Here, we will concisely highlight key steps of the preparation. The kidney is placed into a dry tissue culture dish, decapsulated using two sharpened forceps (see Note 2), and axially cut into thin slices with a sharp razor blade. Kidney slices are immediately placed into a tissue culture dish with ice-cold BSS. Small longitude sectors containing ~20–30 tubules are gently teased out of a kidney slice under a stereomicroscope using watchmaker forceps. Individual sectors are further disassembled into individual tubules under a higher magnification. Distal nephrons can be visually identified by their pale color, coarse shape, and, in many cases, bifurcations. Of note, only a minority (~10–15%) of isolated tubules are distal nephrons. For beginners, we recommend to verify the origin of the isolated segments under a high-magnifying microscope. In contrast to all other segments, distal nephrons have clearly distinguishable cells inside an intact tubule. Isolated distal nephrons are subsequently affixed to the poly-L-lysine-coated 5 × 5 mm coverslips.

3. Split-opening of a distal nephron is another technical challenge to overcome. A coverslip with attached tubule is transferred to a recording/imaging chamber on the stage of an inverted microscope. Tubule is split-opened with two micropipettes controlled with different micromanipulators. Using sharp glass micropipettes with resistances higher than 20 MΩ generally yields the best result. We encourage putting maximal effort towards split-opening a tubule along its length as fully as possible. This will allow to monitor $[Ca^{2+}]_i$ dynamics simultaneously from a greater number of cells within a split-opened area. An example of split-opened distal nephron is shown in Fig. 1a.
4. A coverslip with split-opened tubule is transferred into a 35 mm tissue culture dish with 3 mL BSS solution containing 2 μM of an ester form of a $[Ca^{2+}]_i$-sensitive dye, Fura-2 AM, and incubated for 40–45 min in darkness (see Note 3). It is critical to minimize exposure of the solution to visible light during transferring and incubation since Fura-2 is light sensitive. In a simplest scenario, a tissue culture dish is wrapped with aluminum foil. After incubation, the coverslip is placed back to the recording chamber and washed with bath solution for 10 min to remove an excess of Fura-2 AM.

3.2 Monitoring TRP Channels Function Using $[Ca^{2+}]_i$ Imaging in Distal Nephron Cells

1. The major principle of monitoring $[Ca^{2+}]_i$ in individual cells is based on the phenomenon that fluorescent excitation spectrum of Fura-2 shifts towards shorter wavelengths upon binding Ca^{2+}. Typically, Fura-2 emission is collected at 510 nm in response to repetitive brief excitation at 340 and 380 nm after certain periods of time (see Note 4). The ratio of fluorescent intensities $R = F_{340}/F_{380}$ represents an index of relative changes

in $[Ca^{2+}]_i$. If necessary, the ratio can be converted to actual $[Ca^{2+}]_i$ values after calibration of fluorescent intensities in permeabilized cells (with ionomycin) in solutions with known Ca^{2+} concentration using well-defined protocols (33).

2. Commercially available software (NIS elements, InCa Imaging, etc.) is required for visualization and quantification of Fura-2 fluorescence in split-opened distal nephrons. A snapshot of Fura-2 fluorescent with 340 and 380 excitation is used to adjust exposure times as necessary. Typically, exposure times are in the range of several hundred milliseconds (~300 ms). We recommend monitoring Fura-2 fluorescent signals every 3–5 s. These parameters do not typically lead to more than 10% of decay in Fura-2 signals (due to photobleaching) over 30-min experimental period and are sufficient to reliably monitor changes in $[Ca^{2+}]_i$ in response to activation of TRP channels and membrane G-protein-coupled receptors (such as purinergic and bradykinin receptors). If this is not the case, the exposure times need to be decreased to prevent Fura-2 bleaching. It is important to keep the ratio of the exposure times for 340 and 380 nm constant (e.g., identical exposure times) since otherwise it will complicate comparison of Ca^{2+} dynamics from distinct distal nephrons. Fluorescent signals from individual cells within split-opened area of distal nephron can be easily separated (see Fig. 1a–c). This allows to draw multiple regions of interest (ROI) to monitor changes in $[Ca^{2+}]_i$ simultaneously from many cells. Furthermore, post-experimental fixation and immunohistochemical staining with a selective marker of principal cells, AQP2, enable to probe a role of TRP channels in $[Ca^{2+}]_i$ signaling in different cell types (Fig. 1d; see further Subheading 3.4).

3. Recent pharmacological advances resulted in development of several highly potent agonists for certain TRP channel subtypes. These agonists can be effectively used to probe functional status of TRP channels in distal nephron cells. This is particularly the case for the mechanosensitive TRPV4 channel which can be selectively activated by GSK1016790A (EC_{50} ~ 15 nM) and RN1747 (EC_{50} ~ 1 μM). As exemplified in Fig. 2a, application of 30 nM of GSK1016790A elicits a sustained elevation of $[Ca^{2+}]_i$ in PC to a greater extent than that in IC. This is consistent with a recently reported higher level of TRPV4 expression in PC as was detected with immunohistochemistry (19). In contrast, GSK1016790A has no effect on $[Ca^{2+}]_i$ in mice genetically engineered to lack TRPV4 channel (Fig. 2b).

4. Pharmacological inhibition in combination with existing genetic knockouts of TRP subtypes is instrumental to assess contribution of these channels in generation of $[Ca^{2+}]_i$ signal in response to activation of G-protein-coupled receptors, such as

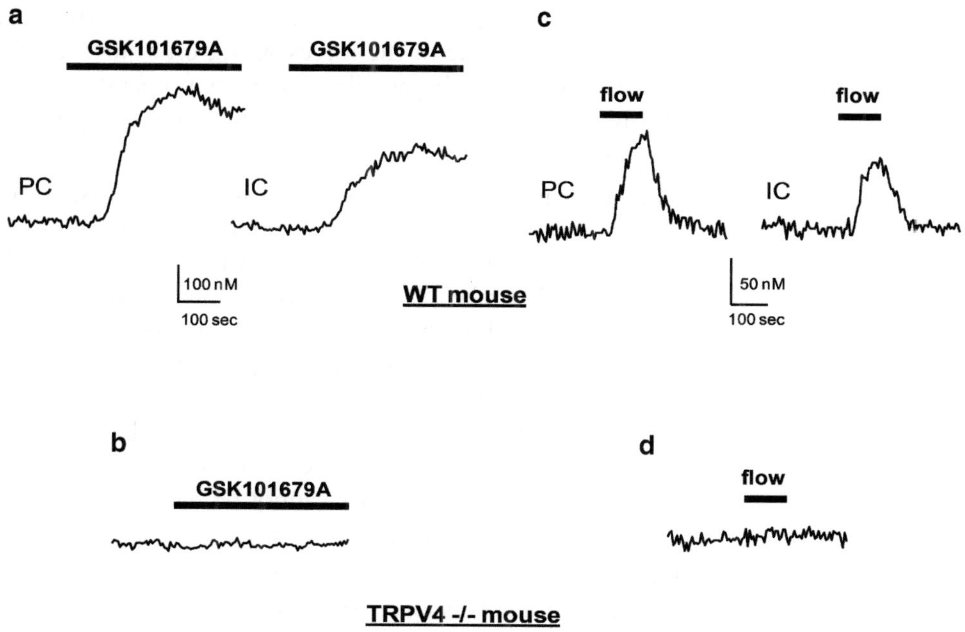

Fig. 2 TRPV4 activity determines flow-mediated $[Ca^{2+}]_i$ responses in both principal and intercalated cells. Representative time courses of changes in $[Ca^{2+}]_i$ in PC and IC (similar to that show in Fig. 1) in response to application of the TRPV4 activator GSK101679A in C57BL/6 wild-type (**a**) and TRPV4$^{-/-}$ (**b**) mice. Changes in $[Ca^{2+}]_i$ in response to an abrupt 10× elevation in perfusion rate from 1.5 to 15 mL/min in PC and IC of wild-type (**c**) and TRPV4$^{-/-}$ (**d**) mice. Respective scale bars for **a**, **b** and **c**, **d** are also shown

P2Y2 (13), and to mechanical stimulation of distal nephron cells, such as elevated flow (19). As an example, a rapid 10× elevation in luminal flow produces a sustained increase in $[Ca^{2+}]_i$ in both PC and IC from wild-type mice (Fig. 2c) but not in distal nephron cells from TRPV4$^{-/-}$ mice (Fig. 2d). This unequivocally suggests a critical role of TRPV4 channel in flow-mediated Ca^{2+} signalization in murine distal nephron. To accurately quantify shear stress which is generated by elevated flow and to determine if the produced mechanical stress fits within the physiological range present in intact distal nephron, a parallel plate chamber is commonly used.

3.3 Using Parallel Plate Chamber to Assess Shear Stress

1. Shear stress is the force per unit area that is generated on the cell surface due to fluid flow over the surface. In the kidney the tubular fluid has a low viscosity, relative to vascular vessels, due to protein-free tubular fluid. Nonetheless, because of high tubular flow rates, the shear stresses in the distal nephron can still be very high with typical values ranging from 0.3 to 25 dyn/cm^2, although even higher values are likely in various pathophysiological states (23, 34). Further, other mechanical stresses may also be at play for a given experimental condition

(e.g., stretching, pressure, cytoskeletal tension) so that a simple shear stress estimate of the forces acting on a surface or channel may not always be possible for many preparations without clearly defining the experimental conditions being employed (35, 36).

2. The application of a defined shear stress to the split-opened tubule requires that the tubule, attached as a flat sheet, be enclosed in a flow chamber of defined geometry and fluid flow properties. A parallel plate chamber is typically used for this type of study where the split-opened tubule, attached to coverslip, is placed on the bottom plate of the parallel plate chamber. An open-top chamber cannot be employed as fluid turbulence and unstirred layers over the tissue will lead to turbulent flow, i.e., non-laminar (see below), and, hence, estimation of the shearing force at the tissue/cell surface cannot be accomplished.

3. A parallel plate chamber can either be constructed from Plexiglas blocks with defined dimensions or purchased from several commercial sources (see Subheading 2.3 for examples). A coverslip forms the bottom portion of the chamber that allows cells to be visualized. This also allows application of fluorescence techniques, such as measurement of intracellular Ca^{2+} while performing shear stress studies. Typical dimensions for such a parallel plate chamber are 250 μm H × 1 cm W × 2–5 cm L (or similar). Fluid enters at one end and is collected at the opposite end. All chamber sides must be parallel and polished so as to achieve laminar, not turbulent, flow (see also (34)). Typical flow rates through the chamber are controlled with a perfusion pump, such as a peristaltic pump, to yield flow rates from 0 to 30 mL/min or more to generate shear stresses from 0 to 30 dyn/cm². For a rectangular chamber, the shear stress at the middle of the chamber (laminar flow) can be calculated as

$$\tau = 6\mu Q / bh^2,$$

where τ is the shear stress (dyn/cm²), μ is the fluid viscosity (using the fluid viscosity of water: 0.01002 P at 20°C and 0.006915 P at 37°C. Note, 1 P = 1 dyn·s/cm²), *Q* is the flow rate (mL/s), *b* is the chamber width (cm), and *h* is the chamber height (cm). With laminar flow the shear stress is linearly dependent upon fluid flow rate. The major determinants of shear stress in this setting are fluid flow rate, fluid viscosity, and temperature (see Note 5).

3.4 Identification of Distal Nephron Cell Types with Immunohistochemistry

1. In order to differentiate $[Ca^{2+}]_i$ signals recorded from different cell types of distal nephron, split-opened tubules are stained for AQP2 (a specific marker of principal cells) following the Ca^{2+} imaging experimentation. A coverslip with a split-opened

tubule is placed into one of the wells of a 12-well microtiter plate (see Note 6) and fixed in 4% paraformaldehyde solution in PBS (pH = 7.4) for 15 min at room temperature. Fixation with paraformaldehyde for more than 15 min is not recommended as the proteins will be cross-linked to the point where the antibody's binding to its epitope may be significantly diminished. The coverslip is then transferred to another well to permeabilize the sample with 0.1% Triton X-100 in PBS for 10 min (see Note 7) and to be subsequently washed with PBS 3 times for 5 min. If necessary the fixed samples can be stored in PBS at +4°C for at least 1 week.

2. Nonspecific binding of the antibodies is blocked with 10% normal goat serum (NGS) in PBS-Tween for 30 min at room temperature. The samples are further incubated with anti-AQP2 antibodies labeled with ATTO-550 (1:100 dilution) in 1% NGS in PBS-Tween for 2 h at room temperature in dark. It is sufficient to cover 12-well microtiter plate with aluminum foil to protect the fluorescent dye from photobleaching. After the incubation with anti-AQP2 antibodies, the samples are washed in PBS three times for 5 min. The samples are then incubated with 4′,6-diamidino-2-phenylindole (DAPI) in PBS (300 nM) for 1 min to visualize nuclei. Alternatively, a mounting medium with DAPI stain can be used. Finally, the samples are dehydrated, covered with a permanent mounting media (one drop on top of the coverslip) and the coverslip is attached to a microscope glass slide. The slide is put in the dark overnight to let the mounting medium dry.

3. Stained tubules are examined with an inverted fluorescent microscope using a 40× objective. Samples are excited with a metal halide lamp through UV-2E/C (for DAPI) and G-2E/C (for ATTO-550) filter sets. We recommend searching for the specimen in epifluorescent mode for the best convenience/visibility. Images are acquired with a CCD camera interfaced to a PC running NIS Elements 4.00 software. The representative micrograph of AQP2 (pseudocolor red) and nuclear DAPI (pseudocolor blue) staining of a split-opened tubule is shown in Fig. 1d. AQP2-positive PC and AQP2-negative IC can be clearly distinguished. This allows further separate quantification of fluorescent $[Ca^{2+}]_i$ signals from these cell types.

4 Notes

1. We recommend using younger (6–8 weeks) mice or rats for experiments as they typically have less connective tissue. It is also considerably easier to visibly identify distal nephrons in these animals. Moreover, the basal membrane, that covers

isolated tubule, is less rigid which makes split-opening much more successful. This, though, does not preclude using older (up to 1 year) animals if necessary for experimentations.

2. We recommend using different sets of forceps for isolation of distal nephrons from murine kidneys. The dullest forceps are necessary to move coverslips (dragging and transferring to another tissue culture dish). The fine forceps are used for teasing out of thin longitude sectors from kidney slices and for providing holding support of individual sectors. The finest forceps are critical for effective pinching and stripping of individual tubules out of a sector. While finest forceps are available from commercial vendors, they are expensive and worn out quickly. With a little practice, an investigator can quickly learn to manually hone dull or worn-out forceps under a stereomicroscope using fine sharpener stones.
3. While this chapter focuses on using Fura-2, several other Ca^{2+}-sensitive dyes, such as indo-1 and fluo-4, are available. They can also be successfully used to monitor Ca^{2+} dynamics in distal nephron cells if necessary.
4. Excitation light at 340 nm does not freely penetrate through glass, particularly through coverslips. This results in decreased intensity of Fura-2 signal at this excitation wavelength. If the signal is low, we recommend using thinner coverslips to affix distal nephrons. Quartz slides do not have this problem, but they are expensive. Alternatively, the fluorescent ratio can be quantified using excitation at 360 nm (Ca^{2+}-insensitive isosbestic point of Fura-2 spectra) and 380 nm but the dynamic range of the ratio will be reduced. The required filters for all wavelengths (340, 360, 380 nm) are available from many vendors (such as Chroma Technology).
5. Mechanosensitive Ca^{2+} responses become greatly potentiated by increasing temperature from 23°C to more physiologically relevant 37°C. Furthermore, gating properties of several TRP channels, including TRPV4, are also temperature dependent. It is a good idea to perform key experiments under both temperatures to bolster experimental conclusions. Several temperature-controlled systems (for example TC-324B/344B from Warner Instruments) are commercially available.
6. Fixation and immunostaining procedures are performed in a 12-well microtiter plate. This enables easy access and transfer of coverslips with forceps while using relatively small volumes of solutions. The volume can be reduced to as low as 500 μL, if the amount of available antibody/reagent is limited. Alternatively 3.5 × 10 mm tissue culture dishes can be used.
7. Triton X-100 is the most widely used detergent to improve the penetration of antibodies. However, it partially dissolves the

membrane lipids and destroys the membranes, thus potentially affecting the staining results. Therefore, the utilization of this detergent for membrane-associated antigens is not always appropriate. There are much milder membrane solubilizers, such as Tween 20, Saponin, Digitonin, and Leucoperm. When used at 0.2–0.5% concentrations for 10–30 min, they make "pores" large enough for antibodies to penetrate through the plasma membrane without dissolving it.

References

1. Eaton DC, Malik B, Saxena NC, Al-Khalili OK, Yue G (2001) Mechanisms of aldosterone's action on epithelial Na^+ transport. J Membr Biol 184:313–319
2. Hummler E (2003) Epithelial sodium channel, salt intake, and hypertension. Curr Hypertens Rep 5:11–18
3. Stockand JD (2010) Vasopressin regulation of renal sodium excretion. Kidney Int 78:849–856
4. Inoue T, Nonoguchi H, Tomita K (2001) Physiological effects of vasopressin and atrial natriuretic peptide in the collecting duct. Cardiovasc Res 51:470–480
5. Satlin LM, Sheng S, Woda CB, Kleyman TR (2001) Epithelial Na^+ channels are regulated by flow. Am J Physiol 280:F1010–F1018
6. Satlin LM, Carattino MD, Liu W, Kleyman R (2006) Regulation of cation transport in the distal nephron by mechanical forces. Am J Physiol 291:F923–F931
7. Woda CB, Bragin A, Kleyman TR, Satlin LM (2001) Flow-dependent K+ secretion in the cortical collecting duct is mediated by a maxi-K channel. Am J Physiol 280:F786–F793
8. Geyti CS, Odgaard E, Overgaard MT, Jensen ME, Leipziger J, Praetorius HA (2008) Slow spontaneous $[Ca^{2+}]_i$ oscillations reflect nucleotide release from renal epithelia. Pflugers Arch 455:1105–1117
9. Hovater MB, Olteanu D, Welty EA, Schwiebert EM (2008) Purinergic signaling in the lumen of a normal nephron and in remodeled PKD encapsulated cysts. Purinergic Signal 4:109–124
10. Leipziger J (2003) Control of epithelial transport via luminal P2 receptors. Am J Physiol 284:F419–F432
11. Liu W, Xu S, Woda C, Kim P, Weinbaum S, Satlin LM (2003) Effect of flow and stretch on the $[Ca^{2+}]_i$ response of principal and intercalated cells in cortical collecting duct. Am J Physiol 285:F998–F1012
12. Liu W, Morimoto T, Woda C, Kleyman TR, Satlin LM (2007) Ca^{2+} dependence of flow-stimulated K secretion in the mammalian cortical collecting duct. Am J Physiol 293:F227–F235
13. Mamenko M, Zaika O, Jin M, O'Neil RG, Pochynyuk O (2011) Purinergic Activation of Ca-Permeable TRPV4 Channels Is Essential for Mechano-Sensitivity in the Aldosterone-Sensitive Distal Nephron. PLoS One 6:e22824
14. Nauli SM, Alenghat FJ, Luo Y, Williams E, Vassilev P, Li X, Elia AE, Lu W, Brown EM, Quinn SJ, Ingber DE, Zhou J (2003) Polycystins 1 and 2 mediate mechanosensation in the primary cilium of kidney cells. Nat Genet 33:129–137
15. Praetorius HA, Leipziger J (2009) Released nucleotides amplify the cilium-dependent, flow-induced $[Ca^{2+}]_i$ response in MDCK cells. Acta Physiol (Oxf) 197:241–251
16. Praetorius HA, Leipziger J (2010) Intrarenal purinergic signaling in the control of renal tubular transport. Annu Rev Physiol 72:377–393
17. Woda CB, Leite M Jr, Rohatgi R, Satlin LM (2002) Effects of luminal flow and nucleotides on $[Ca^{2+}]_i$ in rabbit cortical collecting duct. Am J Physiol 283:F437–F446
18. Liu W, Murcia NS, Duan Y, Weinbaum S, Yoder BK, Schwiebert E, Satlin LM (2005) Mechanoregulation of intracellular Ca^{2+} concentration is attenuated in collecting duct of monocilium-impaired orpk mice. Am J Physiol 289:F978–F988
19. Berrout J, Jin M, Mamenko M, Zaika O, Pochynyuk O, O'Neil RG (2012) Function of TRPV4 as a mechanical transducer in flow-sensitive segments of the renal collecting duct system. J Biol Chem 287:8782–8791
20. Song MY, Yuan JX (2010) Introduction to TRP channels: structure, function, and regulation. Adv Exp Med Biol 661:99–108
21. Goel M, Sinkins WG, Zuo CD, Hopfer U, Schilling WP (2007) Vasopressin-induced membrane trafficking of TRPC3 and AQP2 channels in cells of the rat renal collecting duct. Am J Physiol 293:F1476–F1488
22. Goel M, Sinkins WG, Zuo CD, Estacion M, Schilling WP (2006) Identification and localization of TRPC channels in the rat kidney. Am J Physiol 290:F1241–F1252

23. Wu L, Gao X, Brown RC, Heller S, O'Neil RG (2007) Dual role of the TRPV4 channel as a sensor of flow and osmolality in renal epithelial cells. Am J Physiol 293:F1699–F1713
24. Strotmann R, Harteneck C, Nunnenmacher K, Schultz G, Plant TD (2000) OTRPC4, a nonselective cation channel that confers sensitivity to extracellular osmolarity. Nat Cell Biol 2:695–702
25. Liedtke W, Choe Y, Marti-Renom MA, Bell AM, Denis CS, Sali A, Hudspeth AJ, Friedman JM, Heller S (2000) Vanilloid receptor-related osmotically activated channel (VR-OAC), a candidate vertebrate osmoreceptor. Cell 103:525–535
26. Hartmannsgruber V, Heyken WT, Kacik M, Kaistha A, Grgic I, Harteneck C, Liedtke W, Hoyer J, Kohler R (2007) Arterial response to shear stress critically depends on endothelial TRPV4 expression. PLoS One 2:e827
27. Gao X, Wu L, O'Neil RG (2003) Temperature-modulated diversity of TRPV4 channel gating: activation by physical stresses and phorbol ester derivatives through protein kinase C-dependent and -independent pathways. J Biol Chem 278:27129–27137
28. Du J, Wong WY, Sun L, Huang Y, Yao X (2012) Protein Kinase G Inhibits Flow-Induced Ca^{2+} Entry into Collecting Duct Cells. J Am Soc Nephrol 23:1172–1180
29. Sakmann B, Neher E (2009) Single-channel recording. Springer, New York, NY
30. Burg MB, Grantham J, Abramow M, Orloff J, Schafer JA (1997) Preparation and study of fragments of single rabbit nephrons. J Am Soc Nephrol 8:675–683
31. Burg M, Grantham J, Abramow M, Orloff J (1966) Preparation and study of fragments of single rabbit nephrons. Am J Physiol 210: 1293–1298
32. Weinbaum S, Duan Y, Satlin LM, Wang T, Weinstein AM (2010) Mechanotransduction in the renal tubule. Am J Physiol 299: F1220–F1236
33. Grynkiewicz G, Poenie M, Tsien RY (1985) A new generation of Ca^{2+} indicators with greatly improved fluorescence properties. J Biol Chem 260:3440–3450
34. Jin M, Berrout J, O'Neil RG (2011) Regulation of TRP channels by osmomechanical stresses. In: Zhu MX (ed) Methods in signal transduction series. CRC Press, New York, NY, pp 353–373
35. Resnick N, Yahav H, Shay-Salit A, Shushy M, Schubert S, Zilberman LC, Wofovitz E (2003) Fluid shear stress and the vascular endothelium: for better and for worse. Prog Biophys Mol Biol 81:177–199
36. White CR, Frangos JA (2007) The shear stress of it all: the cell membrane and mechanochemical transduction. Philos Trans R Soc Lond B Biol Sci 362:1459–1467

Chapter 30

Patch-Clamping *Drosophila* Sensory Neurons

Volodymyr Kucher, Benjamin A. Eaton, James D. Stockand, and Nina Boiko

Abstract

Electrophysiological studies provide essential clues about the regulation and physiological function of ion channel proteins. Probing ion channel activity in vivo, though, often is challenging. This can limit the usefulness of such model organisms as *Drosophila* for electrophysiological studies. This is unfortunate because these genetically tractable organisms represent powerful research tools that facilitate elaboration of complex questions of physiology. Here, we describe a recently developed method for recording ion channel activity in *Drosophila* sensory neurons. This approach is based on patch-clamping primary neuron cultures from *Drosophila* embryos. Such cultures allow the study of ion channels in different genetic backgrounds. In addition to describing how to prepare a primary neuronal cell culture from *Drosophila* embryos, we discuss, as an example of utility, analysis of Na^+ currents in cultured class IV multidendritic (md) sensory neurons with the patch clamp technique. Excitability of md sensory neurons, manifested as action potential firing, is revealed with whole-cell current-clamping. Voltage-clamping class IV md neurons revealed the activity of the voltage-gated Na^+ channel, *paralytic*. Moreover, challenging class IV md neurons with acidic pH activates acid-sensing inward Na^+ currents. Genetic manipulation of *Drosophila* combined with this electrophysiological readout of activity identifies pickpocket1 (Ppk1), a member of the Deg/ENaC channel family, as responsible for conducting an acid-sensing Na^+ current in class IV md sensory neurons.

Key words Patch clamping, Primary cell culture, Ion channel, *Drosophila* sensory neurons

1 Introduction

Ion channels are integral membrane proteins that form pores through the lipid bilayer to facilitate ion permeation. This function makes channel proteins important regulators of a variety of biological processes. For instance, they play a vital role in neuronal signal transduction, neurotransmitter release, muscle contraction, cell secretion, and gene transcription. The patch clamp technique allows direct measurement of ion channel function and can provide essential clues about ion channel expression, regulation, and physiological function. However, probing ion

Nikita Gamper (ed.), *Ion Channels: Methods and Protocols*, Methods in Molecular Biology, vol. 998,
DOI 10.1007/978-1-62703-351-0_30,

channels in native cells and tissues often can be challenging and experimentally limiting.

Drosophila melanogaster is a powerful model organism for the study of physiology, including physiological functions of ion channels. The reasons for this are (a) the genetic tractability of this model organism as its complete genome has been decoded and sequenced; (b) many tools are available to modify the genetics of these animals; (c) the structure of the *Drosophila* nervous system is well defined and understood; and (d) the short life span, small size, and simple diet of these animals keep husbandry inexpensive. However it has been impossible, until recently, to take full advantage of the utility of the *Drosophila* model organism for studying ion channels as techniques for recording native ion channels from *Drosophila* neurons were missing. Recent advances in *Drosophila* cell culture and electrophysiology techniques, as described here, have overcome many of these limitations increasing the utility of *Drosophila* as a model organism for the study of ion channel biology and physiology.

Preparation of primary cell cultures from *Drosophila* embryos was first developed in the laboratory of Seecof and colleagues in 1968 (1). Several variations in this technique are available for culturing neuronal as well as nonneuronal cell types (2–4). These cultures are suitable for probing ion channel activity, regulation, and function with the patch clamp method (5, 6). Moreover, neurons in primary neuronal cultures from *Drosophila* retain their native physiological properties: they form functional synaptic connections and exhibit electrical excitability (5, 6). Here we describe the preparation of a primary neuronal cell culture from midgastrula-stage *Drosophila* embryos and detail patch-clamp analysis of class IV multidendritic (md) sensory neurons in this culture to demonstrate their utility for investigating ion channel function and regulation.

2 Materials

2.1 Preparation of Primary Neuronal Cell Culture from Drosophila Embryos

1. The appropriate *Drosophila melanogaster* line preferably with the cell type of interest labeled in some manner (see Note 1).
2. Agar, Granulated (Genesee Scientific; Seattle, WA).
3. Sucrose (Sigma, St. Louis, MO).
4. Apple juice, 100% (Hill Country, can be purchased in any grocery store).
5. Acetic Acid, Glacial (EMD, Gibbstown, NJ).
6. Ethanol, absolute, 200 proof (Sigma St. Louis).
7. Inactive dry yeast (Genesee Scientific; Seattle, WA).
8. Embryo collection cage (Genesee Scientific; Seattle).

9. Graduated plastic centrifuge tubes, 50 mL (Corning, Corning, NY).
10. Mesh, Nitex Nylon 64 μm (Genesee Scientific; Seattle, WA).
11. Paint brush, size three round (Loew Cornell, can be purchased in any art store).
12. Plastic Petri Dishes, 60 mm (Fisher Scientific, Pittsburg, PA).
13. Autoclaved distilled water.
14. Ham's F12/DME Media (DMEM; Irvine Scientific, Santa Ana, CA).
15. Insulin, from bovine pancreas (Sigma, St. Louis, MO).
16. Progesterone (Sigma St. Louis, MO).
17. Transferrin (Sigma St. Louis, MO).
18. Putrescine (Sigma St. Louis, MO).
19. Sodium Selenite (Sigma St. Louis, MO).
20. Tissue culture dishes, 35 and 100 mm (Corning, Corning, NY).
21. Microscope cover glass, 18 × 18 – 2 (Fisher Scientific, Pittsburgh, PA) cut into 5 × 5 mm chips.
22. Borosilicate glass capillaries (World Precision Instr., Sarasota, FL).
23. Dissection stereomicroscope (i.e., Stemi 2000; Carl Zeiss Microscopy, Thornwood, NY).

2.2 Electrophysiological Analysis of Class IV md Sensory Neurons

1. Inverted microscope (we use Eclipse Ti; Nikon, Melville, NY) fitted with epifluorescence and excitation and emission filters for GFP (FITC/RSGFP LP emission; #21012, Chroma Tech. Corp., Rockingham, VT) as needed to identify the cell type of interest.
2. Patch-clamp amplifier (we use Axopatch 200B; Molecular Devices, Sunnyvale, CA; see Note 2).
3. Digitizer, i.e., Digidata 1322A A/D board (Molecular Devices) interfaced with a PC running an appropriate acquisition and analysis software (we use pClamp 9.2 software suite; Molecular Devices).
4. Micromanipulator (i.e., MP-285; Sutter Instr. Co., Novato, CA).
5. Fast exchange recording chamber (i.e., Model RC-22, Warner Instr., Hamden, CT).
6. Perfusion system (we use Valve Bank II pinch valve perfusion system from AutoMate Scientific Inc., San Francisco, CA).
7. Micro-vibration isolation table with Faraday cage.
8. Micropipette puller (we use Model P-97 flaming/brown puller, Sutter Instr. Co.)

9. Micro-forge (we use MF-830; Narishige, East Meadow, NY).
10. Borosilicate glass capillaries (World Precision Instr., Sarasota, FL) pulled and forged to 10–13 mΩ.
11. Extracellular bathing solution: 140 mM NaCl, 5 mM KCl, 1 mM $CaCl_2$, 10 mM d-glucose, 10 mM HEPES, and 10 mM MES (pH 7.4; see Note 3).
12. Intracellular pipette solution for current-clamp experiments: 120 mM K-gluconate, 20 mM NaCl, 0.1 mM $CaCl_2$, 1 EGTA, and 10 mM HEPES (pH adjusted to 7.2 with KOH).
13. Intracellular pipette solution for voltage-clamp experiments: 120 mM Cs-gluconate, 20 mM NaCl, 0.1 mM $CaCl_2$, 1 mM EGTA, and 10 mM HEPES (pH adjusted to 7.2 with CsOH; see Note 3).
14. Tetrodotoxin (TTX).
15. Amiloride.

3 Methods

The *Drosophila* fly line *ppk1-Gal4,UAS-mCD8-GFP* that is used as an example in the subsequent sections is a kind gift from Dr. Darren Williams (King's College London, UK). These flies have targeted expression of green fluorescent protein (GFP) specifically in class IV multidendritic (md) sensory neurons regulated by the *ppk1*-promoter/enhancer GAL4 and upstream activating sequence (UAS) system (7, 8). Class IV md neurons tile the larval body wall and ramify beneath the epidermis (9). These are polymodal sensory neurons involved in proprioception and nociceptive responses (10–12). Figure 1a shows a negative fluorescence micrograph of a dissected *ppk1-Gal4,UAS-mCD8-GFP* larvae stained for GFP expression. The GFP-expressing dorsal da neuron (ddaC, circled) and ventral nerve cord (vnc) are indicated. The fluorescence micrograph in Fig. 1b shows the group of dorsal da neurons co-stained for GFP and neuronal membranes (anti-HRP). The GFP-expressing ddaC neuron indicated in Fig. 1a is shown. Figure 1c demonstrates the identity of dorsal da neurons shown in Fig. 1b (indicated with asterisks). Such GFP expression provides a means to specifically target class IV md neurons in the heterogeneous population of neurons growing in this primary culture. This allowed us to investigate ion channels specifically in this subset of neurons.

3.1 Preparation of a Primary Neuron Culture from Drosophila Embryos

1. Fly stocks are maintained on standard food at 25°C and a 12 h light/dark cycle. Preparation of the primary neuron culture from embryos is performed as previously described (2) with some modifications.

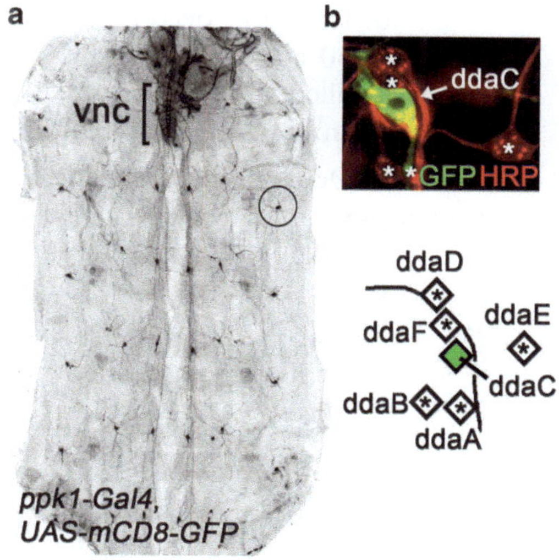

Fig. 1 *Drosophila* class IV md sensory neurons. (**a**) A negative fluorescence micrograph of a dissected *ppk1-Gal4,UAS-mCD8-GFP* larvae stained for GFP expression with anti-GFP antibody and all neurons with HRP. The GFP-expressing dorsal da neuron (ddaC, *circled*) and ventral nerve cord (vnc) are indicated. (**b**) Fluorescence micrograph of a group of dorsal da neurons (from a) co-stained for GFP and neuronal membranes (anti-HRP). The GFP-expressing ddaC neuron *circled* in (**a**) is shown. (**c**) The identity of dorsal da neurons shown in (**b**) is indicated with *asterisks* in this scheme

2. For *Drosophila* embryo collection, prepare embryo collection apple-juice agar plates using the following recipe: boil 375 mL ddH_2O with 125 mL apple juice, 15 g agar, and 6 g sucrose for 6 min. Cool to 50–60°C; add 10 mL ethanol and 5 mL glacial acetic acid. Pour into 100 mm plastic Petri dishes (recipe yields approximately 25 dishes). Cool and store at 4°C.
3. Next place 300–500 young (up to 2 weeks old) flies into a large embryo collection cage. Cover the cage with an embryo collection apple-juice plate containing a small stroke of inactive yeast paste in the center. Make the inactive yeast paste to the consistency of toothpaste by mixing dry inactive yeast with distilled water. Flip the cage and plate where the cage now is standing on the plate. Replace the plate once a day for 2–3 days prior to cell culture preparation (see Note 4).
4. On the third day of cell culture preparation, collect embryos on a fresh apple-juice agar plate containing the inactive yeast paste for 3 h and age them for 4–6 h after egg laying.
5. Add 5 mL of distilled water to the top of the collecting plate and using a fine-tipped paint brush, collect and then transfer the embryos into a dechorionation vessel (see Note 5). After transfer, the embryos should be resting on the mesh netting in the dechorionation vessel.

6. Place the dechorionation vessel containing embryos into a 60 mm plastic Petri dish and gently wash embryos with distilled water three times. The mesh netting should prevent embryos from washing thru.
7. Dechorionate by filling the 60 mm Petri dish with 50% bleach solution to a level above the meshwork containing the embryos for 3 min.
8. Rinse embryos again with sterile (autoclaved) distilled water thoroughly (three to four times) and transfer them to a fresh sterile 60 mm plastic Petri dish by unscrewing the 50 mL conical tube cap and then removing and inverting the mesh netting. Gentle rinsing of the mesh netting can enhance transfer (see Note 6). Dechorionated embryos will stick to the bottom of the plastic dish. Following rinsing, fill the culture dish with 6 mL of culture medium.
9. For cell culture preparation, remove the contents of midgastrula-stage embryos under stereomicroscope (see Note 7). Contents are removed by piercing the dechorionated embryo with a sharpened micropipette (tip diameter 50–100 μm) and gently sucking the content into the pipette leaving the dechorionated embryo husk attached to the Petri dish (see Note 8). The contents of 3–4 embryos are collected into a single pipette.
10. Prepare cover-glass chips prior to seeding with embryonic cells. Chips are prepared by soaking them in 100% ethanol and rinsing with sterile water. Place eight to ten 5 × 5 mm cover glass chips into a sterile 35 mm tissue culture Petri dish. Allow them to air dry and do not cover with media at this time.
11. To each chip add a 10 μL drop of culture medium. Transfer the contents of 3–4 embryos from the collecting pipette (collected in step 9) to each chip by resuspending the contents in this 10 μL drop of culture medium. Let cells settle for 10 min and then fill the culture dish with 3 mL of culture medium.
12. Maintain cultures in a 5% CO_2 incubator with 95% humidity at 23–25°C for 2–3 days (see Note 9).
13. For primary neuronal cell culture, use serum-free Ham's F12/DME media (DMEM) with 20 mM HEPES and 2.5 mM l-Glutamine. Supplement media with insulin from bovine pancreas (50 μg/mL), progesterone (20 ng/mL), transferrin (100 μg/mL), putrescine (100 μM), and sodium selenite (30 nM) (see Note 10).
14. For electrophysiology experiments, cultured neurons can be used 2 days after seeding and for up to 2 weeks (see Note 11).

In these cultures from *Drosophila* midgastrula-stage embryos, neurons arise from neural precursors. Shown in Fig. 2a is a bright field image of midgastrula embryonic cells 1 h after being plated.

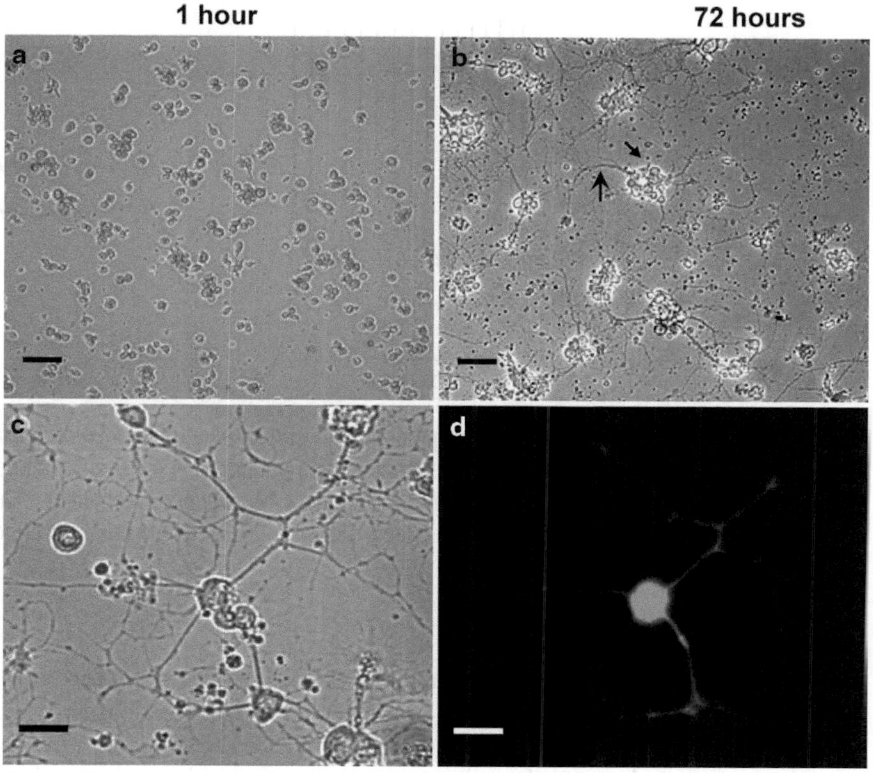

Fig. 2 Morphological development of primary neurons from *Drosophila* embryos plated in cell culture. Bright field images of primary cells from *Drosophila* embryos (**a**) 1 h after plating and (**b**) 72 h after growing in culture. Higher magnification bright field (**c**) and fluorescence (**d**) images of cultured GFP-expressing class IV md sensory neurons. Scale bars are 50 μm (**a**, **b**) and 3 μm (**c**, **d**)

These cells are round and mostly dissociated. Figure 2b–d shows the same cells after 72 h in culture. By this time, the cells have differentiated into neurons containing branching neuronal extensions (big arrow) and forming neuronal clusters (small arrow).

3.2 Electrophysiological Analysis of Drosophila Class IV md Sensory Neurons Held in Culture

1. Patch-clamping *Drosophila* class IV multidendritic sensory neurons follows standard procedures using common equipment (see Note 12).
2. Place a glass chip containing cultured neurons into a fast exchange recording perfusion chamber mounted to the stage of an inverted microscope. Rinse the chip with constant perfusion of the extracellular bath solution to remove culture media and debris.
3. Class IV md sensory neurons are identified in primary neuronal cultures by eGFP emission upon proper excitation. An example of a GFP-expressing md neuron with eGFP emissions is shown in Fig. 2d.

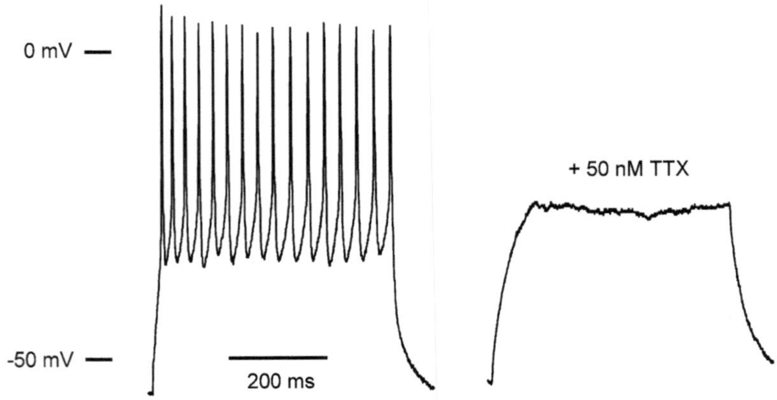

Fig. 3 Recording action potentials in current-clamped *Drosophila* class IV md sensory neurons. Representative action potentials in a current-clamped class IV md sensory neuron evoked by injecting a 40 pA suprathreshold depolarizing current for 600 ms (*left*). The effects of 50 nM TTX on action potential firing is shown to the *right*

4. After compensating offsets, form a high-resistance cell-attached seal on a neuron of interest by lowering a patch pipette (open pipette resistance ~10–13 mΩ) to the cell with a micromanipulator and applying gentle suction. Upon contacting the cell membrane, pipette resistance should increase to ~30–40 mΩ. Seal formation normally occurs when pipette resistance exceeds 1 mΩ (see Note 13). Gentle suction facilitates seal formation. Following seal formation, capacitance is compensated. For whole-cell current- and voltage-clamp experiments, the membrane is ruptured with additional gentle suction to provide access to the intracellular environment (see Note 14). Cell capacitance (~20 pF for md neurons) is determined by integrating the area of capacitive transients from an average of 10 voltage-ramps from −60 to −80 mV. Upon formation of the whole-cell patch, seal resistance decreases below >1 gΩ to ~500–800 mΩ (at a holding potential of −60 mV).
5. Current clamping allows determination of resting membrane potential (the electrical potential across the membrane of neurons at rest). For md neurons this is ~ −56 mV. Stimulation of a neuron with a depolarizing current injection sufficient to bring the neurons to the threshold for the action potential firing evokes a train of action potentials. Figure 3a shows representative action potentials from current-clamped class IV md sensory neurons. These action potentials were evoked with a suprathreshold 40 pA depolarizing current for 600 ms. As shown in Fig. 3b, action potential firing in *Drosophila* md neurons is sensitive to inhibition of voltage-gated sodium channels with tetrodotoxin (TTX).

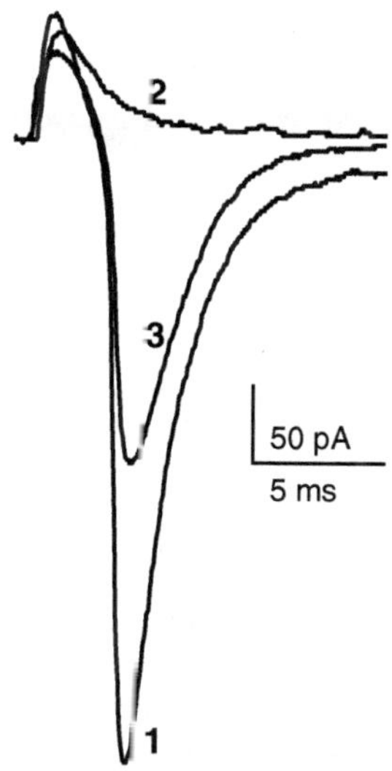

Fig. 4 Recording voltage-gated Na^+ currents conducted by *para* in voltage-clamped *Drosophila* class IV md sensory neurons. Representative macroscopic *para* Na^+ currents in a voltage-clamped md neuron evoked by stepping the holding potential from −60 mV to −40 mV in the presence of pH 7.4 (1) and pH 5.0 (2) and 2 min after returning to pH 7.4 (3). Inward Na^+ current is downwards

6. Macroscopic currents can be quantified with the whole-cell voltage-clamp configuration. Figure 4 shows representative macroscopic currents from the *Drosophila* TTX-sensitive, voltage-gated Na^+ channel *paralytic*. These macroscopic *para* currents are from a whole-cell patch formed on a class IV md neuron. In this example, current was evoked by stepping membrane potential from the holding potential of −60 to −40 mV. The activity of voltage-gated Na^+ channels, as shown in Fig. 4, is also sensitive to extracellular pH, i.e., decrease in current amplitude is observed in acidic (<5.0) pH. Here application of low pH was performed using a computer controlled fast perfusion system.

   In addition to inhibiting *para*, lowering extracellular pH elicits two distinct acid-sensitive Na^+ currents in class IV md neurons: a sustained (I_s) and transient (I_t) current. These are shown in Fig. 5. For this experiment, voltage was clamped to −60 mV with current evoked by rapidly lowering bath pH from 7.4 to 4.5 (noted with hatched gray bar); and for I_t subsequently washing 1 mM amiloride (noted with black bar). Current was recorded in the Clampex gap-free acquisition mode.

7. Combining the patch clamp technique, which is a readout of function, with forward genetics in *Drosophila* provides a powerful tool for studying ion channel physiology. As shown in Fig. 6, genetic knock down of ion channel of interest together

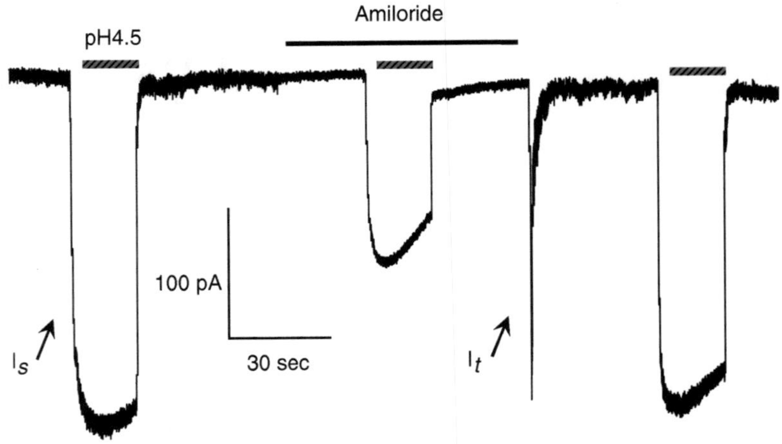

Fig. 5 Recording macroscopic acid-sensing Na^+ currents in voltage-clamped *Drosophila* class IV md sensory neurons. Representative sustained (I_s) and transient (I_t) acid-sensing macroscopic Na^+ currents in a voltage-clamped class IV md neuron. Inward Na^+ current is downwards. This cell was voltage-clamped to −60 mV with current evoked by rapidly lowering bath pH from 7.4 to 4.5 (noted with hatched *gray bars*); and for I_t subsequently washing 1 mM amiloride (noted with *black bar*)

with the rescue of null mutants with a transgene, and overexpression of a gain-of-function mutant channel can be used to identify the specific cellular currents conducted by a particular ion channel. Voltage-clamping class IV md sensory neurons in genetically manipulated flies identified the transient acid-sensing current (I_t) as being conducted by pickpocket1 (Ppk1), a *Drosophila* Deg/ENaC channel family homolog (9). Deletion of the *ppk1* gene abolished I_t but was without effect on I_s. Targeted expression of the wild-type *ppk1* transgene in class IV md neurons of *ppk1* null mutant flies rescued I_t, having no effect on I_s. Moreover, expression of the degenerin gain-of-function $Ppk1^{deg}$ mutant (13, 14) in these neurons resulted in larger and more prolonged I_t.

4 Notes

1. We used the *ppk1-Gal4, UAS-mCD8-GFP* fly line (a gift from Dr. Darren Williams, King's College, London, UK), which restrictively expresses GFP in class IV md neurons (7, 8).
2. There are few different patch-clamp amplifiers, A/D acquisition boards and software, micromanipulators, isolation tables, microscopes, pipette pullers, perfusion chambers, et cetera available for patch-clamp analysis. The user should use those best suited to their experiments and personal preference (see also Chapter 7 of this book).

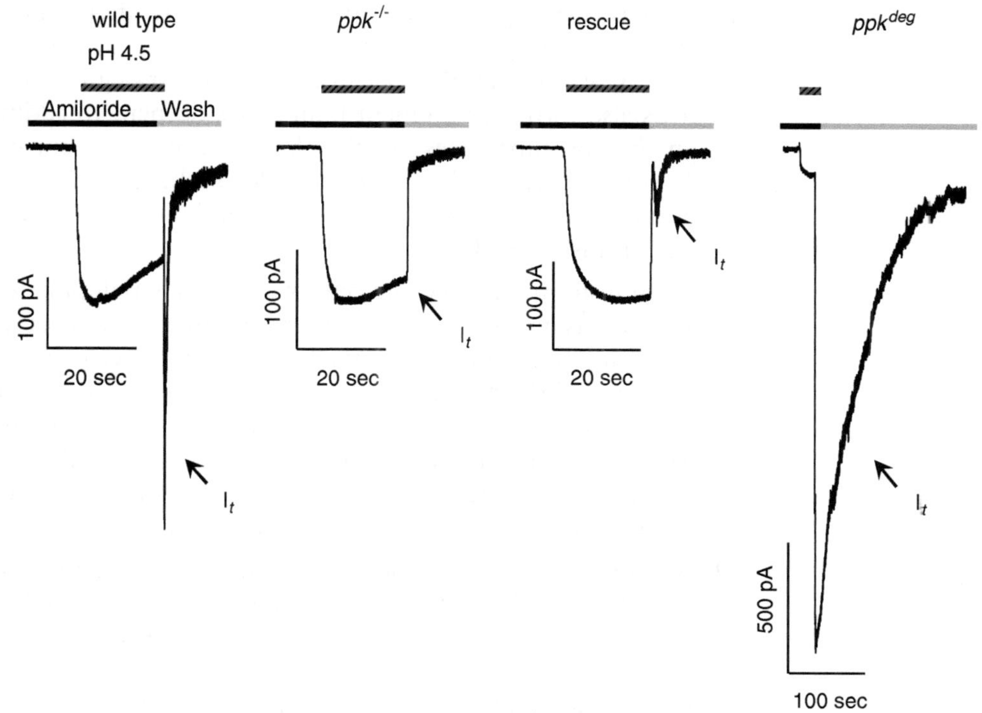

Fig. 6 Identification of the current conducted by Ppk1 in *Drosophila* class IV md sensory neurons. Representative acid-evoked macroscopic Na^+ currents in voltage-clamped (holding potential −60 mV) class IV md neurons from wild type (*left*), *ppk1*$^{-/-}$ null mutants (second from the *left*), *ppk1*$^{-/-}$ mutants rescued with expression of the wild-type *ppk1* transgene in md neurons (third from the *left*), and flies with targeted expression of the gain-of-function *ppk1*^{deg} transgene in class IV md neurons (*right*). In these experiments, neurons were treated with 1 mM amiloride prior to providing the activating acid pulse with blocker and low pH subsequently washed simultaneously to evoke I_t

3. The pipette and extracellular bathing solutions should be appropriate for the channel of interest. The choice of solutions depends on experiments conditions (15).
4. Replacement of embryo collecting plates prior to preparation of the culture allows flies to adapt to the new environment increasing egg laying and the number of eggs available for harvesting.
5. We make the dechorionation vessels by removing the bottom two-thirds of a 50 mL conical centrifuge tube and drilling a ~ 1 in. diameter hole into the cap. The cap is then removed and the top of the tube is covered with 8 × 8 cm mesh netting with the cap then screwed back on. Dechorionation is the removal of the outer protective layer (chorion) from an embryo.
6. Avoid coated tissue culture dishes for dechorionated embryos do not stick to these dishes. We routinely use plastic Petri dishes from Fisher Scientific.

7. Midgastrula-stage embryos appear 5:20–7:20 h after egg laying at 25°C. Embryos in this stage can be identified by having the origins of tracheal placode invaginations. Parasegmental furrows also form and segment boundaries deepen. In addition, gnathal protuberances within the head become apparent. A distinct cleft at the posterior pole of the embryo, which becomes detached from the vitelline membrane, appears at the end of this stage (16).
8. For transferring embryonic cells, we routinely pull pipettes to a fine tip using a P-97 puller (Sutter Instruments) and cut these tips using a ceramic tile cutter under stereomicroscope. We recommend the pipette tips be small enough to prevent sucking whole embryos into the pipette but big enough to avoid damaging individual cells. This should be empirically determined. Cell transfer is accomplished by mouth suction using 40 cm silicon tubing connected to the pipette.
9. We use a standard mammalian incubator with temperature control set at 23°C. The CO_2 level in the incubator was adjusted under control of Bacharach Fyrite Gas Analyzer (Bacharach Inc.).
10. Media not supplemented can be stored at 4°C for up to 2 weeks. The appropriate amount of media should be supplemented just prior to culture preparation. These supplements are designed for in vitro cell growth and differentiation in serum-free medium.
11. For patch-clamp experiments we use the cells growing in culture between 2 and 6 days. The media in the culture is never changed during this time.
12. For a detailed description of the patch clamp method, the reader is directed to the excellent book by B. Sakmann and E. Neher (17) and also to Chapter 7 of this book.
13. A high-resistance seal (>5 gΩ) is formed readily on the membrane of *Drosophila* neurons in culture using minimal suction. For gentle suction we recommend using of 1 mL syringe.
14. Many patch-clamp amplifiers can apply a brief but intense voltage step to facilitate membrane rupture. On the Axopatch 200B this is initiated with the "zap" button. With our conditions, we recommend not using this method for membrane rupture. Rather we prefer to rupture the membrane using additional gentle suction.

References

1. Seecof RL, Unanue RL (1968) Differentiation of embryonic Drosophila cells in vitro. Exp Cell Res 50:654–660
2. Sicaeros B, O'Dowd, DK (2007) Preparation of neuronal cultures from midgastrula stage Drosophila embryos. J Vis Exp 5:226

3. Cross D, Sang J (1978) Cell culture of individual Drosophila embryos: I. Development of wild-type cultures. J Embryol Exp Morphol 45:161–172
4. Shields G, Sang JH (1977) Improved medium for culture of Drosophila embryonic cells. Drosoph Inf Serv 52:161
5. Hodges DD, Lee D, Preston CF, Boswell K, Hall LM, O'Dowd DK (2002) tipE regulates Na^{+}-dependent repetitive firing in Drosophila neurons. Mol Cell Neurosci 19:402–416
6. O'Dowd DK (1995) Voltage-gated currents and firing properties of embryonic Drosophila neurons grown in a chemically defined medium. J Neurobiol 27:113–126
7. Matthews BJ, Kim ME, Flanagan JJ, Hattori D, Clemens JC, Zipursky SL, Grueber WB (2007) Dendrite self-avoidance is controlled by Dscam. Cell 129:593–604
8. Grueber WB, Ye B, Yang CH, Younger S, Borden K, Jan LY, Jan Y-N (2007) Projections of *Drosophila* multidendritic neurons in the central nervous system: links with peripheral dendrite morphology. Development 134:55–64
9. Adams CM, Anderson MG, Motto DG, Price MP, Johnson WA, Welsh MJ (1998) Ripped pocket and pickpocket, novel Drosophila DEG/ENaC subunits expressed in early development and in mechanosensory neurons. J Cell Biol 140:143–152
10. Ainsley JA, Pettus JM, Bosenko D, Gerstein CE, Zinkevich N, Anderson MG, Adams CM, Welsh MJ, Johnson WA (2003) Enhanced Locomotion Caused by Loss of the *Drosophila* DEG/ENaC Protein Pickpocket1. Curr Biol 13:1557–1563
11. Zhong L, Hwang RY, Tracey WD (2010) Pickpocket Is a DEG/ENaC Protein Required for Mechanical Nociception in Drosophila Larvae. Curr Biol 20:1–6
12. Kim SE, Coste B, Chadha A, Cook B, Patapoutian A (2012) The role of *Drosophila Piezo* in mechanical nociception. Nature 483:209–212
13. Hong K, Driscoll M (1994) A transmembrane domain of the putative channel subunit MEC-4 influences mechanotransduction and neurodegeneration in *C. elegans*. Nature 367:470–473
14. Waldmann R, Champigny G, Voilley N, Lauritzen I, Lazdunski M (1996) The mammalian degenerin MDEG, an amiloride-sensitive cation channel activated by mutations causing neurodegeneration in *Caenorhabditis elegans*. J Biol Chem 271:10433–10436
15. Hamill OP, Marty A, Neher E, Sakmann B, Sigworth FJ (1981) Improved patch-clamp techniques for high-resolution current recording from cells and cell-free membrane patches. Pflugers Arch 391:85–100
16. Ashburner M, Golic KG, Hawley RS (2005) Drosophila. A laboratory handbook, Secondth edn. Cold Spring Harbor, New York, NY
17. Sakmann B, Neher E (1983) Single-channel recording. Plenum Press, New York, NY

Part VI

Ion Channels as Research Tools

Chapter 31

Production and Validation of Recombinant Adeno-Associated Virus for Channelrhodopsin Expression in Neurons

John Y. Lin

Abstract

Recent discovery of the light-activated ion channel, channelrhodopsin (ChR), has provided researchers a powerful and convenient tool to manipulate the membrane potential of specific cells with light. With genetic targeting of these channels and illumination of light to a specific location, the experimenter can selectively activate the voltage-gated ion channels (VGICs) of ChR-expressing cells, initiating electrical signaling in temporally and spatially precise manners. In neuroscience research, this can be used to study electrical signal processing within one neuron at the cellular level, or the synaptic connectivity between neurons at the circuitry level. To conduct experiments with ChRs, these exogenous channels need to be introduced into the cells of interest, commonly through a viral approach. This chapter provides an overview of the design, production, and validation of recombinant adeno-associated virus (rAAV) for ChR expression that can be used in vitro or in vivo to infect neurons. The virus produced can be used to conduct "optogenetic" experiments in behaving animals, in vitro preparations and cultured cells, and can be used to study signal transduction and processing at a cellular or circuitry level.

Key words Optogenetics, Channelrhodopsin, Recombinant adeno-associated virus, Optical stimulation, Membrane excitability

1 Introduction

Excitable cells such as neurons and myocytes express voltage-gated ion channels (VGICs) on their plasma membrane. These cells utilize electrical signals to process and convey information between different cells or between different cellular compartments within one cell. The ability to manipulate the electrical signaling in a temporally and spatially precise manner would provide important insights to this complex electrical signaling. The ability to manipulate the cellular electrical signaling with an optical approach provides several advantages, as light is relatively unintrusive in living tissue and can be modulated easily and precisely in spatial and temporal domains with modern optical instrumentations. This "optical manipulation" of the cellular electrical signaling would require the cells of interest to be "light responsive," as most cells do not respond to

Nikita Gamper (ed.), *Ion Channels: Methods and Protocols*, Methods in Molecular Biology, vol. 998,
DOI 10.1007/978-1-62703-351-0_31,

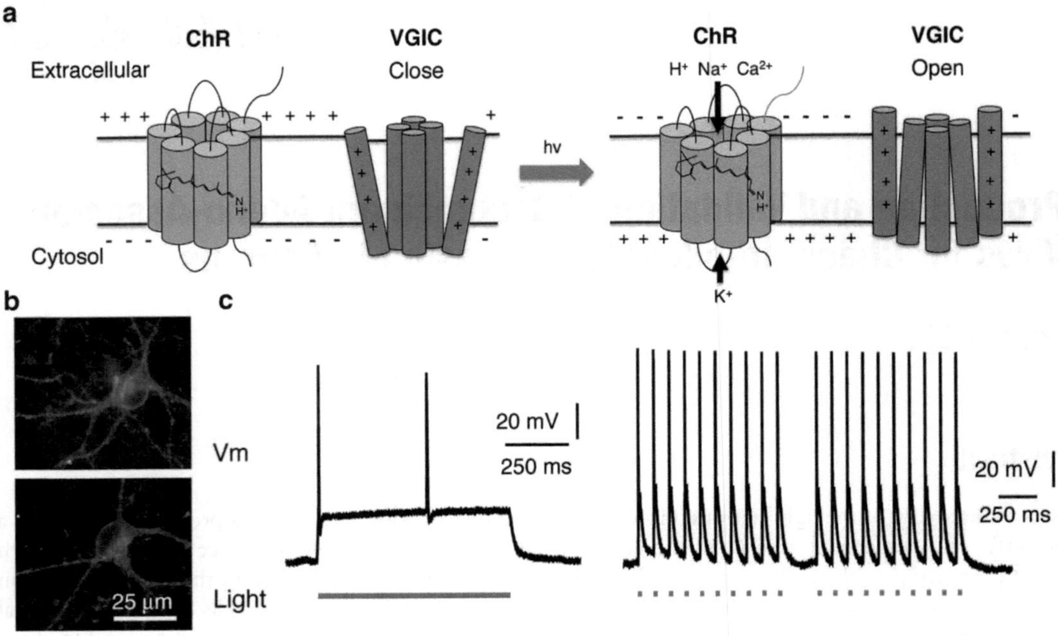

Fig. 1 Schematic of the light-induced activation of voltage-gated ion channel (VGIC) with channelrhodopsin (ChR) and examples of ChR expression in neurons. (**a**) Illumination of plasma membrane expressing ChR leads to the depolarization of the membrane, and the activation of voltage-gated ion channel (VGIC). The opening and closing of ChR is controlled by the photo-isomerization of all-trans retinal incorporated in the ChR protein. (**b**) Examples of neurons infected by recombinant adeno-associated virus (rAAV) expressing ChR-Citrine (eYFP) fusion protein as visualized with the fluorescence of the mCitrine. (**c**) Examples of light-induced depolarization in cultured neurons expressing the ChR variant ChIEF. The *left trace* shows the cell in response to 750 ms light pulse and the *right trace* showed the membrane response to two bursts of 10 Hz light stimulation

light natively. The recent discoveries of light-responsive ion channels, channelrhodopsins (ChRs), from the algae *Chlamydomonas reinhardtii* and *Volvox carteri* have been major breakthroughs in achieving optical manipulation of cellular excitability (1–3). By expressing these exogenous channels in the cells of interest, the experimenters can selectively depolarize the specific cells with light to activate VGICs (Fig. 1). This also creates the opportunity to manipulate selective neurons through genetic targeting. Although other approaches have been previously developed to depolarize membrane potential with light, these alternative approaches often required the introduction of synthetic organic chemicals such as modified azobenzene compounds (4) or caged ligands to act as chromophores (5). These synthetic chemicals are not commercially available and sometimes are difficult to introduce into living tissue. ChRs, on the other hand, utilize all-trans retinal, a naturally occurring precursor of retinoic acid in many tissues (6), which simplifies the approach for many biologists. The genes encoding the functional channel components of ChRs are <1,050 bp, which are small enough to be used in various recombinant viral vectors. As native

ChRs are evolved to provide algae with a simple vision, there are many properties of these proteins that are not ideal for its use as a research tool (7). However, many subsequent developments and engineering of ChRs have improved kinetics, desensitization/inactivation, and membrane trafficking of ChR, with the variant ChIEF being thus far the best general purpose ChR variant with good membrane trafficking, a low level of desensitization, and fast kinetics (7–9). Two aspects that have yet to be improved are the low conductance of ChRs (8, 10, 11), and having different variants with selective ion permeability (2, 8). To compensate for these two properties, a high expression level in the cells of interest is often necessary to achieve significant depolarization.

Neuroscientists have utilized ChRs to selectively excite genetically defined neurons or neurons of a defined location. By expressing ChRs in specific neurons, the experimenters can trigger action potentials in these neurons and artificially induce synaptic release with light in in vitro and in vivo preparations (Fig. 1). The majority of studies utilizing ChRs in rodents achieve targeted expression with recombinant viral vectors. Expressing ChR utilizing a viral vector is a quick, simple, and effective approach compared to alternatives such as generating transgenic animals and in utero electroporation. Of the different recombinant viruses that have been used with ChR, recombinant adeno-associated virus (rAAV) has been the most successful and efficient viral vector for in vivo use. Recombinant AAV is small in size and can be concentrated at a high titer, which can lead to high numbers of infected cells in vivo. At this moment, many core facilities and commercial services can produce high-titer rAAV, although this is typically expensive (~US $1,500/strain) and time-consuming (3–4 weeks). In this chapter, the protocol utilized within our laboratory to generate high-titer rAAV is described. The resulting virus is suitable for use in culture and in vivo in rodents. The protocol typically takes 1–1.5 weeks to complete and generate >200–300 μL of rAAV at titer >10^{12} genome copy (GC)/milliliter. The simplified flow chart of rAAV production is shown in Fig. 2. The main instruments required are cell culture facilities approved for virus generation and ultracentrifuge with suitable rotor, which are available in many research facilities (see Note 1 regarding safety and precautions about handling recombinant virus). Although there are many alternative methods for some of the steps described here (transfection methods, cell line used, virus purification), I have provided the methods that have consistently generated high-titer rAAV in our laboratory in a cost-effective manner.

Prior to generating rAAV for ChR expression, it is important to design the appropriate viral vector for the desired experiment. In the system described here, the maximum genetic insert that can be placed between the standard AAV inverted terminal repeats (ITRs) is ~4.7 kbp, which includes the genetic material for the promoter,

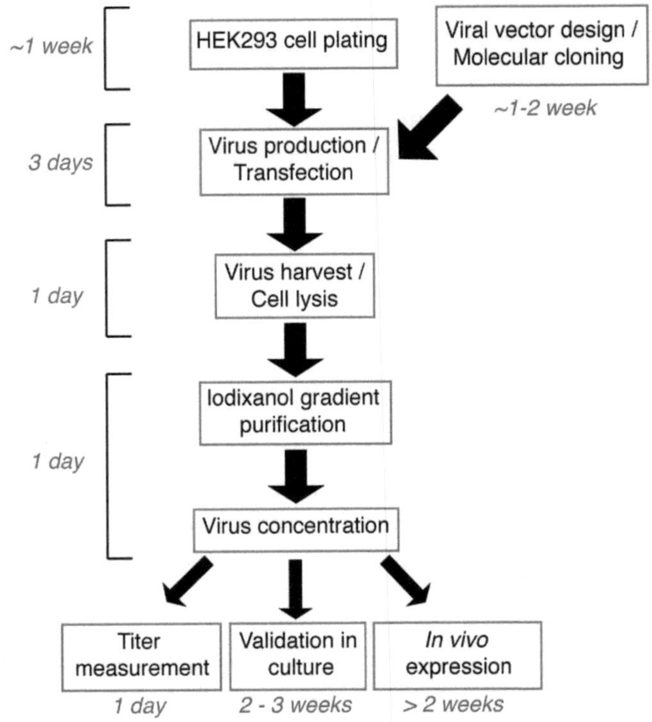

Fig. 2 The flow chart for the production and validation of rAAV for ChR expression. This chart shows the steps and typical time requirement involved in the production and validation of rAAV for ChR experiment. The molecular cloning of the viral vector can be done in parallel to the preparation of cells for transfection

gene encoding ChR, poly(A) sequence, and other features such as fluorescent protein (FP), woodchuck hepatitis virus posttranscriptional regulatory element (WPRE), introns, and lox sites. When utilizing these viral vectors, it is important to understand the functions of these features and the length limitation to produce viable virus for the experiments.

As for the choice of ChR variant, the native ChR2 and VChR1 are now obsolete as they have unfavorable properties. For blue-light-responsive ChR, ChIEF is the variant with thus far the best optimized properties (7, 9). For red-shifted ChR variant, C1V1 with additional trafficking sequence has better membrane expression than VChR1 (12). Both ChIEF and C1V1 are ~1,050 bp in size.

The inclusion of fluorescent protein (FP) allows for the visualization of expressing cells, although it is not essential for the function of ChRs. For direct fusion of FP to the ChRs, monomeric green fluorescent protein (GFP)-based FPs such as eGFP or eYFP/mCitrine are preferred over monomeric red fluorescent protein (mRFP)-based FPs (e.g., mCherry). This is due to the punctate membrane aggregation that is observed when mRFP-based FPs are used as ChR fusion. For RFP-based fusion, tandem dimer Tomato (tdTomato) is the preferred choice, as this design minimizes the aggregation (13). However, tdTomato is ~1.4 kbp in size, twice the size of a monomeric FP, and this difference in size would limit the use of a larger cell-specific promoters. It is also possible to visualize the expressing cells using a non-fusion FP expression system by using an IRES (14) or 2A sequence (15) that will result in cytosolic FP expression in the ChR-expressing cells.

A poly(A) tail is necessary for high expression of the transgene. Commonly used poly(A) tails are the bovine growth hormone (bGH), human growth hormone (hGH), and simian virus 40 (SV40) poly(A) tails. These sequences are typically 0.2 kbp in length.

Several different promoters have been used to express ChRs with rAAV in neurons; the most common ones include the α-CamKII promoter, CAG promoter (cytomegalovirus/CMV early enhancer element and chicken beta-actin promoter), human synapsin promoter, and EF-1α promoter. Although the α-CamKII promoter (1.3 kbp) is reported to be selective for excitatory neurons in lentiviral vectors (16), it failed to confer such selectivity in rAAV, possibly due to the enhancer property of the AAV-ITR sequence (17, 18). We have also found that the α-CamKII promoter in the lentivirus leads to the selective expression of ChR in an unidentified subset of cells in the mouse cortex or primary cell culture. However, the α-CamKII promoter is one of the strongest neuronal promoters available, which can be used to compensate for the poorer properties of some ChR2-based ChR variants. We have found that the human synapsin promoter (~0.5 kbp) gives neuron specificity (e.g., non-expressing in astrocytes or glial cells) and sufficient expression with ChIEF and other variants with improved membrane trafficking. CMV (~0.4 kbp), CAG (~1.6 kbp), and EF-1α (1.2 kbp) promoters are strong ubiquitous promoters. In a standard ChR viral vector containing ChR, monomeric FP, WPRE, and SV40 poly(A) tail, the maximum promoter size is ~2.2 kbp. This is the main limiting factor when utilizing a cell-type specific promoter with rAAV.

Another feature that is often included in the viral vector is WPRE (0.6 kbp) that increases the stability of mRNA and increases the expression level of the transgene by ~ 3-folds (19). However, this feature may not be necessary if a strong promoter is used and high viral titer is achieved. Alternatively, introns can be introduced between the promoter and the transgene or between the transgene and poly(A) tail to increase the expression level (20). The commercially available pAAV vector from Stratagene includes a β-globin intron between the CMV promoter and the multiple-cloning site. To combine the use of rAAV and a Cre-recombinase mouse strain, the FLEX (13) or DIO (21) strategies that place the inverted transgene between two different sets of lox sites allow for the transcription in the correct orientation in Cre-recombinase expressing cells. Figure 3 shows some common

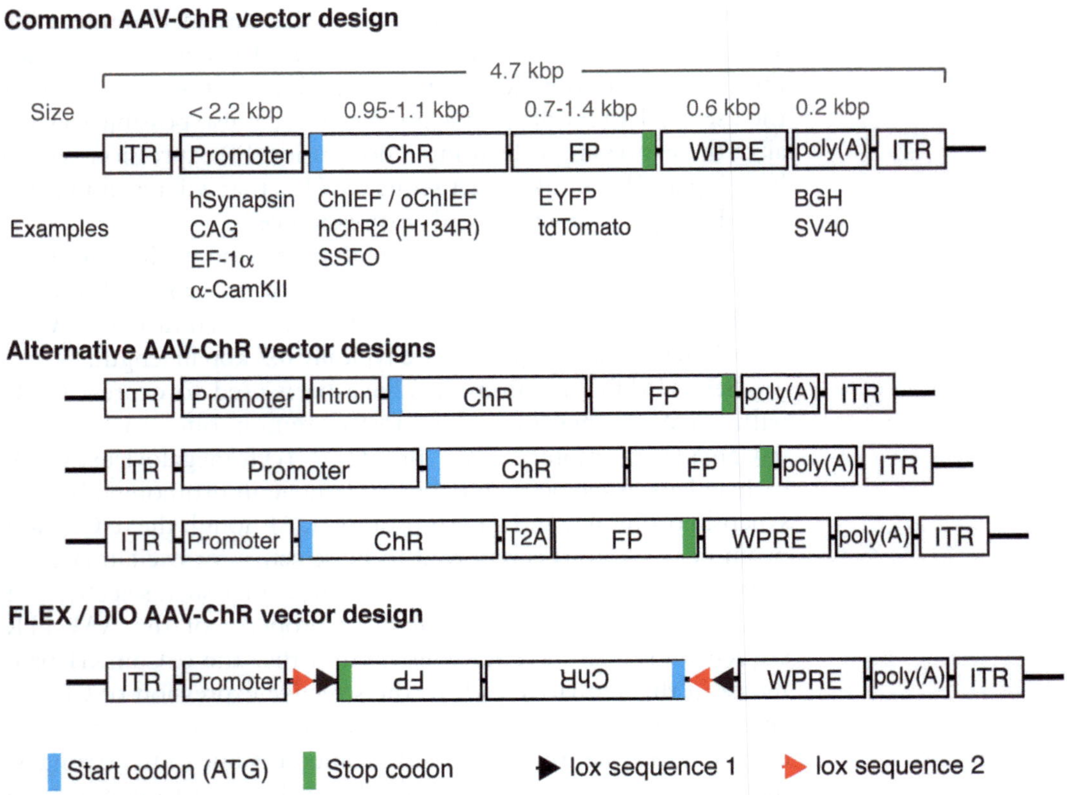

Fig. 3 Common AAV viral vector design for ChR expression. The *top example* shows the most frequently utilized AAV vector design for direct ChR expression which contains a promoter, the transgene-encoding ChR-fluorescent protein (FP) fusion protein, woodchuck hepatitis virus posttranscriptional regulatory element (WPRE), and poly(A) tail between two AAV inverted terminal repeats (ITR). The examples below show alternative designs including the omission of WPRE, utilization of intron to increase expression level, and utilization of a 2A-FP sequence for cytosolic FP expression. The *bottom example* shows the FLEX/DIO approach for targeted ChR expression with a Cre-recombinase-expressing mouse strain where the ChR-FP transgene is inserted in reverse (non-expressing) between two sets of lox sites between the promoter and WPRE. The presence of Cre-recombinase would excise and invert the transgene between the lox sites, leading to expression of ChR-FP

and alternative viral vector designs for expressing ChR with rAAV. The desired components can be assembled with standard molecular cloning procedures into the AAV viral vectors containing the AAV-ITR sites.

2 Materials

2.1 Production of rAAV In Trans

1. Ad-helper plasmid (75 μg), AAV-helper plasmid (30 μg), and AAV-ChR plasmid (25.5 μg). Production of rAAV in trans requires the transfection of three separate plasmids at 1:1:1 molar ratio. We recommend maxiprep-quality plasmid DNA. AAV-helper plasmid contains the *rep* and *cap* genes and deter-

mines the rAAV serotype (e.g., XR8 is used to produce rAAV serotype 8; XR1 is used to produce serotype 1) (see Note 2 regarding choice of serotype). Ad-helper plasmid (XX6-80) contains adenoviral genome region *E2a*, *E4*, and *VA*. AAV-helper and Ad-helper plasmids are available from National Gene Vector Biorepository (https://www.ngvbcc.org/). For the AAV-ChR plasmid, we recommend the AAV2-oChIEF-Citrine or AAV-Flex-rev-oChIEF-tdTomato plasmids. AAV2-oChIEF-Citrine contains mammalian codon-optimized ChIEF (oChIEF) fused to Citrine driven by the human synapsin promoter inserted between the AAV2 ITRs. This construct is suitable for direct expression of oChIEF in neurons. This vector is available from the Roger Tsien lab a University of California, San Diego (http://www.tsienlab.ucsd.edu). The AAV-Flex-rev-oChIEF-tdTomato plasmid can be used with a Cre-recombinase mouse line where the expression of oChIEF-tdTomato is restricted to the Cre-recombinase expressing cells. This plasmid is available as plasmid 30541 from http://www.addgene.org.

2. 293A cell lines (Life Technologies, Carlsbad, California, USA).
3. Complete HEK293 cell culture medium (1 g/mL DMEM + 10% fetal bovine serum + 1% Penicillin/streptomycin) (Life Technologies or Cellgro).
4. Six 15 cm cell culture dishes.
5. CalPhos™ Mammalian Transfection Kit (Clontech, Mountain View, California, USA).
6. Cell Scrapers (Fisher Scientific).
7. –80°C freezer.
8. 37°C water bath.
9. 23-gauge sterile needles.
10. 20 mL syringes.
11. Benzonase DNA nuclease (E1014, Sigma-Aldrich, St. Louis, Missouri, USA).
12. Sorvall Centrifuge RC5-B (Thermo Scientific) or similar centrifuge.
13. 50 mL conical centrifuge tube.

2.2 Purification and Concentration of rAAV

1. L-80 or L-80 XP Ultracentrifuge (Beckman Coulter Inc.).
2. 70 Ti rotor (Beckman Coulter Inc., 337922).
3. Ultra-Clear Quick-Seal tube (Beckman Coulter Inc.).
4. Two Quick-Seal Tube Spacers (Beckman Coulter Inc.).
5. Tube Topper Sealer (Beckman Coulter Inc.).
6. 15 cm metal cannula needles (18–20 gauge) (Hamilton Company or Popper Laboratory/Fisher Scientific).

Table 1
Composition of the 15, 25, 40, and 58% iodixanol gradient solution

	Optiprep (mL)	10× GB (mL)	5 M NaCl (mL)	ddH_2O (mL)	Phenol red (5 mg/mL) (μL)	Total (mL)
15%	10	4	8	18	0	40
25%	11.67	2.8	0	14	70	28.47
40%	16	2.4	0	5.6	0	24
58%	24.17	0.8325	0	0	60	25

7. 18-gauge sterile needles.
8. Syringes (3 mL and 10 mL).
9. Utility clamp and clamp stand.
10. 50 kDa MW Amicon centrifuge filter (Millipore).
11. Iodixanol/Optiprep (D1556, Sigma-Aldrich, St. Louis, Missouri, USA).
12. 10× gradient buffer (GB): 100 mM Tris (pH 7.6), 1.5 M NaCl, 100 mM $MgCl_2$; filter-sterilize using 0.22 μm syringe or vacuum filter. Store at 4°C.
13. 1× gradient buffer (GB): 5 mL 10× GB, 45 mL cell culture grade sterile H_2O.
14. 15, 25, 40, and 58% iodixanol gradient solutions (see Table 1 for composition).

2.3 Validation of rAAV in Cultured Neurons

1. Standard primary cultured neuron preparation (e.g., http://www.invitrogen.com/site/us/en/home/References/protocols/neurobiology/neurobiology-protocols/isolation-culture-and-characterization-of-cortical-and-hippocampal-neurons.html).
2. Glass-bottomed culture dish for recording with an inverted microscope (e.g., MatTek Corporation) or glass coverslips for recording with an upright microscope.
3. Poly-D-lysine solution (100 μg/mL).
4. Complete Neurobasal medium: Neurobasal-A + 1 × B27 + 1 × G lutaMax + 1% Penicillin/streptomycin (Life Technologies).
5. Standard whole-cell patch-clamp recording apparatus and solutions (see Chapter 7 of this book).
6. Epi-fluorescence imaging system with electronic shutter system accepting TTL inputs. This can be a standard xenon arc lamp with a mechanical shutter (e.g., Uniblitz shutter, Vincent Associates, Rochester, NY) or light-emitting diode (LED)-based light source (e.g., Thorlabs, Newton, NJ).

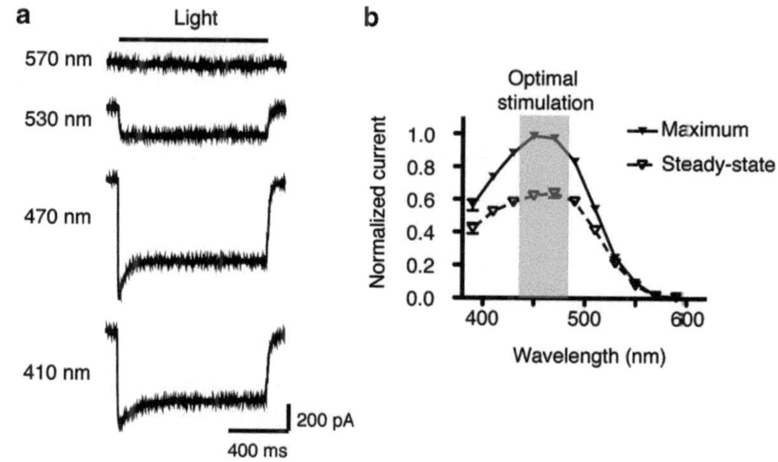

Fig. 4 The responses of ChIEF to different wavelengths of light. (**a**) The representative responses of ChIEF to 570, 530, 470, and 410 nm light of the same photon flux under voltage-clamp. (**b**) The maximal and steady-state (measured at 950–1,000 ms after initiation of illumination) response spectra of ChIEF normalized to its maximal response across the spectra. The optimal excitation wavelength is between 440 and 490 nm (Figure modified with permission from Lin et al. (8))

7. Appropriate filter set for visualizing the fluorescent protein included in the viral vector. See Chroma Technology (www.chroma.com) or Semrock (www.semrock.com) for the appropriate filter sets for the different fluorescent proteins.
8. Appropriate filter set for excitation of ChR with an arc lamp-based system. See Fig. 4 for the spectral response of ChIEF (see Note 3).

3 Methods

3.1 Production of rAAV In Trans

1. Plate 293A cells in six 15 cm dishes with 15 mL of complete HEK293 cell culture medium in each dish (see Note 4 regarding using alternative cell line).
2. Prepare the plasmid DNA mix when cells are 85–90% confluent. For six 15 cm dishes, mix together 30 μg of total AAV-helper/XR(x) plasmid (8 kbp), 75 μg of Ad-helper/XX6-80 plasmid (18 kbp), and 25.5 μg of AAV-oChIEF-Citrine plasmid (~6.4 kbp) which gives molar ratio of 1:1:1.
3. Prepare the calcium phosphate transfection reagent by mixing 1.172 mL of 2 M $CaCl_2$ (from the CalPhos™ kit) and plasmid DNA mix from step 2 in a 50 mL conical centrifuge tube and make up the volume to 9.45 mL with sterile water (see Note 5 and Chapters 2, 4, and 5 regarding alternative transfection methods).

4. Drip 9.45 mL of 2 × HEPES buffered saline (2 × HBS from the CalPhos™ kit) dropwise slowly into the solution from step 3. Tap the tube after each drop. Incubate the mixture for 15 min at room temperature.
5. Add 3.15 mL of the calcium phosphate transfection mixture dropwise into each of the 15 cm cell culture dish. Return the culture dishes into the incubator (see Note 6).
6. One day after transfection, replace the cell culture medium within the dish with fresh complete HEK293 cell culture medium (15 mL/dish).
7. Three days after transfection, remove and discard the media, being careful not to lose the cells in the dish.
8. Add 0.5 mL of 1× GB per plate and scrape the cells off the dish with a cell scraper. Collect the cell suspension off the plates into a 50 mL conical tube (should be ~5–7 mL total).
9. Freeze-thaw the cell suspension by transferring the sample between a −80°C freezer and 37°C water bath (15 min each).
10. Triturate the thawed cell suspension through a 23-gauge needle to lyse the cells and release the virus. Triturate three to four times.
11. Repeat step 9 three times and step 10 twice; lysed cells will become difficult to pass through the needle as intracellular DNA is released. 20- or 18-gauge needle can be used instead if the process is too difficult.
12. Allow the cell lysate to return to 37°C and add Benzonase at 1 μL (250 U) per 5 mL of cell suspension. Mix and incubate the suspension at 37°C for 45 min to 1 h in the 37°C water bath. Gently shake the tube every 15 min.
13. Centrifuge the suspension at 3,000 × *g* for 15 min at 4°C to pellet out the cell debris and DNA aggregates. Extract the rAAV-containing supernatant into a fresh 50 mL tube. The supernatant can be saved at −20°C (for ~1 month) or used directly in the purification step.

3.2 Purification and Concentration of rAAV

1. Build an iodixanol gradient in a Quick-Seal tube by adding 9.7 mL of 15% gradient solution to the bottom of the tube using a 10 mL syringe syringe fitted with a long 18-guage metal cannula needle.
2. Add subsequent layers below the previous layer by extending the syringe needle to the bottom of the tube and slowly injecting the gradient solutions. The remaining compositions are 6.4 mL of 25% gradient solution, followed by 5.4 mL of 40% gradient solution and then 5.4 mL of 58% gradient solution.
3. Add the virus-containing supernatant to the gradient by slowly dripping the solution with an 18-guage needle fitted on a syringe without the plunger onto the top layer of the gradient. Minimize the disruption of the layers.

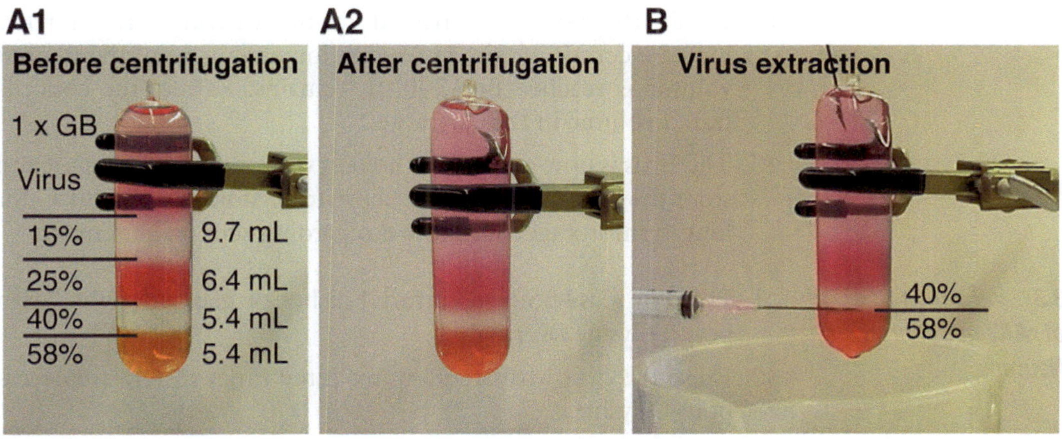

Fig. 5 Example of iodixanol gradients before and after centrifugation. (**a**) The iodixanol gradient preparation showing the layers containing the 1× GB, virus solution, and 15, 25, 40, and 58% gradient solution before (A1) and after (A2) centrifugation. (**b**) The placement of the needle for the extraction of virus solution after centrifugation

4. Fill the remaining empty volume of the tube to the top using 1× GB as in step 3.
5. Weigh the tube and make a counter balance tube of identical weight filled with the same layers (replacing the virus with 1×GB); the differences should be <0.01 g.
6. Seal the tip of the tubes using a Tube Topper Sealer. Gently squeeze the tube to ensure the tips are sealed and airtight. The tube should appear as shown in Fig. 5.
7. Place the two tubes in the opposite spaces of a 70Ti Beckman rotor. Lower the Quick-Seal Tube Spacer on top of the tube prior to sealing the rotor lid. Centrifuge the tube in vacuum at 169,538×*g* for 2 h 10 min at 18°C. The acceleration should be set to "max" and deceleration to "9."
8. Remove the tubes from the rotor and mount the tube containing the virus on a utility clamp (Fig. 5).
9. Insert an 18-gauge needle near the top of the tube as airflow vent. Place a beaker directly underneath the tube to catch any leakage (Fig. 5b).
10. Prepare a 3 mL syringe with an 18 gauge needle for extraction of the virus. Insert the needle at the interface between the 40 and 58% gradient buffer layers with the bevel of the needle facing up (Fig. 5b). Slowly extract ~2 mL of solution, first with the beveled needle opening facing upwards during the first half of extraction and then facing downward for the rest (see Note 7).
11. Add the 1.5 mL of 1× GB to the 2 mL of virus solution and load the solution into a 50 kDa MW Amicon centrifuge filter cartridge. Spin at 1,500×*g* for 15 min.

12. Check the volume remained in the cartridge and further increase centrifuge time in 5-min increments until the desired volume is reached or ~200 μL remains. Extract the concentrated volume in the cartridge.
13. The virus is now ready to be used in vivo or in vitro. Aliquot and store the virus at −20°C for short-term storage and −80°C for long-term storage. See Note 8 regarding titer measurement.

3.3 Validation of rAAV in Cultured Neurons

1. Pre-coat glass-bottom cultured dishes or glass coverslips with poly-D-lysine overnight.
2. Rinse the dish with sterile water three times to remove excess poly-D-lysine.
3. For a glass-bottomed dish with 12–14 mm diameter circular glass area, plate 150×10^3 primary cultured neurons in 2 mL complete Neurobasal medium.
4. One day after plating, add 5 μL rAAV (10^9–10^{10} viral particles) into each dish.
5. Maintain the neuronal culture for 2–3 weeks with standard neuronal culture procedure.
6. On the day of the recording, place the neurons in the recording chamber and visualize the ChR-FP-expressing cells using the epi-fluorescence setting with the correct filter set for the FP. If the FP is fused to ChR, membrane-localized fluorescence should be visible (see Fig. 1).
7. After selecting the ChR-FP-expressing cell for recording, switch the filter set to the one appropriate for activation of ChR (e.g., 480 nm for excitation of ChIEF).
8. Establish whole-cell patch-clamp recording with the ChR-FP-expressing cells using standard techniques.
9. Switch to current-clamp recording and maintain a stable membrane potential with current injection if necessary; start the recording.
10. Send a TTL pulse from the digital out port of the digitizer to open the shutter of the excitation light. We typically use 500–750 ms duration light pulse for prolonged activation and 20 pulses of 15 ms duration at 10 Hz for high-frequency pulsed stimulation (see Note 9). The response of a ChR-expressing neuron should be similar to the one shown in Fig. 1c.

4 Notes

1. It is important to follow the institutional safety guidelines when producing and handling viruses. The procedures should be conducted in approved facilities and the appropriate protective

gears should be used. The personnel should also receive safety training to handle the virus. The instrumentation and reagents coming into contact with the virus should be treated with 10% bleach for 15 min for decontamination.

2. The common rAAV serotypes used for infecting neurons are serotypes 1, 5, 8, 9, and 10. The different serotype has different tropism towards different neuronal types and brain regions. We recommend checking the literature before deciding on the serotype. We have been able to achieve very efficient infections of multiple brain regions with serotype 8.

3. All existing ChR variants (including the red-shifted C1V1-based variants (12) and VChR1 (7)) can be efficiently activated by 470–480 nm light. This can be light either from an arc lamp filtered with a 480 nm interference filter or from a 470 nm LED light source.

4. Although most HEK293 or HEK293t cell lines can be used to produce rAAV, in our laboratory, the 293A cell line (Life Technologies, Carlsbad, California, USA) consistently generates high titer of rAAV over other cell lines tested. The flat appearance of the cells at low density and strong adhesion of the cells to the culture plate at high density are favorable for the transfection and subsequent harvesting of virus post-transfection.

5. Most cell transfection methods (cationic liposomes, poly-ethyleneimine, or electroporation) can be used; however, a high level of transfection and low toxicity are crucial for the production of high-titer rAAV. We have found that the calcium phosphate precipitation method provides a simple, consistent, and cost-effective method of transfection that results in high virus production when used with 293A cells. It is critical to transfect at the right cell density, as low cell density during transfection typically leads to high level of toxicity and low yield of virus.

6. It is possible to visualize the calcium phosphate precipitate with a microscope equipped with phase contrast optics after 1 h. The precipitate will appear as irregular-shaped particles on top of the cells.

7. It is common to have iodixanol contamination in the virus aliquot; this does not affect the use of the virus in vivo in the rodents or in vitro.

8. The titer of rAAV is typically measured in genome copy/milliliter, as one infectious rAAV particle can carry only one copy of viral DNA, and viral particles that failed to incorporate the viral genomic DNA will not lead to transgene expression. There are several methods of measuring rAAV titer, the most accurate being quantitative polymerase chain reaction (PCR) with primers targeting part of viral genome or the transgene. It is also possible to measure the DNA content of the virus with

a non-amplified fluorescence-based assay (e.g., with DNA dye PicoGreen on a fluorescence plate reader). Both approaches would require the lysis of 10–20 μL of viral aliquot to release the viral DNA, followed by measurement of the viral DNA concentration within the lysate against DNA preparations with known concentrations. The number of viral genome copies can be calculated from the DNA concentration if the viral genome size (the DNA length between the AAV ITRs) is known.

9. The mechanical shutter of an arc lamp light is typically limited to 20 Hz of maximal repetitive frequency, with a minimal opening time ~10 ms. This is due to the mechanical speed limitation of the shuttering blade and the underlying electronics. A LED-based light source can achieve much greater temporal resolution (μs scale) as the onset and off rate of light is depending on the speed and quality of the constant-current power supply used to power the LED.

Acknowledgments

J.Y.L. was funded by Foundation of Research, Science and Technology New Zealand and was also supported by National Institutes of Health grants to Roger Y. Tsien (NS027177) and Howard Hughes Medical Institute. I would like to acknowledge Dr. Daniel Gibbs for technical advice regarding rAAV production and Dr. Sharon B. Sann for editorial assistance.

References

1. Nagel G, Ollig D, Fuhrmann M, Kateriya S, Musti AM, Bamberg E, Hegemann P (2002) Channelrhodopsin-1: a light-gated proton channel in green algae. Science 296:2395–2398
2. Nagel G, Szellas T, Huhn W, Kateriya S, Adeishvili N, Berthold P, Ollig D, Hegemann P, Bamberg E (2003) Channelrhodopsin-2, a directly light-gated cation-selective membrane channel. Proc Natl Acad Sci USA 100: 13940–13945
3. Zhang F, Prigge M, Beyriere F, Tsunoda SP, Mattis J, Yizhar O, Hegemann P, Deisseroth K (2008) Red-shifted optogenetic excitation: a tool for fast neural control derived from Volvox carteri. Nat Neurosci 11:631–633
4. Szobota S, Gorostiza P, Del Bene F, Wyart C, Fortin DL, Kolstad KD, Tulyathan O, Volgraf M, Numano R, Aaron HL, Scott EK, Kramer RH, Flannery J, Baier H, Trauner D, Isacoff EY (2007) Remote control of neuronal activity with a light-gated glutamate receptor. Neuron 54:535–545
5. Zemelman BV, Nesnas N, Lee GA, Miesenbock G (2003) Photochemical gating of heterologous ion channels: remote control over genetically designated populations of neurons. Proc Natl Acad Sci USA 100:1352–1357
6. Posch KC, Burns RD, Napoli JL (1992) Biosynthesis of all-trans-retinoic acid from retinal. Recognition of retinal bound to cellular retinol binding protein (type I) as substrate by a purified cytosolic dehydrogenase. J Biol Chem 267:19676–19682
7. Lin JY (2011) A user's guide to channelrhodopsin variants: features, limitations and future developments. Exp Physiol 96:19–25
8. Lin JY, Lin MZ, Steinbach P, Tsien RY (2009) Characterization of engineered channelrhodopsin variants with improved properties and kinetics. Biophys J 96:1803–1814
9. Mattis J, Tye KM, Ferenczi EA, Ramakrishnan C, O'Shea DJ, Prakash R, Gunaydin LA, Hyun M, Fenno LE, Gradinaru V, Yizhar O, Deisseroth K (2012) Principles for applying optogenetic

tools derived from direct comparative analysis of microbial opsins. Nat Methods 9:159–172
10. Feldbauer K, Zimmermann D, Pintschovius V, Spitz J, Bamann C, Bamberg E (2009) Channelrhodopsin-2 is a leaky proton pump. Proc Natl Acad Sci USA 106:12317–12322
11. Gunaydin LA, Yizhar O, Berndt A, Sohal VS, Deisseroth K, Hegemann P (2010) Ultrafast optogenetic control. Nat Neurosci 13:387–392
12. Yizhar O, Fenno LE, Prigge M, Schneider F, Davidson TJ, O'Shea DJ, Sohal VS, Goshen I, Finkelstein J, Paz JT, Stehfest K, Fudim R, Ramakrishnan C, Huguenard JR, Hegemann P, Deisseroth K (2011) Neocortical excitation/inhibition balance in information processing and social dysfunction. Nature 477:171–178
13. Atasoy D, Aponte Y, Su HH, Sternson SM (2008) A FLEX switch targets Channelrhodopsin-2 to multiple cell types for imaging and long-range circuit mapping. J Neurosci 28:7025–7030
14. Pelletier J, Sonenberg N (1988) Internal initiation of translation of eukaryotic mRNA directed by a sequence derived from poliovirus RNA. Nature 334:320–325
15. Osborn MJ, Panoskaltsis-Mortari A, McElmurry RT, Bell SK, Vignali DA, Ryan MD, Wilber AC, McIvor RS, Tolar J, Blazar BR (2005) A picornaviral 2A-like sequence-based tricistronic vector allowing for high-level therapeutic gene expression coupled to a dual-reporter system. Mol Ther 12:569–574
16. Dittgen T, Nimmerjahn A, Komai S, Licznerski P, Waters J, Margrie TW, Helmchen F, Denk W, Brecht M, Osten P (2004) Lentivirus-based genetic manipulations of cortical neurons and their optical and electrophysiological monitoring in vivo. Proc Natl Acad Sci USA 101:18206–18211
17. Beaton A, Palumbo P, Berns KI (1989) Expression from the adeno-associated virus p5 and p19 promoters is negatively regulated in trans by the rep protein. J Virol 63:4450–4454
18. Nathanson JL, Yanagawa Y, Obata K, Callaway EM (2009) Preferential labeling of inhibitory and excitatory cortical neurons by endogenous tropism of adeno-associated virus and lentivirus vectors. Neuroscience 161:441–450
19. Loeb JE, Cordier WS, Harris ME, Weitzman MD, Hope TJ (1999) Enhanced expression of transgenes from adeno-associated virus vectors with the woodchuck hepatitis virus posttranscriptional regulatory element: implications for gene therapy. Hum Gene Ther 10:2295–2305
20. Huang MT, Gorman CM (1990) Intervening sequences increase efficiency of RNA 3′ processing and accumulation of cytoplasmic RNA. Nucleic Acids Res 18:937–947
21. Cardin JA, Carlen M, Meletis K, Knoblich U, Zhang F, Deisseroth K, Tsai LH, Moore CI (2009) Driving fast-spiking cells induces gamma rhythm and controls sensory responses. Nature 459:663–667

Chapter 32

Optical Control of Ligand-Gated Ion Channels

Stephanie Szobota, Catherine McKenzie, and Harald Janovjak

Abstract

In the vibrant field of optogenetics, optics and genetic targeting are combined to commandeer cellular functions, such as the neuronal action potential, by optically stimulating light-sensitive ion channels expressed in the cell membrane. One broadly applicable manifestation of this approach are covalently attached photochromic tethered ligands (PTLs) that allow activating ligand-gated ion channels with outstanding spatial and temporal resolution. Here, we describe all steps towards the successful development and application of PTL-gated ion channels in cell lines and primary cells. The basis for these experiments forms a combination of molecular modeling, genetic engineering, cell culture, and electrophysiology. The light-gated glutamate receptor (LiGluR), which consists of the PTL-functionalized GluK2 receptor, serves as a model.

Key words Optogenetics, Optochemical genetics, Optical switch, Photochromic tethered ligand, Azobenzene, LiGluR, Glutamate receptor

1 Introduction

Biology occurs over a wide range of time and length scales. To understand dynamic biological systems, we require tools for both the spatiotemporal observation *and* perturbation of cellular and molecular events. While the past years have seen rapid growth in optical (e.g., fluorescence-based) real-time reporters of cellular signals (1), the development of means to activate cells on short and small scales has lagged behind. With light being the premier choice for both readout and activation of cellular events, photosensitive molecules have recently enabled us to noninvasively control biological signals with high spatial and temporal resolution. This is achieved in optogenetics and optochemical genetics either by "repurposing" Nature's light-sensing proteins from bacteria, algae, or plants or by engineering synthetic light-gated functionalities (2–4).

Nikita Gamper (ed.), *Ion Channels: Methods and Protocols*, Methods in Molecular Biology, vol. 998, DOI 10.1007/978-1-62703-351-0_32,

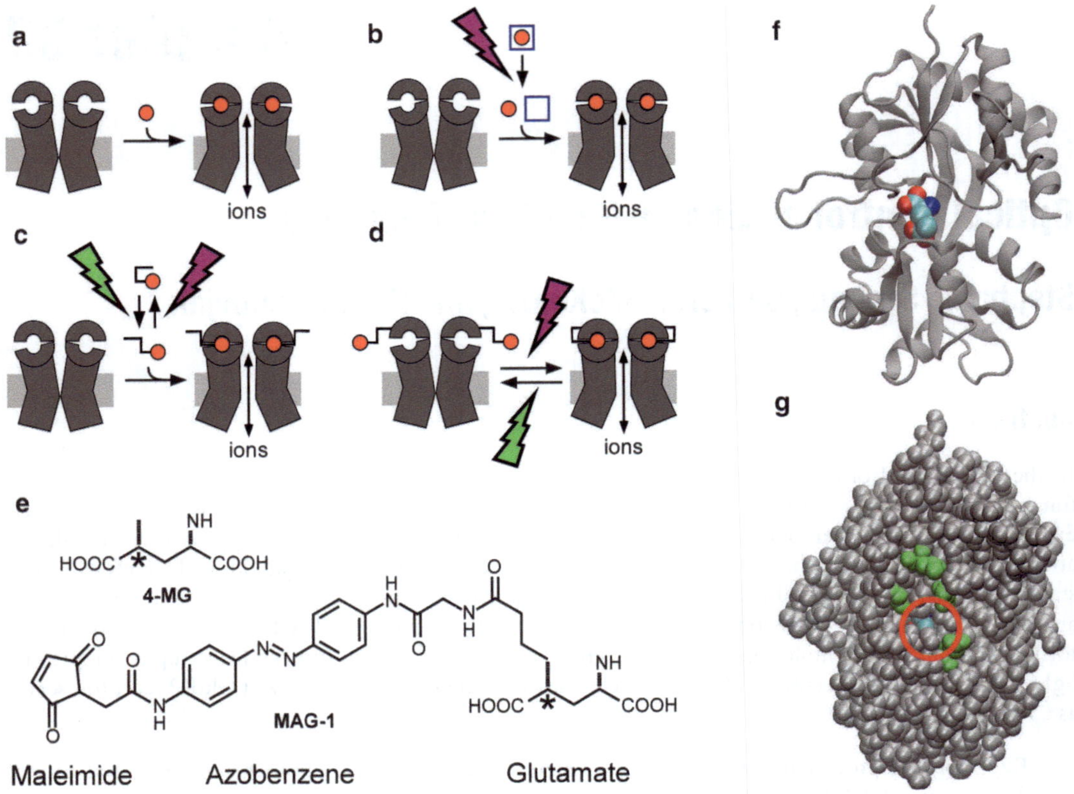

Fig. 1 Synthetic strategies for manipulating ligand-gated ion channels with light. (**a**) In the example of the ionotropic glutamate receptor GluK2, glutamate binding in the extracellular ligand binding domain triggers domain closure and gate opening. (**b**) Gating after ligand uncaging. (**c**) Gating after reversible conversion of a photochromic ligand to a binding-competent conformation. (**d**) Reversible light control by a PTL. (**e**) Chemical structure of MAG1, the prototypical maleimide-azobenzene-glutamate PTL, and 4-MG. (**f**, **g**) Crystal structure of the ligand binding domain of GluK2 in complex with 2*S*, 4*R*-4-methylglutamate (4-MG). 4-MG, residues surrounding the binding site suited for PTL attachment, and exit tunnel are highlighted as described in the text

Three synthetic strategies to light-control signals in cells or in vivo are commonly used: "caged" ligands, photochromic ("reversibly caged") ligands, and photochromic tethered ligands (PTLs) (Fig. 1) (3). "Caged" ligands and photochromic ligands can be optically activated with subcellular resolution and within milliseconds either by releasing the ligand from the cage or by triggering a reversible conformational change in the photochrome. After photoactivation, these molecules act as free ligands on their specific protein targets (Fig. 1b, c). In contrast, PTLs are linear molecules that consist of a ligand moiety, a photochrome in the core of the molecule, and a reactive group that attaches to the protein (Fig. 1d, e). Site-directed attachment is achieved by genetically introducing a cysteine residue near the ligand binding site. After attachment, the agonist or antagonist located at the end of the tether is presented to or retracted from the binding site by photoisomerization of the PTL core (Fig. 1d). Tethered ligands have the unique advantage that they specifically control a selected protein since the cysteine substitution is required for attachment and thus for light sensitivity. Receptors with substitutions can then be genetically targeted to specific cell types or organs in living organisms. PTLs have been developed to light-control several classes of ion channels, including nicotinic acetylcholine receptors (5, 6), glutamate receptors (7), and K^+-channels (8), as well as variant channels with new functionalities derived from molecular studies (9) or evolutionary relationships (10, 11). In these experiments, the tethered ligands were acetylcholine, glutamate, and tetraethylammonium, respectively.

This chapter describes how genetic engineering, cell culture, and electrophysiology are combined to apply PTL-gated ion channels in cell lines and primary cells. In addition, we explain how molecular modeling can be applied to test if a PTL is compatible with a target receptor and how modeling can aid in the choice of PTL attachment sites. In this way, this chapter includes all steps to perform light control of an established as well as a new receptor based on an existing PTL. While procedures for the HEK293 cell line as well as primary neurons are described here, a single cellular model is sufficient for basic functional experiments. To explain these experiments, the light-gated glutamate receptor LiGluR, consisting of the ionotropic glutamate receptor GluK2 (formerly known as iGluR6) mutated at residue L439C and functionalized with MAG1 (maleimide azobenzene glutamate; Fig. 1e), serves as a model.

2 Materials

2.1 Molecular Modeling

1. Desktop computer running visual molecular dynamics (VMD) (12).
2. Protein Data Bank (PBD) file of the ligand-gated ion channel of interest in complex with ligand (here, PDB file 1SD3 containing 2S, 4R-4-methylglutamate (4-MG) co-crystallized in the ligand binding domain of GluK2 (13)).

2.2 Molecular Biology

1. Bacterial cell culture shaker in a 37°C room or 37°C incubator with shaker (~225 rpm).
2. LB medium: 1% bactotryptone, 0.5% yeast extract, 1% NaCl, pH 7.0, autoclaved.
3. Bacteriological agar.
4. Ampicillin stock solution (100 mg/mL in water), stored at −20°C in 500 μL aliquots.

5. QuickChange II Site-Directed Mutagenesis Kit (Agilent, Vienna).
6. QIAprep Spin MiniPrep DNA purification kit (Qiagen, Hilden, Germany).
7. Gene coding for ion channel of choice in a vector suited for mammalian expression (e.g., pcDNA3.1(−)) (see Note 1).
8. Gene coding for yellow fluorescent protein (YFP) in a vector suited for mammalian expression.
9. Gene coding for YFP in a vector suited for neuron-selective expression (see Note 2).
10. Mutagenesis oligonucleotide primers and sequencing oligonucleotide primers ordered custom-made from any of the many commercial vendors (see Note 3).
11. Sanger DNA sequencing provided by any of the many commercial vendors.

2.3 Cell Culture and Transfection of HEK293 Cells

1. Cell culture facility equipped with laminar flow hood, water bath (37°C), incubator (5% CO_2), and cell counting chamber; HEK293 cells (American Tissue Culture Collection, LGC Standards, Teddington, UK) cultured according to the provider's recommendation in 25 cm^2 tissue culture flasks.
2. 500 mL Dulbecco's modified essential medium (DMEM) supplemented with 5% fetal bovine serum (FBS), sterile filtered, and stored at 4°C. Supplement half of the medium with 1% penicillin/streptomycin solution, and use the supplemented DMEM unless noted otherwise.
3. Dulbecco's phosphate buffered saline (DPBS): 8 g NaCl, 200 mg KCl, 200 mg KH_2PO_4, 2.16 g $Na_2HPO_4·7H_2O$, water added to 1 L and sterile filtered.
4. Borate buffer: 1.55 g boric acid, 2.375 g Na^+ borate, water added to 500 mL, pH 8.5, sterile filtered, and stored at 4°C.
5. Opti-MEM I reduced serum medium (LifeTech, Vienna).
6. 0.25% Trypsin-EDTA solution (LifeTech).
7. Lipofectamine 2000 transfection reagent (LifeTech).
8. Poly-L-lysine (PLL) HBr (molecular weight 70,000–150,000). Open PLL container in the laminar flow hood and add water to a concentration of 10 mg/mL. Aliquot in 1.5 mL Eppendorf tubes and store at −20°C.
9. Round cover glasses (18 mm diameter, Carolina Biological, Burlington, NC, USA) (see Note 4).
10. Polystyrene 25 cm^2 cell culture flask with filter lid (0.2 μm pore size, sterile).
11. Polystyrene 12-well plate (sterile).

2.4 Additional Reagents for Cell Culture and Transfection of Hippocampal Neurons

1. Dissection microscope.
2. Dissection tools: Large scissors, large forceps, small scissors, two sharpened forceps, curved spatula, and small scalpel.
3. High-glucose MEM: 12.75 g D-glucose in 50 mL modified essential medium (MEM) sterile filtered, and stored at 4°C.
4. Media: 181 mL MEM, 3 mL high-glucose MEM, 10 mL FBS, and 0.2 mL MITO+ serum extender. Store at 4°C. Once a week, add 1 mL B-27 Serum-Free Supplement (LifeTech) and 500 μL L-glutamine to 48.5 mL media to produce final medium. Store at 4°C.
5. Saline: 1.19 g HEPES, 1.80 g D-glucose, Hank's basic salt solution (HBSS) added to 500 mL, sterile filtered, and stored at 4°C.
6. HBSS washing solution: 7.89 g NaCl, 4.77 g HEPES, 0.30 g KCl, 0.29 g $CaCl_2{\cdot}2H_2O$, 0.20 g $MgCl_2{\cdot}6H_2O$, 0.14 g Na_2HPO_4, 0.18 g D-glucose, water added to 1 L, pH 7.3, sterile filtered, and stored at 4°C.
7. BBS solution: 0.82 g NaCl, 0.53 g BES, 10.7 mg Na_2HPO_4, water added to 50 mL, sterile filtered, and stored at 4°C.
8. $CaCl_2$ solution: 18.38 g $CaCl_2{\cdot}2H_2O$, water added to 50 mL, sterile filtered, and stored at 4°C.
9. Cytosine-β-D-arabinofuranoside (AraC) solution (4 mM), sterile filtered, and stored at −20°C.
10. Two sterile glass Pasteur pipettes: The first pipette is barely flame-polished; the second pipette is flame polished to half its original diameter.
11. Round cover glasses (12 mm diameter, Carolina Biological, Burlington, NC, USA).
12. Polystyrene 6-well plate and 24-well plate (sterile).
13. Cell strainer (40 μm, BD Biosciences).

2.5 Microscopy and Light Activation

1. Inverted microscope equipped with fluorescence condenser, rotating turret for moving filter holders ("filter cubes") into the light path and 20× fluorescence microscopy objective. The microscope is placed on an air table and enclosed by a Faraday cage.
2. Light source capable of producing bright monochromatic illumination (>1 mW/mm^2) that can be computer controlled (i.e., Polychrome V, TILL Photonics, Gräfelfing, Germany) (see Note 5).
3. Power meter with wavelength range from 300 to 800 nm and power range up to 50 mW (PM120VA, Thorlabs, Munich, Germany) (see Note 6). Power range can be extended with neutral density filters (Thorlabs).

4. Standard filter set for visualization of YFP in a filter cube.
5. Total reflectance mirror sized to fit into a filter cube in the diagonal position (e.g., NT43-875, Edmund Optics, Karlsruhe, Germany).
6. Empty filter cube.
7. Standard laboratory spectrophotometer to perform absorbance measurements at a wavelength of 360 nm.
8. MAG1 in solid form.
9. DTT stock solution (10 mM): 15.4 mg DTT in 10 mL extracellular solution, distributed to 1 mL aliquots and stored at −20°C.
10. Concanavalin A solution: 15 mg type VI concanavalin A (Sigma) in 50 mL extracellular solution and stored at 4°C.

2.6 Whole-Cell Patch Clamp Measurements

1. Patch clamp amplifier with data acquisition electronics, electrode holder, and software installed on a desktop computer (e.g., AxoPatch200B, Digidata 1440A, HL-U, and pCLAMP 10, Biberach, Germany).
2. Micromanipulator on rotating base mounted to a tower (e.g., MP-285ROE, 285RBI, and MT-71-9, Sutter Instruments, Science Products, Hofheim, Germany) (see Note 7).
3. Glass micropipette puller (e.g., P-97, Sutter Instruments).
4. Extracellular solution: 145 mM NaCl, 4 mM KCl, 1 mM $MgCl_2$, 2 mM $CaCl_2$, 10 mM HEPES, pH 7.4 (see Note 8).
5. Intracellular solution: 135 mM K-gluconate, 10 mM NaCl, 2 mM $MgCl_2$, 2 mM MgATP, 1 mM EGTA, 10 mM HEPES, pH 7.4.
6. Salt bridge solution: 3 M KCl and 1% electrophoresis-grade agarose in water.
7. Glass capillary tubes (1.5 mm outer diameter, 1.17 mm inner diameter, 100 cm length, Warner Instruments, Hamden, CT).
8. Unplugged glass Pasteur pipettes (length >200 mm).
9. Glass bottom petri dish (glass surface: 25 mm diameter, 0.2 mm thickness, Bioptechs, Butler, PA, USA) (see Note 9).

3 Methods

3.1 Molecular Modeling of PTL Substitution Pattern and Attachment Sites

1. Molecular modeling is applied to test if a PTL is compatible with a target receptor. In the case of MAG1 and GluK2 (Fig. 1), modeling is based on a crystal structure of the ligand binding domain of GluK2 with bound 4-MG (13). Note that MAG1 and 4-MG are both substituted at the 4′ position (the Cγ atom of glutamate; Fig. 1e, highlighted with a star) with identical stereochemistry, and 4-MG therefore is a valid model for

MAG1 with respect to the substitution pattern (14). Furthermore, the closed conformation of the ligand binding domain indicates that the receptor is in an activated conformation (Fig. 1a).

2. To visualize 4-MG in the ligand binding domain, open PDB file 1SD3 in VMD (File/NewMolecule).
3. In the representations panel (Graphics/Representations), reformat the structure by activating drawing method "NewCartoon" and in the "selected atoms" field enter "chain A."
4. Orient the structure such that the ligand faces the front and highlight the ligand by creating a new representation with the drawing method "VDW" and in the "selected atoms" field enter "chain A and resid 998." The resulting view is shown in Fig. 1f.
5. To verify that an extended 4′ substituent of MAG1 will reach to solvent exposed residues, change the drawing methods of "chain A" to "VDW." An exit tunnel becomes visible (Fig. 1g, circle), confirming that the tail of MAG1 may reach the protein surface while still allowing the ligand binding domain to close (7).
6. In the next step, use the molecular model to identify residues that are surface exposed and surround the binding site (Fig. 1g, highlighted residues). Select 4–8 residues and order oligonucleotides to alter them into cysteines. In the case of MAG1, attachment to L439, L482, G486, or E722 of GluK2 has proven successful (15).
7. For every new application of a PTL to a receptor, perform steps 1–5 with a suited crystal structure to verify that an exit tunnel exists and to find sites for attachment.

3.2 Preparation of Growth Plates

1. Solubilize 9 g agar in 600 mL LB medium and autoclave.
2. After autoclaving, let hot LB medium cool to 50°C, add 600 μL ampicillin stock solution and swirl to mix.
3. Pour solution into sterile petri dishes to a depth of ~3 mm.
4. Leave the plates at room temperature and unstacked for 24 h before storing them at 4°C.

3.3 Polymerase Chain Reaction to Generate Cysteine Mutants

1. Dilute the vector containing the ion channel gene to a concentration of 20 ng/μL.
2. Resuspend lyophilized mutagenesis oligonucleotides (forward and reverse) to a concentration of 300 μM by adding X*3.33 μL water to the original tubes (X is the amount of nanomoles delivered in the tube). Vortex for 10 s and spin briefly every 5 min for a total time of 15 min. Store at –20°C.

3. Dilute mutagenesis oligonucleotide primers to 10 μM by adding 2 μL of each oligonucleotide primer to 56 μL water. Vortex, spin briefly and store at −20°C.
4. The PCR is setup following the manufacturer's instructions with 20 ng template vector (1 μL of the dilution) and 10 pmol of each oligonucleotide (1 μL of the mixture). Recommended PCR parameters are 1 cycle at 95°C for 60 s; 18 cycles at 95°C for 30 s, at 56°C for 30 s, and at 72°C for 600 s; 1 cycle at 95°C for 7 min; and 1 cycle at 4°C until the reaction is processed further (see Note 10).
5. After PCR, add 1 μL dpn1 restriction enzyme, pump mix gently, spin briefly, and incubate at 37°C for not less than 90 min.

3.4 Amplification and Selection of Mutagenized Receptors

1. Warm up LB agar plates (one for each PCR) at 37°C for 30 min (see Note 11).
2. Transform PCR without further purification into competent cells. Maintain cells on ice during the transformation and treat cells gently without vortexing or excessive pipetting.
3. Spread bacteria evenly on plates with a sterile glass rod.
4. Place plates at 37°C upside down overnight.
5. On the following day, combine 3 mL LB medium with 3 μL ampicillin solution in glass tubes (four for each PCR).
6. Using a pipette tip, pick four colonies of each plate by lightly touching a single colony with the tip and dropping the pipette tip into a glass tube.
7. Place glass tubes at 37°C with shaking (~225 rpm) overnight.
8. On the following day, extract vector from bacteria with MiniPrep kit following the manufacturer's instructions and send for verification by sequencing.

3.5 MAG1 Reconstitution

1. Prepare ~3 mg MAG1 in a 1.5 mL Eppendorf tube.
2. Add 200 μL DMSO to the tube to yield the MAG1 stock solution (26 mM).
3. Vortex vigorously for 10 s and spin down briefly. Repeat until all solid has dissolved.
4. Prepare 10 μL aliquots and store them at −20°C in the dark (see Note 12).

3.6 Verification of MAG1 Concentration (See Note 13)

1. Add 0.5 μL MAG1 stock solution to 1 mL DPBS, vortex, and spin briefly.
2. Prepare 5 tenfold serial dilutions: Add 100 μL of the first MAG1 dilution to 900 μL PBS to produce a second dilution.

Then add 100 μL of this second dilution to 900 μL PBS to prepare a third dilution. Repeat two more times.

3. In the spectrophotometer, blank absorbance at a wavelength of 360 nm with DPBS and measure absorbance for all serial dilutions.
4. Using measurements in the linear absorbance regime (usually between 0.1 and 1.0 A), determine the concentration of the stock solution. This is achieved by plotting A_{360} against concentration and equating Beer–Lambert's law with the extinction coefficient 0.025 A_{360}/μM and the dilution factor.

3.7 Preparation of Cover Glasses for HEK293 Cells

1. Cell culture should be performed under sterile conditions in a laminar flow hood. Instruments and containers should be disinfected by spraying them with 70% EtOH prior to placing them in the hood.
2. Dilute PLL solution to a concentration of 0.05 mg/mL in borate buffer.
3. Add 2 mL diluted solution to the wells of a 12-well plate.
4. Add a single 18 mm cover glass to each well and incubate cover glasses for 2–8 h (see Note 14).
5. Transfer the desired number of cover glasses (typically 4–8) to a new 12-well plate.
6. Wash cover glasses three times with water (add and aspirate water) and allow to air dry.

3.8 Preparation of HEK293 Cells

1. Warm trypsin solution, DPBS, and DMEM in water bath to 37°C. Warm 9 mL DMEM in a culture flask in the incubator (see Note 15).
2. Add 1.5 mL DMEM to each cover glass prepared previously in the 12-well plate.
3. Remove cells from incubator and aspirate media.
4. Add 10 mL DPBS, let stand for 10 s, and aspirate.
5. Add 3 mL trypsin solution and let stand for 3 min.
6. To detach cells from flask, tap flask against a vertical surface.
7. Add 7 mL media to the flask and pipet up and down with a serological pipette (see Note 16).
8. Transfer 1 mL cell suspension to the culture flask that contains 9 mL pre-warmed media. Place flask in incubator to passage the culture in the future.
9. Transfer the remaining 9 mL cell suspension to a 50 mL conical tube.
10. Centrifuge at 1,000 × *g* for 3 min and aspirate supernatant.

11. Resuspend cells in 2 mL media by carefully pipetting up and down ten times with a serological pipette.
12. Count cells, transfer 80,000 cells to each cover glass, and allow 8–24 h for cells to settle before transfection.

3.9 Transfection of HEK293 Cells

1. Warm DMEM with no antibiotics added in water bath to 37°C.
2. Remove 12-well plate from incubator and replace DMEM on cells with pre-warmed DMEM.
3. For each cover glass to be transfected, combine 25 μL Opti-MEM I, 0.2 μg ion channel vector, and 0.01 μg YFP vector in a 1.5 mL Eppendorf tube (see Note 17).
4. For each cover glass to be transfected, combine 25 μL Opti-MEM I and 1 μL Lipofectamine in a 1.5 mL Eppendorf tube.
5. Let stand for 5 min.
6. Add Lipofectamine solution to vector solution and tap tube gently to mix.
7. Let stand for 20 min.
8. Add 50 μL transfection mix to each cover glass.
9. Allow 18–24 h for expression (see Note 18; refer to Chapters 2, 4, and 5 of this book for alternative transfection techniques).

3.10 Preparation of Cover Glasses for Hippocampal Cell Culture

1. Add 2 mL 70% EtOH to all wells of a 24-well plate.
2. Add a single 12 mm cover glass to each well and incubate for 10 min.
3. Wash three times with water.
4. Treat the cover glasses with PLL as described above.
5. After treatment, wash cover glasses three times with water and allow to soak in water for ~1 h. Wash one more time and let dry.

3.11 Dissection of Hippocampus

1. Obtain *P0–P4* Sprague Dawley rat pups (local ethical permission and other appropriate approvals will be required)
2. Autoclave dissection tools prior to dissection.
3. Warm saline in water bath, warm media (refer to Subheading 2.4) in incubator and bring trypsin solution to room temperature.
4. Fill two 60 mm petri dishes halfway with warm saline.
5. Humanely sacrifice the animal in accordance to the local ethical guidance. Sever neck at base of head with large scissors and place under dissecting scope.
6. Remove skin from back to front with forceps.
7. Slice the skull from back to front with small scissors.

8. Peel back the skull to each side with large forceps, remove brain with curved spatula, and slide it into petri dish containing the saline. Repeat for additional animals and then transfer all brains into a new dish.
9. Remove cerebellum and separate hemispheres with scalpel.
10. Remove meninges from the outer surface of the brain (see Note 19).
11. Identify the hippocampus as a curved structure that is slightly denser and therefore has a dark contrast to the rest of the tissue. Gently cut away the hippocampus being very careful not to damage the tissue.
12. Remove remaining meninges with minimal damage to the hippocampus and clean away any excess tissue still attached to it with small scalpel.
13. Carefully transfer hippocampus to 15 mL conical tube filled with saline.

3.12 Preparation of Hippocampal Cells

1. Remove all but 4.5 mL saline from the tube containing hippocampi, add 500 mL warm trypsin, and invert to mix.
2. Place at 37°C for 8 min.
3. Add 9 mL warm saline to dilute trypsin, invert to mix, and remove solution using serological pipette. Repeat for a total of four washes.
4. Add 1 mL media and triturate six times with large-diameter Pasteur pipette. Remove media with suspended cells and strain into a 50 mL conical tube.
5. Repeat previous step with small-diameter Pasteur pipette if more cells are desired.
6. Count cells and dilute to a concentration of 200,000 cells/mL.
7. Add 500 mL cell suspension to each well and place in incubator.
8. After 15 min, replace media to eliminate debris that has not attached to cover glass.
9. After 4–7 days in culture, AraC is added to a final concentration of 4 μM to prevent growth of glia.
10. To maintain cultures, replace half of the media with fresh media once per week.

3.13 Transfection of Hippocampal Cells

1. For each cover glass, warm 1 mL transfection media (985 μL MEM supplemented with 15 μL high-glucose MEM), 500 μL HBSS, and 500 μL growth media in the wells of a 6-well plate placed in the incubator.
2. Bring $CaCl_2$ solution, BBS, and expression vectors to room temperature.

3. Remove cells from incubator and transfer media from cells to a well of the 6-well plate.
4. Add 500 μL transfection media to cells and aspirate.
5. Add 500 μL transfection media to cells.
6. Mix, in this order, 1.15 μL $CaCl_2$ solution, 1.2 μg ion channel vector, and 0.1 μg YFP vector with neuron-specific promoter and add water to a final volume of 33 μL.
7. To the mixture, add 33 μL BBS dropwise, mix, and transfer 30 μL to the center of each cover glass immediately.
8. Observe cells regularly during the next 6 h. After the shortest amount of time that generates a fine layer of precipitate, aspirate transfection media and incubate cells in HBSS.
9. After the shortest amount of time needed to dissolve precipitate (typically 2–15 min in HBSS), wash cells with 500 mL conditioned media (aspirated from the cells previously) and add both conditioned and fresh media to cells (500 μL each). Refer to Chapters 4 and 5 of this book for alternative methods for neuronal transfection.

3.14 Preparation of Whole-Cell Patch Clamp Measurements

1. Using the flame of a gas burner, bend the end of a Pasteur pipette into a "U" shape where the "sides" and the "bottom" of the U are ~2 cm long (see Note 20).
2. Heat salt bridge solution in a microwave until homogenous. Allow to cool to 50°C.
3. Aspirate solution in a 1 mL pipette tip and fill the U-shaped glass to complete the salt bridge. Store the salt bridge in 3 M KCl at 4°C.
4. Using glass capillary tubing and a multistage puller, prepare micropipettes with taper length of ~4 mm, tip diameter of ~2 μm, and electrical resistance of 2–8 MΩ. Refer to Chapter 7 for further considerations regarding the patch pipette manufacturing.

3.15 Preparation of Microscopy

1. Connect the light source to the data acquisition electronics and configure the software such that the wavelength can be computer controlled.
2. Connect the optical output of the light source to the back port of the microscope.
3. To direct light of all wavelengths to the sample, fix a total reflectance mirror with superglue in the spare filter holder at the position reserved for the dichroic mirror. Place this filter cube in the turret (see Note 21).
4. Place the unmodified filter cube for YFP imaging in the turret.
5. Measure the intensity of the light exiting the objective using the power meter (see Note 6).

3.16 MAG1 Labeling of HEK293 Cells and Neurons (See Note 22)

1. In the laminar flow hood, fill four wells of a 12-well plate with 1 mL extracellular solution and supplement one well with DTT to a final concentration of 1 mM. Fill one well with concanavalin A solution.
2. To remove growth medium, transfer a cover glass with HEK293 cells or neurons from growth medium to extracellular solution.
3. To activate surface cysteines by ensuring that they are reduced, transfer the cover glass to extracellular solution with DTT and incubate for 10 min.
4. To remove DTT, transfer cover glass to extracellular solution and incubate for 2 min.
5. For concanavalin treatment, transfer cover glass to concanavalin A solution and incubate for 10 min (see Note 23).
6. In the meantime, thaw MAG1 stock solution and illuminate with UV light for 1–15 min (see Note 24).
7. Dilute illuminated MAG1 in one of the wells to a final concentration of 10–50 mM (see Note 25).
8. Transfer cover glass to this solution and incubate for 10–30 min. Keep the well plate in the dark during this time.
9. Transfer cover glass to a well with extracellular solution and keep in the dark until the experiment.

3.17 Patch Clamp Measurements and Photoswitching

1. Add 500 μL extracellular solution to glass bottom petri dish and transfer the cover glass to the dish.
2. Connect the glass bottom dish to a dish with the reference electrode using the salt bridge.
3. Adjust the microscope for visualization of YFP by choosing the appropriate filter cube and by switching the light source to 510 nm excitation light.
4. Using the eye pieces, identify a transfected cell (see Note 26).
5. Apply and clamp positive pressure to the micropipette while lowering it into extracellular solution to prevent adsorption of debris at the liquid–air interface.
6. Position the micropipette over the center of the cell (see Note 27).
7. Slowly lower the micropipette until it presses down on the cell surface, causing an indentation.
8. Rotate the modified filter cube that holds the total reflectance mirror into the light path.
9. Adjust the holding potential to −70 mV.
10. Release the positive pressure from the micropipette, observe GΩ seal formation, and use negative pressure to break through

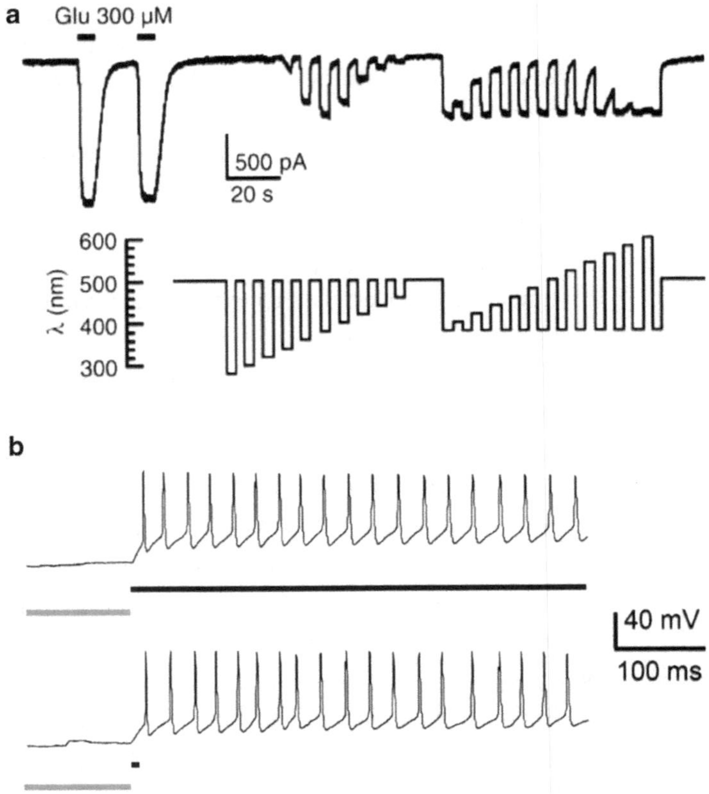

Fig. 2 Optical control of HEK293 cell currents and hippocampal neuron action potentials with LiGluR. (**a**) Whole-cell current in a voltage-clamped HEK293 cells expressing LiGluR (holding potential −60 mV). Current can be photo-controlled with maximum activation at ~380 nm and inactivation at ~500 nm. As a positive control for ion channel expression, glutamate as a free ligand can be added using gravity flow. (**b**) LiGluR activation depolarizes and triggers action potential firing in a current-clamped hippocampal neuron (*top trace*). After a brief pulse, depolarization is stable in the dark (*bottom trace*) (Part **a** is reprinted by permission from Macmillan Publishers Ltd: Nat Chem Biol (7), copyright 2006. Part **b** is reprinted from Neuron, ref. 17 with permission from Elsevier Ltd)

the cell membrane. Allow 5 min to obtain a seal resistance >100 MΩ. Refer to Chapter 7 of this book for more tips on whole-cell recording.

11. Photoswitching can be accomplished by changing illumination wavelengths while recording from the cell in voltage-clamp or current-clamp mode (Fig. 2). Illumination at a wavelength of 380 nm coverts MAG1 into the *cis*-isomer that opens GluK2 and produces an inward (negative) current in voltage-clamp or depolarization in current-clamp. Illumination at a wavelength of 500 nm light coverts MAG1 into the *trans*-isomer, allowing GluK2 to close.

4 Notes

1. In our experience, all mammalian expression vectors containing the cytomegalovirus (CMV) promoter performed well in HEK293 cells and neurons with robust protein expression. Increased expression of ion channels was achieved using the hybrid CMV enhancer/chicken β-actin promoter or with woodchuck hepatitis virus posttranscriptional regulatory element (16). Most ion channel genes can be obtained in expression vectors from Addgene (http://www.addgene.org/).
2. Co-transfection of fluorescent proteins allows selecting transfected cells in patch clamp experiments. Furthermore, fluorescent proteins under the control of cell type-specific promoters can be used to distinguish cell types. To target neuronal cells in hippocampal cell cultures, a modified pcDNA3.1(−) expression vector containing the neuron-specific human synapsin 1 promoter is available from the authors.
3. Mutagenesis oligonucleotide primers can be reliably designed using the PRIMERX website (http://www.bioinformatics.org/primerx/) following the QuickChange parameters. Sequencing oligonucleotide primers should be designed according to the recommendations of the sequencing provider, which also may offer universal oligonucleotide primers that bind to many vectors.
4. In our experience, cover glasses from some other manufacturers require additional cleaning in 70% EtOH solution followed by washing in water.
5. In the past years, the polychrome has become a standard light source for optogenetic experiments as it offers bright illumination with maximum wavelength flexibility. A good alternative is filter-based light sources capable of producing high-intensity illumination (>10 mW/mm^2) with fast switching time between filters (<2 ms) (e.g., DG4, Sutter Instruments, Science Products, Hofheim, Germany).
6. For accurate optogenetic experiments, it is required to measure the light intensity produced in every individual setup. This is commonly done at the very end of the light path by quantifying the light exciting the objective. The total intensity is measured using the power meter, while the illuminated area can be estimated to be the field of view. Latter can be determined using a micrometer-sized grid. For a more elaborate measurement, the illuminated area is controlled using an aperture or pinhole. Most optogenetic applications work well with intensities of >1 mW/mm^2, and MAG1 can be reliably converted with as little intensity as 0.1 mW/mm^2 (11).

7. In the simplest patch clamp setup, the micromanipulator with mounted patch clamp electrode is attached on the optical table using a tower stand, while the cells are placed on the X-Y-table of the inverted microscope. The major advantage of this setup is that localizing fluorescent cells is facilitated as the cells are moved relative to the objective. The major disadvantage is that the field of view or focus cannot be changed after a cell has been patched.
8. In a modified extracellular solution, Na^+ ions can be replaced by N-methyl-D-glucamine to reduce cell toxicity caused by PTLs. PTLs may act as a very weak agonist during labeling and generate Na^+ currents.
9. Glass bottom dishes proved to be an easy and affordable alternative to imaging chambers typically used in patch clamp experiments. They can be reused many times or disposed after single use.
10. We have obtained good results when using 20 ng of template vector and following the manufacturer's protocol, especially for the number of cycles. For the mutagenesis of genes in long or problematic vectors, we recommend supplementing the reaction with additional 40 ng template vector and 3 μL DMSO; see also Chapter 20 of this book for further details of the site-directed mutagenesis approach.
11. Place plates upside down with the base sitting on the ridge of the lid on one side. In this way, condensation will not accumulate on the agar surface and humidity is reduced.
12. The maleimide group of a PTL is sensitive to hydrolysis and precautions are necessary to avoid water contact until the labeling reaction. Store PTL solids in an Eppendorf tube that is placed in a 50 mL conical tube containing ~1 cm of desiccant at the bottom. Place the conical tube in the dark and at −20°C or −80°C. Parafilm wrapping of the Eppendorf tube is neither required nor desired. When resuspending PTLs, use DMSO that was aliquoted from a new stock and stored in individual tubes in a container with desiccant. PTL stock solutions in DMSO are stored as described above for the solid and can be stored for 1 year. Stock solutions should be sufficiently concentrated such that DMSO concentration during labeling does not exceed 1% (at a PTL concentration of up to 200 μM). While it has not shown to be problematic, the number of freeze-thaw cycles should be kept to a minimum.
13. This procedure is important as it verifies the MAG1 concentration and validates the structural integrity of the azobenzene photochrome. It can also be used to determine MAG1 concentration if no scale with mg sensitivity is available or if uncertain amounts of MAG1 were transferred to the Eppendorf tube.

14. While 2–8 h are recommended for treatment, cover glasses can be left in PLL solution for several weeks, and PLL solution can be reused at least six times if kept sterile.
15. Warming media in the incubator is preferred over warming it in the water bath as equilibration with CO^2 and consequently proper pH are ensured.
16. Use of serological pipettes is recommended as narrow 1 mL pipette tips can damage cells.
17. For transfection of most glutamate receptor vectors, the amount of Lipofectamine and vector can be scaled down substantially (up to fivefold) compared to the recommendation in the manufacturer's protocol. It is critical to use miniscule amounts of fluorescent protein vector compared to ion channel vector (typically 10–20 times less fluorescent protein vector than the ion channel vector).
18. For whole-cell patch clamp measurements, HEK293 cells should be sparse at the time of experiment to prevent gap junction formation between adjacent cells. For many other purposes (e.g., imaging) a confluent layer works well and gives the best transfection efficiency and cell health.
19. Meninges will cause microglial growth in the cell culture that is harmful to neuronal growth. When removing meninges, discard damaged hippocampal tissue.
20. In the first step, hold the pipette horizontally and heat it 2 cm from the end until the terminal piece drops by gravity to produce a right angle. Repeat to complete the U shape. Using a glass cutter and appropriate safety equipment, separate the U shape from the rest of the Pasteur pipette.
21. It is not safe to look into the eye pieces when using this modified filter cube as unfiltered stray light may reach to the eye piece. It is at no time safe to expose eyes to unfiltered light from the light source, to light in the UV range, or to bright light of any wavelength: Consider local safety regulations.
22. While this procedure describes MAG1 labeling of cells expressing LiGluR, it is generally applicable to other PTLs and ion channels. Labeling and wash steps can be executed either at room temperature or at 37°C. DTT treatment and UV preillumination are optional steps that yield the greatest efficiency of MAG1 labeling and thus photoswitching.
23. Concanavalin A blocks desensitization of GluK2 and is required for experiments in HEK293 cells but not neurons.
24. Efficiency of MAG1 labeling is increased if MAG1 is converted to the *cis*-isoform in UV light before the labeling. This can be achieved with any handheld UV source, e.g., a UV LED pointer, and at any light intensity. Indicative values: at an

intensity of 0.04 mW/mm^2 (λmax 365 nm) allow 15 min; at an intensity of 5.5 mW/mm^2 (λmax 374 nm) allow 30 s.

25. It is essential to pump mix the solution to fully dissolve MAG1 as DMSO solutions tend to “sink” to the bottom of the well. Prepare the MAG1 dilution fresh for every cover glass to be labeled.
26. Cells with medium bright fluorescence often allow the formation of high-quality whole-cell seals compared to very bright cells.
27. Micropipette positioning can be achieved in the following steps: After focusing on the cell, move the focal plane above the cell by turning the focus wheel on the side of the microscope by ~two turns. In the next step, lower the micropipette until it is in this new focal plane and move it to be centered on the field of view. In the final step, refocus on the cell and then lower the pipette slowly.

Acknowledgments

Our research is supported by grants of the European Union *Seventh Framework* Programme (to H.J.) and the Human Frontier Science Program (to H.J.). We thank D. Fortin for advice on primary cell preparation, transfection, and PTL conjugation, and K. Greenberg for recommendations on promoters and expression elements.

References

1. VanEngelenburg SB, Palmer AE (2008) Fluorescent biosensors of protein function. Curr Opin Chem Biol 12:60–65
2. Fenno L, Yizhar O, Deisseroth K (2011) The development and application of optogenetics. Annu Rev Neurosci 34:389–412
3. Szobota S, Isacoff EY (2010) Optical control of neuronal activity. Annu Rev Biophys 39:329–348
4. Fehrentz T, Schonberger M, Trauner D (2011) Optochemical genetics. Angew Chem Int Ed Engl 50:12156–12182
5. Chabala LD, Lester HA (1986) Activation of acetylcholine receptor channels by covalently bound agonists in cultured rat myoballs. J Physiol 379:83–108
6. Tochitsky I, Banghart MR, Mourot A, Yao JZ, Gaub B, Kramer RH, Trauner D (2012) Optochemical control of genetically engineered neuronal nicotinic acetylcholine receptors. Nat Chem 4:105–111
7. Volgraf M, Gorostiza P, Numano R, Kramer RH, Isacoff EY, Trauner D (2006) Allosteric control of an ionotropic glutamate receptor with an optical switch. Nat Chem Biol 2:47–52
8. Banghart M, Borges K, Isacoff E, Trauner D, Kramer RH (2004) Light-activated ion channels for remote control of neuronal firing. Nat Neurosci 7:1381–1386
9. Chambers JJ, Banghart MR, Trauner D, Kramer RH (2006) Light-induced depolarization of neurons using a modified Shaker K^+ channel and a molecular photoswitch. J Neurophysiol 96:2792–2796
10. Janovjak H, Sandoz G, Isacoff EY (2011) A modern ionotropic glutamate receptor with a K^+ selectivity signature sequence. Nat Commun 2:232
11. Janovjak H, Szobota S, Wyart C, Trauner D, Isacoff EY (2010) A light-gated, potassium-selective glutamate receptor for the optical inhibition of neuronal firing. Nat Neurosci 13:1027–1032
12. Hsin J, Arkhipov A, Yin Y, Stone JE, Schulten K (2008) Using VMD: an introductory tutorial. Curr Protoc Bioinformatics Chapter 5:Unit 5.7

13. Mayer ML (2005) Crystal structures of the GluR5 and GluR6 ligand binding cores: molecular mechanisms underlying kainate receptor selectivity. Neuron 45:539–552
14. Stawski P, Janovjak H, Trauner D (2010) Pharmacology of ionotropic glutamate receptors: A structural perspective. Bioorg Med Chem 18:7759–7772
15. Gorostiza P, Volgraf M, Numano R, Szobota S, Trauner D, Isacoff EY (2007) Mechanisms of photoswitch conjugation and light activation of an ionotropic glutamate receptor. Proc Natl Acad Sci USA 104:10865–10870
16. Donello JE, Loeb JE, Hope TJ (1998) Woodchuck hepatitis virus contains a tripartite posttranscriptional regulatory element. J Virol 72:5085–5092
17. Szobota S, Gorostiza P, Del Bene F, Wyart C, Fortin DL, Kolstad KD, Tulyathan O, Volgraf M, Numano R, Aaron HL, Scott EK, Kramer RH, Flannery J, Baier H, Trauner D, Isacoff EY (2007) Remote control of neuronal activity with a light-gated glutamate receptor. Neuron 54:535–545

Index

Nikita Gamper (ed.), *Ion Channels: Methods and Protocols*, Methods in Molecular Biology, vol. 998,
DOI 10.1007/978-1-62703-351-0, © Springer Science+Business Media, LLC 2013

N

O

P

R

S

T

U

V

W

X

MIX
Papier aus verantwortungsvollen Quellen
Paper from responsible sources
FSC® C105338

If you have any concerns about our products,
you can contact us on
ProductSafety@springernature.com

In case Publisher is established outside the EU,
the EU authorized representative is:
Springer Nature Customer Service Center GmbH
Europaplatz 3, 69115 Heidelberg, Germany

Printed by Libri Plureos GmbH
in Hamburg, Germany